NEHRU

NEHRU

A Tryst with Destiny

STANLEY WOLPERT

New York Oxford

OXFORD UNIVERSITY PRESS

1996

Oxford University Press

Oxford New York
Athens Auckland Bangkok Bogotá Bombay
Buenos Aires Calcutta Cape Town Dar es Salaam
Delhi Florence Hong Kong Istanbul Karachi
Kuala Lumpur Madras Madrid Melbourne
Mexico City Nairobi Paris Singapore
Taipei Tokyo Toronto

and associated companies in
Berlin Ibadan

Published by Oxford University Press, Inc.
198 Madison Avenue, New York, New York 10016

Library of Congress Cataloging-In-Publications Data
Wolpert, Stanley A., 1927–
Nehru: a tryst with destiny / Stanley Wolpert.
p. cm.
Includes bibliographical references and index.
ISBN 0-19-510073-5
1. Nehru, Jawaharlal, 1889–1964.
2. Prime ministers—India—Biography. I. Title.
DS481.N35W65 1996
954.04'2'092—dc20 [B] 95-32743

1 3 5 7 9 8 6 4 2

Printed in the United States of America
on acid-free paper

For Dorothy

my best friend and true love

PREFACE

In 1957, when first I met Jawaharlal Nehru, he was sixty-seven, still vigorous enough to race up some fifty steps to the top of Maharashtra's Pratapgarh in the midday heat, unveil an equestrian bronze of Shivaji, and speak to all of us awaiting his arrival from Delhi that day before he drove back to Pune, where he addressed a much larger audience and opened a new college and clinic before flying off to Calcutta. This was his usual daily routine. Prime Minister Nehru exhausted his younger secretaries and most of the officials around him. Some attributed his powers to yoga, which he practiced every morning, but those who knew him best recognized that he was driven. Was it acute sensitivity to India's race with time to save its starving, suffering millions or the growing consciousness of his own mortality? Was fame the spur or the love of power? Or was this power simply his destiny? Whatever drove Nehru with such relentless fury, he rarely slept more than a few troubled hours and had little daytime rest outside a prison cell.

The second time we met in his office in New Delhi's Parliament. I was a Ford Fellow, completing research for my doctoral dissertation on Tilak and Gokhale. With little hope of success, I telephoned the prime minister's office and asked if I might meet with him to discuss early Indian nationalism. Three days later I was told he would meet me and could spare only fifteen minutes. I was careful to arrive early but found his outer office so jammed with Indians whose business with their prime minister was obviously so much more important and urgent than mine that I expected to be informed that my appointment was canceled. Instead, my wife and I were ushered into his spacious office and remained there more than an hour while Nehru eloquently reflected on the roots of Indian nationalism in its pre-Gandhian era as if he had nothing better to do and no more important matter to worry about. He recalled obscure chronology with amazing accuracy and ranged so widely over broader horizons of history that I was dazzled by his brilliance. We left his office with a clearer understanding of why hundreds of millions of Indians loved their remarkable Panditji.

The third time I saw Nehru was at the opening of New Delhi's Academy of Fine Arts (Lalit Kala Akademi), to which he brought Edwina Lady Mountbatten. I sat just a few rows behind them. I was surprised at how cheerful Nehru appeared that evening and how like adolescent lovers he and Edwina behaved, touching, whispering into each other's ears, laughing, holding hands. "The family line is that they are simply good friends," Lady Mountbatten's grandson Lord Romsey told me thirty-five years later over tea in his sitting room in Broadlands. "Nothing more than that, you see." But I had seen much more, and Lord Mountbatten himself often referred to Nehru's correspondence with his wife as love letters, knowing better than anyone but Nehru how much Edwina adored her handsome "Jawaha," as she lovingly called him. This was why Nehru's daughter Indira hated her, just as she hated her own Auntie "Nan," gracious Madame Vijaya Lakshmi Pandit, who kindly chatted with me for hours in 1985 when we found each other staying at Delhi's Indian International Centre off Lodhi Gardens. We often met for tea or to talk about her brilliant brother and bitter niece.

Romsey suggested that I write his father, Lord Brabourne, who chairs the Broadlands Trust, to request permission to read Nehru's letters to Lady Mountbatten. Brabourne replied: "We made an agreement with Mrs. [Sonia] Gandhi . . . for copies of Nehru's letters to be sent to Mrs. Gandhi and she arranged for copies of Lady Mountbatten's letters to be sent to us. . . . The condition made at that time was that no permission would be given to anybody to read the letters without the agreement of both families." So early in 1994 I flew to India and met with Sonia Gandhi in her Jan Path fortress in New Delhi. "Let me be totally honest with you," she said, after refusing to show me any of the letters. "I don't want you to think we have anything to hide, but . . ."

The deepest passions and fears that drove and tortured Nehru throughout his adult life have always remained hidden, first by himself, then by Indira, and now by those who zealously secrete documents recording his intimate thoughts and concerns during India's first decade of independence, trying more than three decades after his death to perpetuate myths and hoping to hide the true nature of that great man. Nehru, however, played too important a role in recent Indian history to be kept locked away in closets and trunks in the basements of Broadlands, Jan Path, and Teen Murti. For almost a decade, foolish British bureaucrats kept Jawaharlal Nehru behind bars, and his equally foolish heirs and self-appointed guardians have locked up his mind and heart for three times as long.

Many good friends, fine scholars, and excellent librarians helped me find the elusive keys to the secret chambers of Nehru's personality over the last quarter century of my research into this most written about yet least understood of all Indian leaders. I am doubly indebted to Nehru's sister Nan and to her daughter Rita, whose recent early death was more shocking than Madame Pandit's demise at ninety. The jurist Nagendra Singh, whose death also came much too soon, talked with me at length about Nehru as well as Indira. My closest Pune neighbors and dear friends, Rao Sahib Patwardhan and his brothers Achyut and Pama, all of whom knew and admired Nehru, taught me much about both his weaknesses and his strengths. Friends in Delhi have also confided in me over the years and helped sharpen my appreciation and understanding of Nehru's complex character. I thank them all, most especially Raja Dinesh Singh, Attorney General Milon K. Banerji, the Honorable Inder Gujral, Khushwant Singh, Dr. P. N. Chopra, Prakash Tandon, Patwant Singh, and Sheila Kalia. My warmest thanks to Ambassador John Kenneth Galbraith and his gracious wife Kitty for so generously sharing their vivid memories of Nehru with me in their lovely living room in Cambridge. My old friend Janki Ganju, who died soon after my wife and I enjoyed one of his delicious Kashmiri dinners in Washington, taught me several things about his fellow Kashmiri, Pandit Nehru, that I had only guessed before. I am also grateful to Ambassador and Mrs. S. S. Ray for their kind help in my search and to Mrs. Ray's sister Mrs. Chaya Ray and her husband in London for their kind hospitality and cogent reminiscences. Many earlier Indian ambassadors to Washington, starting with B. K. Nehru, Keval Singh, Shankar Bajpai, Prakash Kaul, Karan Singh, and Abid Hussain, all have been most helpful and kind. I met Krishna Menon several times, and Indira as well, but both were always as secretive and carefully guarded in every way as was Nehru himself.

Jaya Prakash Narayan and Morari Desai, when the latter was prime minister, taught me many truths about Nehru and Indira that I had long disbelieved. The morning I spent interviewing Lord Mountbatten in London in the summer of 1978 proved as illuminating concerning Nehru as it was about Jinnah, whose biography I was then writing. I thank the present Countess Mountbatten for her gracious hospitality in London and her son Lord Romsey for his hospitality in Broadlands. I am most grateful to Janet, Lady Balfour of Burleigh, the only person granted permission by Brabourne to read the love letters while she was writing her biography of Edwina, for speaking with me at such length in London about our mutual friends and their affairs. Janet's good friend Roger Bolton was most helpful and more than generous in helping me find many photos of Nehru, several of which are reproduced in this book. My dear old friend, Professor William Fishman of London, kindly arranged for me to meet Harrow archivist A. D. K. Hawkyard, thanks to whom I learned many things about Nehru's first home in England. I thank Earl Listowel for his hospitality and help and Lord Wyatt of Weeford for his most hospitable kindness and astute insights into the public schoolboy Nehru and his later political nature. The archivist of London's Inner Temple led me to his basement records of Nehru's admission papers and applications for the bar, which he so kindly photocopied for me. Thanks to the librarian and staff of Friends House on Euston Road in London, I was given full access to the Agatha Harrison Papers with her correspondence with Nehru and several fine photos of Nehru, one with his grandsons. Thanks also to Dr. C. M. Woolgar and his staff, I had access to all the Mountbatten papers now stored in the University of Southampton's fine Hartley Library. I also thank the excellent and efficient staff of the National Registry of Archives in Chancery Lane, the fine librarians at London's Public Record Office in Kew, and those who assisted me in finding the Nehru Papers at Cambridge and Oxford.

Dr. Richard Bingle of the British Library's India Office Library and Records has helped me so often—thanks to his unique knowledge of the many important collections of public and private papers he oversees and to his unfailing kindness and cordial cooperation whenever I call on him—that I find myself again most deeply in his debt. I thank my splendid typists, Jo Perugini and Jane Bitar, for their careful, speedy transformation of my messy typescript into WordPerfect disks.

My excellent Oxford editor, Nancy Lane, has been midwife for eight of my books over the last sixteen years! Nancy, I am so grateful for your criticism and support that I hardly know how to thank you enough.

My dear Dorothy first met Nehru when I did, and I have long suspected that he might not have kept me alone in his office for more than five minutes, but he seemed happy to talk all hour as he smiled and nodded in Dorothy's direction. I think she intuitively understood Nehru long before I did, and I thank her for inspiring me to write this book and for her editorial acuity.

Los Angeles
July 1995

S. W.

CONTENTS

1	Tryst with Destiny, 1889–1905	3
2	Master Joe, 1905–1911	11
3	Reluctant Reentry, 1912–1917	28
4	The Coming of Gandhi, 1918–1920	39
5	In and Out of Prison, 1921–1922	47
6	To Change or Not to Change, 1923–1925	57
7	At Sea, 1925–1927	67
8	Toward Complete Independence, 1927–1928	77
9	Return to Revolution, 1929–1930	93
10	Prison and Round Tables, 1930–1931	112
11	Summit in Delhi, 1931	124
12	Rigorous Imprisonment, 1932–1933	134
13	Out on India's Left Wing, 1933–1934	153

Contents

14	Whither Nehru?, 1934–1935	164
15	European Refresher, 1935–1936	188
16	Reclaiming the Crown, 1936–1937	201
17	Provincial Powers, 1937–1938	221
18	Western Thunder, 1938–1939	237
19	War Again, 1939–1940	255
20	On the Eve of Revolt, 1940–1941	266
21	War Behind Bars, 1941–1942	281
22	From Cripps's Mission to "Quit India," 1942	301
23	Last and Longest Incarceration, 1942–1945	318
24	From Prison to Power, 1945–1946	345
25	Provisional Prime Minister, 1946–1947	369
26	Interim Raj, 1947	384
27	Freedom and Partition, 1947	397
28	Days of Darkness, 1947–1948	408
29	Pandit Prime Minister, 1948–1950	433
30	Himself at the Top, 1950–1956	457
31	Decline and Fall, 1957–1962	468
32	Good Night, Sweet Prince, 1962–1964	484
	Footnotes	499
	Bibliography	536
	Index	539

NEHRU

1

Tryst with Destiny

1889–1905

"LONG YEARS AGO, we made a tryst with destiny," the handsome fifty-seven-year-old prime minister told India's Constituent Assembly in New Delhi that mid-August midnight in 1947, "and now the time comes when we shall redeem our pledge."[1]

Nehru's high-pitched, Harrow–Cambridge nasal tone was strangely soothing and comforting to that perspiring audience of India's elected elite, chosen by ballot as well as birth to represent some 300 million people, most of them peasants—Hindu, Muslim, Sikh, and Parsi—all of them Indians now, his adoring countrymen. A thin fringe of silver was visible below the cotton Gandhi cap set atop Nehru's otherwise bald head. His red rose, replenished several times daily from his garden, adorned the chest button of his golden silk jacket.

"[N]ot wholly or in full measure," a shadow darkening his face as he thought of Pakistan—Jinnah's fragmented, truncated Muslim child born only yesterday, torn so violently by partition from both sides of his own beautiful India—"but very substantially." The shadow passed swiftly, for Jawaharlal Nehru was never one to brood. Adversity only strengthened his resolve and often cheered him. As a rule, he loved danger and pain more than life's placid pleasures. "At the stroke of the midnight hour, when the world sleeps, India will awake to life and freedom. A moment comes, which comes but rarely in history, when we step out from the old to the new, when an age ends, and when the soul of a nation, long suppressed, finds utterance."

No tears for what lay behind, no fears for tomorrow's mayhem and madness. This was the moment Nehru had lived and worked for—all his life it seemed, but forty years at least. For nine of those he was locked behind British bars, yet all that was behind him. Now he had freedom and the power to carve out a new destiny, not only for himself, but also for India, and Asia, and all the world. It was, after all, an era of rapid change, revolutionary in every sense of the word Nehru loved so well.

"It is fitting that at this solemn moment we take the pledge of dedication to the service of India and her people and to the still larger cause of humanity. At the dawn of history India started on her unending quest. . . . Through good and ill fortune alike she has never lost sight of that quest or forgotten the ideals which gave her strength."

Nehru's Darling Indu sat in the front row, smiling up at him. His only child, Indira, was not yet thirty but already the "old soul" he knew her to be.[2] She would carry the torch when his fingers faltered and his body failed. And she had given him two grandsons to continue after his own body returned to ash and dust: Rajiv, her heir, and Sanjay, the "spare," both Nehru boys despite their surname of Gandhi. All three were his progeny, in the best and truest sense, all Nehrus, destined to rule. India would long remain in sound and safe hands after his own aging body had withered away, of that he could now rest assured.

"We end today a period of ill fortune and India discovers herself again. The achievement we celebrate today is but a step, an opening of opportunity, to the greater triumphs and achievements that await us. Are we brave enough and wise enough to grasp this opportunity and accept the challenge of the future?" *Are we brave enough?* His voice grew faint. He shuddered at the rumbling of distant thunder out of Punjab, and the prospect of war in Kashmir. Murder and mayhem had already drenched dusty roads out of Pindi and Peshawar with the blood of Pathans and Sikh Jats, whose scimitars and kirpins would never dry all this long year of brutal civil war that fertilized Indo–Pakistani soil from Lahore to Amritsar. Even now the drums were beating, he well knew, in the eastern districts of rural Bihar and Bengal, where blood lust fed its own fever as Muslim fanatics slaughtered innocent Hindus, who picked up rocks and sturdy sticks to crush neighbors with whom they had lived in peace for generations. *Are we wise enough?*

"Freedom and power bring responsibility," he told them. How many of the weary old men in this crowded chamber tried to teach him that over the past three decades! Yet somehow it never sounded quite right until now. He had always hated paternalism and cautionary advice, which he considered cowardly or reactionary, but now words such as *responsibility* sounded much wiser. His own voice seemed to sweeten them, imparting a different meaning perhaps, a more salubrious nuance. "That responsibility rests upon this Assembly, a sovereign body representing the sovereign people of India. Before the birth of freedom we have endured all the pains of labour and our hearts are heavy with the memory of this sorrow. Some of those pains continue even now. Nevertheless the past is over and it is the future that beckons to us now."[3]

Nehru loved thinking about the future. Planning was one of his passions, for his mind always raced ahead, often flying high above the boring wasteland of the moment, liberating him from prison-cell isolation or a dull routine of any sort. His vision of the future was of one world of socialist humanism, in which each person would create or labor to the limits of his or her ability, and all would receive as much education as they were capable of absorbing and whatever goods or services they needed from a government enlightened enough—thanks to his or Indira's leadership—to provide it. George Bernard Shaw first taught him that; Harold Laski believed it; so did Lenin and Trotsky and Krishna Menon, and even as conservative a Republican as Wendell Wilkie thought some of it was possible. And the Mahatma believed it as well.

But why had Gandhi abandoned him today? Why had he rushed away from Delhi just on the eve of his triumph and greatest glory? He should have been sitting up here on the platform; his naked legs crossed; the hand-spun shawl covering his otherwise naked brown torso; his toothless grin cheering them all; the rimless glasses perched low on his nose; and those Mickey Mouse ears protruding from his bald head, so "ugly" as to be beautiful. For Gandhi's beauty was inside that "great soul," *Mahatma*.

"That future is not one of ease or resting but of incessant striving so that we might fulfill the pledges we have so often taken and the one we shall take today," Nehru reminded his audience, remembering Gandhi. "The service of India means the service of the millions who suffer. It means the ending of poverty and ignorance and disease and inequality of opportunity. The ambition of the greatest man of our generation has been to wipe every tear from every eye. That may be beyond us but as long as there are tears and suffering . . . our work will not be over."[4]

But why had Gandhi abandoned him? Why the sudden urgency to walk barefoot through blood and mud in Bengal and Bihar? Was that more important than being here in this great assembly, from which Nehru's words were broadcast to the world, preserved by the miracle of recording science, so that generations of young Indians, young Asians, world youth, would someday memorize this message, Nehru's first official message as prime minister of independent India, in the way that young Americans memorized the words of George Washington and English youth remembered the stirring words of that other great Harrovian prime minister, Winston Churchill? Was pleading with Calcutta's riff-raff and illiterate peasants, possibly at the risk of losing his own life, more important to the

Mahatma than to hear him now, to congratulate him, to embrace him, to cheer him? For surely Gandhi knew how depressed he, Nehru, also felt at the waste of life, the slaughter that continued and threatened to wash away all they had won. But Gandhi's primary duty was to be here, to address this audience, underscoring this moment for history! Was that not more important to India's future than tramping through blood? Why had he become so strange? So bitter? Was it senility? Surely he was not jealous? After all, Nehru had long been Mahatma Gandhi's disciple. He had once been like a son to that "little father," his *bapu*. And Gandhi was guru to all the Nehru family! But how unpredictable he was! Telling Dickie Mountbatten to ask Jinnah to be India's first prime minister! Hateful, horrible Jinnah, the cause of all this killing and the creator of Pakistan! How could Bapu even suggest such a thing? "He's out of touch," Nehru explained, when Dickie asked his opinion of that idea. He said it sadly, in his self-effacing Cambridge-don style, no bluster or bombast. "Bapu's been away from Delhi too long, you see. I'm afraid he's lost touch with political reality."[5]

How else to explain such madness? Mohammad Ali Jinnah had, after all, said nothing positive, entertained no progressive, no modernist, no useful idea in the last forty years of his bitter, dry, lawyer's affluent selfish life. He had gained more political mileage out of his miserable band of reactionary Muslims in his narrow-minded Muslim League by saying no to every offer ever made to him by Congress or any number of viceroys and secretaries of state for India, in New Delhi, Simla, and London, at conferences large and small. Did Gandhi honestly believe that such a naysayer could suddenly reverse himself and say yes to ruling India? And if ever he did, what would Jinnah's rule be like? Total abject surrender to every reactionary force on earth! Worse than the British Raj! And where would that leave all of us, moreover? Would any Member of a Nehru Cabinet ever be invited to serve in a Government run by Jinnah?

"And so we have to labour and to work and work hard to give reality to our dreams. Those dreams are for India, but they are also for the world. . . . This is no time for petty and destructive criticism, no time for ill will or blaming others. We have to build the noble mansion of free India where all her children may dwell."

His father, Motilal Nehru, had built a mansion in Allahabad. He called it Anand Bhawan, "Abode of Bliss." A strange astrologer once prophesied that none of them would live very long inside that huge white British bungalow and that those who did would never enjoy much "bliss." Nehru never believed in such nonsense, though his cousin Birju swore that he'd seen the "ghost." His mother, of course, always claimed to hear "her" from her wing of the big house, but father pooh-poohed such things, roaring derisively whenever someone mentioned hearing "her" last night! *She* was supposedly the English wife of the former owner, who had been murdered by her husband.[6] As a child he used to look for the ghost, but he never glimpsed her gray figure or floating robes that Birju had so vividly described. Indira once told him that she had seen the ghost, but he always suspected it was her sickly mother, Kamala, who put her onto that. Kamala believed everything anyone told her about ghosts, spirits, voices, visions, auras, and past and future lives. What a sad, lonely creature she was, Kamala. If ever two people were mismatched . . . yet he had tried his best to avert that disaster. But his father would hear none of his protests, overruling each of his objections as he overruled any and every opponent in the High Court. What else could he do but consent? Kamala had seemed so robust as a child bride, yet how quickly she deteriorated, how fast that dread disease consumed her lungs! Even the purest Swiss air and the best European doctors could not save her! They had spared no expense. But now Kamala was "no more," though Indira's sad eyes often reminded him of her.[7]

Motilal had moved them into their "Abode of Bliss" when Jawaharlal was ten, early in 1900. He couldn't remember much about the house in which he had been born, a more modest place in the heart of what was now Allahabad's red light district. Only his father's study remained vivid to his mind, with its large mahogany desk, heavy plush curtains, book-lined walls, Persian carpets, crystal decanters of red wine (which he thought was blood), the fine leather-bound blotter on top of the always neat desk, and the inkstand. And there were two pens. The five- or six-year-old Jawahar thought that his "father could not require both at the same time," so he borrowed one of them to help him practice his alphabets.[8] His new English governess was teaching him both to read and write. Nehru was so deeply immersed in his work that he hardly heard them at first, shouting "Theft!" Not until the servants started racing about, banging doors and drawers, searching every room, did he realize what they were combing the house to find. Then he panicked, trying to hide the pen under his bed instead of returning it at once. Then he ran off, trying to hide in a closet. But the house was too small and the servants were too terrified of the wrath of Motilal Nehru. When his majordomo found the purloined pen, eagerly explaining to "Burra Saab" where he had found it and where he found "Chota Saab" cowering, Motilal's fury fell with a mighty hand on the bare bottom of his trembling, terrified, tiny son.

"Father was very angry, and he gave me a tremendous thrashing," Jawaharlal recalled much later. "Almost blind with pain and mortification at my disgrace, I rushed to mother, and for several days various creams and ointments were applied to my aching and quivering little body." That traumatic experience left deep scars on Nehru's sensitive psyche, affecting young Jawahar in ways he never fully realized. "I do not remember bearing any ill will toward my father because of this punishment," he wrote. "I think I must have felt that it was a just punishment, though perhaps overdone. But though my admiration and affection for him remained as strong as ever, fear formed a part of them." For the remaining quarter century of Motilal's life, his son was very careful never again to rouse his wrath to similar retribution.[9]

When he reached Harrow, Jawahar learned that bare-bottom "birching" remained the customary punishment in the best English schools, as well as in the best family in British India. Nor did Nehru himself ever hesitate to thrash a lazy or wayward servant or a peasant or other simpleton who pushed too close to him in those crowded meetings for which India's nationalist struggle became so famous.

But young Jawahar felt no fear of Swarup Rani, his tiny unlettered mother, hardly taller than himself at the time of that trauma. "I had no fear of her, for I knew that she would condone everything I did and, because of her excessive and indiscriminating love for me, I tried to dominate over her a little."[10] Swarup's upper-caste Hindu mother's approach was to do whatever lay in her power to alleviate the slightest pain or suffering that her darling son—whom she loved much more than her husband—ever felt or feared. For hours each day she knelt at his bedside, covering every bruise with the soothing creams and ointments. She did not permit herself a moment's rest or sleep as long as he remained awake, crying and quivering. He soon sensed how easy it would be for him "to dominate over her" and later often enlisted her as his ally in many of the struggles he waged against his father, especially over his perennial need for more money while living in Cambridge and London. Whether the pain of Motilal's thrashing or the pleasure of Swarup Rani's delicate application of ointments to his "quivering little body" left more potent traces on young Nehru's psyche it is impossible to assess. Both had at least as much impact, however, on his mind and psychic memory as they did on his bottom and back. And both left enduring attitudes toward gender difference on his consciousness.

Men like Nehru's father were to be feared and obeyed, or at least mollified. Women like his mother were condemned for their excessive and indiscriminating love and could easily be dominated, as so many of the women in Jawaharlal's later life were, as indiscriminating sources of pleasure enjoyed for the moment but then dismissed as soon as their demands for closer companionship or love became excessive. In his maturity, Nehru never succumbed to female tears or seductive pleas that might easily have moved a less cold or aloof person. The only two women who affected him differently—moving him in ways that only his closest male comrades ever did, touching the innermost recesses of his mind and spirit—were his daughter Indira and Edwina Mountbatten, each in her own way as courageous, brave, and masculine as any man he had loved.

The first young man to whom Jawahar developed an early attachment was his French–Irish tutor, Ferdinand T. Brooks.[11] Motilal hired Brooks in 1902 when the family moved into Anand Bhawan, and he remained there as Jawahar's closest friend and constant companion for three years, until fifteen-year-old Nehru's first trip to England. "In many ways he influenced me greatly," Jawaharlal recalled. "F. T. Brooks developed in me a taste for reading . . . I also developed a liking for poetry. . . . Brooks also initiated me into the mysteries of science. We rigged up a little laboratory. . . . Apart from my studies, F. T. Brooks brought a new influence to bear upon me which affected me powerfully. . . . This was theosophy. He used to have weekly meetings of theosophists in his rooms, and I attended them. . . . There were metaphysical arguments, and discussions about reincarnation and the astral and other supernatural bodies, and auras . . . it all sounded very mysterious and fascinating, and I felt that here was the key to the secrets of the universe . . . I dreamed of astral bodies and imagined myself flying vast distances. This dream . . . has indeed been a frequent one throughout my life."[12]

Before coming to India, the handsome young Brooks had been a disciple and lover of "elder brother" Charles Webster Leadbeater,[13] a renegade Anglican curate who wrote extensively about "the astral plane" and "the unseen world" of "auras" and after 1906 was accused of child molestation and pederasty on several continents. Eloquent Annie Besant,[14] Leadbeater's leading defender on the board of the Theosophical Society and "the greatest orator in England,"[15] converted Motilal Nehru to Theosophy and initiated Jawaharlal as well. Through Annie, Brooks was hired by Motilal, who obviously knew nothing about his relationship with Leadbeater. Motilal was too busy to read any of Leadbeater's books and had no idea of his open advocacy of "mutual masturbation" with his "younger brothers" as the best way to "help them grow strong and manly."[16] Leadbeater insisted that he could tell from a boy's "aura" when he needed sex, offering to "relieve the pressure" by masturbating him while encouraging the boy to relieve him in the same way. "Glad sensation is so pleasant, Darling," Leadbeater told his boys, the most famous of whom was Krishnamurti, whose "aura" Leadbeater hailed as a "reincarnation of Jesus Christ."[17] Several of Leadbeater's "boys," including Brooks, had severe headaches and breakdowns, and Brooks committed suicide, drowning himself in France a few years after Jawahar left for England.[18]

Nehru was initiated into the Theosophical Society at its annual meeting in Benares (Varanasi) in December 1902, by Annie herself. He recalled that she gave him "good advice and instruction in some mysterious signs" and that "I was thrilled."[19] Leadbeater attended that Benares meeting at the peak of his powers, having just completed a triumphant lecture tour of America and four best-selling books. Nehru never mentions him, however, in his autobiography: "Soon after F. T. Brooks left me I lost touch with theosophy . . . I am afraid that theosophists have since then gone down in my estimation. Instead of the chosen ones they seem to be very ordinary folk."[20]

His admiration for Annie remained warm, however, as did his fond memories of Brooks. Later in life only one other man ever got as close to Nehru as Ferdinand Brooks had, his alter ego (or dark side), V. K. Krishna Menon.[21] The brilliant, Kerala-born Krishna had been recruited into theosophy by Leadbeater's disciple George Arundale and was sponsored by Annie to travel to England to teach at the society's boys school in Letchworth, in 1924. Krishna Menon also soon abandoned theosophy, however, moving to the secular London School of Economics, where he became a disciple of Harold Laski, later one of Nehru's heroes as well. Krishna Menon was also a member of London's Inner Temple bar.

From the first day they met in London, Jawaharlal and Krishna Menon recognized each other as soul mates, brothers in much more than the fraternity of London's bar. Nehru was seven years older than Krishna Menon, who volunteered to serve him in many ways, at London's India League, with Britain's Labour Party, in publishing his books, and finally as free India's first high commissioner in London, where he always proved useful, ready to receive his prime minister at any airport at any hour, to drive him down to Broadlands, Edwina's secluded country house. It was one of only two places on earth where Jawahar and Edwina could enjoy complete privacy.

"At this solemn moment when the people of India through suffering and sacrifice, have secured freedom, I Jawaharlal Nehru," he swore aloud, leading all the other elected members of this assembly in their oath as the conch shell's midnight cry faded to echoes of thunder in this modern chamber, "do dedicate myself in all humility to the service of India and her people to the end that this ancient land attain her rightful place in the world and make her full and willing contribution to the promotion of world peace and the welfare of mankind."[22]

"In all humility" had begun to trouble him more of late, but Indians liked that phrase, always talking of themselves as a "humble people." Or is it "ourselves?" he wondered, never quite sure any more. "A queer mixture of the East and the West," he'd once called himself.[23]

2

Master Joe

1905–1911

"MASTER JOE" NEHRU was enrolled at Harrow in the fourth form, in the Trinity (fall) term of 1905.[1] His father, mother, and sister Nan (later Vijaya Lakshmi Pandit) sailed with him from Bombay in early May aboard the P & O's steamship *Macedonia*. Reaching Dover, they entrained for London "in high good humor."[2]

Motilal had hoped that his son would go to Eton but found no vacancy there. The headmaster of Harrow, Joseph Wood, proved more flexible, agreeing to admit Jawahar after his father's generous pleading, provided the boy could bone up on Latin the preceding summer. There were several "crammers" in London, and young Jawahar went to one run by solicitor Atkins in Dalston Lane while living with the Rev. and Mrs. Tanner at their rectory in Highgate.[3]

The elder Nehrus and Nan went to Germany that summer, seeking curative waters for Motilal's lumbago and pregnant Swarup Rani's morning sickness. From Cologne they journeyed to Bad Hamburg and then to Bad Ems, but none of those waters brought relief for what ailed the ex-wrestler Motilal or his tiny wife, whose pregnancies always proved difficult. By September they were back in London to escort their son to Harrow, where he was given a small drafty room in the headmaster's old Victorian house facing High Street.

Five Indians were enrolled at Harrow when Nehru joined that second most prestigious of England's exclusive public schools, where Prime Ministers Palmerston, Peel, and Churchill all learned early lessons in politics, Byron penned his first poems, and Trollope wrote his first essays. The princely son of the gaekwar of Baroda occupied the best room in Head Master's House, and when he left for Harvard in 1906, the heir of the maharaja of Kapurthala moved into it.[4] He and Nehru became friends and his younger sister, Rajkumari Amrit Kaur, later joined the National Congress and was the only woman in Nehru's first cabinet, serving as his minister of health.

Master Joe soon got used to Harrow and even liked the place, although he played no "footer" (football) and was obliged to "fag" for upper classmen, drawing their baths, lighting their fires and "toshes" (foot pans), and performing all those petty, demeaning acts of personal servitude required of new boys in every fraternity as rites of passage. From his grown-up vantage point at Cambridge several years later, Nehru looked back on his early Harrow years as "the shackles of boyhood."[5] He rarely complained, however, and his weekly letters home, usually addressed to "My dear Father," always closed "With love, from your ever loving son, Jawahar."[6]

Motilal and Swarup Rani were back at Anand Bhawan before November, where a second son was born to them on the same day and month as Jawaharlal's birth, November 14, only sixteen years later. The frail infant lived less than three weeks. The "curse" and "ghost" of Anand Bhawan were much on many Nehru minds at this time, and Motilal felt so depressed after his second son's death that he would have sold the "haunted" mansion if it had not be for Jawaharlal's surprise at the prospect and strong opposition. His wife sided with her son, reminding her husband that an earlier first boy had been born to them, also dying in his first month, long before they moved to this house. And Motilal well knew that he had first married another Kashmiri child bride in her teens who bore them a son, but both died within a year.

Headmaster Wood soon assured Motilal that he was "*fully* satisfied" with his son, making Master Joe "my special care," giving him "my best advice," especially on "the vital question of clothing."[7] Jawahar emerged from his term examinations first in his form of twenty-nine pupils. He won a prize for his excellent work, a copy of Lamb's *Essays of Elia,* which he found "very nicely bound" but said nothing about its contents. He felt more lonely at Harrow, however, than he had been at Anand Bhawan, not having yet found a close friend to replace Ferdinand Brooks. For Christmas Nehru returned to the rectory at Highgate but reported that London was "dull." It was also cold. When walking outside, despite his warm clothing, sixteen-year-old Jawahar recalled "cutting winds went right through me," indeed, "almost paralysed me . . . my hands and feet were quite frozen." However, he described that sensation as "not altogether unpleasant."[8]

Early in 1906, Jawahar acquired his first English girlfriend and wrote his Father about her and how excited he felt at her fondness for him. "You must not confuse real love, with a passing passion, or a feeling of pleasure in the society of a girl," cautioned Motilal. "You know all the arguments against Indians marrying English women." Accordingly, Motilal embarked at this time on a strenuous search for the "right" Kashmiri Brahman girl for his son to marry. Despite Motilal Nehru's English attire and Western habits and tastes, he considered himself first of all a Kashmiri Brahman Pandit and intended to arrange his son's marriage to a daughter of his own Hindu upper-caste community.[9]

Almost two centuries earlier, Motilal's great-grandfather, Raj Kaul, had moved permanently from the princely state of Kashmir with his family to the Mughal capital of Delhi, yet Nehrus never ceased looking to Kashmir as their ancestral home. As Hindu Brahmans, they felt obliged to marry within their own caste, to avoid pollution or eternal damnation. Many Kashmiri Brahmans remained in Srinagar where some of them ran the Hindu maharaja's state government, but with the ever expanding demand for the services of these clever, reliable, resourceful Pandits, more and more of them moved south, first serving the Muslim Mughal emperors and later the British Raj, as Motilal's advocate *(vakil)* grandfather, Lakshmi Narayan Nehru, and police chief *(kotwal)* father, Ganga Dhar Nehru, both did in Delhi.

The Indian revolt (the British called it the mutiny) of 1857 left Delhi in flames and the Nehru family fled south to Agra, where Motilal's father died of tuberculosis in 1861, three months before his youngest son was born. All three Nehru boys studied law. Motilal's eldest brother joined the British Judicial Service, and Motilal and his middle brother Nand Lal became advocates before the provincial high court in Allahabad. In 1887, Nand Lal, then only forty-two, died, leaving the entire family and their lucrative growing practice in the hands of twenty-six-year-old Motilal. But his energy, ambition, and intelligence helped him wrestle his way to the top of Allahabad's bar by the time his second tiny Kashmiri bride bore their first healthy son and heir, on November 14, 1889. Motilal had no intention of now "losing" this precious son to any English woman!

Master Joe's fascination with and attraction to English women was not permanently frustrated by his best-intentioned father, though at seventeen Jawahar was hardly ready to defy or disobey him. "I must have expressed myself very clumsily," Jawahar wrote home. "I did not mean that I wanted to be engaged as soon as possible. On the contrary I had a vague dread [of it]." He assured Motilal that since his father knew so "much more about the persons concerned than I do," he was "far better qualified to make a choice." Motilal had gathered astrological charts as well as detailed family and financial information about each of the young girls considered as appropriate candidates for marriage to his son. All proper Kashmiri marriages were arranged by the parents, with help from intermediaries. The future husband and bride took no direct part until all preliminary arrangements were complete, contracts had been drawn up, and dowry agreed on. The young couple might then actually be permitted to meet. If the shock proved too great for either party, the arrangement might then be called off.[10]

"I wish you all success in your endeavors to find a 'real gem,' " Jawahar wrote from the comfortably remote distance of Harrow. "As for looks, who can help not feeling keen enjoyment at the sight of a beautiful creature? And yet . . . Beauty is after all skin-deep," he added, hoping perhaps to placate Motilal with platitudes.[11]

Master Joe's constant companions at Harrow were, of course, all boys. In February 1907 he went for a run with two of them, and taking their ice skates along they ventured out on a pond covered with thin ice. "The ice cracked like anything and gave way completely at places," Jawahar reported, "but we managed to return unharmed." It was the sort of adventure he sought to replicate many times in later life, skating out on thin ice or jumping into an icy river, testing himself against harsh elements and trying his capacity to endure extremes of heat and cold or his climbing equipment, to see whether the single rope that linked him to life would hold. Or was it self-death he thus sought, like the one that claimed Ferdinand Brooks?[12]

Master Joe's classmates took little or no interest in India; indeed, most of them were indifferent even to British politics. Only Nehru had been able to name Campbell-Bannerman's Liberal cabinet in 1907, when "Honest" John Morley served as secretary of state for India. Motilal presided that year over the first provincial conference of India's National Congress held in Allahabad. Jawahar was anxious to see what his father would say about the suddenly excited state of Indian politics, roused in the wake of the unpopular first partition of Bengal, which Viceroy Lord Curzon's government had approved over the angry opposition of Calcutta's Hindu majority. Bengali leaders like Surendra Nath ("Surrender Not") Bannerjee and Rabindra Nath Tagore were opposed to Curzon's "divide-and-rule" tactics in splitting their Bengali motherland, separating its mostly Hindu western half from the poorer Muslim eastern half. The boycott of all British goods and services swept the land from Calcutta to Bombay, from Kashmir to Madras, bringing equally strong alternative demands for the production and purchase of only *swadeshi* (India-made) goods, as well as for *swaraj* (home rule or freedom).[13]

Motilal disappointed his son, however, by describing the Bengali advocates of the boycott and swadeshi as "oily *babus.*" *Babu* was a British term of denigration used for most Bengalis, meaning "boys," and the noxious addition of "oily" made Motilal sound like the worst Tory reactionaries who filled the British service clubs around India. "Their erratic methods have made me respect them," Jawahar replied, defending the radical Bengalis, whose courage he admired more than the inert condition of most Allahabad leaders. Foremost among the latter was the provincial president, Motilal Nehru, who argued from his rostrum, "I firmly believe John Bull means well." After less than two years living in Britain's bracing intellectual climate, Master Joe expressed much less confidence in John Bull's good faith than his father had.[14]

During his last year at Harrow, Jawahar enjoyed a field-day mock battle against Oxford, whose team had cavalry and a maxim gun. Nonetheless, "We licked them hollow."[15] Nehru learned to handle a gun but never much enjoyed the "sport" of killing birds, to which he was invited by several of his friends. By now he played both cricket and footer, rooting for Harrow's eleven against Eton at Lord's that July 1907, marking the close of his Harrow career. He cried the night before leaving the school he had learned to love, as he did for at least one of the friends he left behind. "I had grown rather fond of the place,"[16] he admitted later. Many years later, after becoming prime minister, he returned to Harrow with that other most famous old boy, Winston Churchill, with whom he enjoyed a quiet lunch. They both wore their blue and white ties and reminisced about the wild "cock-house" matches in which they had fought and the fagging and birching they had survived. Then before that first and only reunion of a British and an Indian prime minister, who had shared the same early school memories, was over, they joined the new boys inside the crowded speech room, on opposite walls of which their portraits still face each other, singing *Forty Years On,* Harrow's oldest song:

> Forty years on, when afar and asunder
>   Parted are those who are singing today
> When you look back, and forgetfully wonder
>   What you were like in your work and your play,
> Then, it may be, there will often come o'er you
>   Glimpses of notes like the catch of a song—
> Visions of boyhood shall float them before you,
>   Echoes of dreamland shall bear them along.[17]

In 1907, of course, neither of the two greatest Harrovians of the twentieth century dreamed it possible that in just "forty years on" one would lead India's nation out of the empire over which the other had tenaciously presided, vowing never to permit that "brightest jewel" to leave Britain's Crown.

The summer before starting at Trinity College, Cambridge, Nehru toured Ireland with his cousin Birju (Brijlal Nehru), raising Motilal's blood pressure at the thought of his son's being in Belfast when the army opened fire there in August 1907. "I would have dearly liked to have been there," the intrepid Jawahar replied from Dublin, where he had attended a Sinn Fein rally that almost brought the tramway workers out on strike.[18] What he saw and heard in Ireland did more to revolutionize his mind than anything he had ever seen or learned about India. Morever, in Cambridge the first memorable lecture he heard was given by George Bernard Shaw and entitled "Socialism and the University Man."[19] Soon after that he started reading whatever he could find by Shaw, who remained one of his lifelong heroes. Two years after India's independence, in the last year of Shaw's life, Nehru was driven from Claridge's Hotel with his private secretary, M. O. "Mac" Mathai, in Shaw's Rolls Royce to Ayot St. Lawrence, with Krishna Menon following them alone in the High Commission's Rolls, in which the three bachelors later drove down to Broadlands. Mac recalled that Shaw told Nehru that he and Stalin were "the only hope" for the world. Shaw was ninety-three at the time. He misspelled Nehru's first name in the *Sixteen Self-Sketches* he gave him, writing "Jawaharial" and insisting, after Mac pointed out the mistake, "keep it like that; it sounds better!"[20]

"I am getting used to Cambridge," Jawahar wrote his father in October 1907. He had put his name down on the "tubbing list," and being the "lightest person" to turn out at the boathouse, they made him "cox" of the practice eight. He retained tiller command of his crew for the next few years. He also attended a meeting of the Indian undergraduate association, Majlis, that first month, despite Motilal's warning to stay clear of any "extremist" Indian clubs. Jawahar insisted, however, that he "failed to discover anything very reprehensible" in the Majlis, which he continued to attend more or less regularly. There was another "Native" Club, which he had hoped might be more politically radical, for it also was started by Indians at Cambridge, but he soon learned that a "native" in Cambridge meant a "raw oyster." He informed Motilal that quite unjustly, "Cambridge Indians have got a bad name on account of the doings of a very small number of gentlemen."[21]

For his first year at Cambridge, Jawahar rented spacious and comfortable lodgings at 40 Green Street, an easy walk to the main gate of Trinity, in the thick stone wall that surrounded the handsome quad of the college's grounds and dorms. The intellectual atmosphere of Cambridge stimulated his mind and made him more daring in his letters to his father, whom he tried to convince that India's "extremist movement" was no worse than

Ireland's Sinn Fein, arguing: "Their policy is not to beg for favours but to wrest them."[22] Motilal, however, became more conservative as his son got more radical, fearing that the bonfires of boycott and swadeshi threatened to burn down all he had built in Allahabad, personally as well as professionally. He wrote an article for the English *Pioneer*[23] urging a return to the politics of moderation and loyalty to the British Raj, which had done so much good for India, helping advance its poor people in so many ways, bringing all the arts and sciences of the Western Enlightenment to India's previously dim and troubled shores. Jawahar read his father's article and wrote that "I do not like it at all . . . I had till now an idea that you were not so very moderate. . . . But the article almost makes me think that you are 'immoderately moderate'. . . with strong 'loyalist' tendencies."[24]

Motilal was enraged when he read that critique, for it accurately described his political persuasion. At the peak of his professional powers, head of his extended family as well as of Allahabad's bar, and one of Congress's most respected leaders, whose dinner guests and friends included British governors and judges, collectors and commissioners, and members of the viceroy's Executive Council, forty-seven-year-old Motilal Nehru was not accustomed to such condescending criticism. And especially from a young man who wired him less than a week later for more money to cover the costs of a long Christmas holiday that he and Birju were enjoying at the Cairn Hydro in Harrogate.

"We have had a lot of entertainments here during the past week, mostly dances," Jawahar wrote, explaining why he had spent his last pound. "There was one fancy dress ball and some Cinderellas," he added, reporting in greater detail about a most interesting staff ball in which all the servants danced with all the visitors. If anything could upset Motilal more than his earlier letter, it was this remarkably frank report. "I . . . waited on the girl who serves us at table. She was the best dressed and the prettiest of the lot. . . . Birju Bhai ("brother") asked one of them [waitresses] for a dance and she answered 'Yes, Sir.' It was so funny, her saying 'Sir.' " Motilal did not find it the least bit funny. He intensified his search for the proper Kashmiri bride for his wayward son, more anxious than ever to bring Jawahar home soon so that he could marry and start practicing before the High Court as his junior partner.[25]

"I felt elated at being an undergraduate with a great deal of freedom," Jawaharlal recalled. "I had got out of the shackles of boyhood and felt at last that I could claim to be a grown-up." He and his friends discussed Nietzsche and Shaw and the "latest" by Lowes Dickinson. "We considered ourselves very sophisticated and talked of sex and morality in a superior way," although looking back he confessed that "most of us were rather timid where sex was concerned. At any rate I was." He admitted to "a certain shyness . . . as well as a distaste for the usual methods." He thought perhaps that he remained "a shy lad . . . because of my lonely childhood." In addition to Shaw and G. Lowes Dickinson, Nehru read and admired Oscar Wilde and Walter Pater, labeling his "general attitude to life" a "vague kind of Cyrenaicism." He enjoyed "a soft life and pleasant experiences," emerging from Cambridge "a queer mixture of the East and the West."[26]

Motilal was angry enough to stop his monetary bailouts and to bring his prodigal son home, but the gentle Swarup spoke up in defense of him. She understood her husband's temper well, but fierce as the thunder might be, she knew the fury of his wrath was fleeting. At last she saw him nod slowly: "Boys must be boys, I suppose." Then he sent the £100 cheque Jawahar needed "most desperately" on the eve of his "expulsion" from Green Street, where he owed £36 in back rent, with a current balance of £2 in his Cambridge bank account. Neither his cousin Birju nor his other Cambridge cousin, Shridhar, could lend him enough to cover his deficit, though a third Nehru cousin, Krishnalal (Kishan Bhai), then studying medicine at Edinburgh University, did come through with some funds, which Jawahar expended with his usual alacrity, off to Paris for an Easter week holiday with Birju.[27]

Much to the Nehru cousins' chagrin, however, their Paris hotel "proved rather expensive," as Jawahar reported to his father upon his penurious return to Birju's London lodgings. The "hotel bill," he explained, "staggered us." It was fifty francs "more than we had expected. . . . The mistake arose from our not having included a day in our estimation."[28] Poor Paris servants had to be left without "tipping," and worse pity, hungry young Jawahar and Birju could not afford "a substantial meal" till they reached foggy London.

Motilal suffered a severe "asthmatic flareup" when he read that letter and went into a depression whose source none of his doctors could diagnose. Motilal Nehru had always been the most cheerful, optimistic, convivial bon vivant in Allahabad. His laughter was legendary, so loud and prolonged that it could almost be "heard in Lucknow," some of his dinner guests at Anand Bhawan used to say. And when he opened his mouth to argue a case in court, no judge blinked an eye or nodded off to sleep, for Motilal had as powerful a voice as his wrestler's biceps. Yet now, suddenly, he could barely breathe, and his hair had grown perceptibly whiter, almost overnight. The more superstitious servants blamed Burra Saab's illness on the ghost, but this time Swarup Rani knew that the cause of his malady was more distant yet intimate. Jawahar now wrote his mother to say how "glad" he was to learn "from father's letter yesterday that he has decided that I should come to India this year."[29]

His earlier letters had urged his father to bring his mother and sisters Nan and Betty to London for the summer, but now Motilal wanted him home for a prenuptial "screening" at least, if not the actual marriage. Motilal had begun to fear that he was losing his only son to the high life of London, Paris, and Cambridge. Wife and work were the cures, Motilal believed, for most adolescent ailments.

"I am very sorry to learn that father is unwell," young Jawahar wrote his mother. "I am particularly distressed to hear that he is depressed and might not have liked something I wrote to him in my letter a few weeks ago. What I wrote to him was written purely in fun and it never occurred to me that it could offend him. I was surprised [at] . . . such a silly thing." He promised to write his next letter directly to father, to apologize. "So father thinks I have changed in Cambridge. I cannot understand this," Jawahar added, but then he seemed to grasp it a bit, reflecting perhaps on how wise his parents were. "Well, everyone slightly changes when he is a student and I think I am no exception. But I do not think I have changed more than that. . . . You have written that I am more interested in games and do not pay enough attention to my studies. To some extent, this is. . . . true. I did not study much during the last two terms. . . . I used to spend a lot of time every day on the river. I did this only because father desired it. . . . I may play tennis occasionally. You misunderstand about my dancing. I do not go to dances every day. . . . Father himself wanted me to learn it." He was only trying to do what his father wanted, and now they kept reminding him of his cousin Shridhar's remarkable success at Cambridge. Shridhar was a grind, who never stopped studying. After taking a first at Cambridge, Shridhar Nehru went on to do his doctorate at Heidelberg and returned home in 1913. He joined the prestigious British Indian Civil Service, serving as collector, magistrate, and commissioner in the United Provinces, where his first cousin spent so many years behind British prison bars while Shridhar held the keys in his pocket.[30]

A week after sending his diplomatic softener to his mother, Jawahar wrote his father saying how "very sorry" he was to "hear that you were unwell and had the toothache again. I suppose . . . that you have been overworking yourself as usual. I know well that you can hardly help working hard and yet I think that you ought to take better care of your health and have the right amount of exercise every day. . . . I shudder to think what the consequences would be if you had a breakdown more serious than the last. One of my reasons for not wishing to go to India was the thought that your coming here would do you a great deal."[31] "I have been told that you did not like something I wrote to you a few weeks ago," Jawahar continued. "I was rather surprised to know this as what I wrote to you was written purely in fun. All the same I was right sorry I ever wrote that or thought it. But what amazes me is that you should have ever thought me capable of being guilty. . . . I am sure you will pardon me for an offense which I did not intend to commit."[32]

Had the traumatic aftermath of taking his father's pen returned to haunt him before Jawahar finished that letter? Motilal wrote no more about the insult to his integrity but wanted his son to sail home in early June. Jawahar argued that his tutor had adamantly "refused to let me go down a single hour before the end of full term." This meant that he could not possibly leave before mid-June, and anyway he had "tons of things" to do in London, especially buying gifts to bring home, and he wanted to visit Oxford as well. He hoped in a year or two to transfer to Oxford, to continue his studies for several more years there. He had met some congenial Oxford students and attended a few balls there, and the thought of returning home to marry and settle down again in Allahabad was too dreadful to contemplate. He had also invited his best friend at Harrow to come and stay with him in Cambridge for the last week in May.[33]

The summer at home in 1908 proved a disaster from Jawahar's point of view and a failure for his parents, who showed him at every Kashmiri wedding in Delhi, Agra, and Allahabad, introducing him to every available Kashmiri beauty in north India, but none captured his heart.[34] Nehru could hardly wait to go back to what had by now become his real home—England. When he returned in early October he knew, however, that the only way he could stay in England with his father's continued support was to enroll in one of London's temples of the bar, although he had no more interest in the law than he did in marriage. But both his mother and father had made it quite clear to him that with Motilal's health deteriorating as quickly as it was, it would be imperative for his son to enter the practice as soon as possible unless he wanted to throw away everything his father had worked so hard all his life to acquire. Jawahar could hardly admit, of course, that he did not really care what happened to his father's business as long as Motilal continued to support him in the proper style to which he had grown accustomed. One bothersome change Motilal demanded was that he abandon his fine lodgings on Green Street and move into a room in Trinity's shabbiest dormitory, Whehwell's Court, outside the main courtyard gate on Trinity Street. It seemed a paltry saving for so substantial a fall in the comfort level of his daily and nocturnal life over the next year.

Jawahar had to go to London to apply for admission to the Inner Temple on December 1, 1908, after his Trinity tutor, William Fletcher, introduced him to his old barrister friends Henry Bond and W. Rowse Ball. Following their interview, Bond and Ball certified that "Mr. Nehru" was a man of "good character" and "gentlemanly behavior" and was "likely to be popular and successful" at the bar.[35] Then for the not inconsiderable fee of £40 16s 3p, Jawaharlal Nehru gained admission to the London Inn of Law established on land acquired by the Military Order of Knights

Hospitaller in the twelfth century, whose graduates included Prime Minister George Grenville and seven lord chancellors, as well as one of Britain's most eminent judges, Sir James Stephen.[36]

Two weeks after his admission to the Inner Temple, from which he could graduate only after passing his exams and keeping twelve terms by attending at least three dinners each term in the temple's hall, Jawahar was off for his Christmas vacation at Harrogate's hydro, of which Leadbeater and Brooks had been so fond. The hot tubs there kept him quite warm during what had been one of London's severest winters. From Harrogate he went with another of his English emigré cousins, Jivan Lal Katju, to Heidelberg to visit his cousin Shridhar. Back in Cambridge by mid-February, he played a good deal of tennis and did some riding. He was quite comfortable on horseback now, as he would remain for the rest of his life.

His mother and father kept pressing him to accept one of a fast-diminishing number of young Kashmiri women still awaiting his nod. "You and Father are unnecessarily worried about my marriage," Jawahar wrote his mother, arguing that it was "not essential for me that I should marry a Kashmiri." After giving the question of arranged Hindu-caste marriages considerable thought, Jawahar was more convinced than ever that "everyone in India should marry outside his or her community." His enlightened Western approach would, to his later chagrin, be accepted and followed by his daughter, but it was not appreciated by his own father or mother. Motilal, in fact, threatened to leave everything and rush to England immediately to bring his wayward son home.[37]

Jawahar now wrote to say how sorry he was "for all the trouble I am giving you over such a detail as Marriage." He knew, of course, that to his aging parents it was no detail, but the primary concern of their lives. Yet he had to try, before it was too late, to explain his free spirit and cloistered heart. "Personally, and I mean it, I am not violently looking forward to the prospect of being married to anybody," he wrote his father. "The joys of matrimony appear to me wholly imaginary, and do not at all appeal to me." As important as the fact that Nehru's seven most vital adolescent years were spent in England or elsewhere away from home was that they were also lived virtually without the benefit of any intimate or consistent female companionship. Between the ages of fifteen and twenty-two there were really no important women in Jawaharlal Nehru's life. Perhaps even more significant, he never seemed to mind it.[38]

What he did mind, however, and still feared most was losing his father's support by rousing his wrath. "You, doubtless, will object to my remaining in the unblessed ranks of the unmarried," he added in his most rapid about-face, "and that settles it. . . . As regards the girl if marriage it is to be—I am, on the whole, in favor of your latest choice. My opinion, of course, will coincide with yours. . . . In any case, my opinions are of secondary importance. You are the man on the spot and are best able to judge. There is just one point which I should like to remind you of and that is the possibility of my marrying outside the Kashmiri community. If you ever consider this practicable you will find me strongly in favor of it."[39]

By so frankly opening his letter, admitting that he really had no desire to marry "anybody," Jawarhar was equally honest when he ended it by giving his father a free hand to select any bride for him he wished. What difference, after all, did the girl make to Jawahar? He wanted none of them. To Motilal and Swarup, pedigree mattered most, for to Brahmans nothing was more important than choosing one's mate from one's own caste, which really meant one's own community. That alone could ensure future rebirth or, better still, the ultimate "release" *(moksha),* which to every good Hindu meant salvation. To Motilal and his wife, the mere idea of even looking outside their community for their only son's wife was unthinkable.

After two years at Cambridge, however, Jawaharlal's frame of mind, dress, and manners were those of an upper-class Englishman, rather than of any caste of Indian, recalled Syed Mahmud, who met him at the Cambridge Union in 1909. Nehru wore "a pink shirt, with socks and tie to match," and was remarkably "soft spoken" and "quiet."[40] The Irish radical Fenner (later Lord) Brockway also met Nehru at the Cambridge Union in the same period, finding him "almost European." As he did whenever offered a platform, the fearless Fenner attacked British imperialism, in what young Nehru considered an "extremist" speech. Nor was Nehru ever to lose that air of upper-class English politesse and slightly disdainful ennui. When Ian Stephens first met him in 1931, he felt "as if I was back in Cambridge talking to a young Cambridge Don."[41] A decade and a half later, Woodrow Wyatt (Lord Wyatt of Weeford), who first met him on a cabinet mission, felt instinctively at ease with him, for Wyatt had also gone to Harrow and saw Nehru as "an English public schoolboy."[42]

Nehru certainly reacted as an upper-class Englishman to the death of King Edward VII in 1910, reporting to his father from Cambridge in May that the past week "has not been a very cheerful one. . . . [O]n Friday last . . . we first heard of the King's illness. And that night he died. . . . I was sorry to hear of it for he was not a bad fellow. And the suddenness of his death had an element of the tragic about it. . . . On Tuesday last the new King [George V] was proclaimed here. The Vice-Chancellor read out the communication. . . . A couple of gorgeously attired trumpeters were present but unhappily their command of the trumpets was of the most limited kind and consequently the result was not of as dignified a character as the occasion merited . . . the saddest . . . death of Edward the VII of Blessed and Glorious memory."[43]

After much anxiety that he might not pass his tripos (final exams) at Trinity, Jawahar managed to reach what he called the "Olympian heights" of a second class, which was lower than his father's expectations but better than the gentleman's third, with which young Nehru would have been "very content."[44] Wearing his most elegant evening attire with ermine-collared cape and white tie, Nehru knelt before Cambridge's vice-chancellor, hands joined straight-fingered together Indian fashion. "The V.C. muttered something about the grace of God in Latin and lo and behold!" he reported to his father, he was now a graduate of Cambridge University.[45]

His mother asked whether he would still treat her "with respect" after he returned home in so exalted a status. "What is the use of a man being educated if he does not know how to behave towards his parents?"[46] Jawaharlal reassured her in Hindi. Motilal now urged him to study law at Cambridge while keeping his requisite dinner terms at the Inner Temple, so that he might return home in April 1912 with an LL.D. as well as his barrister's wig and gown. But Jawahar was "tired of Cambridge" and wrote again about hoping to spend "the next year or so at Oxford instead of Cambridge." His reasons were three: "I am getting rather bored with Cambridge," he candidly explained. "I can afford to spend a year or even two years at Oxford," and the latter would mean having to wait only two months more than his Easter 1912 date for completing his temple bar terms, as Oxford's graduation was in June. "My chief reason for wishing to go to Oxford is," however, Jawahar added, that "Cambridge is becoming too full of Indians."[47]

The impact of this letter on Motilal's health can only be imagined. It is small wonder that Swarup Rani had asked him the timid question she did! *Too full of Indians*? What did it say about his self-image? ("Galbraith, I'm the last Englishman to rule India," Nehru much later confided to another old Trinity, Cambridge, man, U.S. Ambassador John Kenneth Galbraith.) As to his being able to "afford" spending a year or even two at Oxford, Motilal may have reflected, "Why shouldn't he?" And what difference could it make to Jawahar whether he came home in April or late June? Why rush simply to get married and see two old parents?[48]

Jawahar did rush, however, to embark on a sudden trip to Norway that July, one that almost took his life. His traveling companion was an unnamed Englishman, possibly the Harrow companion who had lived with him in Cambridge for a week two years earlier, one of his Trinity crew, or perhaps a new Oxford friend, which might help explain his eagerness to move there. In his autobiography, Nehru's cryptic account of his "accident" in Norway refers only to "my companion, the Englishman," also described as "a young Englishman," but never named. Nehru's reluctance to name him seems almost more strange than his account of the"heroic" role played by that young Englishman in "saving" his life.

Jawahar's earlier letter home in August 1910 was, however, written from his unnamed friend's "private family hydro" in Maidenhead, recounting their trip and the "accident." They had left their cruise ship at Visnes for "an overland excursion" and found "magnificent scenery on our way," then "eventually spent the night at a primitive kind of hotel about four or five thousand feet above sea level. I being one of the most energetic members of the party arrived at this place with another some time before the others. Both of us felt so hot after our tedious climb that we decided to have a bath immediately. The hotel people were much surprised by this strange request. . . . After some thought, however, they informed us that some times, when people wanted to wash themselves they went down to the river. We therefore went to the rushing mountain torrent, which was called the river, and plunged in . . . the water was of course icy cold. We were both quite numbed entering. After a while my foot slipped and I completely lost control of myself. I was being merrily borne along with the current when my companion stopped me. The exciting part of it lies in the fact that there was a mighty water-fall of about 400 ft. quite near. I am sure I could not have stopped myself before I reached the bottom of the fall."[49]

A quarter century later, when Nehru wrote his autobiography in a British Indian prison, his companion of this letter, also merely referred to as "another," acquired his "Englishman" identity, thus clearly differentiating him from the Indian nationalist leader Nehru had by then become. At the time of the accident, however, Nehru obviously felt no "national" distance from his intimate friend with whom he must have raced nude into the mountain torrent. In his autobiography, Nehru adds that "armed with table napkins or perhaps small face towels, which the hotel generously gave, two of us, a young Englishman and I, went to this roaring torrent which was coming from a glacier near by. I entered the water; it was not deep but it was freezing and the bottom was terribly slippery. I slipped and fell and the ice-cold water numbed me and made me lose all sensation or power of controlling my limbs. I could not regain my foothold and was swept rapidly along by the torrent. My companion, the Englishman, however, managed to get out and he ran along the side and ultimately, succeeding in catching my leg, dragged me out. . . . Later we realized the danger . . . " and he tells of the waterfall "two or three hundred yards ahead."[50]

Can Jawahar's strange accident in Norway be read as his own carefully doctored metaphoric confession of a passionate, "hot" and "icy cold"—

indeed "numbing"—love-affair with a young Englishman too important for him to name, too dear to forget, his heroic other? That resourceful companion had great strength and a firm grip, which saved Jawahar when he had "completely lost control of myself," in what might otherwise have been a fatal, or at least life-threatening "fall," the sort of death his wretched tutor Brooks had chosen.

A fortnight later he was back in London, still hoping to enroll in one of the three Oxford colleges to which he had written, with Oriel being his first preference, then Trinity, and New College. He had not gone to Oxford to interview, fearing that that might be a problem, since he had never mentioned in his letters to the heads of those colleges that he was Indian, as he explained to his father, hoping that one might promise to take him, and then find it difficult to get out of his promise when he saw Nehru's skin color. Motilal refused, however, to encourage any move to Oxford, urging his son to remain in London if he could no longer abide Cambridge.[51]

Jawahar seems to have found even London dull by now, however, hurrying off to Buxton for a week or so, planning another trip to the Continent, first to Paris and then Dieppe. In late September he wrote his father from the casino at Dieppe, again referring to an anonymous companion by starting with "We came here from Paris yesterday," planning to stay another four days, and then shifting to the first person with "I hope to go back to London on Sunday, and thus end my wanderings for the summer of 1910." Such travel, of course, was "far from inexpensive," and he thanked his father for his "welcome cheque." By now Motilal had not only decided against Oxford but also had given his son one of two choices: either live in London or come home immediately.[52]

"I have already decided to remain in London," Jawahar replied, remembering his father's temper. "I intend doing Law, and law alone, for the next few months." One additional subject interested him, however, "if I think I can spare the time," and that was economic philosophy, which he had heard was taught brilliantly at the London School of Economics. But is was another full year before he found any time.

For most of the winter of 1910 Jawahar was kept busy preparing a gala "At Home" reception show for the father of his old Harrow friend, the gaekwar of Baroda, who was visiting London in late November. The gaekwar's son, Maharajkumar Jaisinghrao, had finished at Harvard and was back in London, which helped cheer up Jawahar and his several cousins who also were there. Nehru had "to labour greatly" to make the gaekwar's At Home "a success," as he explained to his Father; "The *piece de resistance* on our program will be a series of tableaux from Omar Khayyam. They will be taken from [Edmund] Dulac's illustrations to Omar. I believe you have got that book and so you can form some idea of what guys we are going to make ourselves. . . . We have to be rehearsing most of the time and the spare time is filled up by visiting Mr. Clarkson [a famous couturier in London] to arrange about the wigs & make-ups. There was a great scarcity of men willing to take part in the tableaux and so I had to offer myself most unwillingly."[53]

Wearing his wig, made up with lipstick, powder, and eye shadow, his body draped in silks and satins, Jawahar "most unwillingly" offered himself up night after night to those "endless" rehearsals for the gaekwar's At Home as a beautiful young girl, holding out her jug of wine and loaf seductively to her poet lover, Omar. Nor was that the only time he used those expensive silks and wigs. Artistic "tableaux" performers were very much in demand for those seductively dim velvet-draped Victorian sitting rooms owned by his aristocratic Cyrenaicist companions all over London, and there was always "a great scarcity of men" willing or able to take the woman's part in the clever games they all loved to play. "I enjoyed life and I refused to see why I should consider it a thing of sin," Nehru wrote of his odd behavior and hedonism at this time. Looking back, however, he later confessed, "I was a bit of a prig with little to commend me."[54]

Jawahar enjoyed his last year in London as much as possible, dancing the *Chocolate Soldier* and the *Quaker Girl* waltzes, playing tennis, and attending the processions associated with George V's coronation. He was, of course, "foolish enough to run up some bills," he wrote his Father in June 1911, requesting another £150 less than a month after he had acknowledged the earlier receipt of £100 urgently needed. Motilal's patience proved as deep as his pockets, and even though his health worsened with each new letter from his son, he never cut him off, resolving only to conclude the marital arrangements and bring him home as soon as he was called to the bar.

"I do not, and cannot possibly, look forward with relish to the idea of marrying a girl whom I do not know," Jawahar wrote that December. "At the same time . . . [i]f you are intent on my getting engaged to the girl you mention I have no objection . . . I shall abide by your decision."[55] Kamala Kaul was then only twelve, living in the huge old Delhi compound owned by her Kashmiri grandfather, Pandit Kishan Lall Atal, whose fifth son, Jawaharmul, was Kamala's father. Motilal called her "a little beauty," and although she was quite plain in appearance, she seemed "very healthy" when the elder Nehrus first met her that summer in Delhi at another Kashmiri wedding. The contract was, therefore, drawn up, and all dowry and other details were agreed on, contingent only on Jawahar's final approval after he returned from London. As a British barrister as well as a Cambridge degree holder, the sole heir to Motilal's legal practice and accumulated fortune in real property, there was no better "catch" in all of north India, not to speak of the tiny Kashmiri Brahman community. Kamala's parents and grandparents were overjoyed at their "great good fortune." None of them dreamed of how tragic a trial this marriage would prove to their little daughter.[56]

Jawahar made one last effort to get himself out of the marriage, writing to his mother from Beacon Hotel in Sussex, where he had gone for a brief holiday in March 1912: "My only fault is that I do not wish to marry a total stranger. Would you like me to marry a girl who I may not like for the rest of my life or who may not like me? Rather than marry in that way it would be better for me not to marry at all." He meant it and was quite right; but there was no escape, for either of them. Parents knew best, as every Indian child was taught from the day he or she was born, and parents had the most experience in such matters and so were best able to choose the proper mates for their children. "I accept that any girl selected by you and father would be good in many respects, but still, I may not be able to get along well with her," Jawahar tried desperately to explain. "In my opinion, unless there is a degree of mutual understanding, marriage should not take place. I think it is unjust and cruel that a life should be wasted merely in producing children."[57]

Swarup Rani had no soothing balm this time for her son's emotional pain, for hers had been precisely that sort of "wasted," life, devoted "merely" to her children. She had no reply for that letter, but Motilal answered for them both. "Your last letter pained and surprised me very much," Barrister Jawahar (called to the bar on June 13, 1912) replied, admitting that he had "spent far too much money" and had not given nearly as much attention to his studies as he should have done. One of his aristocratic English friends had "lost" a £40 loan from him, for which "I repented of my folly." By demanding an account of his expenditures during the last six months, however, Motilal appears to have been challenging his barrister son's "reputation for veracity." Jawahar answered, "I suppose you have ceased to trust me altogether . . . the very idea of furnishing accounts is anathema . . . I am not desirous of staying in England or anywhere else a single day under these conditions." Motilal repented for having wounded his son's pride, wired more money, and urged Jawahar to stay in London through August. So he booked passage to Bombay on the ship that would bring him home after more than seven years abroad.[58]

3

Reluctant Reentry

1912–1917

MOTILAL NEHRU rented a bungalow in Mussoorie, the favorite cool British hill retreat in the United Provinces in the summer of 1912. From Bombay, Jawaharlal took the train to Dehra Dun, the nearest station, where his father and the servants were waiting with the horses, on which they galloped uphill to his mother and sisters. "I can still see them," his sister Betty recalled half a century later, "Father in his superbly cut white breeches and polished boots on a big bay horse, roaring with happy laughter, and beside him a slim, handsome young man with rather delicate features. . . . He slipped out of the saddle throwing the reins to a groom, and ran lightly to hug my mother, lifting her right off the ground. Then he was engulfed in a multicolored wave of female relatives . . . looking over their heads he spotted me. . . . I was almost frightened as he came toward me and swung me up in his arms . . . and kissed me." Betty was only five at the time and "did not like my brother much at first," for he was "the most awful tease" and she was afraid of him, never knowing what his next prank might be. When she mounted her pony he might "sneak up" and "whip" it, making the shy creature "rear or bolt." Once the pony ran away with the terrified child "clinging to his back," making Motilal angry at his son, but Jawahar "only laughed and said it was good for me." The "meanest thing" he ever did to his sister was to grab "my leg and throw me into deep dark water" in the indoor swimming pool at Anand Bhawan, shortly after they returned to Allahabad. "I forgot everything I knew about swimming and went down like a rock. Of course Bhai dove to the rescue and soon had me up . . . dumping the water out of me." But it was years before Betty lost her fear of water or of her brother.[1]

Nan, however, was seven years older than Betty, and she was the beauty of the family who had for years waited eagerly for the adored brother to return home. "For me it was, in a very real sense, an awakening," Nan recollected. "He had time to ride with me, read to me, and encourage me to discuss things with him. . . . To me Bhai was a knight *sans peur et sans reproche.*"[2] Nan was eleven years younger than Jawahar, almost exactly Kamala's age. She became, in fact, the first adult woman Jawaharlal loved, in a truly platonic sense, for she was beautiful, bright, sweet, charming, and good, all the things he always imagined women should be, and since she was his sister there was never that wretched, awkward sexual tension he always felt with Kamala and hated. For years Nan remained the family member closest to Jawahar, and when he became prime minister she was his first ambassador to Moscow, then Washington, and finally his high commissioner in London. She was the first woman elected president of the United Nations General Assembly. After her brother's death, although she lived another quarter century, Nan was never appointed to anything by her niece, Indira, who had so long vied with her for Nehru's primary affection.

Returning to Allahabad from Mussoorie, the twenty-three-year-old barrister Nehru joined the High Court, and soon "I felt that I was being engulfed in a dull routine of a pointless and futile existence." He found the "intellectual and cultural torpor" of British India so "insipid" that he knew exactly how his Cambridge idol, G. Lowes Dickinson, later felt, as reported to E. M. Forster and quoted by Nehru in the British edition of his autobiography: "And *why* can't the races meet? Simply because the Indians *bore* the English. *That* is the simple adamantine fact."[3] What Jawahar found, moreover, was that the old colonial Englishmen he met in Allahabad were as boring as the Indians. "Each bores the other and is glad to get away from him to breathe freely and move naturally again."

The only diversion Jawahar could find for his unused energy and intellect was Indian politics. He never really took to the law or legal practice, the way Motilal did, or as did the brilliant barrister Jinnah, one of his father's earlier juniors, now fast becoming a leader of Bombay's bar as well as of India's National Congress and soon of the Muslim League. Nehru was still much too shy to get to his feet in the High Court and speak up, trembling and muttering whenever he felt himself so obliged. It would be some time before he found his public voice. Politics at least offered him an arena of conflict, one sufficiently diverse to invite intellectual dissent, indeed, even to reward the maverick outsider who felt perpetually dissatisfied and alienated by every social and political institution confronting him in this impossible imperial Raj.

Jawahar went with Motilal to his first meeting of the Indian National Congress at Bankipore, Bihar, in December 1912. Backward Bankipore almost made Allahabad seem like Cambridge, but worse than its remote provincial venue was the ultra moderate character of those "English-knowing upper-class" Indians, who came there dressed in "morning coats and well-pressed trousers."[4] Five years earlier, at the peak of the boycott by India's National Congress against the British Raj, after the first hated partition of Bengal in 1905, Congress had split into two "parties,"[5] this loyalist moderate rump that now met in Bankipore and a "revolutionary" new party led by the fiery young Bipin Pal of Bengal, who emigrated to a cold-water flat in London, and the older Lokamanya ("friend of the people") Bal Tilak of Maharashtra, who was jailed for sedition and transported to Mandalay prison, where he remained until the eve of World War I.

"British rule is recognized by all rational and thoughtful persons to be a Providential dispensation, destined to contribute to the material, moral and political elevation of this land," the Bankipore Congress president R. N. Mudholkar told his silent, slightly somnolent audience. Motilal, of course, agreed with Mudholkar, as did the ailing Gopal Gokhale, the brightest and hardest-working moderate politician, hailed after his death in 1915 by Mahatma Gandhi as "my political Guru."[6] Jawahar was "impressed by him," finding Gokhale "one of the few persons present who took politics and public affairs seriously."[7] He also obviously agreed with the goal expressed by Mudholkar for Indians, to be claimed as "British subjects" with the "full rights of British citizenship."[8] What Nehru did not appreciate, however, was the timid moderate loyalist insistence that the only way to reach that goal was by "peaceful methods."

"Politics . . . to me," Nehru insisted, "meant aggressive nationalist activity against foreign rule." He had listened to Sinn Fein agitators in Dublin, heard Bipin Pal address his Cambridge majlis, and audited lectures by George Bernard Shaw. The idea of attending sessions of Congress once a year simply to hear old men in tails caution him against violence and "unconstitutional" methods of agitation was not only deadly dull to Jawahar but also highly insulting. If politics was to be his game, he had no intention of playing it according to rules set by dim-witted old boys who knew nothing either of London's latest fashions or the intellectual radicalism of modern Cambridge, Oxford, and Dublin.[9]

"When he came back to us from England he was more West than East," his sister Betty recalled, "with his superbly tailored clothes from Savile Row and his head full of radical ideas that the young intellectuals at Cambridge were tossing around. They had made him a convinced Socialist. . . . In some ways he was a utopian Marxist until he died."[10]

Less than two years after Bankipore, the shot fired at Sarajevo was heard round the world, and when Great Britain entered World War I on August 4, 1914, Viceroy Lord Hardinge notified British India that it too was at war. The "war to end all wars" had thus begun, and India's initial response was universal, hearty support for King George and the empire. Even fifty-eighty-year-old Tilak, just released from Mandalay prison, cabled his "loyal support" to George V. Forty-five-year-old barrister Mohandas Gandhi, back in London after winning a short-lived "victory" over Smuts in South Africa, urged every Indian there to "think imperially" and recruited Indians for a field ambulance training corps. He then launched a mini-Satyagraha ("hold onto the truth") campaign against its British colonel before they were shipped to the front in France, thus saving his own life for more important work in India. All of India's 570 princes wired their support of the defense of the empire, and the twenty-seven largest princely states immediately placed their well-trained Imperial Service troops at the disposal of the viceroy. The *Loyalty,* a hospital ship paid for by several princes, steamed off to France with the first of many Indian expeditionary forces, initially totaling fifty thousand Anglo-Indian troops. They left Karachi in late August, just in time to reach the Western Front and to defend it against the German assault at Ypres. By year's end more than seven thousand Indian troops lay dead in Flanders field, and by the war's end nearly 100,000 of the more than one million Indian troops sent abroad had fallen in Mesopotamia, at Gallipoli, and on the Western Front.[11]

Most Indians, like most Englishmen, expected the war to end swiftly with an Allied victory, in a year with luck, or possibly two. No one but Kitchener dreamed it would last as long as four. India's political leaders all believed that their "loyal support" for Great Britain's world struggle against Prussian forces of "evil" and "chaos" would bring India its just reward of political freedom. Dominion status—if not immediately after war's end, then a few years later—was almost universally anticipated. As the war dragged on, however, and the casualties mounted, with rising prices and growing shortages of every food, deeper resentments and hatreds surfaced in India. Nehru sensed "little sympathy with the British in spite of loud professions of loyalty. Moderate and extremist alike learnt with satisfaction of German victories. There was no love for Germany, of course, only the desire to see our own rulers humbled. It was the weak and helpless man's idea of vicarious revenge."[12]

He well knew that "weak and helpless man's" feeling, for Motilal and his astrologers had set Jawahar's wedding date for the "festival of spring"

(Vasanta Panchami), which would arrive early in February 1916. "The house was full of jewelers, merchants and tailors coming in and out throughout the day," Betty recalled, remembering how loudly her father and Jawahar argued, "getting very agitated." There was no escape, Jawahar knew, when Kamala moved to her uncle's home in Allahabad a full year before the marriage date, so that she could come to Anand Bhawan, almost daily it seemed to him, for special tutoring in English, and in order to dine with them, to learn how to use Western silver, and perhaps to allow her to talk to Nehru and for them to grow more fond of each other. For Kamala that may have been the case, but not for Jawahar. The more he saw of this strange illiterate child, the less he liked her and the more outraged and frustrated he became.[13]

"Father had been closely watching my growing drift towards Extremism, my continual criticism of the politics of talk and my insistent demand for action," Jawahar recalled of this stressful interlude. "Father imagined that I was heading straight for the violent courses adopted by some of the young men of Bengal. This worried him very much. As a matter of fact I was not attracted that way, but the idea that we must not tamely submit to existing conditions and that something must be done began to obsess me more and more. . . . I felt that both individual and national honor demanded a more aggressive and fighting attitude to foreign rule."[14]

Had he gone to Oxford or stayed in London, it is hard to imagine young Jawahar feeling as he now did on the eve of a wedding he did not want, but could do nothing to stop. His impotence enraged him, until he became obsessed more and more with the feeling that personal honor at least "demanded a more aggressive and fighting attitude," if not against his father then against foreign rule. Those two tyrants soon merged in his mind, for loyalist Congress moderates—Gokhale's liberal Anglophile wing of gentlemen like Motilal—politely petitioned the British Raj for the redress of grievances and invited Englishmen to dine and drink at home, convivially laughing and joking with them. Motilal thought he could control his son by bringing him home to marry a proper Kashmiri maiden, he himself had chosen. Jawahar resisted as long and hard as he dared or could for the right to choose his own mate, to live his own life in a style he had learned to love and enjoy, but his father was too powerful. Jawahar was forced to accept a loveless marriage, saddled with a bride he never knew or wanted. Now he longed for the "vicarious revenge" that he would soon take against his father and foreign rule, by adopting a new, more militant, saintly Hindu "little father" (bapu), choosing to follow the revolutionary leadership of that "great soul" (Mahatma) rather than the more timid path of the moderate Motilal.

"The major event in our life in 1916 was Bhai's wedding," remembered Nan. "The wedding was lavish to the point of ostentation . . . one of those contradictions in Father's personality difficult to reconcile with the kind of man he really was. His love for my brother was deep—nothing in the world was too good for him, and his bride must have the best of everything. Many of the ornaments given to her were designed by Father, the precious stones chosen by him and mounted in the house where a regular little goldsmith's shop had been set up and where Father spent every minute he could spare from his professional work. . . . Bhai kept himself severely aloof." [15]

Motilal hired a special train to bring his more than three hundred guests from Allahabad to Delhi's "Nehru wedding camp," a small tented town erected just outside the heavy wall surrounding old Delhi, capital of the Mughals, whose loyal servants many members of the Nehru clan had been. A few miles south, New Delhi was slowly rising on the plain of red stone and brick that had hosted nine previous incarnations of the capitals of once mighty empires, where dozens of monarchs had ruled north India for more than three thousand years. In little more than a decade the British Raj moved into its sumptuous new palaces, large enough, it was hoped, to accommodate the British viceroys and their army of servants for at least another few hundred years. But just thirty-one years from now, the shy and angry young man whose wedding would lavishly be celebrated here for more than a week in 1916 redeemed a pledge he had not yet made with destiny.

Every Kashmiri Brahman in north India came to pay respects to the wealthy Motilal and his son, whose beige brocade *sherwani* and silken turban shimmered as he rode to his bride's resplendently lit and flower-decked Atal House on a white Arabian horse that first midnight in the spring, February 8, 1916. Kamala wore a pink sari with floral bangles, necklaces, and ornaments made to approximate the shape and color of the jewels Motilal would give her next day, to prevent the Brahmans who chanted Vedic mantras at the wedding and poured ghee onto the divine Agni's fire from taking them all as "tips." The officiating Brahmans could claim only the two gold bangles worn by the bride. Tied together, the young couple took the seven steps prescribed by Brahmanic ritual around the sacred flames, thus binding themselves to each other as man and wife.

Jawahar's mother applied red tika dots of kumkum to Kamala's and her son's foreheads. The second evening the slight, seventeen-year-old bride received all her real jewels, unable to stand without stooping under the weight of so many diamonds, emeralds, pearls, and rubies. Festive eating and drinking, talking and dancing, continued all week. After the Delhi parties ended, the garlanded train full of Nehrus and their friends returned to Allahabad, when the real celebrations began, for Nan at least, evening after evening of gaiety at Anand Bhawan, to which the whole province seemed to have been invited. Younger sister Betty was, however, so exhausted by then that she considered having survived the Delhi festivities and the weary ride home in itself "a tribute to our sturdy constitutions."[16]

Nehru's own terse account of his wedding in his autobiography reflects how depressed he was about the grand celebration and his child bride. "My marriage [he does not mention Kamala] took place in 1916 in the city of Delhi," he wrote, adding that it was the first day of spring. "That summer we spent some months in Kashmir. I left my family in the valley and, together with a cousin of mine, wandered for several weeks in the mountains." One of the things he did during those weeks of wandering was to shoot and kill a bear, the first and only bear he killed in his life. Was that to prove his masculine prowess? Another was nearly to kill himself, again in an icy "accident," which from his own report of it almost sounds like attempted suicide.[17]

"The wind was cold and bitter but the sun was warm . . . the loneliness grew . . . not even trees or vegetation to keep us company—only the bare rock and the snow and ice. . . . Yet I found a strange satisfaction in these wild and desolate haunts . . . I was full of energy and a feeling of exaltation. . . . At one place on our march beyond the Zoji-la pass . . . we were told that the cave of Amaranath was only eight miles away . . . an enormous mountain all covered with ice and snow lay in between and had to be crossed, but what did that matter? Eight miles seemed so little. In our enthusiasm and inexperience we decided to make the attempt. . . . We crossed and climbed several glaciers, roping ourselves up . . . some of our porters . . . began to bring up blood. It began to snow and the glaciers became terribly slippery; we were fagged out and every step meant a special effort. But still we persisted in our foolhardy attempt. . . . We had now to cross this ice-field . . . very tired but in good humor. . . . It was a tricky business and there were many crevasses and the fresh snow often covered a dangerous spot. It was this fresh snow that almost proved to be my undoing, for as I stepped upon it and it gave way, down I went (into) a huge and yawning crevasse. . . . But the rope held and I clutched to the side of the crevasse and was pulled out. We were shaken up by all this but still we persisted. . . . The crevasses, however, increased in number and width and we had no equipment or means of crossing some . . . so at last we turned back, weary and disappointed . . . the cave of Amaranath remained unvisited . . . I resolved to come back. . . . That was eighteen years ago, and . . . I have not even been to visit Kashmir again, much as I have longed to . . . ever more and more I have got entangled in the coils of politics and public affairs. Instead of going up mountains or crossing the seas I have to satisfy my wanderlust by coming to prison."[18] The "strange satisfaction" of what he called the "exciting experience" is reminiscent of his earlier accident in Norway. Once again he tested the limits of his stamina, courage, or self-hatred, and survived.

As Jawahar now daily told his father and would soon try to teach Kamala, "we must not tamely submit to existing conditions"! The idea that "something must be done" had already begun to obsess him more and more. The only question that remained unanswered was what to do and how. He rushed back to Allahabad, ostensibly to work, though Motilal remained a month longer with Kamala and the rest of the family in Kashmir. But the law offered him no solace, none of the release that his "individual . . . honor demanded" and that Kamala's love would certainly never bring him, for in accepting her as his bride he proved to himself only that he was his father's "weak and helpless man." What he craved was aggressive political action, the vicarious revenge that a true revolution might bring, liberating him from his sense of impotence.

Motilal wrote to Jawahar from Srinagar that Kamala began complaining of headaches soon after her husband left, but he attributed them to "nervousness." He prescribed homeopathic remedies, several of which he had been taking himself for his lumbago and a persistent cough. Later he decided that 80 percent of the cause of Kamala's headaches was "hysteria." The poor child bride must have felt guilty for having driven away her handsome husband because of her own inadequacy and lack of sexual experience. But the headaches only got worse, and soon a deeper depression brought a fatal weakness to her lungs, relieved only briefly during her pregnancy and the birth of Indira, in November 1917. Nehru makes no reference to their daughter's birth in his autobiography, nor does he write anything about her childhood or early adolescence. Kamala appears to have contracted tuberculosis by the time their second child, a son, was born prematurely and died only two days after his birth in 1924. The last ten years of Kamala Nehru's life were spent in Swiss and other sanatoria, futilely seeking a cure.[19]

As World War I passed its midpoint in December 1916, India's nationalist movement reached its peak of unprecedented unity. Both wings of Congress agreed to reunite at a historic annual session in Lucknow. "We have found our luck in Lucknow," Tilak proclaimed. The Muslim League, which also met in that old capital of provincial Mughal power, under Jinnah's presidency, adopted the same platform as the Congress did, the historic Lucknow Pact, demanding a fair share of special electoral representation on every council in British India. The pact, drafted in final form by Jinnah and Motilal Nehru, with Tilak's approval, was Gokhale's political legacy to India, having outlined its points as his final act on the eve of his early death in 1915. Jinnah, hailed at Lucknow as the "Muslim Gokhale," called for an early "transfer of power from the bureaucracy to democracy," urging all Hindus and Muslims to "stand united."[20] Congress's venerable president, Mazumdar of Bengal, agreed: "Blessed are the peacemakers," he told his large and cheering audience. "I break no secret when I announce to you that the Hindu–Moslem question has been settled and the Hindus and Mussalmans have agreed to make a united demand for Self-Government."[21]

Such optimism was premature, of course, for India's Hindu–Muslim conflict would never be resolved, not even after the birth of Pakistan, thirty-one hard-fought and bloody years later. Jinnah, who was forty at the time, lived long enough to become first governor-general of that Muslim nation carved out of both wings of what had been British India, but most of the others were long since gone by then. Only Nehru and Gandhi, who first met at that Lucknow session of Congress, also survived. Mahatma Gandhi had won his reputation as a saintly revolutionary in South Africa, where he had initially gone in 1893 to work as a barrister for a Muslim Indian firm in Durban and remained to serve and inspire the entire Indian community of South Africa in its struggle for racial equality and basic human rights. Gandhi's method of political revolution proved most inspiring, first to Indians and later to oppressed peoples the world over, for its technique was to wrest concessions, political or otherwise, by means of nonviolent unrelenting self-sacrifice. Gandhi's genius was his fearlessness and seeming simplicity. He had abandoned his Savile Row suits in South Africa, after being thrown out of a first-class coach for which he had bought a ticket, and now dressed as a Gujarati peasant, attending the Congress barefoot, his naked limbs covered with a heavy shawl and his shaved head wrapped in a loosely wound turban.

"He seemed very distant and different and unpolitical to many of us young men," Nehru remembered as his first impression of the little man soon to become his political guru and surrogate father. To Gandhi, moreover, this young "Englishman" was Motilal Nehru's son—whose big marriage *(tamasha)* had been noted in all the Delhi newspapers for weeks—and nothing more. Neither as yet guessed how important each of them was to be to India, to the world, and to each other.[22]

For Jawahar the most inspiring "Indian" political leader at this time was old Annie Besant, whose sudden emergence at the center of India's political stage proved as startling and exciting as most of the other remarkable things she had done in her many "prior incarnations."[23] Annie, who won notoriety as the radical Charles Bradlaugh's comrade, helped him defend Dr. Knowlton's banned "dirty, filthy book" on birth control in 1877, later joined Bernard Shaw's Fabian Socialist Society, led the first women matchmakers' strike in London, and spoke eloquently in favor of Irish home rule and every other good cause, before meeting Madame Blavatsky in London in 1889 and converting to theosophy. For the next quarter century Annie focused on theosophy alone, which she believed was "divine wisdom," starting chapters, writing books, opening schools and colleges in north and south India, bringing in converts like Motilal Nehru and his son, but taking little political interest in Congress or Indian politics until 1915. She then decided to start an Indian home rule league, hoping to bring the sadly dispirited, divided Congress back together, personally visiting both Tilak and Gokhale, as well as Jinnah and all the other major leaders, urging them to unite and win home rule or dominion status from the British government at war's end. Her energy was inexhaustible, her powers of speech mesmerizing, and her optimism contagious, rallying Indians wherever she spoke to cheer her, calling her "the mother," a reincarnated Hindu mother goddess in Irish skin and English voice. British officials hardly knew what to make of her or how to deal with her, for she was much smarter than most of them yet looked rather like Queen Victoria. They panicked. It was, after all, wartime, so they had given themselves emergency powers to arrest anyone for sedition if they disliked anything that was said or written about the Raj. So early in 1917, the idiot governor of Madras ordered the arrest of sixty-nine-year-old Annie Besant, who overnight became the mother martyr to young India and was, by year's end, shortly after her release from detention, the first woman and the only English woman elected to preside over India's National Congress.

Annie's arrest and detention in Ootacamund Hill Station triggered Jawahar's activist political consciousness, inspiring him in June 1917 to write his first letter to the editor of Allahabad's *Leader*. Anxious perhaps at how his father would react to so public a splash into political waters, Jawahar launched his otherwise undisguised attack with Latin he had learned at Harrow. "*Qui Deus vult perdere prius dementat*" [Whom the gods would destroy, first they make mad]. . . . We are seeing the proverb justify itself in our ancient land. . . . But with my sorrow at the treatment accorded to Mrs. Besant is mingled joy and gladness that this day has come. For madness has fallen on the bureaucracy and that is the surest presage of their coming fall. Home Rule has come and we have but to take it if we stand up like men and falter not."[24]

The Madras government's arrest of Annie Wood Besant thus served to make the boy she had initiated into theosophy fifteen years earlier stand up like a man and begin his fight against foreign rule and his father. The next day a protest meeting was called in Allahabad by Mother Besant's Home Rule League branch, and Jawahar wrote to make sure that more than speeches and resolutions would emerge from it, arguing that "ours have been the politics of cowards and opium-eaters long enough and it is time we thought and acted like live men and women who place the honor and interests of their country above the frowns and smiles of every Tom, Dick and Harry who has I.C.S. attached to his name. . . . Every one of us who holds an honorary position under the Government should resign it and refuse to have anything to do with the bureaucracy. I am aware that many will not be prepared to do this, that they would sooner go to the devil than offend the collector. Of the likes of such we have no need, we want no faint hearts or wobblers in the Home Rule League . . . let them part company with us and continue to kow-tow before and worship at the feet of our masters from across the seas."[25]

The political activism and boycott he had learned from the Sinn Fein in Dublin and from Bipin Pal and Bernard Shaw in Cambridge and London. But the joy and gladness he felt reflected Jawahar's personal pleasure in asserting his political independence. Annie Besant's detention had ignited in his mind the spark of courage to speak out and stand up in public against all his father had always defended in their nightly arguments when Jawahar felt himself to be a "coward and opium-eater." Now he could tell them all, first Motilal and then the rest, to go to the devil, and part company with us if they still feared to offend some Tom, Dick or Harry Collector! For although he lacked the courage to abandon the marriage he never desired or leave the bar and legal practice he found boring and dull, at last he had found a legitimate reason to rebel. Political activism and revolution gave Nehru the strength to stand up and, when required, to satisfy his wanderlust for many long years behind prison bars. Soon he would follow his little father, Bapu, much poorer and humbler than Motilal, but with political courage and daring as intense and as great as his own.[26]

4

The Coming of Gandhi

1918–1920

"SOME HIDDEN violence in himself," Erik Erikson suggests, may have helped Gandhi win younger Indians to his revolutionary method of attacking the roots of British rule by destroying its popular support. Jawahar became one of those younger converts to Gandhi's method of revolution.[1]

In 1918, however, it was ironically Gandhi who worked himself sick recruiting Indians for the British army, fruitlessly walking behind a British drummer from village to village in Gujarat, urging young peasants to join up. "You are the votary of Ahimsa [non-violence], how can you ask us to take up arms?" the peasants asked him. "My optimism received a rude shock," Gandhi confessed, suffering so severe a breakdown that he was obliged to take to his bed. He was still in bed, fearing his illness was "bound to be prolonged and possibly fatal," when Sardar Vallabhbhai Patel, his leading Gujarati lieutenant, brought him the news of Germany's surrender. The realization that the war was over, and no further recruitment would be required of him came as "a very great relief," Gandhi reported, swiftly recovering from what his doctor diagnosed as "a nervous breakdown."[2]

For India, however, War's end brought no freedom, no home rule, not even its distant promise, only sharper swords of British repressive legislation and action. Half a million Indian troops returned home from the various fronts abroad, having enjoyed the love of European women and the respect of the Englishmen with whom they had served, to find themselves treated again as "nigger natives" in British India, where painfully little had changed. And with the returning soldiers came influenza, which claimed no fewer than twelve million lives before 1919 ended. Political India was shocked to learn that its immediate reward for supporting the war effort would be a renewal of martial law and official terror throughout Punjab.

First came the Black Acts. Justice Rowlatt tabled two bills at the first postwar meeting of the viceroy's Legislative Council, calling for a six-month extension of the emergency wartime Defence of India Criminal Regulations that had been introduced in 1915, suspending all civil liberties and judicial due process during the war. "There was no precedent or parallel in the legal history of any civilized country to the enactment of such laws," Barrister Jinnah warned Viceroy Chelmsford. Such measures would "create unprecedented discontent" and have the "most disastrous effect upon the relations between the Government and the people." His wise warning went unheeded, and Rowlatt's bills were enacted over the unanimous opposition of all Indians on the viceroy's council. Jinnah was the first to resign his seat, protesting that "the fundamental principles of justice have been uprooted and the constitutional rights of the people have been violated at a time when there is no real danger to the State." [3]

Mahatma Gandhi by now had recovered his health, and chose April 6, 1919, as the "sacred" start of "National Week" to protest the Black Acts, beginning with a strike *(hartal)* shutting down all Indian businesses and the public taking of vows by Indians everywhere "to refuse civilly to obey" those repressive laws. Gandhi established the Satyagraha Sabha (society for nonviolent noncooperation), which attracted many young members, including Jawaharlal Nehru. Members were to pledge to "court gaol [jail] openly and deliberately" in public.

"When I first read about this proposal . . . my reaction was one of tremendous relief," Jawahar wrote. "Here at last was a way out of the tangle, a method of action which was straight and open and possibly effective. I was afire with enthusiasm . . . I hardly thought of the consequences—law-breaking, gaol-going, etc.—and if I thought of them I did not care." At last a "way out of the tangle"—that dull routine between Allahabad's bar and Anand Bhawan, where Kamala waited all day long to welcome him home with her news of what baby Indira did or failed to do. [4]

His father, of course, was "dead against this new idea." Motilal Nehru "was not in the habit of being swept away by new proposals; he thought carefully . . . and the more he thought of the Satyagraha Sabha and its program, the less he liked it. What good would gaol-going of a number of individuals do, what pressure could it bring on the Government? . . . It seemed to him preposterous that I should go to prison." [5]

"Father was appalled," Betty recalled. "All his training, his inclination to compromise and to find a solution within constitutional limits, his veneration for the rule of law . . . made him distrust *Satyagraha.*"[6] But none of Motilal's arguments, not even his violently raised voice or the pounding of his formidable fist on his once-terrifying desktop could weaken his son's resolve to follow the new path of revolution. If anything, Motilal's frustration and rage at this "stupid idiocy" and the "futile waste of time and energy" involved in such tactics against the British Raj sufficed to convince Jawahar of the wisdom of Gandhi's solution. Despite everything his father said, "I had no doubt in my mind that I had to go the way of Satyagraha."[7] Jawahar later discovered that Motilal actually tried sleeping on the floor, to find out what it might feel like in prison.

Finally, at his wit's end, Motilal invited Gandhi to visit Anand Bhawan, hoping to dissuade him from revolution before it was too late. "We went down to the station to meet him in two cars," Betty wrote. "My eyes searched the handsome first-class carriage for a glimpse of the great man. Then I saw Bhai and Father greeting a strange person who got out of a third-class carriage. He was small and wizened, and wore only a loincloth and a white shawl draped over his bare shoulder. . . . . 'Is this insignificant person going to make everybody go to prison, and overthrow the British?' " Betty decided that it was not possible. Nor would she soon change her mind, though all week she watched everyone rushing around to bring fruit and goat's milk to the strange visitor, finding it all "terribly affected." Jawahar, however, and Kamala, Motilal, and Swarup Rani attended the prayer meetings conducted daily by Gandhi on their lawn, the only parts of which Betty enjoyed were singing hymns such as *Abide with Me,* and *Lead Kindly Light.*[8]

Motilal had long talks with Gandhi, from which Jawahar was purposely excluded, for his father was now wrestling with the Mahatma for his only son's heart, mind, and soul. Wealthy and powerful advocate though he was, the elder Nehru proved no match for the "great soul" whose yogic powers would soon shake the British Raj. After their last private round in Motilal's study, "Gandhiji advised me not to precipitate matters or do anything which might upset Father," the rebellious Jawahar recalled. "I was not happy at this." Motilal was even less happy.[9]

"You have taken my son," he reproached Gandhi near the end of their final talk on the eve of the Mahatma's departure, "but I have a great law practice in the British courts. If you will permit me to continue it, I will pour great sums of the money I make into your movement. Your cause will profit far more than if I give it up to follow you. "No," Gandhiji said. "No! I don't want money. I want you—and every member of your family."[10]

That April, Satyagraha Week triggered violent confrontations with British police and soldiers in many Indian cities. Most dreadful, however, was the massacre that turned the heart of Punjab's Sikh capital, Amritsar, into the cradle of India's nationhood. The deportation of several of Gandhi's Punjabi lieutenants led to a mass protest march toward the collector's house, which was protected by British troops, who opened fire. Several Indians were killed in that first round, and the mob raced off in all directions, burning British banks, attacking Englishmen and two Western women. British troops, under Brigadier R. E. H. Dyer, were called in to help restore order in the city. On the afternoon of April 13, 1919, Dyer unleashed his massacre in Jallianwala Bagh, marching his soldiers into that otherwise totally enclosed garden *(bagh)*, where thousands of unarmed Indians had gathered, many innocent peasants who had come to celebrate the spring holiday. Without a word of warning, Dyer ordered his men to open fire. Sixteen hundred rounds were shot point-blank into that desperately fleeing crowd, many of whom plunged into a deep stone well to their deaths, with others trying to climb the surrounding walls. None of those bullets was wasted. Four hundred people died, and twelve hundred were wounded in less than ten minutes, as the sun set over Amritsar and moderate Indian nationalism. Thousands of loyal Indians who had never attended a meeting of Congress were converted into ardent nationalists as soon as they learned what had happened in the garden at Amritsar.[11]

"We must do reverence to the sacred memory of the dead who were killed in Amritsar," President Motilal Nehru solemnly told the huge gathering of India's National Congress, which met in Amritsar that December. "India will never forget the sacrifice and the sufferings of these children of hers . . . at the hands of the alien and reactionary bureaucracy. . . . [T]he bogey of the frontier is exploited to the uttermost and the proposals made by the 'man on the spot' seldom fail to secure acceptance at the hands of the higher authorities . . . but repression and terrorism have never yet killed the life of a nation; they but increase the disaffection and drive it underground to pursue an unhealthy course of . . . violence. And this brings further repression and so the vicious circle goes on. No one can but deplore violence and political crime. But let us not forget that this . . . is due to the perversity of the executive which blinds itself to the cause of the discontent and, like a mad bull, goes about attacking all who dare to stand up against it."[12]

The equally moderate, law-abiding Jinnah was similarly outraged. As president of the Muslim League, which met in Calcutta that September, Jinnah denounced those "celebrated crimes" of Punjab, "which neither the words of men nor the tears of women can wash away. 'An error of judgment,' they [the British] call it . . . error of judgment it is and they shall have to pay for it, if not to-day then tomorrow. . . . [T]his Government must go and give place to a completely responsible Government." But Jinnah was reluctant to support Gandhi's call for a nationwide Satyagraha to attain "complete *swaraj* [self-rule] for India according to the wishes of the Indian people." [13]

He was not sure of how those "wishes" were to be ascertained, or by whom. As a minority Muslim, moreover, Jinnah knew that there were many different "Indian people." Although he was not a devout Muslim, his secular British style and Western tastes in food and drink and his pre-occupation with jurisprudence in both London and Bombay made him distrust Gandhi's constant use of Hindu symbolism, Hindu prayers, and Jain–Hindu vows in his revolutionary movement. Nor was Jinnah the only Congress leader to distrust the Mahatma. Bengal's greatest barrister, C. R. Das, also doubted him, as did the Bengali radical Bipin Pal and Annie Besant, who now spoke of Gandhi's method as a "channel of hatred." Several of Bombay's moderate Parsi leaders considered Gandhi "fanciful" or a "madman." But Motilal Nehru supported him because Jawahar had won over his father to Gandhi's side, though not without the help of Britain's right-wing bureaucracy. [14]

Gandhi's Satyagraha resolution was passed at a special meeting of Congress in Calcutta the following year (1920), when Satyagraha was launched nationwide. Without Motilal's support, Gandhi could never have come so swiftly to a position of such prominence and power in the National Congress and over the future course of India's nationalist movement. Although Motilal was never completely convinced of either the wisdom or the rationality of Gandhi's style of politics, he knew that by now Jawahar was sufficiently attracted to this radical "saint" and his Satyagraha to fear that he would lose his son unless he played along with the Mahatma. It was also thanks to Motilal that Gandhi took control of Bombay's Home Rule League, whose name he changed to Swaraj Sabha. Jinnah and his allies resigned in the wake of that change, after which Gandhi wrote to invite him to return to "share in the new life that has opened up before the country."

"If by 'new life' you mean your methods and your program," Jinnah replied, "I cannot accept them; for I am fully convinced that it must lead to disaster. . . . [F]aced with a Government that pays no heed to the grievances, feelings and sentiments of the people . . . your methods have already caused split and division in almost every institution that you have approached . . . and in the public life of the country not only amongst Hindus and Muslims but between Hindus and Hindus . . . and even between fathers and sons; people generally are desperate all over the country and your extreme program has for the moment struck the imagination mostly of the inexperienced youth and the ignorant and the illiterate. All this means complete disorganization and chaos. What the consequence of this may be, I shudder to contemplate . . . I do not wish my countrymen to be dragged to the brink of a precipice in order to be shattered." [15]

That was Jinnah's first apocalyptic vision of British India's "shattered" partition, more than a quarter century before it occurred, when the Muslim quarter of his compatriots followed their *Quaid-i-Azam* (great leader) to Pakistan. Gandhi had embraced the radical brothers Shaukat and Mohammad Ali and their popular Pan-Islamic Khilafat movement supporting the restoration of Turkish territories to the defeated Ottoman caliph as the first "plank" in his Satyagraha platform. Jinnah never trusted "Hindu *bania*" (merchant) Gandhi to represent impartially Muslim interests or demands any more than he believed the Mahatma could keep his mass Satyagraha nonviolent, stirring up as it did millions of desperate, "ignorant," and "illiterate" followers. When Jinnah wrote of the "inexperienced youth," whose "imagination" was "struck" by Gandhi's revolutionary zeal, he meant, among others, Jawaharlal Nehru, thirteen years his junior, whom he later called "Peter Pan." Was Jinnah's vision of so many Indians soon to be "dragged to the brink of a precipice" by a Mahatma who had just usurped his mantle of Congress leadership prophetic or was it, as many of his opponents insisted, the revenge he dreamed of after losing his political power?[16]

For young Nehru there was never any doubt. He remembered one of Gandhi's first appearances before "a very tame gathering of timid, middle-aged folk" of the Muslim League in early 1920, where Gandhi and Shaukat Ali went to enlist support for Satyagraha and the Khilafat movement. "Gandhiji . . . spoke well in his best dictatorial vein. He was humble but also clear-cut and hard as a diamond. . . . . This is going to be a great struggle, he said, with a very powerful adversary. If you want to take it up, you must be prepared to lose everything, and you must subject yourself to the strictest non-violence and discipline. When war is declared martial law prevails, and in our non-violent struggle there will also have to be dictatorship and martial law on our side, if we are to win. You have every right to kick me out, to demand my head, or to punish me whenever and howsoever you choose. But so long as you choose to keep me as your leader you must accept my conditions, you must accept dictatorship and the discipline of martial law."[17]

This sort of appeal stirred Jawahar's heart. He accompanied his new little father to Bombay on August 1, 1920, the day that the militant old Lokamanya Tilak died. "The whole of Bombay's million population seemed to have poured out to do reverence to the great leader."[18]

Suddenly the enormous crowd found itself being addressed by a new leader, a nonviolent "dictator" who announced his launching of his first nationwide Satyagraha that very day on Chaupati Beach in Bombay. Next month Jawahar wrote again to the editor of *The Leader,* urging a boycott of all British imports and institutions, including government councils, describing himself "an ardent believer" in "noncooperation" as "laid down by Mahatma Gandhi . . . and I am firmly convinced that noncooperation and no other course will bring us victory. That victory may not come in a day or a year, but come it will and must, *ruat coelum* [though the heavens fall]."[19]

Gandhi promised that India would win *swaraj* (freedom) within one year of the actual launching of his Satyagraha campaign. Many of his followers, less learned in Latin than Jawahar, believed him. Others, neither so simple nor so sophisticated, such as the old servants and women at Anand Bhawan, knew that "bad times" and "sad times" had come for India, for their giant king cobra was dead, killed by a stupid new servant.

The cobra lived in the woodshed out behind Anand Bhawan itself but close enough to slither in and out of the great house whenever it wished, for cobras in India are magical creatures, almost divine in their wondrous powers, and good luck to every household they grace with their hooded presence, for they bring healthy and happy progeny and guard the family and servants of the lucky house they inhabit. Most Hindus worship cobras as *naga-raja* (king snake), royalty incarnate, with the largest king cobras believed to choose only the homes of kings to inhabit. So all of the older servants and most of the Nehrus believed that one day a Nehru would, indeed, be king. Then a new servant panicked when he saw the giant snake slithering past his feet one night in 1920, so he picked up the axe outside the woodshed and broke the back of the king cobra, killing him and striking terror into the hearts of Swarup Rani, her daughters, and Kamala. Tearfully, Swarup Rani ran to her husband when the stupid servant proudly told her what he had done.

" 'The cobra has been killed,' she said. 'Ill luck will come to us.'

'My dear,' Father said [Betty recalled], 'ill luck is already here, if you consider it that; for I have joined Mahatma Gandhi. I am going to give away this house and will have to sell all my horses and my wine cellar. . . .'

'Must you?' Mother asked.

'Our son has joined Gandhiji. . . . . Would you want me not to go with him?'

'No,' Mother said. 'I am glad you are with your son, and we shall all join the movement together. But the cobra is dead.' "[20]

Jawahar's rational, Western-educated mind feared no such nonsense. Myth and superstition was the curse of India! He had found his cause in the service of his new father, Bapu, who had taken him to see the real India, village India, where impoverished peasants lived in huts of mud or wattle, eating less in a week than was thrown away every day at Anand Bhawan, where even the servants had more to eat than their stomachs could hold.

Nehru was thirty before he visited village India during that fateful summer of 1920. "That visit was a revelation to me. We found the whole countryside afire with enthusiasm and full of a strange excitement. Enormous gatherings would take place at the briefest notice by word of mouth. One village would communicate with another, and the second with the third . . . and all over the fields there would be men and women and children on the march to the meeting-place. . . . They were in miserable rags, men and women, but their faces were full of excitement and their eyes glistened, and seemed to expect strange happenings which would, as if by a miracle, put an end to their long misery. They showered their affection on us and looked on us with loving and hopeful eyes, as if we were the bearers of good tidings, the guides who were to lead them to the promised land. Looking at them and their misery and overflowing gratitude, I was filled with shame and sorrow, shame at my own easy-going and comfortable life and our petty politics of the city which ignored this vast multitude of semi-naked sons and daughters of India, sorrow at the degradation and overwhelming poverty of India. A new picture of India seemed to rise before me, naked, starving, crushed and utterly miserable. And their faith in us, casual visitors from the distant city, embarrassed me and filled me with a new responsibility that frightened me." This was Jawahar's moral awakening, his "discovery of India," but, equally important, his discovery of himself, of his calling in life.[21] Until now he had been searching over snow-covered mountains and icy ponds and treacherous streams, in Hydros and Casinos, and tableaux-decorated sitting rooms, bored to tears, taking orders and doing things he hated, bored, bored, bored! Yet here at last he found a "whole countryside afire with enthusiasm." And the "strange excitement" of those "miserable rag-clad men and women," with glistening eyes was caused by *his* being there, *his* coming! He and Bapu. Filled to overflowing with shame and sorrow, Jawahar was also afire. For the faith of that adoring army "filled me with a new responsibility."

Now he *knew* why he'd gone abroad and learned so much, yet came back and seemed to have lost it all, giving up the high life of London and Cambridge and Paris for Allahabad's Bar and Kamala! It was for *this,* these "faces full of excitement," this "vast multitude of semi-naked sons and daughters of India," who looked up to *him,* to perform the miracle that would save them! Soon he would find words in which to express his mission so eloquently that all of India's National Congress would echo his promise, follow his passionate lead.

5

In and Out of Prison

1921–1922

NEHRU'S AWAKENING to the plight of India's peasant population coming not only as late as it did in his life but also after so many foreign adventures and so much self-indulgence imprinted his mind and heart with the power of a sudden religious conversion. Sheltered in Motilal's palatial grounds as a boy, spirited away to even grander private schools of higher learning designed to stimulate grandiose dreams in the hearts of aristocratic boys and men, Jawahar returned home to find himself, in much the way Prince Siddharta Gautama had, face to face with poverty, misery, social sickness, and death.

He had, of course, seen many peasants before June 1920, when he first spent three days walking through the village hinterland of Partabgarh and the Rae Bareli district of the United Provinces, which became his family seat in the parliament. "I had often passed through villages, stopped there and talked to the peasants," he later recalled. "But somehow I had not fully realized what they were and what they meant to India. Like most of us, I took them for granted" until his conversion, awakening, or psychic illumination—whatever it could be called—opened his mind, heart, and spirit, as well as his eyes. "Ever since then my mental picture of India always contains this naked, hungry mass . . . perhaps I was in a receptive frame of mind and the pictures I saw . . . were indelibly impressed. . . . These peasants took away the shyness from me and taught me to speak in public. Till then I hardly spoke at a public gathering; I was frightened at the prospect, especially if the speaking was to be done in Hindustani, as it almost always was. But I could not possibly avoid addressing these peasant gatherings, and how could I be shy of these poor unsophisticated people? I did not know the arts of oratory and so I spoke to them, man to man, and told them what I had in my mind and in my heart."[1] These poor peasants, unsophisticated and, illiterate, gave Jawaharlal Nehru his voice.

"The shadow on our lives was Kamala's failing health," his sister Betty recalled of this time. "Ever since Indira's birth [1917] she had seemed frail and easily tired; now she began running a low temperature and feeling miserable. Of course Bhai and Father got the best doctors for her but it was a long time before they diagnosed her illness."[2]

Jawahar's sister Nan was married in 1921. But not to the man she loved, for he was Muslim, the handsome, brilliant, English-educated Syed Hossain, hired by Motilal to edit a nationalist newspaper, *The Independent,* he had started in 1917 in Allahabad. What did it matter to the young lovers that he had been born a Muslim, and she a Kashmiri Brahman? Weren't Bhai and Gandhiji always saying that "Hindu-Muslim unity was the first prerequisite to India's Freedom"? What better antidote for communal hatred and conflict than Love? But to Motilal and Swarup Rani, Hossain's being Muslim mattered as much as Jawahar's wanting to marry a British barmaid. All of them, including Jawahar, tried talking sense into Nan. Nothing they said, however, brought any response but silence and tears. Finally, Motilal called in Gandhi, who came to Anand Bhawan in 1920 to take Nan with him to his Gujarati *ashram* (village community) in Sabarmati, on the outskirts of Ahmedabad in west India.

"My heart sank when I first saw the place," she later wrote. "Everything was so utterly drab and so unpleasing to the eye. I wondered how long I could survive there. . . . Life in the ashram was austere beyond belief. Rising at 4 A.M. for prayers, we went on to the chores of the day, which consisted of sweeping and cleaning our living quarters, washing our clothes in the river, cleaning the latrines . . . to let me down gently I was not required to clean the latrines, for which I offered thanks to the Almighty!" Nan's ashram therapy was, of course, meant to dampen her desire for Syed, and even though she never completely forgot him, she returned home less than a year later prepared to accept the Hindu Brahman, Ranjit Pandit, whom her mother and father and Mahatma Gandhi had found and arranged for her to marry.[3]

Ranjit Pandit had been to Oxford with Nan's cousin Jivan. He was also a barrister, tall and a good athlete who played tennis and polo. He sang arias in French and German, read Sanskrit, and had just written an article, published in Calcutta's *Modern Review,* hailing Jawaharlal's "rising star" in the nationalist movement, entitled "At the Feet of the Guru." His one major drawback, however, was that instead of belonging to the Kashmiri Brahman community, he came from a Maharashtrian Saraswati

Brahman family, devoutly Hindu and very rich, but still not acceptable to the oldest Kashmiri families in Delhi and Srinagar.

"The Kashmiri community as a whole had announced its decision to boycott the wedding," Vijaya Lakshmi noted, "because, according to them, I was marrying 'outside the community.' . . . The boycott was quite ridiculous but caused tremendous comment at the time. . . . Gandhiji had been a close friend of Ranjit's father, and this wedding had his consent and approval." But when Gandhi also insisted that instead of her beautiful sari and jewels, the bride must wear khadi and no jewels. "Mother could not have been more angry. She . . . could not understand his politics, and certainly did not think he had the right to advise the family on personal matters. His growing influence over Bhai disturbed her greatly. She felt intuitively that this man was the enemy of her home. . . . Khadi at that time was not only coarse, it was . . . very ugly." But Motilal, happy to have his elder daughter speaking to him again, gave her 101 silk saris to make up for the coarse hand-spun cotton one she was forced to wear at her wedding, and many beautiful jewels, the last of his treasured supply.[4]

The number of political leaders of the new young India now emerging from the shell of the old Raj and more ancient Hindu past, who flocked to the wedding made up for the missing Kashmiris. Jawahar and Motilal and all the other Congress leaders filled the Nehru garden on May 10, 1921—the most auspicious date to marry, according to many Hindu astrologers—coincidentally the fiftieth anniversary of the start of the 1857 "Mutiny." That made many I.C.S. officers and British brigadiers fear that another revolt was about to begin here and now in Allahabad. After seeing that army of Gandhi-capped and khadi-clad politicos flock to Allahabad station and march up every road toward Anand Bhawan, some British officers packed off their wives and children to Delhi or Lucknow. No shots were fired, however, at the gala wedding; no new mutiny occurred.

"Many of us who worked for the Congress program lived in a kind of intoxication during the year 1921," Jawahar wrote. "We sensed the happiness of a person crusading for a cause. We were not troubled with doubts or hesitation; our path seemed to lie clear in front of us and we marched ahead, lifted up by the enthusiasm of others. . . . We worked hard, harder than we had ever done before. . . . Above all, we had a sense of freedom and a pride in that freedom. The old feeling of oppression and frustration was completely gone."[5]

"I became wholly absorbed and wrapt in the movement . . . I gave up all my other associations and contacts, old friends, books, even newspapers, except in so far as they dealt with the work in hand . . . . In spite of the strength of my family bonds, I almost forgot my family, my wife, my daughter. It was only long afterwards that I realized what a burden and a trial I must have been to them in those days, and what amazing patience and tolerance my wife had shown towards me. I lived in offices and committee meetings and crowds . . . I experienced the thrill of mass-feeling, the power of influencing the mass . . . and I felt at home in the dust and discomfort, the pushing and jostling of large gatherings, though their want of discipline often irritated me. . . . I took to the crowd and the crowd took to me, and yet I never lost myself in it; always I felt apart from it."[6]

This was Nehru's first autobiographical reference to his daughter, who would be four in November 1921. By now the crowd had displaced his wife and daughter in his life, yet even from the masses he loved so much more than his immediate family, Jawaharlal always "felt apart," explaining: "From my separate mental perch I looked at it [the crowd] critically, and I never ceased to wonder how I, who was so different in every way from those thousands who surrounded me, different in habits, in desires, in mental and spiritual outlook, how I had managed to gain goodwill and a measure of confidence from these people. Was it because they took me for something other than I was? Would they bear with me when they knew me better? Was I gaining their goodwill under false pretenses? I tried to be frank and straightforward to them; I even spoke harshly to them sometimes and criticized many of their pet beliefs and customs, but still they put up with me. And yet I could not get rid of the idea that their affection was meant not for me as I was, but for some fanciful image of me that they had formed. How long could that false image endure? . . . And when it fell down and they saw the reality, what then?"[7]

Nehru often feared that one day he would be "unmasked," his carefully crafted image unveiled, revealing his true persona to the crowd that loved him because they never really knew him. So fearful was he that he quickly added, "I am vain enough in many ways, but there could be no question of vanity with these crowds of simple folk. There was no posing about them, no vulgarity, as in the case of many of us of the middle classes who consider ourselves their betters. They were dull certainly, uninteresting individually, but in the mass they produced a feeling of overwhelming pity and a sense of ever-impending tragedy."[8]

As noncooperation gathered momentum in 1921, more peasants and city-born Congress followers turned out, ready to boycott everything British, from cloth to courts, titles, and council chambers. Great Britain's home government found it all puzzling and tried to appease its brightest imperial jewel by sending the Prince of Wales to India that November. A royal visit was usually cause for jubilation among the "natives." The Prince of Wales, however "dreaded" going out to that "queer . . . country"[9] but decided it might be bearable if his favorite cousin, Lord Louis Mountbatten (nicknamed Dickie), came along to cheer him up. This handsome sailor, Louis Francis Albert Victor Nicholas Mountbatten, had just turned twenty-one and was at the time courting the flapper debutante Edwina Ashley, so Dickie in turn invited her. Edwina's German Jewish grandfather, Sir Ernest Cassel, had just died, leaving her most of his vast fortune, which she would inherit upon coming of age in 1922. India's only Jewish viceroy, Rufus Isaacs, Lord Reading, had just begun his turbulent half decade as viceroy in Simla and New Delhi. Lord Reading had been a good friend of Sir Ernest, so when Edwina wired to ask him if she might come to India, he immediately invited her to stay with him and Lady Reading whenever she liked and, for as long as she wished. Edwina reached India in December 1921, a month after Dickie and his royal cousin Edward, known in the family as David. That was the same month that Jawaharlal Nehru, of whom neither Dickie nor Edwina had yet ever heard, was first jailed by the British.

Black-flag protests greeted the prince, and general strikes *(hartals)* called by Congress shut down most businesses in the big cities he visited, including Allahabad. India's usually bustling bazaars and main streets had become ghost towns, lined only by police and soldiers as David and Dickie drove through. Jawahar was general secretary of the United Provinces Congress Committee, over which Motilal presided, and they planned to "shut down" Allahabad on December 9, 1921, when the Prince was expected to arrive, on his way to Calcutta. But three days earlier, both Nehrus were arrested at home, the first of many such traumatic events at Anand Bhawan.

Jawahar received a message from Motilal in his Congress office in the city that the police were searching Anand Bhawan and that they were waiting there to arrest them both. "I motored back . . . at about 6:30 P.M. Found the family gathered together in father's room. Father having his evening bread and milk. . . . On the whole every one was taking matters very cheerfully. Only mother had a lost expression. Kamala behaved admirably. Indira made quite a nuisance of herself objecting to her food and generally getting on people's nerves." [10]

A large crowd had gathered outside, and both Motilal and his son addressed them briefly before they were driven off to the district jail. "The idea of our loved ones being put behind prison bars worried us," his sister Betty remembered. "It was hardest of all for my mother, to whom the past few months of change had been a sort of nightmare she had not quite fathomed. But she was a brave wife and a still more brave mother. . . . Mother and Kamala smiled bravely as they parted from their husbands, but though their smile was courageous, there was sadness and loneliness in their hearts." Motilal was tried the next day before the young district magistrate Knox of the I.C.S., whose face reddened as he nervously sentenced the senior member of Allahabad's bar to six months in jail for having added his name to the top of a list of noncooperating "volunteers" that his *Independent* newspaper had recently published. More embarrassed, however, was Justice Banerjea, an old Bengali friend of Motilal's, who had so often enjoyed his regal hospitality yet now was obliged to state the requisite charges against the elder Nehru, who sat silent throughout the brief proceedings, with his only granddaughter, Indu, perched on his khadi-covered knee.[11]

Jawahar had written a message for his father that was published the next day in his newspaper: "Having served you to the best of my ability while working among you, it is now my high privilege to serve the motherland by going to jail with my only son. . . . I have only one parting word to say—continue non-violent non-co-operation without a break until Swaraj is attained . . . Adieu."[12]

The younger Nehru was taken the night before to Lucknow by a European inspector and five police guards, and they were met at Lucknow Station by the superintendent of police. "An offensive person," Jawahar noted in his prison diary. He was driven in a prison motor van with twelve police guards to Lucknow's district jail, where he met many of his comrades, already locked in their barracks. Now he, too, issued a message, published together with the one he had written for his father, in the December 8 *Independent:* "Some days ago you did me the high honor of appointing me the General Secretary of the Committee. . . . Today a higher honor and a greater service await me and I welcome it with the conviction that you will carry on the work of the Committee more vigorously and successfully. . . . It has pleased Providence to give this province a chance of leading the fight for liberty. May you, the representatives of the people prove worthy of this high trust. . . . [W]orkers from the districts must come to the headquarters and keep the flag flying at the citadel. . . . May God guide their deliberations and give them strength and wisdom. One thing I would have you remember. There can be no compromise or parlaying with evil. This struggle must and can only end in complete victory for the people. Any weakening . . . will be a betrayal of the thousands who have given of their best for the cause. Au Revoir! We meet again, I hope, as free men."[13]

Sir Harcourt Butler, lieutenant-governor of the United Provinces at this time, was an old friend of Motilal's, who had often dined at his table and drunk much of his champagne. He ordered that his friend be kept with his son in a special prison "shed about twenty by sixteen feet . . . allowed to order meals from outside, have all the books he wanted and write letters." It was not Anand Bhawan, of course, but fifty-two-year-old Motilal was wise enough by now to know that there would be little comfort and fewer pleasures left for him to enjoy during the last decade of his life. Jawahar at least was there to assist him in many ways, cooking and cleaning up and nursing him when he fell ill, as now more often happened.[14]

More than thirty thousand Congress protesters were arrested and kept behind bars before the end of 1921, yet the British cautiously refrained from jailing Gandhi. They were afraid of inciting a more violent revolt, knowing how many Indians viewed their little father more as a divinity than a political leader. The Mahatma was, therefore, allowed to remain free, even as his leading lieutenants were locked up, and every British jail throughout much of India was filled with defiant Indians, more cheerful than most of the police sent to arrest them. "I go to jail with the greatest pleasure, and with the fullest conviction that therein lies the achievement of our goal," Jawaharlal assured the "citizens of Allahabad," as he addressed his friends in a second message run in the family newspaper. "The most important thing is to preserve complete peace and an atmosphere of non-violence," he cautioned, paradoxically adding, "I trust you will always be in the firing line in the battle of Swaraj."[15]

Gandhi invited Swarup Rani and Kamala to his Sabarmati ashram that December, to attend the annual Congress meeting held during Christmas week of 1921 in Ahmedabad. The Nehru women decided to go, taking little Indira along. They traveled third class in the new spirit of national austerity. "Sitting on the hard wooden seats in the stifling railway carriage for hours on end was agony for my delicate little mother and for Kamala," Betty wrote. "Their fortitude amazed me." Betty found the ascetic routine at the ashram as trying as her elder sister had the year before. It was cold that December, and rising at 4 A.M. to follow Gandhi to the Sabarmati riverbank for prayers was hardly a wealthy young girl's ideal holiday. But Kamala found the spiritual atmosphere appealing, her own ascetic predisposition resonating to Mahatma Gandhi's austerity and the monastic routine of ashram life. Swarup Rani had a room to herself, but the others had to sleep in a large "sort of students' hostel." It was Indira's baptism in the national struggle, and it left a deep imprint, hardening her heart against the British enemy that had caused her mother and all the family so much anguish and pain.[16]

Part of noncooperation was never to argue against an arrest or sentence in court and never to pay any fine imposed by a judge, as happened in the first arrest of both Nehrus. They refused to pay anything, so the police went to Anand Bhawan to confiscate goods worth much more than the total rupee amount of both fines. As the police rolled up one fine Persian carpet, Indira stamped her tiny foot and shouted, "You can't take that away. . . . It . . . belongs to Mommy and Father." Then she kicked and had to be held by Kamala, crying bitterly. "When he heard of it, Bhai [Jawaharlal] was afraid she would hate the British forever." Not quite forever, but much more than her father ever could. For Indira's hatred was not only forged earlier than Jawaharlal's but also hardened without the benefit of any prior experience with or knowledge of those pleasant days at Harrow and Cambridge. She was, in most respects, "never a child," growing old before she had a chance to grow up.[17]

Gandhi tried his best to keep his Satyagraha troops under the strict moral discipline of *ahimsa* (nonviolence), which he insisted was "the highest religion" of ancient India, essential to waging any war of truth. But in Bombay, as elsewhere throughout India, many impassioned young enthusiasts raced through the streets, demanding Swaraj and proclaiming Hindu–Muslim unity yet, in their frenzy, picking up brickbats and breaking shop windows and heads. To Mahatma Gandhi, such violence "stinks in my nostrils." After learning about a rampage on November 17 through Bombay's busy bazaar, he prayed for divine guidance in this struggle he had launched and believed he still led. Were the troops who claimed to serve under him truly dedicated to Ahimsa? Patiently, Gandhi waited, listening for his "inner voice" to tell him what to do, whether to carry on this movement that was getting away from him, or call it off.

Then in February 1922 Gandhi learned that twenty-two policemen had been burned up inside their station house in Chauri Chaura, Bihar, trapped there by a mob of killers and arsonists, who called themselves Satyagrahis. "God has been abundantly kind to me," Gandhi wrote of the crime of Chauri Chaura. "He has warned me the third time that there is not as yet in India that truthful and non-violent atmosphere which, and which alone, can justify mass disobedience. . . . God spoke clearly through Chauri Chaura. I understand that constables who were so brutally hacked to death had given much provocation . . . gone back upon the word, just given by the Inspector . . . stragglers . . . abused by the constables. The former cried out for help. The mob returned . . . constables opened fire. The little ammunition they had was exhausted and they retired to the Thana [police station house] for safety. The mob, my informant tells me, therefore set fire to the Thana. The self-imprisoned constables had to come out for dear life, and as they did so, they were hacked to pieces and the mangled remains were thrown into the raging flames." So Gandhi called a halt to the movement he had started, much to the chagrin and amazement of most of his followers, especially those behind prison bars.[18]

"We were angry when we learnt of this stoppage of our struggle at a time when we seemed to be consolidating our position and advancing on all fronts," Jawahar wrote. "But our disappointment and anger in prison could do little good to any one, and civil resistance stopped and non-co-operation wilted away."[19] The British knew that arresting Gandhi now would cause less concern among his troops, many of whom had begun to call him "an old fool" or "a British stooge" or "crazy." For in trying to justify his decision, he had written, "The only virtue I want to claim is Truth and Non-violence. I lay no claim to superhuman powers. . . . I wear the same corruptible flesh that the weakest of my fellow beings wears, and am therefore as liable to err."[20]

To Nehru, Gandhi's response to Chauri Chaura came as a shock from which he never fully recovered for the remaining quarter century of his adopted little father's life. Suddenly Jawahar saw that even Mahatma Gandhi was made of clay. "The sudden suspension of our movement . . . was resented. Our mounting hopes tumbled to the ground. . . . What troubled us even more were the reasons given for this suspension and the consequences that seemed to flow from them. Chauri Chaura may have been and was a deplorable occurrence . . . but were a remote village and a mob of excited peasants in an out-of-the-way place going to put an end . . . to our national struggle for freedom?"[21]

It was a basic difference between them. Gandhi always insisted that unethical means could never elicit desired ends. The result of violence would only be more violence, not truth or *ahimsa*. Nehru was attracted to Gandhi's Satyagraha movement not because of its ethical philosophy but for what he believed to be its practical purpose, to win India's freedom. Unarmed India, after all, had no hope of ever defeating the well-armed British Raj, more powerful than all of Europe's Central Powers combined. Even the mutiny of Sepoy soldiers had failed. Therefore, India's modern revolt must be nonviolent in order to succeed, rallying the masses to rise up without arousing enough British wrath to replicate ten thousand times the massacre at Jallianwala Bagh. But once India's army was up and marching, all prison cells filled to bursting with happy, self-assured national troops, why call off the war? To Nehru, it was little short of madness, exposing a flaw in Gandhi's mind and method that he had never seen before. The aftermath of Chauri Chaura, however, taught Jawahar much about the weakness of his leader and, perhaps in equal measure, about his own motivation for political action. He had accepted Gandhi's "method . . . because of a belief in its effectiveness." If it were no longer effective, he would abandon it for one better suited to reality, or realpolitik.

On March 10, 1922, Gandhi was arrested and sentenced to six years for sedition. No one protested, and no street revolution followed to tear down the Raj. A few days earlier Jawaharlal himself had just as suddenly been released from prison. "I do not know why I have been released," he told the press. "My father, who is suffering from asthma, and many hundreds of my comrades are still in jail. I have only this to say—keep on fighting, keep on working for independent India and do not rest. Do not forsake your ideals. Do not go in for false compromises." He returned to work in his Allahabad office, issuing manifestos and instructions to the workers of the Congress committee he ran, urging disciplined "humility and toleration" and telling every one who worked with him now what he would later tell his nation: "We must talk less and do more."[22]

Nehru now led a campaign to close down the business of every cloth merchant in Allahabad who continued to stock and sell foreign cloth, most of which was imported from Manchester. He organized volunteer pickets that marched outside those shops, urging people not to shop there. And he personally approached the shop owners, demanding money to compensate other cloth merchants who adhered to the Congress program and sold Swadeshi clothing. As a result, in May 1922, six weeks after his release from prison, he was tried and returned to jail, sentenced to two concurrent terms of nineteen months behind bars. Although he refused to plead either guilty or not guilty, Jawaharlal did read a statement of principle to Allahabad's court:

"I stand here charged with criminal intimidation and abetment of an attempt to extort. . . . Less than ten years ago I returned from England after a lengthy stay. . . . I had imbibed most of the prejudices of Harrow and Cambridge and in my likes and dislikes I was perhaps more an Englishman than an Indian. . . . And so I returned to India as much prejudiced in favor of England and the English as it was possible for an Indian to be. . . . Today . . . I stand here in the dock . . . an ex-convict who has been to jail once already for a political offense, and a rebel against the present system of Government in India. . . . Today sedition against the present Government in India has become the creed of the Indian people, to preach and practice disaffection against the evil which it represents has become their chief occupation. . . . England is a mighty country with her armies and her navies, but today she is confronted with something that is mightier. . . . We are fighting for our freedom, for the freedom of our country and faith . . . India will be free; of that there is no doubt, but if England seeks the friendship of a free India she must repent and purge herself of her many sins, so that she may be worthy. . . . I shall go to jail again most willingly and joyfully. Jail has indeed become a heaven for us, a holy place of pilgrimage since our saintly and beloved leader was sentenced. . . . One feels almost lonely outside the jail, and selfishness prompts a quick return. . . . I shall go with the conviction that I shall come out to greet Swaraj in India."[23]

That impassioned statement reflected the depth of Jawaharlal Nehru's disappointment in the British Raj and his own dreams for India's national struggle. India was now fighting for its Freedom. There could be no turning back. India would be free. Swaraj was much closer than any Englishman imagined. It was also more remote than Jawahar believed, but his faith in its ultimate realization was never to waver. That faith and hope helped sustain him through all his dark prison years ahead. From now on, going to jail would become the highest form of Nehru's Indian national service.

6

To Change or Not to Change

1923–1925

EVEN THOUGH GANDHI called off his Satyagraha campaign early in 1922, he did not abandon the boycott of most things British, including Manchester cloth and the new councils elected under the more liberal Government of India Act passed by Parliament in 1919. Motilal Nehru, released at the end of his prison term in June 1922, viewed Gandhi's boycott of elections and legislative councils as self-defeating for India's political aspirations. It was one thing to boycott every British institution as long as the revolt was gaining momentum, but with the inevitable drop in morale that followed Gandhi's about-face in the wake of Chauri Chaura, his failure to take advantage of every opportunity to engage the British in political debate and possibly to defeat them within council chambers, seemed suicidal to Motilal. He convinced his Bengali barrister friend, C. R. Das, that it was time for a change. So they joined forces and early in 1923 organized the Swaraj Party to contest every local election. They hoped to use the enlarged British Indian legislature to attack and undermine the Raj from within, rather than to sit passively outside its formidable walls hoping they might come tumbling down thanks to mantras muttered by a Mahatma.

Gandhi recognized the challenge to his national leadership posed by so formidable an alliance of Congress's leading United Provinces and Bengali lawyers in favor of change. Although the Mahatma remained behind bars, he skillfully guided his most loyal lieutenants—Gujarati Sardar Vallabhbhai Patel, Madras Brahman C. Rajagopalachari, and Bihari Dr. Rajendra

Prasad—to carry his "no-change" flag to every Congress meeting. In December 1922 the "No-changers" soundly defeated the Swaraj Party "pro-changers" at the Gaya Congress by almost two to one, a major setback for the then Congress president, C. R. Das. Neither Das nor Motilal lost faith, however, in the soundness of their strategy, for both men were not only realists but also brilliant lawyers, who knew Viceroy Reading well enough from his own legal career to believe that their arguments inside his council chamber would have a significant impact on the nature and conduct of his Raj. They also had doubts about Mahatma Gandhi, whose "inner voice" and often unpredictable course of action or rationales for inaction confused them nearly as much as they did Secretary of State Edwin Montagu, who considered Gandhi either "naive" or a "Bolshevik."

By now, Motilal's asthma was so bothersome that he asked Das to accept the presidency of their new party, with himself serving as secretary. He would not, however, be able to convince his son of the value of joining the Swaraj Party, when in January 1923 Jawaharlal was again released from prison well before his term had ended.

"There is always a feeling of relief and a sense of glad excitement in coming out of the prison gate," Jawahar wrote. "We felt exhilarated, but this was a passing sensation, for the state of Congress politics was discouraging enough. In the place of ideals there were intrigues, and various cliques were trying to capture the Congress machinery. . . . My own inclination was wholly against Council entry, because this seemed to lead inevitably to compromising tactics and to a continuing watering down of our objective." Nothing Motilal wished or President Das argued in defense of their Swaraj Party changed young Nehru's mind. Nor was Jawahar much happier with the "constructive" program of spinning and social reform work on which his adopted *bapu*'s "no-changer" lieutenants focused their energies, mostly spinning and weaving khadi cloth in village ashrams, in order to "uplift" India's rural majority. The Swarajists seemed too reformist and constitutional to Jawaharlal, the no-changers too passive and pious. He yearned to play a more activist role and for a few months tried to bridge the gap separating the camps, but none of the older leaders really listened. C. R. Das, who was now closer to Motilal than anyone other than his wife, found Jawahar surprisingly "cold-blooded."[1]

"I suppose he was right," young Nehru reflected. "Compared to many of my friends and colleagues I am cold-blooded. And yet I have always been afraid of being submerged in or swept away by too much sentiment or emotion or temper. For years I have tried my hardest to become 'cold-blooded.' " Bored by the pro-changer/no-changer conflict and intrigues in both camps of Congress, Jawahar resigned from his centrist post on the Congress Working Committee, and in September 1923 he "had a strange and unexpected adventure." He was free, but felt frustrated. Although he could not go to Paris or Norway, the princely state of Nabha was not far from Delhi, only a day away, and the British administrator running that state had become the magnet symbol of imperial tyranny to many youthful activists, including Jawaharlal.[2]

Nabha, one of the major princely Sikh states of Punjab, had been taken almost directly under British control during World War I when the old maharaja was deposed and replaced by his young son, who was carefully watched by his British mentors, educated in the best British martial tradition and, by World War II, had became one of British India's leading generals. Militant Sikhs, some loyal to the deposed maharaja and others eager to "cleanse" Sikh temples, called *Gurdwaras* (the Guru's door), of their corrupt Hindu *mahants* (managers), marched in nonviolent bands *(jathas)* across the Nabha state border toward Jaito, where the Gurdwara struggle in Nabha was centered.[3]

At a special session of Congress in Delhi in September 1923, Jawahar learned of a protest march into Nabha that was to be launched that month and decided to join it. Since 1857, when loyalist Sikhs, those bearded "lions" *(singhs)* of Punjab, played so important a role in helping the British recapture Delhi from the "mutinous" Bengali Sepoys, Sikhs had served as one of the sword arms of the British Raj. The other sword was held by the Muslims.

After the mutiny, British social Darwinists developed a "martial race" category for their spurious theory of India's pluralistic population, classifying loyal Indians, like the Sikhs, Gurkhas, and some of the Muslims (for others had rebelled) as *martial races.* Other Indians, especially the upper-caste Hindu Brahmans and Kshatriyas of Bengal's Sepoy army, which had revolted, were called *nonmartial peoples.* Only the martial races were recruited for the reorganized British Indian army after 1858; all nonmartial peoples were declared unfit for military service of any kind and were disqualified from ever enlisting or bearing arms. This theory rationalized the ever present feeling of post-1857 British minds in India that high-caste Hindu "nigger natives," especially Bengali *babus,* could never be trusted with any weapon. They all had been "untrue to their salt," murdering British officers at Meerut in May 1857 and then killing, raping, and butchering every English man, woman, and child they could find all the way down the Gangetic plain to Lucknow. One could therefore never trust a Hindu or turn one's back on one. But the Sikhs were "brave" and "strong as lions" and hence were encouraged to join up and fight for the Raj they loyally served, and they were permitted to wear the "sword" or "knife" *(kirpin)* that bearded, turbaned Sikhs always kept within reach.

But the massacre at Jallianwala Bagh and the popular postwar anti-Rowlatt Act, Satyagraha, planted doubts in British minds as to the Sikhs' loyalty. During the war German agents had subverted Sikhs abroad, especially in San Francisco, where the Ghadr ("Mutiny") Party of Sikh militants emerged as a threat to the British war effort, as they sought to send arms to India, where they hoped to trigger a revolt. In 1920 the Akali ("Immortal") Party of devout Sikhs, who daily recited the prayers of their *Guru Dev* (divine teacher) Nanak, founder of the Sikh faith, emerged to cleanse and purify the Gurdwaras that had been polluted by corrupt Hindu Brahmans and their servants. Nehru admired the Sikhs' courage, egalitarian ethic, and hard work. There was a "fearlessness"—some labeled it "foolhardiness"—about Sikhs that appealed to most Western minds, including young Jawaharlal's, so he joined the Jatha headed for Jaito that September and was arrested for illegal entry into Nabha state. The British administrator of Nabha, J. Wilson Johnston, had signed an order, which was served on Nehru by Nabha's border police, prohibiting his entry and stating that if he had already entered when served, he must "leave immediately." He refused and so was then handcuffed to a Congress colleague, and both of them were led by chain to Nabha's jail.[4]

"This march of ours down the streets of Jaito town reminded me forcibly of a dog being led on by a chain. We felt somewhat irritated to begin with, but the humour of the situation dawned upon us, and on the whole we enjoyed the experience. We did not enjoy the night that followed. . . . Neither of us could move at all without the other's co-operation . . . not an experience I should like to repeat. . . . Nabha Gaol . . . was small and damp, with a low ceiling . . . we slept on the floor, and I would wake up with a start, full of horror, to find that a rat or a mouse had just passed over my face." It was thus hardly surprising that after leaving Nabha two weeks later (his two-year prison sentence having been suspended by Johnston), Jawahar came down with so severe a case of typhoid fever that he had to remain bedridden at home for almost a month.[5]

Motilal was so distressed when he learned of his son's arrest in Nabha that he contacted Lord Reading to complain, urging his immediate release. The viceroy was sympathetic and advised Johnston to take "special care" of his prisoner, who was not only a Cambridge graduate and barrister but whose father happened to be the viceroy's old friend. Motilal was given the viceroy's permission to go to Nabha himself to visit his son in jail, indeed, to bring him home as soon as he promised not to repeat his illegal entry into that state. Administrator Johnston was properly deferential, of course, for was there an English civil servant in India, young or old, who did not know the name of, and bow his head before, the august presence of Pandit Motilal Nehru? Nowadays, in free-flowing white khadi, he looked more like a tribune of Rome than the Savile Row aristocrat of his earlier days. He sailed through the prison, to the innermost cell where his son sat silent, unsmiling, greeting him more with surprise in coldly questioning silence than with love or relief.

"My dear J," Motilal Nehru wrote his son the next day. "I was pained to find that instead of affording you any relief my visit of yesterday only had the effect of disturbing the even tenor of your happy jail life. After much anxious thinking I have come to the conclusion that I can do no good either to you or to myself by repeating my visits. I can stand with a clear conscience before God and man for what I have done so far after your arrest but as you think differently it is no use trying to make opposites meet. . . . For the present I hardly know what to do with myself and shall wait here for a couple of days or so. . . . Please do not think that I have written this letter either in anger or in sorrow. I have tried my best after an almost all-night consideration to take a calm and practical view of the position. I wish you not to have the impression that you have offended me as I honestly believed that the position has been forced upon both of us by circumstances over which neither has any control."[6]

What a long and lonely night of painful reflection that must have been for the sixty-two-year-old Motilal Nehru. His son, for whom he had worked so hard and planned so carefully, would rather remain locked up in Nabha's vermin-infested jail as a prisoner than accept the viceroy's pardon from his own father's hand. "The position has been forced upon both of us by circumstances over which neither has any control," he finally decided as dawn's gray light etched the dim walls of the dismal state guest house in which he had reflected all night, worrying, wondering, and trying to understand what had happened to him, to his son, to India, to his British "friends"—now suddenly become "enemies"—and to all he had believed all his life.

Motilal left his new assistant, a young lawyer named Malaviya, to watch the proceedings in Nabha's court and help out if Jawaharlal would let him say or do anything, or at any rate to report to him on what was happening after he went home. His asthma returned, as did his lumbago and the other strange aches and pains that wracked his fast failing body. Nonetheless, Motilal continued to try to organize and lead a more rational, if not successful, political movement than the one Gandhi had launched and then abandoned. His son returned home too sick with fever to stand up, not to mention argue, either in court or outside for elections. Fresh elections were scheduled to be held in November throughout British India, and the month before, Motilal invited all the leaders of his Swaraj Party to join him and President Das at Anand Bhawan to draft the text of their election manifesto.

Their long-range party goal was "the attainment of Swaraj," but its immediate objective was more practical: the "speedy attainment of full Dominion status."[7] That had been essentially the goal of both Congress and the Muslim League in Lucknow as early as 1916. Motilal and Das were simply trying to return India's nationalist movement to the rational political track from which it had veered off after the massacre at Jallianwala Bagh. Before spelling out their ways of attaining that immediate goal, the Swarajists tried to win the greatest measure of nationalist support by insisting that theirs was "a party within the Congress"; indeed, they thought of themselves as "an integral part of the Congress." Both Motilal and Das had been presidents of Congress, and both remained the most powerful leaders of their provincial Congress committees. Yet to "no-changers" like Vallabhbhai Patel and Rajendra Prasad, they were now outsiders. Even Jawaharlal viewed them as so close to "loyalists" or "moderates" that he refused to join them and wanted no help from his father in expediting his release from Nabha.

Mahatma Gandhi was not to be released from prison for four more months, but then his appendicitis gave Reading the face-saving opportunity he needed to approve Gandhi's early liberation. Reading, like Motilal and Das, now hoped that once Gandhi recovered from his operation and enjoyed the fresh air of freedom, he might relent a bit and that all of them, Englishmen of liberal goodwill and Indians of liberal constitutional predisposition, might sit down together at a long or round table and talk calmly about how best to formulate a constitution for India.

This was the stated immediate goal of the Swaraj Party. If it could also become the goal of the entire Congress, all would be well. Dominion status within the British Empire might be achieved within a few years or as soon as both houses of Parliament back in Westminster carried the legislation from its first drafts and readings as a bill to its final act. With Lord Reading, a former lord chief justice of England, in command of Simla–Delhi and Motilal and Das running the Congress, India might well have achieved full dominion status in 1924 or soon after. Such a Dominion, moreover, would have included all of South Asia, for Jinnah allied himself with Motilal Nehru's Swaraj bloc in the viceroy's Legislative Council, voting with them as he would readily have governed with them in a free, united dominion of India. Such reasoning reckoned, however, without the piety of Mahatma Gandhi or the passion of Jawaharlal Nehru.

"I feel that a public reception to the Viceroy is a shameful thing for anyone to whom the honour of India is dear and precious," Jawaharlal wrote to Allahabad's Municipal Board from his sickbed in October 1923. "I wish no personal discourtesy to Lord Reading. None of us wish him any discourtesy. But we are human, with eyes to see and often weep, and ears to hear, and hearts to feel, and feel the more because our arms are weak and there is no strength in us to stand upright and protect our own. . . . Lord Reading's Government in the arrogance of power have started a crusade against a brave and gallant people whose chief sin in the past has been a too great loyalty to the British Government. Their highest and their bravest have been arrested, and are being arrested by the hundred.

Every Sikh with a black turban is being hunted down. . . . Is it this that we are asked to celebrate? . . . [F]or which we wish to honor Lord Reading? I can be no party to this infamy. . . . I am weak and powerless but I too have a little pride—the pride of the weak perhaps it may be."[8]

Jawaharlal's Cambridge mock humility was hardly the pride of the weak. Thanks to his name alone, he currently chaired Allahabad's Municipal Board and was regularly consulted by the district magistrate, Knox, and by Allahabad's new chief justice, Sir Grimwood Mears, also a Cambridge man and temple barrister, whose "cordiality" had "a little surprised" young Nehru when he was early released from prison.[9] "I would sooner be trampled by Lord Reading's soldiery and ground to dust, rather than bow down to welcome a person [the viceroy] who was responsible for so much sorrow to my country."[10]

Sir Grimwood tried at this time to lure Jawahar back to what had so long been his friendship for and trust of all Englishmen and their methods of governance as well as education and play. "He had an idea—he told me so later—that I would go far, and he wanted to be a wholesome influence on me to make me appreciate the British view-point. His method was subtle. . . . Sir Grimwood's idea was . . . friendly intercourse and frank and courteous treatment. I saw him several times . . . and then he led up the conversation to the new Councils and their Ministers and the opportunities these Ministers had for serving their country. Education was one of the most vital problems before the country. . . . Suppose, he went on, a man like you, with intelligence, character, ideals, and the energy to push them through, was in charge of education for the province, could you not perform wonders? And he assured me, adding that he had seen the Governor recently, that I would be given perfect freedom to work out my own policy. Then realizing, perhaps that he had gone too far, he said that he could not, of course, commit anybody officially. . . . That idea of my associating myself with the Government as a Minister was unthinkable for me; indeed, it was hateful to me . . . such is our fate, that we can only reach the land where we can build after passing through the deserts of conflict and destruction."[11]

Had Jawahar agreed in 1923 to accept Sir Grimwood's generous offer, not only would the next two decades of his own life have been both easier and more productive, but all of India might have benefited, accelerating its transition less tragically from imperial dependence to full dominion status. The Government of India Act in 1921 had introduced several reforms that made the constitution of British India far more liberal than it had been by the end of John Morley's era in 1910. The franchise was extended to more than five million Indians—landowning, taxpaying, or literate—empowered to elect their own representatives to widely expanded provincial councils, and one-fifth that number could vote for the more powerful members of the Central Legislative Assembly, India's nascent lower house of Parliament. A majority of one hundred of the 140 members would thus be elected.

More than forty of these new members of the assembly belonged to Motilal's Swaraj Party; others, like Jinnah, were independent Muslim members, who readily joined forces with Motilal and Das to defeat any measure they considered reactionary or anti-Indian, including the viceroy's annual budget. The viceroy's Executive Council thereafter had at least three Indian members on its powerful administrative body, and a number of departments of the central government of India were from this time on to be headed by elected Indians. Education, sanitation, and other less generously budgeted matters of state—but in some respects the most important ones—were thus transferred to elected Indian control, with the rest still "reserved" to British power. But nonetheless the Congress "no-changers" continued to boycott all elections and councils.

Jawaharlal did not vote or run for anything that year but presided over the first All-Indian Volunteers Conference held in Kakinada during Christmas week. "I thank you for the honour you have done me . . . [to] put me, a humble worker, in this Presidential Chair. I bow to your decision. But I come to you as a mere soldier and not as an officer. Perhaps many of you are not aware that years ago I was a member of the Officers Training Corps in my school in England. But I ended my career in the Territorial Army of England as a private and was not even awarded any badge. . . . We meet as soldiers of freedom and as soldiers we must be men of action rather than of words. . . . Let us remember that noble Hector, King Priam's son, as he sallied out of the plains of Troy to fight the mighty Achilles, being urged by his brother to bow to unfavorable omens and retreat, replied:—'One omen is best—to fight for our country.' "[12]

It is unlikely that many members of his Kakinada audience had heard of Hector or read the *Iliad,* but to any who had, Nehru's allusion might have seemed unfortunate in light not only of Hector's dreadful fate but also because his arrogance and bluster have long associated his name with bullying rather than discipline. Jawahar's militant hectoring at any rate clearly reflected his own desire to fight for India's total freedom from British dominion rather than accept the half-loaf of Council entry his father was ready to settle for, or even the larger portion of administrative power Sir Grimwood had offered. "It is a matter of joy to the Congress that an All Indian Volunteer Organization is coming into being," Nehru told his followers at Kakinada. "It is proclaimed all along that this movement is nonviolent. . . . Nevertheless you will understand that this is war. . . . We are fighting the British now. If you are capable of governing your country you must be ready to offer yourselves at the altar of your Motherland. You will be able to defeat your opponents the moment you outdo them in the immensity of your sacrifice. . . . In the Punjab . . . the whole of the Sikh community is opposed to the Government." Twenty years after Nehru's death, Sikhs would still be fighting the government in Delhi, an Indian government headed by his only child, Indira, who would be assassinated by two of her trusted Sikh guards.[13]

Soon after Gandhi's early release to treat his appendicitis, he went to Bombay's Juhu Beach to recuperate. Motilal rented a beach cottage next to the Mahatma's, for "Father wanted to explain to him the Swarajist position, and to gain his passive co-operation." Gandhi now focused his energy on spinning, the first plank of his "constructive" program, seeking also to foster Hindu–Muslim unity and abolish Untouchability. Motilal had never liked to spin cotton, though Jawaharlal learned to enjoy it during his months in jail. The elder Nehru feared that the opportunity of winning real concessions from the new British Labour government led by Ramsay MacDonald was being lost through sheer indifference to political reality and Gandhi's lack of awareness of the sympathy for India felt by many of Britain's most powerful leaders at this time.[14]

"Pandit Motilalji and I have had prolonged discussions about propriety of non-co-operators entering the Legislative bodies," Gandhi reported early in April 1924. "There is an honest and fundamental difference between the Swarajists and myself." So Congress remained divided. The gulf between Motilal and the Mahatma appeared unbridgeable, reflecting the distance between modern Western society and ancient village India. Jawaharlal remained almost equally torn between his intellectual attraction to everything modern and Western and his love for Bapu and his country's peasant populace.[15]

But the rift in Congress did not pose as great a problem as did the growing violence that divided Hindus from Muslims during the post-Khilafat breakdown. Mustafa Kemal Ataturk, the modernist president of the Turkish Republic, abolished the caliphate along with the fez in 1924, leaving millions of Indian Muslims—roused to a fever pitch of political expectation by the brothers Ali and Gandhi's insistence on making Khilafat "liberation" the primary plank of his Satyagraha movement—feeling ill used and abandoned. First they had tried to march through Afghanistan to liberate the caliph, but the Afghans closed their border to the ragtag army of Muslim brothers, who brought nothing but faith in Pan-Islam to feed themselves and their families. Most of these Pan-Islamic pilgrims had abandoned their homes in India, leaving real property behind in their religious zeal and politically motivated march toward what they were told would be Pan-Islamic conquest and vindication. Turned back at the Khyber and Bolan frontier passes by armed Afghans, who had hardly enough food for themselves and showed no fraternal interest in feeding a million Indians, the hungry, weary Khilafat army wended its way home, only to find that their Hindu neighbors had swiftly appropriated their abandoned homes and fields. Communal battles were then joined, the Hindus fighting to ward off the enraged Muslims. More than two decades of growing Hindu–Muslim hatred and communal conflict, in Malabar's villages as well as cities like Bombay and Lahore, Lucknow and Calcutta, continued to plague every province of British India from 1924 until the partition of 1947 that gave birth to Muslim Pakistan.

"I consider the religious principles of an adulterous Muslim to be better than Mahatma Gandhi's religious principles," Maulana Mohammad Ali told an angry Muslim audience in Lucknow at this time. "In my humble opinion," Gandhi responded, when this was called to his attention, "the Maulana has proved the purity of his heart and his faith in his own religion by expressing his view." Many Hindus as well as Muslims now began to blame Gandhi for misleading them, first promising freedom in one year, then "robbing" them of all they owned, and now denying them access even to councils, where they might at least try to improve themselves and the lives of all Indians.[16]

The Swarajists and their independent allies led by Jinnah meanwhile carried a resolution by seventy six to forty eight through the Central Legislative Assembly, calling on the viceroy to summon a round table conference to draw up a constitution for India that would lead to full responsible government. "We then waited for the response," Motilal reported at a public meeting in Bombay that April, but when it proved "highly unsatisfactory," the Swaraj Party voted down every official budgetary request for grants, whether for customs, income taxes, salt, or opium, forcing the viceroy to overrule his assembly. Motilal reminded Prime Minister MacDonald of his earlier expressed hope for India, a "fully responsible" Viceroy's Council and "control over her own finances."[17] But even though Labour's Ramsey MacDonald was prime minister at the moment, his government was a Liberal–Tory coalition destined a year later to fall before Stanley Baldwin's mounting Tory majority. Liberal India's false dawn of reform then disappeared as the dark clouds of Britain's harshest imperial reaction rolled east, blowing away all Swarajist dreams of dominion glory, validating, it seemed, in their storms of repression, Mahatma Gandhi's worst fears of Western tyranny and vindicating Jawaharlal's growing radical antipathy to all the things British that he once had loved.

7

At Sea

1925–1927

LATE IN 1924 GANDHI wrote several pleas to Motilal on behalf of Jawaharlal. "He is one of the loneliest young men of my acquaintance in India. The idea of your mental desertion of him hurt me."[1] Typically, Gandhi, never hesitating to intrude in the most intimate of relationships, was certain he knew best how to resolve every conflict. To Motilal, however, who well understood and had long known that he had lost his son to this Mahatma, the impertinence of such a letter—making it sound as though Gandhi really cared more about Jawahar than he did—was too much. Motilal's rage reached his son's ears first, and Jawaharlal wired Gandhi, asking him to plead no more with his Father. "I had heard from Father already," Gandhi replied with equally typical contrition. "I am so sorry. I thought I was writing a harmless letter showing the depth of my feeling."[2]

Jawaharlal's depression and frustration, nonetheless, his belief in Motilal's "mental desertion" was quite real. India's wealthiest lawyer had given up his practice not to follow Gandhi but to march with his son, to stand by him in any struggle, to face any hazard at his side, whether in the courtroom or the prison cell. Yet now that Motilal had taken his own plunge into politics, organizing with Das their thriving Swaraj Party, did Jawahar join them? Did he stand for a seat, any one of which he could have captured without contest? Did he say one word or volunteer an hour of service to the Swaraj Party cause? No. He offered no response, no help, no encouragement. The mental desertion felt by both father and son was mutual.

"Gandhiji," Jawaharlal wrote about his adopted *bapu,* "had made of his life an artistic whole. Every gesture had meaning and grace, without a false touch. There were no rough edges or sharp corners about him, no trace of vulgarity. . . . Having found an inner peace, he radiated it to others and marched through life's tortuous ways with firm and undaunted step. How different was my father from him! . . . He was neither meek nor mild, and . . . he seldom spared those who differed from him. Consciously imperious, he evoked great loyalty as well as bitter opposition. It was difficult to feel neutral about him; one had to like him or dislike him."[3]

Fueling Jawaharlal's growing dislike of his father was his continued dependence on Motilal's generous support, not simply for himself, but for his wife and daughter as well. They all continued to live rent free in their second-floor apartment in Anand Bhawan. Jawaharlal received no pay for any of his Congress work, though he was by now the semipermanent general secretary of the All-India Congress Committee. At thirty-five he had lost most of his hair and suffered from piles. His ailing wife, pregnant again, gave birth prematurely to a son that November, only to watch him die two days later.[4]

Kamala's condition deteriorated in 1925, and her doctors finally diagnosed tuberculosis. There was as yet no cure for it, but Switzerland was generally believed to be the most comfortable atmosphere in which to die of it. Nor was sleepy Allahabad any longer immune to communal violence, with Hindu–Muslim riots erupting within earshot of Anand Bhawan during 1924–1925. In fact, these riots grew so severe that Gandhi announced a three-week fast to end them. Because Nan's first daughter had just been born in Bombay, she and Ranjit also were eager to travel abroad. "We booked our passage . . . in March 1926," Nan recalled, "Kamala had not been well . . . Bhai too was very tired and badly needed a change. Finally we all left India together. Indira, then aged nine, accompanied her parents."[5]

Indu, as Jawahar and Kamala's daughter was nicknamed, was a frail, introspective, brooding child, reared in a world of silent suffering and deeply felt hatreds. Her Aunt Betty recalled seeing her at this age talking to herself as she clutched a pillar on the railing of Anand Bhawan, her other hand raised high. "What in the world are you trying to do?" Betty asked. "She looked at me solemnly . . . her dark eyes burning, and said, 'I'm practicing being Joan of Arc. I have just been reading about her, and some day I am going to lead my people to freedom just as Joan of Arc did.' "[6] "I was fascinated by Joan of Arc because she fought the British," Indira later wrote after becoming prime minister, "and because being a girl, she seemed closer to me than other freedom fighters . . . the sort of person I liked reading about best."[7]

Jawahar rented a small apartment in Geneva, and Betty came out to take care of it for him, caring for Kamala as well while he went to visit many friends, old and new, among them Romain Rolland in his Villa Olga in Villenueve. "I had never lived in anything so small before," Betty reflected on their Geneva apartment. "Jawahar presented me with a map of Geneva and an English–French Dictionary as well as a book containing bus and tram coupons. I was told that that was all I required to get along by myself and . . . as Kamala was unwell, I should have to do the housekeeping . . . I was rather taken aback by my brother's ultimatum but I knew it was no good arguing with him."[8] At first Indira was enrolled in L'Ecole internationale, later leaving Geneva to go to the Ecole nouvelle in Bex, which she liked much better. Indira's recollection was that "my father went off and left me in a flat with my mother who was not well. I had to deal with the maid, order the food and so on. This is how I learnt French—the hard way or maybe the easiest . . . I was more or less in charge of the household. My aunt was there but I think she used to go with my father to various places."[9] Whether it was Indu or Betty who actually ran the flat, both agreed that Jawahar had little interest in doing so and seems to have spent less time there than any of the others did.

"I have been leading a fairly quiet life here," Jawahar wrote from Geneva to his Cambridge friend Syed Mahmud, who had become a close comrade back home. "Geneva is developing into a big international centre and there is always something going on here. The International Labour Conference will begin day after tomorrow. . . . Meanwhile I propose to go to Paris next week and to Italy next month for a week or two. . . . News from India is most distressing . . . the communal frenzy is awful to contemplate. We seem to have been caught in a whirlpool of mutual hatred and we go round and round and down into the abyss. Rioting appears to have become a permanent feature of Calcutta life."[10] Two months later he wrote Mahmud again, feeling "more inclined to think that the only remedy [for communal conflict] is to scotch our so-called religion and secularize our intelligentsia at least. How long that will take I cannot say but religion in India will kill that country and its peoples if it is not subdued."[11]

A few months in Europe, far from Gandhi, thus sufficed to revitalize Nehru's faith in socialism and secularism. Nor would he ever again abandon either of those beliefs as a way to resolve India's problems of poverty and pluralistic conflict. He remained optimistic, moreover, despite the news from India, that "England cannot hold on to India for long." Both Syed Mahmud and Dr. M. A. Ansari had joined the Pan-Islamic Khilafat movement at its inception, but they now resigned, focusing their talents and political energy instead on working for freedom through Congress, as nationalist Muslims. Nehru later spoke of himself as India's leading "nationalist Muslim," but many Muslims like Mahmud and Ansari and the more famous Maulana Abul Kalam Azad remained Congress stalwarts all their lives.[12]

In August Jawaharlal wrote that Kamala was "doing well," but a month later she relapsed, never shaking the persistent fever that killed her in less than a decade. Bertrand Russell's *On Education* convinced Jawahar of "how utterly wrongly we bring up our children," and "as for our treatment of women," he wrote Mahmud, "Is it any wonder that we have continually to contend against bodily ailments and mental troubles? Do you know of any single family . . . in India which is not always faced with disease? . . . We are always hobnobbing with doctors and *hakims* but it does not strike us that the way to combat disease is to . . . remove the causes. These causes are largely interwoven with our unfortunate social system and customs . . . Kamala, I am sorry to say, has not been keeping very well." But two days later he was off to London.[13]

It was Nehru's first visit back in fourteen years. A coal miners' strike had left London "in semi-darkness." The general strike launched earlier in the year had by then collapsed, as the new Tory government cracked down with a heavy hand on all labor movements, which Jawahar felt "as a personal blow," sensing no doubt what India would soon confront in the most bitter decades of its struggle still ahead. He visited Cambridge and Derbyshire but found few of his old friends, the war having taken so many of them. Yet now he started to forge new alliances in England with radical members of the Independent Labour Party like Fenner Brockway and Indian exiles, especially young Krishna Menon, who ran the India League from a "tiny office" in the city, packed with every radical journal and newspaper in five languages, and his ever boiling teapot.[14]

It was here that Krishna Menon first captured Jawahar's mind, his brilliant, bitter wit, igniting Nehru's spirit at first sight. Here at last he found a soul mate, darker skinned, gaunt, ugly, impoverished at birth, and yet better informed about the Marxist–Leninist dialectic he had taught himself and the Fabianism he had learned at the London School of Economics than any Indian Nehru had ever met. It was Krishna Menon who introduced Nehru to Harold Laski, as "one of us." The long nights of "tea and more tea, and grilled tomatoes on toast," and dialectic that Jawahar spent with Krishna Menon were his awakening to foreign affairs. Just as Gandhi taught him to see India's peasant village realities, Krishna Menon now helped expand Nehru's knowledge of foreign policy and diplomacy, which soon became his favorite intellectual retreat.[15]

Krishna Menon told Nehru about the forthcoming Congress of Oppressed Nationalities that was being organized at this time in Berlin, to be held in Brussels during February 1927. Although Germany had lost the war, radical German intellectuals now hoped to win the peace. A generation of Western Europe's best and brightest young men were dead, and tsarist Russia had fallen to revolution. At this time Jawahar knew only that India was an "oppressed nationality." Therefore, it seemed to him fortunate to be in Europe when the first such Congress was organized. He was eager to attend as an official delegate of India's National Congress, and he wrote to the Congress secretary, Rangaswami Iyengar, requesting permission and the funds to cover his travel to and board in Brussels. He received £50, becoming India's only accredited delegate to the Congress, at which he delivered his first international speech.

"We in India have felt the full weight of imperialism . . . since the British arrived . . . they have adopted the old policy of 'Divide and Rule.' I regret to say that that policy is still very much in evidence . . . it is terribly common for innumerable comrades and friends of ours to go to jail with or without trial. . . . But the real injury that the British have done to India . . . is the systematic way in which they have crushed the workers and peasants of India, and made India what she is today. . . . Now it is frequently stated and made much of in the English Press, that the Indians are fighting against each other, the Hindu against the Moslem, etc. We must remember, apart from the fact that these troubles are greatly exaggerated that it is the policy of the British to create these troubles. . . . The noble example of the Chinese nationalists has filled us with hope, and we earnestly want as soon as we can to be able to emulate them and follow in their footsteps [long applause]. We desire the fullest freedom for our country . . . [and] the freedom to develop such relations with our neighbors and other countries as we may desire." [16]

Soon after the Brussels Congress ended in mid-February, Jawaharlal sent two reports home to India, one for general publication and distribution to all members of the All-India Congress Committee (A.I.C.C.) and the other for Congress's Working Committee. "The Brussels Congress . . . was an event of first class importance and it is likely to have far-reaching results," he assured his Working Committee colleagues, including Mahatma Gandhi. Although the English press gave "little or no publicity to it," as proof of its importance, Jawahar listed the delegations from Asia, including China, and the "Negro and White" delegates from South Africa and the United States, as well as European socialist and labor leaders.

In Brussels Nehru met the talented German Jewish dramatist Ernst Toller. He later recalled: "We were attracted to each other and I met him subsequently . . . in London. . . . He was very sensitive to human suffering, and it was a matter of grief for me to learn of . . . his unfortunate death." Toller, who initially fought for Germany during World War I, organized strikes against the war, and was later elected to preside over a swiftly crushed "Bavarian Soviet Republic" and then jailed for five years. After Hitler came to power, Toller's books were burned, and his greatest play, *Maschinensturmer* (The machine wreckers), was never staged again until after World War II. He fled Germany to the United States, where he committed suicide in 1939, feeling hopelessly "at odds" with the world all around him, much as Nehru so often felt.[17]

In India, the elections to the provincial councils as well as the Central Assembly were held in 1926, and Motilal was forced to bear alone the full burden of the Swarajist campaign, for his dear friend and president of the party, C. R. Das, had died the previous year at the age of fifty-five. "We have come out strong though not a majority in Madras and Bengal," Motilal wrote to his son that December, after returning home from an exhausting election tour. "Our strength in the Assembly is likely to be somewhat greater than it was during the last three years but there is a debacle in the U.P. [United Provinces] Council . . . I had hardly any workers worth the name to help me in my own Province." Not only had Jawahar deserted him, but former friends and old allies, Pandit Madan Mohan Malaviya and Lala Lajpat Rai, both ex-Congress presidents, founded their own Nationalist Party that year, a Hindu-first communal party, which won many seats in the United Provinces and elsewhere.[18]

"It was simply beyond me to meet the kind of propaganda started against me under the auspices of the Malaviya–Lala gang," wrote Motilal. "Publicly I was denounced as an anti-Hindu and pro-Mohammedan . . . almost every individual voter was told that I was a beef-eater in league with the Mohammedans to legalize cow slaughter." Malaviya was president of the Hindu Mahasabha, a conservative communal society that focused on saving cows from slaughter by Muslims while trying to force the conversion of Muslims to Hinduism, arguing that most of India's Muslim population had originally been Hindus but had forcibly been converted to Islam during some five hundred years of Muslim rule. One of the most militant popular Hindu communal leaders of that "reconversion" *(shuddhi)* movement, Swami Shraddhanand, was assassinated in Delhi that December by a Muslim extremist. The swami, like Lala Lajpat Rai, belonged to another fundamentalist Hindu society, the Arya Samaj, which advocated turning back India's history more than three thousand years to an ancient Aryan tribal polity, reflected in Vedic scripture, when Brahmans and cows were treated as gods on earth.[19]

"Communal hatred and heavy bribing of the voters was the order of the day, "Motilal reported to Jawahar in Geneva. "I am thoroughly disgusted and am now seriously thinking of retiring from public life. What is worrying me is how to occupy my time. . . . The Malaviya–Lala gang aided by Birla's money are making frantic efforts to capture the Congress. They will probably succeed . . . in the present state of communal tension my voice will be a cry in the wilderness. I shall consult Gandhiji but as you know his hobbies do not interest me. . . . The elections have left me thoroughly exhausted, but there is no peace for me yet."[20]

Motilal felt that his only hope now of bringing some semblance of rationality as well as unity to India's fast-deteriorating "nationalist" movement was to lure Jawahar back to his side and that the best way to bring his son home was to reconcile his rift with Gandhi, despite how negatively he felt about the Mahatma's "hobbies" of spinning, fasting, and protecting cows. At sixty-five, Motilal felt much older and knew that little time remained for him to achieve all he had dreamed of doing. Still, he might be able to leave what remained of his dream to his son, if only he could bring him to his senses. He first had to convince Gandhi, however, that Jawahar was ready to take the helm of Congress. He continued, therefore, not only to send his son ample funds (an additional £300 that November) but also to keep him fully informed of the political infighting that now threatened to tear apart the movement. So in the summer of 1927, Motilal himself returned to Europe to spend time with his son and family.

"There is some talk of your being chosen as President for the coming Congress," Gandhi wrote Jawaharlal that May 1927. "I am in correspondence with Father about it. The outlook here is not at all happy. . . . I do not know whether the process of breaking heads will in any way be checked. We have lost hold upon the masses, and it seems to me that if you become President, you will be lost for one year at any rate to the masses. . . . Someone has got to do it; but there are many who are willing and eager."[21]

Jawahar decided, however, that it would "be better if I am not chosen President." His year abroad had revitalized him politically, and the Brussels Congress and its aftermath of continued contacts with a number of delegates he had befriended made him "feel strongly that a *laissez faire* attitude and trusting to Providence to right matters is a very feeble way of combating evil . . . about the Presidentship. . . . Personally I should have thought that Ansari would be the best choice," he wrote to Gandhi. "I fancy you have got a wrong impression about my idea of the utility of the League against Imperialism. I do not expect much from it. . . . But I welcome all legitimate methods of getting in touch with other countries and peoples so that we are able to understand their viewpoint and world politics generally. I do not think it is desirable nor indeed is it possible for India to plough a lonely furrow now or in the future. . . . Our salvation can of course come only from the internal strength that we may evolve but one of the methods of evolving such strength should be study of other people and their ideas."[22]

Gandhi no longer found him the "loneliest" young man he was when he left India little more than a year before. Nehru had gained much strength from contacts in Europe with other people and their ideas, and he was now ready to reassert himself as a newly independent, internationally integrated person. Having jettisoned his father long ago in all but material and formal familial ways, he now felt strong enough to do battle with, if not yet totally abandon, his adopted *bapu,* whose trust in "Providence" Jawahar found "very feeble." Having discovered Krishna Menon, he now no longer needed any father, natural or adopted, so at least it seemed to him on the eve of Motilal's visit that summer of 1927. Remote from India's reality, he also now thought it should be "simple enough" for Hindus and Muslims to "live peacefully." "Hindus and Muslims in India are steeped in superstition and religion. . . . As soon as we shed our religion we shall . . . behave better. Europe has got rid of religion by mass education which followed industrialism. . . . This process is bound to be repeated in India. I am convinced that the religion of our masses is skin deep," he naively assured Syed Mahmud. "What is distressing is the behavior of our prominent politicians."[23]

Motilal arrived in Venice in midsummer, and his children were there to meet him. "Retired or not . . . he was the same *grand seigneur* in his manner of traveling," Betty remembered. "Suddenly we were going first class, with chauffeured limousines to meet us, and staying at the *grande luxe* hotels." From Italy to London and then Paris, even Kamala appeared to be a bit better. From Paris to Berlin, where a Comintern invitation was waiting for Jawaharlal Nehru to come to Moscow, as an "honored guest" of the Soviet government for its tenth anniversary of the Russian Revolution. Almost every Communist in Europe knew the name of Nehru by now, and although the invitation was only for the younger one, Motilal had no intention of letting his son go that far away all alone. Kamala and Betty were, moreover, eager to see what the revolution's "brave new world" looked like. That November, Jawahar wrote from Moscow to Nan, the sister he named twenty years later as India's ambassador to Moscow.[24]

"We are in a topsy turvy land. All one's old values get upset and life wears a strange aspect here. Everybody is *tovarish* [comrade]. We have to address the waiter or porter or cabby as *tovarish* . . . it takes some getting used to. Moscow is in some ways like any big city, and yet. . . . The comparative poverty of the people is obvious. There are very few cars. . . . The roads and pavements are crowded with pedestrians. The shops . . . are poor in comparison with shops elsewhere. But the spirit of equality is rampant, and pride in the revolution. We arrived a day too late and missed the great celebration . . . a grand parade of the army and a march-past of a million and a half people in front of Lenin's tomb . . . Moscow is full of flags and bunting. The hammer and sickle . . . everywhere. . . .

And of course Lenin's pictures. One cannot move or turn . . . without gazing at Lenin. . . . He is the God of the Russians."[25]

The Nehrus stayed only three days in Moscow, leaving for Germany through Poland, on November 12, but Jawaharlal was most impressed by what he saw, especially "the absence of different classes," finding workers without ties or jackets seated near them in the beautiful Bolshoi Theater, where the ballet "was exceedingly good," he reported to Nan, "the best I have seen." They were also taken to a large central peasants' house, where the very latest agricultural implements were on exhibit and free legal advice, as well as food, was available to all resident peasants. He was told that 350 other such large houses were located all over the Soviet Union and that electrification was proceeding "very rapidly," with "dozens of very big power stations scattered all over."[26]

These seeds of economic planning and development sown in Jawahar's fertile mind bore fruit less than two decades later, as India's first five-year plan. "There can be no doubt," he concluded, "that the conditions of the peasantry and the workers in Russia, in spite of the general poverty and low wages, are far superior to those elsewhere. What is more, their whole moral stature has improved and they look forward with hope and confidence to the future." That vivid impression of Russian "hope and confidence" remained in Nehru's mind until his death and was passed along to his daughter, who inherited his sociopolitical faith together with his power. Nehru viewed Lenin's corpse with admiration, "his great forehead calm and peaceful, satisfied, as it were, with the work he had done." He met Lenin's widow, Commissar Krupskaya, President Kalinin, and Comrade Bukharin—then still one of Stalin's trusted allies—and Madame Sun Yat Sen, who particularly impressed him. "She is delightful, looks twenty five and is full of life and energy . . . speaks English with an American accent, but not unpleasantly."[27]

With Brussels, Berlin, and Moscow firmly imprinted on his mind, Jawahar felt ready for a grand reentry into India's sadly divided National Congress. The Muslim League was at this time of continued communal conflict poised on the brink of opting for separate nationhood while Hindu extremists were gathering strength from their numbers to appeal to their militant majority to return Bharat (India) to the mythical golden age of its pre-Muslim Brahmanic past. Motilal had found Moscow "grim and drab." He could barely tolerate the habits of Stalin's stooges, all of whom adopted the timetable of their dictator, insisting, as he did, on meeting every foreign visitor in the middle of the night because he could sleep only during the day. When Motilal was informed that his appointment with the Soviet foreign minister, Chicherin, was scheduled for 4 A.M., he angrily refused to accept it. The audience was, therefore, rescheduled for 1 A.M.!

Motilal took his family back to London from Berlin. Sir John Simon, then Britain's leading barrister, was to appear before the Privy Council late in November on behalf of Motilal's first client, a large landowner of the United Provinces whose suit Motilal had initiated in 1894, which was finally to be heard by Britain's highest court of appeal. Simon had also just been invited by Prime Minister Baldwin to head a royal commission to India to consider the next step in British India's constitutional "reforms." Motilal visited Simon often in his chambers near the Temple of Justice to discuss their client's appeal. "I accompanied my father on one occasion," Jawaharlal recalled, "to Sir John's chambers for a consultation." If they also consulted each other about India, it may not have been a very pleasant meeting, for Simon was as close to the extreme right of British imperial policy as Jawaharlal was by now to the left of India's nationalist spectrum. But each of them used the opportunity to gauge the other's metal and mind before the battle each led against the other was joined. Simon's all-British, white commission reached India to find itself faced with millions of protestors shouting "Simon, go back!" A forest of black flags welcomed Sir John and his "seven dwarfs," symbolizing a nationalist boycott even more extensive than that greeting the Prince of Wales. This time it was organized in great measure by the more radical, revolutionary young Nehru on the eve of his rise to the peak of Congress popularity and power.[28]

On December 2, Jawahar, Kamala, Betty, and Indu left Venice for Colombo and then on to Madras. Nan had been carefully instructed to await them in Ceylon with the khadi clothing needed by Jawaharlal for his proper reentry into Congress's political fold. The nationalist faithful gathered by the khadi-clad thousands along the banks of Adyar River, within sight of Annie Besant's Theosophical Society that Christmas of 1927. India's future national savior must, after all, be appropriately dressed when he arrived. Having seen Europe's empires crumble and having glimpsed Soviet Russia's promised land, Nehru felt at last ready to lead India "Toward Freedom."

8

Toward Complete Independence

1927–1928

"THIS CONGRESS declares the goal of the Indian people to be independence with full control over the defence forces of the country, the financial and economic policy and the relations with foreign countries." This was Nehru's first resolution when Congress met in December 1927, adding the "demand" of an early and "complete withdrawal of the alien army of occupation."[1]

Mahatma Gandhi did not interrupt his spinning to attend the Subjects Committee meeting at which that resolution was first announced, but his faithful follower Dr. Rajendra Prasad, destined to be India's first president, called it "silly," fearing it would only turn Congress into "the laughing stock of the world."[2]

"This attitude," Jawahar shot back, only showed how timid and provincial Congress had become. "Swaraj within the British Empire was nothing." He felt it to be "degrading for this country to be within the British Empire. A great country like ours must be a free country." Just back from Moscow, after visiting London, Brussels, and Berlin, he suddenly felt himself dragged back to an earlier century, to a strange planet where time stood still. His frustration and rage returned at the sight of and cautious sounds from so many ancient heads and slow-moving minds, the only one of which he found tolerably intelligent being Annie Besant's. She spoke up in favor of Nehru's resolution, providing he agreed to simplify it to read that the goal of the Indian people should be "complete national independence." This was the goal that Nehru also wanted, so he accepted it, and it passed the Congress majority. He also drafted resolutions boycotting the Simon Commission, supporting his league against imperialism (which emerged from the Brussels Congress), against sending any Indian troops to China, and calling for Hindu–Muslim unity.[3]

The day after all his Congress resolutions had passed in Madras, Jawaharlal presided over a new radical Republican Congress, which held its first and what turned out to be its only session. He thanked the "workers and comrades" who organized this mostly Bengali congress, which its critics dubbed a "Bengali Soviet." "Some people said that now that the National Congress declared its goal to be full national independence, there was no necessity for a Republican Congress," President Nehru noted, insisting, however, that the duty of this organization was to get things done in order to implement the Indian National Congress resolution of independence, to keep from "losing the support of the masses."[4]

"You are going too fast," Gandhi wrote to warn him a few days later. "You should have taken time to think and become acclimatized. Most of the resolutions you framed and got carried could have been delayed. . . . Your plunging into the 'republican army' was a hasty step. But I do not mind these acts of yours so much as I mind your encouraging mischief-makers and hooligans."[5] Having declared his complete independence, however, Jawaharlal was now ready to spell it out to Gandhi as well as to the world. He wrote so angry a response that even the Mahatma was sufficiently provoked to "destroy it."[6] A few days later, however, Jawahar calmed down and wrote to him again, though "loath to inflict another letter on you." He felt "much troubled by your criticism of the Congress resolutions," knowing how "very careful," Gandhi always was with his words. "It amazes me all the more to find you using language which appears to me wholly unjustified. . . . You have referred to discipline and to the Working Committee as the National Cabinet. May I remind you that you are a member of the Working Committee . . . ? You have described the independence resolution as 'hastily conceived and thoughtlessly passed.' . . . But I attach more importance to it than to almost anything else. I have thought over every word you said the other day in Madras on this question and it has merely confirmed me in my opinion."[7]

This was written exactly two decades before Gandhi's assassination. It was the first open rift between Mahatma Gandhi and Jawaharlal Nehru, a gulf that grew wider and deeper, though both of them repeatedly attempted to patch it up publicly in order to hide their basic differences. Jawahar now viewed Gandhi as no less conservative and bourgeois than his father had long appeared to him. To Nehru's radical international outlook Gandhi represented Hindu village conservatism, and Motilal epitomized urban India's Westernized elite. But each of them was rooted in the past, whereas he and his newly found comrades belonged to the future. "It passes my comprehension how a national organization can have as its ideal and goal Dominion Status," the National Congress secretary Jawaharlal added in this declaration of independence to Gandhi. "The very idea suffocates and strangles me. . . . You know how intensely I have admired you and believed in you as a leader who can lead this country to victory and freedom. I have done so in spite of the fact that I hardly agreed with anything that some of your previous publications . . . contained. . . . During the noncooperation period you were supreme. . . . But since you came out of prison something seems to have gone wrong. . . . I have asked you many times what you expected to do in the future and . . . [a]ll you have said has been that within a year or 18 months you expected the *khadi* movement to spread rapidly . . . and then some direct action in the political field might be indulged in—several years and 18 months have passed since then and the miracle has not happened. . . . [B]ut faith in your amazing capacity to bring off the improbable kept us in an expectant mood. . . . [F]aith for an irreligious person like me is a poor reed to rely on. . . . You have stated . . . that India has nothing to learn from the West and that she has reached a pinnacle of wisdom in the past. I entirely disagree . . . I think the western or rather industrial civilization is bound to conquer India . . . you have stated that in your opinion there is no necessary conflict between Capital and Labour. I think that under the capitalist system this conflict is unavoidable." [8]

Gandhi was shocked by this unexpected revelation of Jawahar's true feelings and his clear statement of how basic their differences were about virtually everything. "Though I was beginning to detect some differences of viewpoint between you and me," Gandhi replied, "I had no notion whatsoever of the terrible extent of these differences. Whilst you were heroically suppressing yourself for the sake of the nation and in the belief that by working with and under me in spite of yourself, you would serve the nation and come out scatheless, you were chafing under the burden of this unnatural self-suppression. . . . I see quite clearly that you must carry on open warfare against me and my views. For, if I am wrong I am evidently doing irreparable harm to the country and it is your duty . . . to rise in revolt against me. . . . I can't conceal from you my grief that I should lose a comrade so valiant, so faithful, so able and so honest as you have always been; but in serving a cause, comradeships have got to be sacrificed." Nonetheless, Gandhi closed his letter "With Love" and invited his adopted son to come and stay with him at his ashram so that they might discuss their differences at greater length. He was never one to cut off discussion, seeking to resolve every conflict by patient, persistent argument. [9]

Jawahar was not yet ready, however, to engage in face-to-face "open warfare" with the Mahatma and denied reports of their rift. The press wrote of his referring to Gandhi as "effete and fossilised" in a recent speech in Benares. "What I wished to say was that we must develop the spirit of youth and not become unchanging embodiments of age," Jawahar protested to the editor. "An extraordinary inference has been drawn that I was criticizing Mahatma Gandhi. This is a monstrous notion. . . . He is the supreme example of latter day India, of all that is good in youth—energy and action, courage and daring, perseverance and resolution. . . . We who talk of youth and the call of youth are pigmies before his giant and irrepressible youth." [10]

Nehru found Gandhi's letter "painful reading," denying that he had ever thought of "open warfare" or of unfurling any "banner" against his *bapu*. "Is any assurance from me necessary that nothing . . . can alter or lessen my deep regard and affection for you? . . . No one has moved me and inspired me more than you and I can never forget your exceeding kindness to me. There can be no question of our personal relations suffering. But even in the wider sphere am I not your child in politics, though perhaps a truant and errant child?" [11] The bracing air of London, Brussels, and Moscow was swiftly smothered by India's political heat and humidity. Home less than a month, he felt himself a "pigmy" and "truant child."

Motilal lingered on in London, too sick and tired after his Moscow journey to rush home. He had gone to see many doctors, for his eyes were dimmer, his heart weaker, and his lumbago worse than ever. He wrote to Jawahar of his "approaching end" and also informed him that he would be sailing to Bombay in early February. He arrived on the same ship that brought Sir John Simon and his commission, all of whom were greeted with black flags and shouts of "Go back!" Jinnah had organized a universal boycott in Bombay of Simon's commission, outraged at the arrogance of the pigheaded Tory secretary of state Lord Birkenhead for not having invited a single Indian to join it. Jawaharlal took the train to welcome his father home, and both of them joined the boycott rally in Bombay, west India's last great show of Hindu–Muslim unity, soon to be shattered beyond repair.

One of the resolutions passed at the previous December's Congress called for convening an all-parties conference to try to draft a constitution for India acceptable to the Muslim League, liberals, and communal Hindu parties, as well as the Centrist National Congress. It was to be India's national answer to Birkenhead and Simon. The conference would meet in Delhi, chaired by the Congress president Ansari and attended by Motilal, Jawaharlal, Jinnah, Malaviya, Lala Lajpat Rai, and all the other major leaders of India's political movement, except for Gandhi, who continued quietly to spin khadi at his ashram, as he had little faith in councils or conferences.

The first question tackled by the conference in Delhi was to decide on the basic nature of the constitution desired. Should India be a dominion or a republic? Most of those attending, including Motilal and Jinnah, preferred the former. But Jawaharlal insisted that since Congress had just resolved in favor of complete national independence, any mention of dominion status was unacceptable. It took several days of heated argument before they finally agreed to call what they wanted "the establishment of full responsible government." The following week was consumed in more angry and even more futile debate over Muslim rights and special privileges and the question of just how many seats should be reserved for other minorities, especially Sikhs and Christians, on various elective bodies, provincial as well as central. "After ten days of it the strain was too great for me and I fled to avoid riot and insurrection!" Jawaharlal wrote to Gandhi the day after he fled from Delhi to Allahabad. He still had not visited Gandhi's Sabarmati ashram but hoped to find time to go there soon.[12]

Jinnah played a constructive role at this time, bringing nearly all the factions of India's Muslim minority to agree to a creative scheme he had devised that called for establishing a new province, Sind, then part of Bombay, and upgrading to full provincial status the still centrally administered northwest frontier and Baluchistan. This plan would have ensured the Muslims of at least three new majority provinces while demanding that not less than one-third of all central legislative seats be reserved for separately elected Muslim members. It was a bold plan that, if accepted, might have avoided partition. None of the Hindu communal leaders would listen, however, trusting Jinnah no more than Birkenhead did. "It does not do to take these people too seriously," Birkenhead advised the new viceroy, Lord Irwin (later Lord Halifax), who was ready to encourage Jinnah's brilliant scheme.[13]

Jawaharlal saw no real value in the All-Parties Conference, though he returned to Delhi early in March to announce that he could support a constitution only to establish "a democratic socialist republic in India," whose representatives would be elected by "economic units" rather than on any territorial basis.[14] The Soviet model of central democracy remained his most favored system, with each trade union or peasant collective sending its chosen members to sit on higher collective councils. Jawahar found the All-Parties Conference "a very trying affair," he told his friend Mahmud, "a battle of a few extremists. . . . Jinnah and his group on the one side and the Hindu Mahasabha on the other."[15] His inability to acknowledge Jinnah's legal brilliance and integrity, dismissing him as an "extremist," reflected Jawahar's own impatience with more conservative elders. Moreover, Jinnah's elegant dress and success at the bar resembled Motilal's. Had Nehru been able to appreciate just how much India and Congress would have gained by working closely with Jinnah and his own father at this time to draft a constitution acceptable both to Muslims and moderate Hindus, the tragedy of partition and its dreadful toll might have been averted.

The golden opportunity passed, however, in March 1928. Jinnah left for London and Congress later regrouped its Working Committee to try to decide how to win greater Muslim support. In May, Ansari persuaded Motilal to chair a new commission, known as the Nehru Commission, charged with the impossible task of producing by July 1, 1928, an Indian constitution acceptable to all parties. Ansari and Motilal hoped they could lure Jinnah back and reach an agreement, in order to prove to Simon and his own equally frustrated commission that Indians were wise enough to understand their own needs and could solve their own problems, reconciling communal differences for the nation's greater good. But they labored in vain. Now, Jawaharlal joined his father, inviting others to Anand Bhawan to sound out "new proposals." One of the other young men, who had entered the political arena with charismatic potential, was the thirty-year-old Subhas Chandra Bose of Bengal. The greatest disciple of C. R. Das, Bose was destined to preside over Congress in a decade and, during World War II to lead the Indian National Army, sponsored by Japan, from Malaya to the eastern border of British India. Bose came to the Nehru Commission meeting in Allahabad on July 6, bringing with him Bengal's nationalist Muslim Maulana Abul Kalam Azad. But agreement remained "as far off as ever," Jawaharlal confessed to Gandhi.

He also reported that Kamala's health had shown no improvement over the past few years. She had gone with Indira to Mussoorie's Savoy Hotel for the summer, where they both had been examined by several doctors. Indira seemed almost as sick as her mother, incapable, it appeared, of adding an ounce to her bony body, her eyes deep set behind shadowy rings that belied her youth, mirroring the depression of her dying mother. The girl's lungs, however, were clear, and even though she never fully emerged from the dark shadow of her mother's fatal illness, she did escape her disease. "Papu" (father) as Jawaharlal was called by his "darling Indu," promised to try to come soon to visit her in the hills, "But when I come I wish to find you looking well . . . go into the garden of the hotel every day and play there." He knew that his eleven-year-old daughter spent most of her days at her sick mother's bedside.[16]

Before mid-July Motilal felt that his committee's proceedings had reached their "final stages," and he asked Gandhi to pass judgment on what he had drafted. The Mahatma now urged Motilal to accept "the crown" of Congress president, in view of all he had undertaken and was still doing for Congress and the nation. But Motilal was reluctant, suggesting his son instead, unless Vallabhbhai Patel, who had just led a heroic and successful Satyagraha campaign in Bardoli district of Gujarat, wanted to take command. The "Sardar," as the tough, landowning lawyer Vallabhbhai was popularly called throughout Gujarat, later became India's first deputy prime minister, remaining a loyal lieutenant of Mahatma Gandhi. But Sardar Patel's rural powers were most effective when dealing with peasants or as the tight-lipped boss of a political organization, not as a nation's international spokesman. The elder Nehru, Gandhi insisted, was "a man for honourable compromise," India's greatest need at that time. He also felt, even more strongly, that the younger Nehru was "not ready" to put on a "crown of thorns."

Subhas Bose, who chaired Bengal's provincial Congress, pressed Motilal more vigorously than Gandhi did, arguing that "the situation in the country today is such . . . that we can think of nobody else who can rise to the occasion." None of the alternative names would be acceptable to Bengal's Congress, which was to host the session. "If for any reason you decline the Presidentship," Subhas wrote, "the effect would be so disastrous in this province that it will seriously affect the success of the Congress session." That session was to convene in Calcutta during Christmas week of 1928. Subhas respected Motilal almost as much as he had his Bengali guru Das, but he viewed Vallabhbhai as little more than a country bumpkin. He thought of Jawahar much the way he did of himself—though he was eight years younger—as still belonging to the second echelon of Congress leadership.[17]

Motilal alone knew how weak he felt, hence he tried his best to elevate his son to the Congress presidency before his fast-deteriorating health gave out. To Gandhi he conceded that Jawahar "has no doubt frightened many of our goody goodies by his plain talk. But the time has come when the more energetic and determined workers should have their own way of guiding the political activities of the country. . . . Our race is fast dying out and the struggle will sooner or later have to be continued by men of Jawahar's type."[18]

To Annie Besant, who also feared that Jawahar's "complete independence" rhetoric was rousing a ragtag army of youthful revolutionaries who would be impossible to control, Motilal wrote reassuringly, "I have no fear from this group which have at their head an earnest patriot [Jawahar] always willing to look at the other side of the shield as is evidenced from the fact that in spite of his raging tearing propaganda in favour of complete independence Jawahar is sparing no pains to make the All Parties decisions a complete success."[19] Just as he had not cut off his funds no matter how hedonistic Master Joe became in Cambridge and London, Motilal did not now give up on his son, despite Jawahar's romance with Marxist–Leninism and his demand for full freedom, though he could not support himself or his wife and daughter. Wise old man that he was, Motilal knew that his generation was dying and that he himself was "more or less a spent force."[20] Because his son had refused to inherit his lucrative legal practice, he had instead taken up his son's profession of politics.

"As for my presidentship of the Congress," Jawahar wrote his comrade Mahmud in midyear, "don't worry. I had rather not preside. The real objection to me is not youth or jealousy but fear of my radical ideas. I do not propose to tone down my ideas for the presidentship." Motilal accepted it, of course, knowing that unless he agreed, Gandhi would have to allow the crown to be placed again by acclamation on his own head, or Calcutta would withdraw its invitation to host the Congress. And if Mahatmaji accepted it this year, then his most trusted Gujerati lieutenant, "Sardarji" Vallabhbhai probably would take up the burden in the next session. But once Motilal agreed, he could, in India's best time-honored dynastic tradition, hand his political regalia to his son, who, despite his denial, coveted the prospect of becoming the president of his nation's premier party before he turned forty.[21]

The All-Parties Conference reconvened that August in Lucknow in the palatial garden of the raja of Mahmudabad. Jawaharlal and Subhas Bose founded their own Independence for India League, which drew support from the many youth and workers and peasants organizations that had sprung up in several provinces during the last few years, inspired by the Soviet success story and Marxist–Leninist ideas. At Lucknow Jawaharlal served as spokesman for this radical youthful group of dissidents, who opposed the Nehru report, which he had drafted at his father's instruction. "What is the meaning of the resolution that we are considering?" he asked the gathering. "The preamble tells us that it is open to us to carry on activity and propaganda for independence. But this is a mere flourish. The speeches in support of the motion . . . embody an ideology of a past age utterly out of touch with facts and realities of today. We were told of the injustice in not having Indian governors, Indians in the Service and in the Railway Board. . . . Is this our idea of freedom? It seems to me that we are drifting back from the 20th century to the ways . . . of the 19th. . . . They are dead as Queen Anne, as Charles I, as Louis XVI of France and as the last Tsar of Russia. . . . What does the British Commonwealth of Nations . . . stand for? It stands for one part domineering over and exploiting the other. . . . Are we prepared to be tied to the chariot-wheels of England . . . ? The only practical goal is . . . independence. . . . I do submit that it would be a wrong thing and a fatal thing for India to make Dominion Status as our objective. Those of us who think with me have carefully considered this resolution and we have definitely come to the conclusion that we cannot support it."[22]

The report passed, despite Jawaharlal's attacks, yet its fatal weakness remained its elimination of separate electorate representation and Muslim minority weight beyond the actual proportion of the Muslim community in every province where Muslims remained a minority. Such communal minority "safeguards" had been enshrined in the last Government of India Act, which adopted them from the Indian Councils Act of 1909. The Ali brothers had by now totally lost faith in Congress, owing to the demise of the Khilafat movement and the subsequent Hindu–Muslim riots. Jinnah's Muslim League also refused to support the Nehru report, which they viewed more as a victory for Hindu communal leaders than for "reactionary" British imperialists. Motilal tried his best to win over Jinnah, but Jinnah was not his son, and Jinnah would never trust Jawaharlal or be as accommodating to his rapidly shifting moods and mind as his father was.

"Do you want or do you not want the Muslim India to go along with you?" Jinnah soon asked Motilal and his Hindu Mahasbaha supporters, who rejected Jinnah's demand that no less than one-third of all Central Assembly seats be reserved for Muslims. "Minorities cannot give anything to the majority. It is, therefore, no use asking me not to press for what you call 'these small points'. . . . If they are small points, why not concede? It is up to the majority, and majority alone can give. I am asking you for this adjustment because I think it is best and fair to the Musalmans. . . . We are all sons of this land. We have to live together. . . . Believe me there is no progress of India until the Musalmans and Hindus are united . . . and nothing will make me more happy than to see a Hindu–Muslim union." But that was not to be.[23]

Jawahar believed that Hindu–Muslim conflict was caused primarily by economic disparities or clashing "class interests" rather than dogmatic religious beliefs and incompatible customs. He also viewed most of the continuing conflict as the product of British initiative in keeping with its imperialist "divide and rule" policy, which dictated such tactics. Annie Besant now felt that "Congress is becoming an intolerable tyranny by denial of free speech to the minority."[24] As Congress secretary, Jawahar was offended by her remark, countering: "If Dr. Besant wants the majority of the Congress to bow down to the minority or to an individual, that surely would be some kind of tyranny and the tyranny of the minority or of an individual is worse than any other tyranny."[25] He had convened a meeting of his new Independence for India League in Delhi and now announced that it "represents the left wing element in the country and the Congress." Many such leagues sprang up in major cities of most provinces, keeping Jawahar busier than he had been in a long time, for as secretary to both this new league and the old Congress he was the key organizer of every meeting and had to travel ceaselessly, keeping in close touch with his comrades wherever he went.

On October 30, 1928, a large protest parade in Lahore against the Simon Commission's arrival in that capital of Punjab was attacked by the police, who wielded long metal-tipped sticks, called *lathis,* against the peaceful protesters, led by Lala Lajpat Rai. The sixty-four-year-old *lala* (teacher) was struck hard twice, on his chest and shoulder and had to be taken to hospital, where he died a few weeks later of a heart attack. "Lahore brought matters to a head and suddenly sent a thrill of indignation throughout the country," Jawaharlal recalled. "How helpless we were, how despicable when we could not even protect the honour of our chosen leaders!" The Simon Commission was expected later that month in Lucknow, and as soon as Jawaharlal learned that they had been invited to attend the university convocation, he issued his battle cry to all Lucknow students: "Every school boy knows that India has resented the . . . coming of the Simon Commission and has boycotted it. Has no whisper of this reached the academic ears of the Lucknow University authorities? Have they not heard or felt in their *boudoirs* the deep rumble of a nation in anger? Do they not know how one of the greatest of India's sons was treated in Lahore city less than a month ago . . . ? And yet the authorities of Lucknow have invited this Commission to their convocation? And yet, they have dared![26]

"Young men and women of Lucknow, what say you to this infamous and insolent challenge? Is the honour of your country nought that it can be kicked in the dust by a policeman clad in the livery of our alien rulers? But the poor policeman was but an instrument in other hands. What of your University authorities who of their own free will dare to honour those whom the nation has repudiated and boycotted? . . . This is the measure of their nationalism, of their love of India and her honour. This is the tribute they pay to the memory of Lajpat Rai. Will you too forget soon this fragrant memory and this shining example? They have dared! What dare you? . . . Will you not take full part in the boycott demonstrations in Lucknow, and, above all, boycott the convocation to which your University authorities have invited the Simon Commission? They have dared! Do you also dare?"[27]

Jawaharlal had now reached his revolutionary stride in rhetoric redolent of both the French and Russian Revolutions, leading young demonstrators through the streets of Lucknow, shouting, "Simon, go back!" "It was in this connection that . . . my body felt the baton and *lathi* blows of the police," Jawaharlal later wrote. "I saw the horses charging down upon us; it was a discouraging sight. But then, I suppose, some other instinct held me to my place and I survived the first charge, which had been checked by the volunteers behind me. Suddenly I found myself alone in the middle of the road; a few yards away from me . . . were the policemen beating down our volunteers. Automatically, I began moving slowly to the side . . . but again I stopped and had a little argument with myself

. . . prompted by my pride, I . . . could not tolerate the idea of my behaving like a coward . . . a mounted policeman was trotting up to me, brandishing his long new baton. I told him to go ahead, and turned my head away. . . . He gave me two resounding blows on the back. I felt stunned, and my body quivered all over but, to my surprise and satisfaction, I found that I was still standing. . . . Our volunteers gathered together again, many of them bleeding and with split skulls . . . and all of us sat down facing the police . . . and it became dark." Lucknow officials had by then arrived at the scene and, seeing who was there, allowed the demonstrators to go their peaceful way without further obstacle.[28]

"The bodily pain I felt was quite forgotten in a feeling of exhilaration," Jawahar then recalled. Motilal was back home in Allahabad, but although Jawahar felt no pain he decided to call his father, since as he put it, "I was afraid that the news of the assault on me, when he read about it in the next morning's papers, would upset him. . . . So I telephoned to him late in the evening to assure him that all was well, and that he should not worry. But he did worry and, finding it difficult to sleep over it, he decided at about midnight to come over to Lucknow." The last train had gone by then, however, so weary, worried old Motilal started off to cover those 150 dismal miles by motor car in the dead of India's night. No motor journey anywhere in India was either common or easy at the time, for there were no petrol stations, no services, no road lights, and few roads without frequent and dangerous pot-holes. Tired out and exhausted, Motilal drove into Lucknow at 5 A.M. By now Jawaharlal had thousands at his back, marching four abreast and arms firmly locked, shouting, "Simon, go back!"[29]

"We were stopped by the police as we approached the station. . . . The place was full of foot and mounted police, as well as the military. The crowd of sympathetic onlookers swelled up. . . . Suddenly we saw in the far distance a moving mass . . . galloping down towards us, and striking and riding down the numerous stragglers. . . . That charge of galloping horsemen was a fine sight, but for the tragedies that were being enacted on the way . . . the horsemen were soon upon us. . . . We held our ground, and, as we appeared to be unyielding, the horses had to pull up at the last moment . . . their front hoofs quivering in the air over our heads. And then began a beating of us, and battering with *lathis*. . . . It was a tremendous hammering, and . . . [a]ll I knew was that I had to stay where I was, and must not yield or go back. I felt half blinded with the blows, and sometimes a dull anger seized me and . . . I thought how easy it would be to pull down the police officer in front of me from his horse and to mount up myself, but . . . discipline held and I did not raise a hand, except to protect my face from a blow. . . . I was suddenly lifted off my feet from behind and carried off, to my great annoyance. Some of my young colleagues, thinking that a dead-set was being made at me, had decided to protect me."[30]

After another hour or so, Jawahar made his way back to the Congress office, where his father was anxiously waiting. "I was covered with contused wounds and marks of blows. But fortunately I was not injured in any vital spit," Jawaharlal recalled. "I emerged with a somewhat greater conceit of my physical condition and powers of endurance. But the memory that endures with me, far more than that of the beating itself, is that of many of the faces of those policemen. . . . Most of the real beating and battering was done by European sergeants. . . . And those faces, full of hate and blood-lust, almost mad, with no trace of sympathy or touch of humanity! Probably the faces on our side just then were equally hateful to look at. . . . And yet, we had no grievance against each other; no quarrel that was personal, no ill-will. . . . Blindly we struggled, not knowing what we struggled for and whither we went. The excitement of action held us; but as it passed, immediately the question arose: To what end was all this? To what end?" No doubt this was a question Motilal asked as well, although for him the end was much nearer now.[31]

After his Lucknow beating, Jawaharlal emerged more heroic than he had ever appeared before and more popular among the younger members of Congress as well as his own Independence League, who viewed him with awe and admiration. Simple prison, after all, was one thing, but surviving a *lathi* charge was quite another level of bravery, self-sacrifice, and national honor. He had now served time and sustained severe blows yet seemed remarkably cheerful, for nothing excited him as much as action, physical danger, the quivering of his body as it endured shattering blows or any sort of thrashing.

"So far as I am concerned, I have not the slightest grievance against the government or the officials in Lucknow," Nehru informed the press a week later. "The Boycott Committee ought to give them honourable badges for the tremendous help that it received owing to their stupidity in making the boycott demonstrations a magnificent success. . . . It was painful of course . . . to see the extraordinarily callous and brutal behaviour of the police, but I suppose that is the price which we shall often have to pay . . . but in their stupidity . . . they had completely played into the hands of the boycotters. They have brought the real issue before the people of the country bereft of all sophistries and legal quibbles. That issue is that British rule in India means the policeman's baton and the bayonet and the real problem is how to overcome them. Logic and reason is unhappily lost on the baton and the bayonet. They will be overcome by the strength and the sanctions that the country develops. In this process . . . we can have no greater allies than British officials of . . . Lucknow."[32]

Revitalized by Lucknow, Jawahar journeyed in mid-December to Poona to deliver the presidential address to Bombay's Youth Conference. "My attraction to a conference of youth remains, for it is so unlike the gatherings of older folk," he told the audience. "Many of you also, it may be, when you grow older . . . fall into the ancient ruts and forget the spirit of adventure and dare-devilry which was yours. . . . But today you are young . . . and I, with the years creeping on me, have come to you to be a sharer in your abounding hope and courage and to take back with me . . . some measure of your faith and enthusiasm. . . . Why have you met here today? . . . Not simply to play a prominent part in the political or social arena, to become a celebrity and be intoxicated by the applause of the multitude." Was he asking that of the audience or himself?[33]

"You have met here, I take it, because you are not content with things as they are and seek to change them. . . . If this is the urge that has brought you here, then you have met well. . . . But if you are not dissatisfied with existing conditions, if you have not felt this urge which makes you restless and drives and lashes you to action, then wherein do you differ from . . . older people who talk and debate and argue much and act little? . . . It is not the sleek and shiny people having more than their share of this world's goods who are the apostles of change. The world changes and progresses because of those who are disaffected and who are not prepared to tolerate the evils and injustice of things as they are. . . . What do we find in this world of ours today? Utter misery is the lot of vast numbers of people. . . . Wars and conflicts ravage the world and . . . what of our own unhappy country? Foreign rule has reduced her to utmost poverty and misery and . . . has sapped the life out of her. . . . And this . . . imperialism is the direct outcome of a system of society which . . . is called capitalism. . . . We must aim, therefore, at the destruction of all imperialism and the reconstruction of society on another basis . . . one of cooperation, and that is another name for socialism. Our national ideal must, therefore, be the establishment of a cooperative socialist commonwealth and our international ideal, a world federation of socialist states."[34]

This remained Nehru's lifelong ideal and his oft-repeated goal. He had just turned thirty-nine and finally felt he knew what he truly wanted, not only for himself, but also for India and, beyond that, for the world. "Religion has in the past often been used as an opiate to dull men's desire for freedom. Kings and emperors have exploited it for their own benefit and led people to believe in their divine right to rule. . . . And with the aid of religion the masses have been told that their miseries are due to *kismet* or the sins of a former age. Women have been and are still kept down and in the name of religion in many places are made to submit to that barbarous relic of an earlier age—the *purdah* system. The depressed or the suppressed classes cry out to the world how infamously religion has been exploited to keep them down. . . . The problem before us is the problem of the conquest of power. . . . Every Indian knows the crimes that have been and are being committed in the name of law and order. . . . Law and order are the last refuge of the reactionary, of the tyrant and of him who has power and refuses to part with it."[35]

Motilal told his son, as Gandhi well knew, that unless his report was accepted by Congress he would not preside over that body for a single day, not to speak of an entire year. But Jawaharlal remained so adamantly opposed to dominion status, by any definition, as the goal of India's national movement that it seemed possible for him to scuttle the entire Nehru report if he were to oppose it openly. Gandhi, as usual, devised what he considered a suitable mediating formula, suggesting that if Jawaharlal and his cohorts agreed to support the Nehru report for one year, then he would join them after December 31, 1929, in nonviolent noncooperation if Great Britain's Parliament had not by then enacted a constitution for India based on the Nehru report. Jawahar and Bose knew, of course, that Britain's Parliament rarely enacted anything within a year, much less major legislation for India. Speaking of the stated goal of dominion status, even for a single year, Jawaharlal insisted that all his "energy oozes out of me at the very thought of Dominion Status." The real conflict, he argued, was between two ideals, with the real question being which one to raise before the country. "This is a conflict between imperialism and all that is not imperialism," young Nehru insisted. "You cannot for one moment think of Dominion Status so long as Great Britain has the empire around her." Gandhi made further concessions, and Motilal wondered whether the price he had been ready to pay for his son's cooperation might not be too high. Still, he said nothing, and the resolution to present the Nehru report to the entire Congress passed through the Subjects Committee by a vote of 118 to 45. Then on the floor in open session next day, Subhas Bose offered an amendment to compromise all he had agreed on, and Jawaharlal strongly supported him.[36]

"You may take the name of independence on your lips," Gandhi told them, now feeling betrayed, "but all that muttering will be an empty formula if there is no honour behind it. If you are not prepared to stand by your words, where will independence be? Independence is a thing made of sterner stuff. It is not made by the juggling of words." Gandhi's argument and resolution carried the Congress. Finally Motilal spoke. "Both Subhas and Jawahar have told you . . . that, in their opinion, we old-age men are no good, are not strong enough and are much behind the times. . . . It is common in this world that the young always regard aged men as behind times. I would give you one word of advice. Erase from your mind from today those two terms borrowed from foreign language, namely, independence and dominion status, and take the two words swaraj and *azadi* [freedom]. Let us work for swaraj by whatever name we might call it. . . . One year is nothing in the history of the nation. I have not the least doubt that the next Congress will see us united and taking another forward step." His good sense and optimism were roundly applauded, and his prediction proved correct, though the Congress's union would no longer suffice to speak for one Indian nation. More and more of India's Muslim quarter came to believe that *their* interests and *their* concerns were being ignored by the Nehru report. And even though the future might belong to Jawaharlal, many Hindus of the past also lost faith in his "muttering" an "empty formula," fearing, indeed, that there was "no honour behind it." [37]

President Motilal Nehru told his Calcutta Congress on December 29, 1928: "I am for complete independence—as complete as it can be—but I am not against full dominion status . . . provided I get it before it loses all attraction. . . . My views may not be acceptable to . . . the younger men. I quite appreciate their impatience." Forty-three white horses had pulled his ornate presidential carriage from Howrah station to the beautiful giant tent of bright lights and Congress flags erected on the lush green of Calcutta's most spacious public park, the site of this grand forty-third annual session of Congress. Subhas Bose, dressed in the military uniform he had designed and in which his bespectacled moon-faced image would soon be framed in every Bengali home, pranced about on his white stallion, ordering an army of his Bengali volunteers to keep the crowd of thousands eager to attend this Congress from crushing the delegates. The entire Nehru family—India's First Family not simply for this session of Congress but from this date forward—were there, royally received, cushioned, and pillowed in the front row. And with them was Mahatma Gandhi, looking more than ever like Mickey Mouse, as India's national poet, Sarojini Naidu, called him, silently spinning his plans for the nationwide Satyagraha he would launch little more than a year later. [38]

"We need both patience and impatience," President Motilal continued. "Patience with those who differ from us, impatience with ourselves. I have no quarrel with the ideals of the younger men. . . . [A]ll exploitations must cease and all imperialism must go. But the way to do it is a long and dreary one." His voice all but failed him now. The cough that had begun to plague him day and night refused to leave even for this moment of political pomp and glory. Little more than two years remained before the advanced fibrosis that soon filled his right lung would silence Motilal's "voice of freedom." [39]

"Let the younger men by all means preserve their own mentality, but let them not, for the sake of the very motherland they seek to serve, divide the country into more factions and parties than there are already," Motilal cautioned his son as well as Subhas Bose and the militant band that cheered their every cry for "Complete independence! Not tomorrow, but now!" Few of the younger men listened to Motilal's sober speech, however, or feared his warnings, for as Jawaharlal put it in attacking the very Nehru report that Congress had just resolved to support, "If you are prepared to pull down the flag of independence then. . . . [w]ords do not carry us very far. . . . By accepting Dominion Status you show to the world that you are prepared to accept the psychology of imperialism."[40]

"Now that I have shaken the dust of Calcutta I wish to apologize for all the trouble I gave you," Motilal wrote to Gandhi soon after he returned home early in January 1929. For it was in response to his personal plea that the Mahatma had journeyed so far from his ashram retreat. "You have saved a complete fiasco."[41]

"No apology whatsoever is necessary," Gandhi replied, confessing that attending the Calcutta session had given him "an insight into the present working of the Congress," wisely concluding, "And after all, we have to battle both within and without."[42]

9

Return to Revolution

1929–1930

"THE WORLD IS moved by ideas," Jawaharlal told a cheering young audience in Delhi early in 1929. "The British Government is afraid of Soviet principles and is trying its best to uproot socialist ideas and not to allow any communist to preach his gospel. . . . But the government should know that ideas can never be checked by cruel laws. . . . Youth should not be slaves of tradition."[1]

During 1929 Jawaharlal was catapulted to the pinnacle of Congress power, snatching its crown from several older rivals thanks to Motilal's intervention with Gandhi. Ideologically he remained the darling of India's youth, and because of his earlier close encounters with comrades of the Second and Third Internationales, he became one of the leaders of the slowly emerging Indian labor movement that had started to strike in several large cities and mill towns after World War I. Nehru's patrician heritage and aristocratic English education thus blended with the Marxist–Leninist ideas and new formulas for revolution he had learned in Ireland and Moscow, lifting him above all his political contemporaries, at least in argument and public action. He was watched, of course, by many British officials, some of whom were positive that he was nothing more nor less than a communist, with others finding him more dangerously elusive, the potential national leader of a mass Indian revolt, who had none of the Mahatma's scruples about violence, indeed, few, if any, of Gandhi's religious principles.

Gandhi had long planned another trip abroad. He had been invited to the United States as well as most of the capitals of Western Europe by many friends and admirers. Motilal pleaded with him, however, not to go overseas at this critical juncture, feeling himself too weak to keep Congress nonviolent in the face of anticipated British provocation. He specially feared Jawaharlal's arrest, knowing too well what he said morning, noon, and night at home as well as in public meetings. Ironically, Gandhi himself was arrested that March in Calcutta for setting fire to a pyre of foreign cloth, but his "one rupee fine" was paid by an anonymous benefactor, so the Mahatma eluded imprisonment for one more year. Both Nehrus also remained free all year, and Motilal devoted some of his remaining time to designing and supervising the building of a new house for himself and his family, a more modern, equally grand palatial mansion on his vast estate in Allahabad, which he initially called "The Cottage" and later renamed Anand Bhawan. Between the ghost and the dead cobra that cursed older residence (donated to Congress) had lost its appeal. He hoped, of course, for happier days in his new home but an astrologer predicted that its owner would not live much longer to enjoy it. Jawahar, Kamala, and Indira were again to occupy the second-floor apartment, though Jawaharlal roundly berated his father for spending too much on his house, making it too fancy, with too many marble floors and too much elegant furniture. But their new home remained the Nehru family's residence in Allahabad until Indira donated it in 1970 to the nation as a Nehru museum.[2]

Motilal continued to lead his party in the Central Legislative Assembly and spoke vigorously against the government's annual budgetary request for funds, arguing in favor of the most recently passed Congress resolution, in support of his committee's constitution. "It has been described as an ultimatum," Motilal informed Viceroy Lord Irwin and his council of mostly bored British and anxious Indian advisers. "Well, in one sense, you may call it an ultimatum, but it is really an invitation to you to make up your minds within the time given in the Resolution . . . to fulfil your oft-repeated promise to put India on her own legs. . . . [W]e shall accept full responsible government within that time. But if you show no inclination . . . at the end of the year . . . we shall try to follow our own programme. We are willing to keep up the British connection only on honourable terms and those honourable terms are stated in the Resolution as being the same as dominion status. . . . You may continue your present system . . . but you cannot continue it for ever. Nemesis may be slow, but it is always sure."[3]

Jinnah, however, was no longer willing to vote as Motilal advised. "I know, the Nehru Report is my Honourable friend's pet child, but I am speaking dispassionately and I want him to realise, and the sooner he realises it the better—that it is not acceptable to the Muslims." Soon after that, a bill of the finance member, Sir George Shuster, passed, with Jinnah and his Muslim bloc voting with the solid British official majority in favor of it. A month earlier Jinnah had buried his estranged young wife; now he was burying his own faith in the Hindu–Muslim unity, of which he had once been "India's best ambassador" a title that Sarojini Naidu had bestowed on him. The Lucknow Pact of 1916 had been Jinnah's "pet child," but that too was part of India's prewar past, dead and buried.[4]

Throughout 1929 Jawaharlal listened to the drum of revolution as the only salvation for himself and India. But he knew that Great Britain could never succumb to any Indian Congress demand for even so paltry a thing as dominion status in one year. He knew that British rulers—whether Tory, Liberal, or Labour by party—never surrendered one inch of imperial soil until they were forced out of it. Why should India expect better treatment? "In India, millions of people are underfed and underclothed and the mere discussion or talk of freedom cannot bring them freedom," Jawaharlal told his cheering audience in Punjab's capital of Lahore that February. "I disapprove of the spirit of fatalism. . . . [W]henever we attempt to create a new order of society, the government immediately gets nervous and sees 'red.' . . . The whole aim of the government is to shut out all new ideas . . . particularly Russian literature, . . . but if nationalism is combined with socialism the day of freedom would be hastened."[5]

One of Jawaharlal's Congress comrades at this time was Acharya Narendra Deva, who later presided over the Socialist Party of India, which he started with Nehru's support in 1934. Deva now wrote him about their floundering Independence for India League: "We lack in our midst a body of earnest men of deep convictions who have a living faith in some economic programme. . . . The ideas of most of us are vague and indefinite and most of us do not know how to proceed about the business. . . . I think the apathy that we see around us is, more or less, due to want of any intellectual convictions. . . . [T]he League should run a weekly paper, if necessary funds are forthcoming and have a bookshop of its own . . . open study circles and produce cheap literature in Indian languages. . . . You know, when I accepted the office of the Secretary, I made it quite clear that my present duties . . . will not leave me time to go about the country. I can only correspond from here." It was to become a familiar complaint, one that Nehru himself often used, explaining to Gandhi why he spent so little time among the peasants.[6]

"On my return from Europe [in December 1927] I had the fixed intention of spending a few months at least in some village areas," Nehru recalled. "I wanted to try to organise them according to my own ideas but even more so I wanted to educate myself and try to get at the back of the mind of the villager. The Congress secretary duties and repeated and continuous calls for other work prevented me. . . . Somehow or other I cannot extricate myself from the tangle of activities I have got into." He was growing restless again, but party obligations kept him tied down.[7]

He had begun to smoke too much, more than a pack of cigarettes daily, making his head ache, so he decided to stop. One Nehru coughing the way Motilal did nowadays was enough. Kamala also got worse, suffering suddenly from "appendicitis." She was rushed to Calcutta to have her "appendix" removed. A third pregnancy at this time had resulted in "a miscarriage in the third month"[8] according to a "Note on Mrs. Kamala Nehru's Case," written later "by the patient's husband." While Kamala was recovering from her emergency surgery in Calcutta, she decided secretly to join the Ramakrishna Mission there, taking vows of celibacy as a disciple of Swami Sivananda, to whom she turned for solace as she prayed night and day for her final "release" *(moksha)* that would take almost seven more years of pain and anguish. Indira now attended the Convent School in Allahabad, where she felt very "unhappy, and I didn't learn anything because most of the girls there were Anglo-Indians and very pro-British. We were always being scolded."[9] Jawahar's mother, badly bitten by mosquitoes, came down with dengue fever.

Labor unrest intensified, from a jute-mill strike near Calcutta that had begun in late 1928 and grown more vigorous for much of 1929, to a general strike called in Bombay. The British hoped to discourage labor discontent by arresting thirty-two leaders of several unions of Bengal and Punjab as well as Bombay, charging all of them as communists with sedition and bringing them to Meerut for a trial that lasted more than five years. "A defence committee was formed for the Meerut accused, of which my father was chairman," Jawahar recalled. "We had a difficult task. Money was not easy to collect. . . . [M]oneyed people had no great sympathy for communists and socialists and labour agitators. And lawyers would only sell their services for a full pound of somebody's flesh. . . . We had to collect money, often in coppers from the poorest workers, and pay out fat cheques to lawyers."[10] Jawahar himself gave little of his time to the Meerut Trial, never developing any real interest in or talent for the law. He was, however, elected to preside over the Trade Union Congress that year, though he assured one "Comrade Kulkarni," who ran against him for that job, that "I accepted the presidentship of the T.U.C. much against my will . . . I feel that the presidentship should go to an active unionist and worker like you and not to a person like me who has not been connected with trade unionism before. But owing to my absence I was elected. . . . My only difficulty is want of time . . . I am exceedingly busy and I shall be touring about . . . for many days."[11]

Britain's Labour Party won a narrow victory that May, returning Ramsay MacDonald to power over a Liberal–Labour coalition cabinet in Westminster and appointing William Wedgwood Benn (later Lord Stansgate) as his secretary of state for India. "With the Conservatives in power the issue as regards India is usually a fairly straight one, but with Labour there is so much empty and pious talk that some minds are apt to be confused,"[12] Jawaharlal told Allahabad's press, feeling no faith in any British party at this time. "India's prospect depends not on any government in power in England but only on the organised strength of the Indian people," he added a few days later in Calcutta. "The Labour Party dare not conciliate various groups in India. It is quite possible they may adopt an aggressive anti-Indian attitude."[13] He sounded as if he were debating the question before the Cambridge Union. He was anxious to be back in London "for a month or two," as he confessed to Fenner Brockway, "but I can expect no such luck."[14]

Each day Nehru became more and more entangled, by both the National Congress and his Trade Union Congress. Each of British India's eighteen provincial Congress committees voted by midsummer for its most favored candidate to preside over the annual session of Congress to be held that December in Lahore. Ten of the provincial committees proposed Gandhi; five voted for Vallabhbhai Patel; and only three favored Jawaharlal Nehru. For Motilal, however, the choice was a foregone conclusion: "I have been thinking hard on the matter," Motilal wrote Gandhi that mid-July. "It appears to me that, leaving one awkward element in the case, all reasons point to your accepting. That element was present in my case. It consists in our apparent stinginess in parting with power and keeping the younger set out of it. . . . The revolt of youth has become an accomplished fact. . . . It would be sheer flattery to say that you have today the same influence as you had on the youth of the country some years ago. . . . All this would indicate that the need of the hour is the head of Gandhi and the voice of Jawahar. . . . There are strong reasons for either you or Jawahar to wear the 'crown,' and if you and Jawahar stand together, as to which there is no doubt in my mind, it does not really matter who it is that stands in front and who behind."[15]

Gandhi, of course, bowed to Motilal's pressure, having no craving or need for crowns, and wrote immediately to Jawaharlal to offer him the presidency. "About the presidentship," Jawahar replied, "My own personal inclination is not to be shackled down to any office. I prefer to be free and to have time to act according to my own inclinations." That was true enough. "So far as I am concerned the presidentship will . . . be a burden to me . . . and a nuisance. You will not of course be deceived by various odd persons . . . expressing their confidence in me. . . . I have won the goodwill of a number of people but . . . I represent nobody but myself. I have not the politician's flair for forming groups and parties." Here again was the mock humility Nehru had learned at Cambridge, in which neither he nor Gandhi believed. Such humility, however, was generally viewed by gentlemen as well as mahatmas as a virtue, especially among politicians.[16]

"If there is going to be effective action next year," Jawahar continued in this masterfully "nonpolitical" letter, "it will make all the difference in the world whether you lead it or not. Of course you can lead, as you have done in the past, without being Congress President. But it would help matters certainly if you are also the official head of the organisation. I feel that it would be a great gain if you would preside. That would strike the imagination of the country and other countries. If I have the misfortune to be President you will see that the very people who put me there, or many of them, will be prepared to cast me to the wolves."[17] That August Jawaharlal visited Gandhi in his ashram and reported next day to his father that Gandhi felt that "few people are prepared to follow him although they demand loudly for his presidentship. If he was convinced that he was seriously wanted to lead and not merely to be exploited, he would, I think, agree to preside."[18] Neither Nehru now pressed the reluctant Mahatma, however, to wear the Congress crown once more.

On September 28, 1929, the All-India Congress Committee met in Lucknow to choose its next president. Most of the delegates had traveled several nights and many days to cast their votes for Mahatma Gandhi, although many from Gujarat Province were even more vigorously supportive of Sardar Vallabhbhai Patel, still their "hero of Bardoli." Yet out of the blue, it seemed, the name of Jawaharlal Nehru emerged as the newly anointed one, the young prince whose head was now to be measured for a proper crown of khadi. "So Gandhiji was recommended for the presidentship by the Provincial Committees," Nehru later reflected over that remarkable event. "But he would have none of it. His refusal, though emphatic, seemed to leave some room for argument, and it was hoped that he would reconsider it . . . and almost to the last hour all of us thought that he would agree. But he would not do so, and at the last moment he pressed my name forward. The A.I.C.C. [All-India Congress Committee] was somewhat taken aback by his refusal, and a little irritated. . . . For want of any other person, and in a spirit of resignation, they finally elected me. I have seldom felt quite so annoyed and humiliated. . . . It was not that I was not sensible of the honour, for it was a great honour. . . . But I did not come to it by the main entrance or even a side entrance; I appeared suddenly by a trap-door and bewildered the audience into acceptance. They put a brave face on it, and, like a necessary pill, swallowed me. My pride was hurt, and almost I felt like handing back the honour. Fortunately I restrained myself . . . and stole away with a heavy heart."[19]

What a bitter "pill" for him to swallow, finding himself pushed into the presidency not by the masses or the workers of whom he spoke, for there were no peasants in that politically smoke-filled caucus, nor was it even the provincial Congress's legal and landed elite who chose him. It was, of course, as he knew too well, his own dear father.

"My beloved Jawahar," Sarojini Naidu wrote him the next day, "I wonder if in the whole of India there was yesterday a prouder heart than your father's or a heavier heart than yours. Mine was the peculiar position of sharing in almost equal measure both his pride and your pain. I lay awake until late into the night thinking of the significance of the words I had used so often in reference to you, that you were predestined to a splendid martyrdom. As I watched your face while you were being given the rousing ovation on your election, I felt I was envisaging both the Coronation and the Crucifixion—indeed the two are inseparable and almost synonymous in some circumstances and some situations: they are synonyms today especially for you, because you are so sensitive and so fastidious in your spiritual response and reaction and you will suffer a hundred fold more poignantly than men and women of less fine fibre and less vivid perception and apprehension, in dealing with the ugliness of weakness, falsehood, backsliding, betrayal . . . all the inevitable attributes of weakness that seeks to hide its poverty by aggressive and bombastic sound. . . . I have an abiding faith in your incorruptible sincerity and passion for liberty and though you said to me that you felt you had neither the personal strength nor a sufficient backing to put your own ideas and ideals into effect under the turmoils of so burdensome an office, I feel that you have been given a challenge as well as offered a tribute: and it is the challenge that will transmute and transfigure all your noblest qualities into dynamic force, courage and vision and wisdom. I have no fear in my faith . . . I believe that the invincible faith of one's spirit kindles the flame of another in radiance that illumines the world." Sarojini signed it—and always would remain—his "loving friend and sister," just as her exiled brother, "Chatto" Chattopadhyaya, remained one of Nehru's closest European communist comrades, but it was her adoring daughter Padmaja, not Sarojini, whose love he later accepted and reciprocated.[20]

Jinnah viewed Labour's return to power at Westminster much more positively than Nehru did. He went to Simla when the news reached India, for "a long personal talk" with Viceroy Irwin. Soon afterward the viceroy sailed home to confer with his new Liberal secretary of state and with Prime Minister Ramsay MacDonald, urging both of them that the time was ripe to reaffirm Britain's faith in India, thus giving heart to moderates like Jinnah and Motilal, who were otherwise fast losing hope as well as popular support. Lord Irwin by now deeply regretted having bowed to the previous pigheaded Tory secretary's insistence on keeping Indians out of his Simon Commission. Birkenhead's assumption that all of India's "Muslims would rally" to welcome Simon and testify "wholeheartedly" before his commission had proved as inaccurate as his own knowledge of things Indian was totally outdated and racially misguided. Irwin was wiser now and listened with considerable sympathy to everything Jinnah told him and to most of the harsher words of criticism he so often heard from Motilal.[21]

Jinnah also knew MacDonald personally, having served with him previously on a royal commission, and wrote to urge the prime minister to launch a positive initiative to break the political deadlock in India before it was too late. To restore Indian confidence in Britain, Jinnah advised a declaration from Whitehall "without delay" that Great Britain remained "unequivocally pledged" to its previously (1917) articulated policy of eventually granting India "full responsible Government with Dominion status." Jinnah believed that a round table conference convened by the prime minister in London and attended by political leaders of India like himself, who could "deliver the goods," was the only way of avoiding conflict and chaos. Viceroy Irwin agreed, conveying much the same message when he reached Whitehall that summer. The Liberal Simon and Tory lords like Birkenhead were outraged at the prospect of having their report shelved before it was born, but Wedgwood Benn and Irwin bolstered MacDonald's conviction that Jinnah and Motilal were right. "The report of the Simon Commission you need have no hesitation in assuming was never intended to be anything more than advice given for the guidance of the Government," Ramsay MacDonald wrote Jinnah that mid-August, reassuring him that "we want India to enjoy Dominion status."[22]

Lord Irwin returned to India in September and conveyed the same hopeful message to his Muslim confidant, who as president of the Muslim League was soon to emerge as the most powerful spokesman for almost one-quarter of British India's population. The same message, moreover, was conveyed personally by the viceroy to Motilal Nehru, who as president of Congress still represented the majority of politically conscious India, even though his mandate and time limit to win British assent for his constitutional report were rapidly expiring. The other Nehru, Irwin understood only too well, would not be nearly as easy to reach an agreement with as Motilal, whom even Simon admired. Motilal ordered Jawahar to wire every member of the Working Committee, inviting them to Delhi on November 1, to consider what Congress's response should be to the viceroy's impending official pronouncement, which he had been shown in confidence that October.

On October 31, 1929, Viceroy Irwin announced that His Majesty's Government planned to convene a round table conference next year in London to resolve the "broad question of British Indian constitutional advance" along lines, which, as "implicit in the declaration of 1917" would lead to "the attainment of Dominion Status." It was as close to accepting of the Nehru report and the last Congress's resolution deadline as Great

Britain would ever come. In Bombay, Jinnah—joined by Sarojini Naidu and sixteen other moderate national leaders of several parties—issued an enthusiastic public response, welcoming the viceroy's statement as a "fundamental change of procedure . . . which might carry the willing assent of political India." In Delhi, Congress's Working Committee agreed on a "manifesto" response the next day, which Motilal had initially written and which he and Gandhi finally convinced Jawaharlal to sign after accepting his amendments of it, for as president-elect of Congress as well as its general secretary he had to agree with his own Working Committee's otherwise unanimous position.[23]

"We appreciate the sincerity underlying the [viceroy's] declaration, as also the desire of the British Government to placate Indian opinion," Congress's Delhi manifesto began. "We hope to be able to tender our cooperation to His Majesty's Government in their effort to evolve a scheme of Dominion constitution suitable for India's needs. . . . We understand, however, that the conference is to meet not to discuss when Dominion Status is to be established but to frame a scheme of Dominion constitution for India. We hope that we are not mistaken in thus interpreting the import and implication of the weighty pronouncement of His Excellency the Viceroy. Until the new constitution comes into existence, we think it necessary that a more liberal spirit should be infused in the government of the country." This was Motilal's last great triumph. Not only had Congress resolved to accept his report, and his son for its next president, but Jawaharlal, who had so often spoken so angrily, so forcefully, against ever accepting dominion status as his goal, had signed the manifesto. The old man almost felt young again. His heart beat strong, and he could take a deep breath without coughing up blood.[24]

For Jawahar, however, that night and the next few days brought nothing but sleepless hours of agony, heartache, and "brain fever" that tortured him. He felt he had betrayed not only his comrades abroad and in India but also himself. Not since he had agreed to marry Kamala had he felt so weak, so stupid, so angry. His only hope now was to resign, first as general secretary, and then as president-elect. So after two days and sleepless nights, he wrote Gandhi, "I can take, I think, a calmer view of the situation than I could two days ago but the fever in my brain has not left me.

"Your appeal to me on the ground of discipline could not be ignored. . . . And yet I suppose there can be too much of discipline. Something seems to have snapped inside me evening before last and I am unable to piece it together. As General Secretary of the Congress I owe allegiance to it and must subject myself to its discipline. I have other . . . allegiances. I am president of the Indian Trade Union Congress, secretary of the Independence for India League and am intimately connected with the youth movement. What shall I do with the allegiance I owe to these . . . ? I realise now . . . that it is not possible to ride a number of horses at the same time. . . . [I]t is hard enough to ride one. . . . [T]he conviction has grown stronger that I acted wrongly day before yesterday. . . . I am afraid we differ fundamentally. . . . I believe the statement to be injurious and a wholly inadequate reply to the Labour Government's declaration. I believe that in our attempts to soothe and retain a few estimable gentlemen we have ruffled and practically turned out of our camp many others who were far more worth having. I believe that we have fallen into a dangerous trap." He therefore resigned as general secretary and begged to be excused from becoming the next president of Congress, urging Gandhi to take up that burden himself.[25]

Gandhi, of course, knew exactly how to bring his adopted child's brain fever back to normal. "How shall I console you? . . . I said to myself 'Have I been guilty of putting undue pressure on you?' I have always believed you to be above undue pressure. I have always honoured your resistance. It has always been honourable. Acting under that belief I pressed my suit. Let this incident be a lesson. Resist me always when my suggestion does not appeal to your head or heart. I shall not love you the less for that resistance. But why are you dejected? I hope there is no fear of public opinion in you. . . . [W]hy dejection? The ideal of independence is not in conflict with greater freedom. As an executive officer now and President for the coming year, you could not keep yourself away from a collected act of the majority of your colleagues. In my opinion your signature was logical, wise and otherwise correct. I hope therefore that you will get over your dejection and resume your unfailing cheerfulness . . . [T]here is no hurry. . . . If you feel like talking things over with me, do not hesitate. . . . I hope to see Kamala hale and hearty when I reach Allahabad. If you can do wire that the blues are over."[26]

Gandhi's medicine of patient reassurance worked well enough to keep Jawaharlal from resigning immediately. But four days later, after another tortured letter from Jawahar, Bapu wrote again, more firmly this time: "You must not resign now . . . it will affect the national cause. There is no hurry and no principle at stake. About the crown, no one else can wear it. It never was to be a crown of roses. Let it be all thorns. . . . But let us reserve the whole of this for calm and detached discussion when we meet. . . . [M]ay God give you peace."[27]

But no peace would calm Nehru's troubled spirit. The mere thought that the British were ready to settle or agree to what his father wanted, and Gandhi was willing to accept, all but terrified him. For then there could be no revolution, no continuing struggle, no Soviets emerging from India's impoverished soil and soul. It seemed like a dreadful prospect. Yet he could not bolt the Congress and reject the high honor just bestowed on him. "Whatever your reasons may be to explain away your surrender to the traitors who are negotiating for their own class interests," his friend Chatto reproached him, "I myself cannot see why you did not prefer immediate resignation. That would have . . . rallied all the youth, the workers and peasants to your side and you would have been able to defeat easily the compromisers in the Congress. . . . It is a fundamental political error to think that unity in the Congress is more important than the vital interests of the masses. . . . [Y]our signature to the Delhi Manifesto was a betrayal of the Indian masses in the struggle for independence."[28]

Chatto wanted Jawahar to resign from Congress and organize instead an "All India Anti-Imperialist Federation" to lead India's "masses" in a Soviet-style revolution. It was, indeed, what part of Jawaharlal yearned to do himself. But he knew how easy it was for Chatto to instruct him in the international communist line from the Berlin Secretariat of their League Against Imperialism, and he knew much better than Chatto just how weak and vacillating he was. He tried to harness his ambivalence, first addressing his Trade Union Congress on November 30, 1929, in central India's Nagpur. "The fact that I stand here today and address you as your president is itself a sign of your weakness. . . . I shall welcome the day when the worker from the mine and the factory and the field stands in this place . . . only then will the true voice of the worker be heard. . . . It is the system that is wrong . . . based on the exploitation of the few and the prostitution of labour. . . . [Y]ou will have to root out both capitalism and imperialism and substitute a saner and healthier order. . . . It will not profit you much if there is a change in your masters and your miseries continue . . . if a handful of Indians become high officers of the state or draw bigger dividends, and your miserable conditions remain, and your body breaks down through incessant toil and starvation and the lamp of your soul goes out. You want a living wage and not a dying wage. . . . Is it the socialist or the communist who separates the classes and preaches discontent or the capitalist and imperialist who by his policy and methods has reduced the great majority of mankind into wage slaves who are worse even in many ways than the slaves of old? The class war is none of our creation. It is the creation of capitalism and so long as capitalism endures it will endure."[29]

It is hardly surprising that India's leading financiers and industrial mill owners, the Birlas, Tatas, and other major supporters of Mahatma Gandhi and Motilal's Congress found Jawaharlal far less appealing. Indeed, it would be difficult to say at this point whom the younger Nehru alarmed more, the British imperialists or the Indian industrialists. The only mitigating factor in the grave concerns felt by both those powerful groups about Jawaharlal was their awareness that many of his comrades now viewed him as a traitor to world communism. By mid-December, as preparations for the Lahore Congress moved into high gear, Lord Irwin met with Motilal, Jinnah, Gandhi, and other moderate leaders in New Delhi. Jawaharlal had not been invited to the viceroy's house. "I do not know who took the initiative in arranging this interview," he later wrote. "Gandhiji and my father were present . . . representing the Congress view-point. . . . The interview came to nothing; there was no common ground, and the two main parties—the Government and Congress—were far apart from each other. So now nothing remained but for the Congress to go ahead. The year of grace given at Calcutta was ending."[30]

Lord Irwin told Gandhi, Motilal, Jinnah, and the others that afternoon that if they agreed to accept the government's invitation to a round table conference, it would offer them the chance "of doing something big," but if they did not, they would lose "a great opportunity." Jinnah and the others agreed, but Gandhi and Motilal were constrained by Jawaharlal's threat of resignation to adhere to the letter of the Calcutta Resolution of 1928. Irwin was equally constrained by right-wing imperialists in the British cabinet to limit what he might concede to the letter of his vaguely worded October 31 "promise" of dominion status sometime in the future. Thus the gap, though now narrow enough for hands of hope and trust to span, would never be bridged. By New Year's Eve, a week later, a new tricolor flag of saffron, white, and green would be raised in Lahore, and a younger drummer's beat would be heard from the Congress podium.[31]

"I have just unfurled the national flag of Hindustan," President Nehru told his Congress colleagues. "I want you to take a vow that you will have sufficient strength to protect this flag, and that you are ready to sacrifice your lives for freedom. The flag under which you stand today and which you have just now saluted does not belong to any community. It is the flag of the country. . . . All those who stand today under this flag are Indians, not Hindus, not Muslims, but Indians. The volunteers, who have saluted the flag today, should be prepared to lay down their lives for its honour. Remember . . . now that this flag is unfurled, it must not be lowered so long as a single Indian, man, woman, or child lives in India."[32] Twelve-year-old Indira was there with her ailing mother. "It was bitterly cold," Indira recalled, "and we all stayed in tents. As the Congress President, he led the procession and went on a white horse. We were standing, my grandmother and all the family, in a shop or in somebody's house, on the first floor, so that we could get a good view. Close by, there was a band playing . . . 'God save the King!' "[33] In 1947 Lahore would become the capital of Pakistan's Punjab, not of India's.

"The whole world today is one vast question mark, and every country and every people is in the melting pot," President Nehru told his Congress followers. "The age of faith, with the comfort and stability it brings, is past, and there is questioning about everything. . . . Old established ideas of liberty, justice, property and even the family are being attacked. . . . We appear to be in a dissolving period of history, when the world is in labour and, out of her travail, will give birth to a new order. . . . Europe has ceased to be the centre. . . . The future lies with America and Asia." He then gave a brief outline of India's history, stressing its "wonderful stability" and powers to "absorb" and "tolerate" invaders and foreigners, though he admitted that with the "coming of the Muslims the equilibrium was disturbed." India was about to "restore" its ancient "harmony" when "the British came and we fell."[34]

Jawaharlal often spoke of himself as India's "best historian" and later had ample time in prison to write about the past. Indeed, his nationalist reconstruction of India's ancient glory became the most popular text for several generations of young Indians who followed his quest for the *Discovery of India.* He had a Marxist view of world history, seeing most problems as rooted primarily in economic class conflicts. Nehru continued to believe that Hindu–Muslim communalism was caused only by economic inequality and British imperial duplicity. "Do we want outsiders, who are not of us and who have kept us in bondage, to be the protectors of our little rights and privileges, when they deny us the very right of freedom? No majority can crush a determined minority, and no minority can be sufficiently protected by a little addition to its seats in legislatures. . . . The politics of a subject race are largely based on fear and hatred. . . . So far as I am concerned I would gladly ask our Muslim and Sikh friends to take what they will without protest or argument. . . . Meanwhile . . . the All Parties Report has to be put aside." The year was about to end; Jawaharlal had heard no satisfactory response from Britain, saw no easing of the martial "iron grip" of British repression by bayonet and baton.

"We stand, therefore, today for the fullest freedom of India. This Congress . . . will not acknowledge the right of the British Parliament to dictate to us in any way. . . . We are very conscious of our weakness, and there is no boasting in us or pride of strength. But let no one, least of all England, mistake or underrate the meaning or strength of our resolve . . . there will be no turning back. . . . We are weary of strife and hunger for peace and opportunity to work constructively for our country. Do we enjoy the breaking up of our homes and the sight of our brave young men going to prison or facing the halter? . . . I am a socialist and a republican, and am no believer in kings or princes, or in the order which produces the modern kings of industry . . . whose methods are as predatory as those of the old feudal aristocracy. . . . [W]e must realise that the philosophy of socialism has . . . [spread] the world over. . . . India will have to go that way, too, if she seeks to end her poverty. . . . All these are pious hopes till we gain power, and the real problem, therefore, before us is the conquest of power."[35]

As the midnight hour struck that New Year's Eve along the bitterly cold banks of Lahore's River Ravi, Nehru led his National Party to vow and resolve to dedicate themselves "to the attainment of Complete Independence *[Purna Swaraj]* for India." The first step was for all congressmen to "abstain from participating directly or indirectly" in any future elections and to order every current Congress member of any assembly or council to "resign their seats. This Congress appeals to the Nation zealously to prosecute the constructive programme of the Congress, and authorises the All-India Congress Committee, whenever it deems fit, to launch upon a programme of Civil Disobedience, including non-payment of taxes."[36]

"We had burned our boats and could not go back," Jawahar recalled. The Nehru report was cast aside, as was Motilal's Swaraj Party and the dream he had shared with Das. Lahore's cold further weakened Motilal's precarious health, and so he did not hesitate to resign from the Viceroy's Council. As his Congress crown passed to his son, Motilal surrendered his grip over Congress policy, trusting that Jawahar's voice and Gandhi's mind would soon lead the nation to the freedom that he would not live to see.[37]

To retain credibility with Jawahar and his comrades, Gandhi knew he would soon have to launch a new Satyagraha campaign, but first he insisted on testing the waters of popular support for Congress's new goal of *Purna Swaraj,* so he called for a national day of celebration on Sunday, January 26, 1930, on which in every city, town, and rural market all over India a public pledge would be taken as the Congress flag was unfurled, a declaration of independence that Jawaharlal drafted, with help from Thomas Jefferson, and read aloud.

"We believe that it is the inalienable right of the Indian people, as of any other people, to have freedom and to enjoy the fruits of their toil and have the necessities of life. . . . We believe also that if any government deprives a people of these rights and oppresses them, the people have a further right to alter it or to abolish it. The British Government in India has not only deprived the Indian people of their freedom but has based itself on the exploitation of the masses, and has ruined India economically, politically, culturally, and spiritually. We believe, therefore, that India must sever the British connection and attain *Purna Swaraj* or Complete Independence. . . .

"We hold it to be a crime against man and God to submit any longer to a rule that has caused this fourfold disaster to our country. . . . We will therefore prepare ourselves by withdrawing, so far as we can, all voluntary association from the British Government, and will prepare for civil disobedience, including non-payment of taxes."[38]

Thousands of crowded meetings were held, with millions of Indians solemnly taking the pledge, raising their fists after this proclamation of national freedom was read out to them in every language of the land. This was the first public articulation of the "tryst with destiny" that Nehru and India made, but it took seventeen more years to redeem that pledge, though "not wholly or in full measure." Each year thereafter, India's Independence Day has been celebrated on January 26. After 1950, when the Republic of India's constitution was adopted, this national holiday was renamed Republic Day.

For Jawahar personally, however, Lahore was more than a mere political triumph. And for the unlettered masses of India it was a coronation. Hindus have always worshipped youthful handsome gods. Lord Rama is not only a god but a king, while Lord Krishna is both god and earthly lover. Jawaharlal was viewed by millions of Indians as a reincarnation of these sacred heroes. He even had something of the Buddha about him, austere, often melancholy in manner, rejecting the luxury to which he had been born, for freedom's struggle and the hardships of prison life.

"I was in Allahabad during the early part of January," Jawahar recalled, "my father was mostly away. It was the time of the great annual fair, the Magh Mela; probably it was the special Kumbh year, and hundreds of thousands of men and women were continually streaming into Allahabad . . . holy Prayag, as it was to the pilgrims . . . chiefly peasants, also labourers, shopkeepers, artisans, merchants . . . indeed, it was a cross-section of Hindu India. . . . I wondered how they would react to the call for civil resistance. . . . How many of them knew or cared for the Lahore decisions? How amazingly powerful was that faith which had for thousands of years brought them . . . from every corner of India to bathe in the holy Ganga [Ganges River]! Could they not divert some of this tremendous energy to political and economic action to better their own lot? Or were their minds too full of the trappings and traditions of their religion to leave room for other thought?" he reflected, having by now himself become one of their icons.[39]

"Our house attracted crowds of pilgrims . . . and on the days of the mela an endless stream of visitors would come to us from dawn to dusk. . . . Our political slogans they knew well, and all day the house resounded with them. I started the day by saying a few words to each group . . . but soon this proved an impossible undertaking, and I silently saluted them when they came. There was a limit to this, too, and then I tried to hide myself. It was all in vain . . . each door and window had a collection of prying eyes. . . . This was not only embarrassing, it was annoying and irritating. Yet there they were, these people looking up with shining eyes full of affection, with generations of poverty and suffering behind them, and still pouring out their gratitude and love and asking for little in return, except fellow-feeling and sympathy. It was impossible not to feel humbled and awed by this abundance of affection and devotion. . . . I had achieved, almost accidentally as it were, an unusual degree of popularity with the masses. . . . I was a bit of a hero, and a halo of romance seemed to surround me in their eyes. Songs had been written about me, and the most impossible and ridiculous legends had grown up. . . . Only a saint, perhaps, or an inhuman monster could survive all this, unscathed and unaffected. . . . It went to my head, intoxicated me a little, and gave me confidence and strength. I became . . . just a little bit autocratic in my ways, just a shade dictatorial." Wondering about the real reasons for his unique popularity, Nehru confessed that "my reputation as a hero is entirely a bogus one," cogently adding, however, that "there is the idea of mixing in high society and living a life of luxury and then renouncing it all, and renunciation has always appealed to the Indian mind."[40]

"Bhai was riding high," his sister Betty recalled. "The Indian press went flat out, loading Bhai with extravagant praise, and people addressed him by such magniloquent phrases as '*Bharat Bhushan*' [jewel of India] and '*Tyagamurti*' [o embodiment of sacrifice]. . . . Kamala and I played our part in keeping his feet on the ground. When Bhai came down to breakfast we bowed deeply and asked how the Jewel of India had slept, or if the Embodiment of Sacrifice would like some bacon and eggs. Even little Indu joined in the game. Bhai blushed and smiled sheepishly."[41]

That March Gandhi decided to focus the civil disobedience movement he was ready to launch on the British salt tax. Although the tax was just a few annas per pound, salt was used in quantity by every peasant of India, especially during the hot months—most of each year—simply to keep alive. Indeed, since the late nineteenth century the cumulative annual salt tax had been third largest source of revenue in British India's budget, exceeded only by the taxes on land and opium. It was small wonder then that British merchants insisted on monopolizing its production and sale, taxing every ounce packet sold in government monopoly salt stores. Yet salt could be found on beaches along hundreds of miles of India's coastline, drying in the sun. What an ingeniously simple, brilliant plan of Gandhi's, to launch a "salt march" from his Sabarmati ashram to the sea, promising at its end to "break the British law" by picking up untaxed salt from the beach, encouraging millions of Indians to follow his example and to break the law in exactly the same way.

"Before embarking on Civil Disobedience," Gandhi now wrote to Lord Irwin, "I would fain approach you and find a way out." He explained that while holding "British rule to be a curse," he intended no "harm" to any Englishman, "claiming many Englishmen as dearest friends." He viewed

British rule as a "curse" because "it has impoverished the dumb millions by a system of progressive exploitation and by a ruinously expensive military and civil administration which the country can never afford. It has reduced us politically to serfdom. It has sapped the foundations of our culture. And . . . it has degraded us spiritually. . . . I had hugged the fond hope that the proposed Round Table Conference might furnish a solution. But, when you said plainly that you could not give any assurance that you or the British Cabinet would pledge yourselves to support a scheme of full Dominion Status, the Round Table Conference could not possibly furnish the solution for which vocal India is consciously, and the dumb millions are unconsciously, thirsting. . . . I know that in embarking on non-violence, I shall be running what might fairly be termed a mad risk. But the victories of Truth have never been won without risks. . . . Conversion of a Nation that has consciously or unconsciously preyed upon another far more numerous, far more ancient and no less cultured than itself, is worth any amount of risk."[42]

Gandhi left his Sabarmati ashram after prayers on the morning of March 11, 1930, followed initially by fewer than eighty disciples, marching staff in hand toward the sea at Dandi, some 240 miles away. The eyes of the world's press were on this walking saint, worshiped as the Mahatma by countless millions of Hindus and viewed as a latter-day Jesus or Moses throughout the Judeo-Christian world. At every inch of his long journey peasants bowed and prayed, sprinkling water on his bare legs, first hundreds and soon thousands joining his band of Satyagrahis on that most publicized and peaceful of all modern Indian protest marches. "The air is thick with the rumour that I shall be arrested," Gandhi wrote Jawaharlal that first night, reporting that otherwise all went "extraordinarily well," with "volunteers . . . pouring in."[43]

Lord Irwin, good Christian that he was, gave orders not to arrest Gandhi or to inhibit his progress as long as he broke no law. So the march continued peacefully, exciting all of India and magnetizing the world's attention. As Gandhi marched, Jawaharlal both spoke and wrote: "The spark that was lit by our leader on the banks of the Sabarmati is already spreading like a prairie fire throughout the country and soon the whole land will try to redeem that pledge. Many a tragic scene will be enacted and many an actor on this vast stage will suffer torment before the curtain rings down finally on a free India." Subhas Bose and Vallabhbhai Patel had already been arrested, together with several Congress comrades in Bengal and Gujarat. The British officers now wasted no time on trials, sentencing each congressman to three or six months as soon as he was taken into custody and also to fines of one hundred or more rupees.[44]

Jawaharlal and Motilal went by train and car to Ahmedabad in mid-March to meet with their colleagues on Congress's Working Committee, to decide how best to continue the struggle as each of them was arrested. After that meeting, both Nehrus went alone to the village of Jambusar in Broach to meet with Gandhi, who had stopped for the night there en route to Dandi. "We spent a few hours with him there," Jawahar recalled, "and then saw him stride away . . . my last glimpse of him then as I saw him, staff in hand, marching along at the head of his followers, with firm step and a peaceful but undaunted look."[45]

Gandhi walked into the sea on the morning of April 6, the eleventh anniversary of the launching of his first Satyagraha's "national week" in 1919. He picked up salt lying on the shore, and Sarojini Naidu, who had come to Dandi a day earlier, shouted, "Hail, Deliverer!" There were many other shouts of "Jai!" (victory) to Mahatma Gandhi and to India, to Nehru, and to Congress, but no arrests as yet. Gandhi now urged all Indians to break the salt law openly. "This salt being manufactured by Nature in creeks and pits near sea-shore, let them use it for themselves and their cattle. . . . This war against the Salt Tax should be continued during the National Week. . . . Those who are not engaged in this sacred work should themselves do vigorous propaganda for the boycott of foreign cloth . . . and the prohibition of liquor." Small packets of the salt Gandhi picked off the sands of Dandi beach that day were soon selling for "the upset price" of a thousand rupees. Hundreds of thousands of Indians busied themselves all week long finding or manufacturing salt by boiling seawater. Soon tens of thousands filled every prison in British India.[46]

Jawaharlal was arrested on April 14, 1930. He admitted to Allahabad's magistrate that "I have deliberately broken the salt laws" and was sentenced to six months in prison. "Keep smiling," he called out to his family in the courtroom as he was led away, "fight on and see it through."[47] From Naini Prison he telephoned Motilal, asking him to convey the following message to Gandhi, who was then still free: "I have stolen a march over you." On May 5, Gandhi was arrested and taken to Yeravda Prison near Poona. His last message before his own arrest was that "Swaraj obtained without sacrifice, never endures. . . . In real sacrifice there is only one-sided suffering, that is, without killing others one has to die. May India accomplish this ideal. At present the self-respect and everything of India are concealed in a handful of salt. The fist may be broken, but it should never be opened. . . . The conductor of this fight is God and not I. He dwells in the heart of all. . . . Our path is fixed. Whole villages should come forward to pick or manufacture salt. Women should picket liquor and opium shops and foreign cloth shops. . . . There should be bonfires of foreign cloth. Hindus should regard none as untouchables. Hindus, Muslims, Parsees and Christians, all should heartily embrace one another. . . . Students should leave Government schools, and Government servants should resign and be employed in the service of the people. . . . Thus shall we easily complete Swaraj."[48]

"So Bapu has been arrested!" Jawahar wrote in his prison diary that Monday night. "*Ça marche bien*. It is full-blooded war to the bitter end. Good."[49]

10

Prison and Round Tables

1930–1931

MOTILAL TOOK BACK the Congress presidency the day his son was jailed, but his blood pressure remained so high, his asthma so intense, that he could barely chair the Working Committee meeting convened at Anand Bhawan on May 12, 1930. They resolved to continue the Boycott of foreign cloth and to inaugurate a no-tax campaign in several provinces. Motilal tried to prepare his son for the inevitable, reporting on how precarious his health had become. "Your reference . . . to the possibility of a physical and nervous breakdown has alarmed me," Jawahar replied.[1]

Motilal also worried about Indira's education, and from prison Jawahar agreed that a "good private school for little boys and girls is very necessary in Allahabad. . . . I suppose . . . English women should not be employed, but . . . I should suggest getting a really competent American girl out if necessary." Jawahar himself soon started writing "Glimpses of World History" as weekly letters to his daughter from his prison cell. The letters were published by Krishna Menon in 1934 and remained India's most popular and often republished world history. Jawahar amused himself by reading Spengler's *Decline of the West.*[2]

After two months in jail, Jawahar wrote to his favorite sister, "Nan dear. . . . What am I to write about? Life in prison is not meant to be exciting—it is about as uneventful as the existence of the average turnip. . . . Inside the jail—and outside—the massive gates and the high walls of the jail separate two worlds . . . and none love the jail so much as to wish to remain here. Two worlds! You could also compare the two to the animal and the vegetable kingdoms. The object of jail appears to be first to remove such traces of humanity as a man might possess and then to subdue even the animal element in him so that ultimately he might become the perfect vegetable! Soil-bound, cut off from the world and its activity, nothing to look forward to, blind obedience the only 'virtue' . . . spirit considered the great sin—is it any wonder that the prisoner approximates to the plant? . . . You will be surprised to learn that I often get up at 3.30 in the morning—I am myself amazed at this . . . for I have never been guilty of it before. . . . I read and write and spin and weave and run and walk—indeed I am always running a race with time. Why this unseemly haste, you will ask, as if the Devil was after me? I want to create tension, to live in it. Perhaps it is opposed to the Indian mentality and outlook. . . . But I believe in it as a method of keeping in form and as an incentive to continuous effort . . . my way of keeping myself in training, so that when I bid *au revoir* to this delectable spot I may get out as a human animal, and not as a plant. . . . For some weeks I have done less and less reading. . . . I spent more time on spinning . . . and weaving . . . I found this, and specially spinning, soothing to the nerves. . . . [S]emi-automatic occupations . . . allow one to think gently at the same time. To some extent the effect is similar to having a pipe in your mouth!"[3]

A few years later, in another prison cell, when writing his autobiography, Jawahar confessed that "my real conflict lay within me, a conflict of ideas, desires and loyalties, of subconscious depths struggling with outer circumstances, of an inner hunger unsatisfied. I became a battleground, where various forces struggled for mastery. I sought escape from this; I tried to find harmony and equilibrium, and in this attempt I rushed into action. That gave me some peace; outer conflict relieved the strain of the inner struggle. . . . The quest is still the same, in prison or outside, and I write down my past feelings and experiences in the hope that this may bring me some peace and psychic satisfaction."[4]

Motilal journeyed to Bombay with his frail wife and Kamala during the latter part of June. Bombay was still the heart of the revolt, where riots, arrests, strikes, marches, and defiance kept the British in a state of constant alert and ever deepening anxiety. Peshawar had been silenced by bullets and bayonets; martial law had been imposed on the entire frontier; and Lucknow had been beaten numb with lathis and batons; but greater Bombay proved the most difficult to subdue, perhaps because of its location midway between Sabarmati and Poona, the beginning of Gandhi's salty pilgrimage and his prison "temple" as the cell of his incarceration was popularly called. While Motilal was in Bombay, some attempts were made to approach him indirectly on behalf of the government, as Jawaharlal learned from Motilal on June 28, when his father, mother, Kamala, and Indu all came to visit him in prison.

London's *Daily Herald*'s correspondent, George Slocombe, had asked Motilal what his "attitude" would be if he were now invited to a round table conference in London. "We must be masters in our own household," Motilal answered, "but are ready to agree to reasonable terms for the period of transfer of power. . . . We must meet the British people in order to discuss these terms as nation to nation on an equal footing."[5] Slocombe had written with moving accuracy of the brutal attacks against the unarmed and unresisting Satyagrahis who had marched to the British salt factory near Bombay and were mercilessly beaten by "mounted European" police, who galloped down upon them, wielding lathis. "It was humiliating for an Englishman to stand . . . and watch the representatives of the country's administration engaged in this ludicrous, embarrassing business," Slocombe reported.[6] He took the bold initiative of interviewing Gandhi in prison and reported that "even at this critical hour, a settlement is possible and Mr. Gandhi is prepared to recommend to the Congress a suspension of the Civil Disobedience movement and co-operation with the R.T.C. [round table conference]." Such reports electrified India's supporters in the House of Commons, enraged the Tories, and embarrassed Lord Irwin and every liberal Christian.

After their meeting, Jawaharlal confided in his prison diary, "Father's account of Bombay most comforting. . . . Indu does not want to go to Convent [school] any more—Have told her she need not go."[7]

Two days later Motilal and Syed Mahmud, then serving as secretary of the Congress, were arrested for sedition, sentenced to six months, and brought to Jawahar's central prison in Naini. "Father shares my cell with me and my bathroom, and Mahmud shares Narmada Prasad's. Each of our cells is about 11 ft square. One of these is our pantry, baggage room, bathroom, lavatory . . . cram-full . . . The place is horribly uncomfortable for father. . . . Naturally all of us try to be of some service to father. At his age [69] to have to put up with these discomforts . . . is a great trial."[8] There was, however, "plenty of fruit" and other good things to eat, now that Motilal had arrived, his servants bringing "an abundance" of fresh supplies every day. But Motilal ran a low fever now without respite and had little appetite for all the mangoes and melons brought to his tiny cell. Jawahar took care of him as best he could, but a tumor in Motilal's right lung had spread to a kidney, and his enlarged heart beat irregularly. "I hate myself for being so lazy and taking up such a lot of Jawahar's time," Motilal wrote his daughter from jail that July. "But he anticipates everything and leaves nothing for me to do. I wish there were many fathers to boast of such sons."[9]

Lord Irwin hoped that two trusted leaders of India's Liberal Party,

Kashmiri Brahman Tej Bahadur Sapru and Maharashtrian Brahman M. R. Jayakar, might be able to help him lure the Congress leaders out of prison to attend a round table conference that Prime Minister Ramsay MacDonald was ready to convene in London before year's end. Sapru and Jayakar met with Gandhi in Poona's jail on July 23–24, and then went to Naini to meet with Motilal and Jawahar a few days later. "My dear Bapuji," Jawahar wrote on July 28, 1930, "Dr. Sapru and Mr. Jayakar came yesterday and had a long interview with father and me. Today they are coming again."

Gandhi had informed the intermediaries that he was willing to attend a round table conference provided that Congress "raise the issue of independence" there, discussing "safeguards for the transitional period only," and that he would be willing for the interim to call off civil disobedience but insisted on continuing the "peaceful picketing" of foreign cloth and liquor shops and free "salt manufacture" without punishment, as well as the release of all Satyagrahi prisoners, the restoration of confiscated property, the refunding of fines, and the reinstatement of all dismissed Indian officers. "I must confess that your point (1) regarding the 'Constitutional Issue' has not won me over," Jawahar wrote. "I do not see how it fits in with our position or our pledges. . . . Father and I entirely agree with you that we can be 'no parties to any truce that would undo the position at which we have arrived today.' It is because of this that the fullest consideration is essential before any final decision is arrived at. I must confess that I do not see any appreciable advance yet from the other side and I greatly fear a false or weak move on our part. I am expressing myself moderately. For myself, I delight in warfare. It makes me feel that I am alive. The events of the last few months in India have gladdened my heart and have made me prouder of Indian men, women and even children than I have ever been. . . . May I congratulate you on the new India that you have created by your magic touch? . . . Father has been unwell. . . . He has grown very weak. The long interview last night tired him out."[10]

Gandhi knew from the moment he read that letter that it would not be possible for him to go to London with the younger Nehru and rightly feared that the more moderate older Nehru might die before he could attempt so arduous a voyage. Nonetheless, the doomed peace negotiations continued, for Sapru and Jayakar were patient mediators. "Sapru & Jayakar have had two long interviews with father and me," Jawahar wrote in his prison diary on August 1. "Lot of time wasted in listening to what Irwin said & Jinnah did and Fazl-i-Husain intrigued. Nothing important or hopeful. Bapu's note disappointing. . . . We have not suggested anything. Have said that without full consultation with colleagues, specially Gandhiji, we cannot make any suggestions. We have not committed ourselves in any way although we have hinted at a stiff attitude. I have been worried however for last two or three days. I wish we had definitely stopped all talk of 'peace.' It is harmful and diverts attention. There is no peace anywhere in sight."[11]

Sapru then returned to say that the viceroy agreed to their seeing Gandhi in Poona's prison if they liked. "We told him we had and could have no objection to going to see Gandhiji but under the circumstances there was no chance of us three deciding anything." The entire Working Committee would not be allowed to meet. A special train rushed them from Naini to Poona on August 12, and for most of August 13–15 they all conferred. By the end of this ordeal Motilal was running a high fever and coughing up blood. The foregone conclusion, that Jawaharlal reached before any talks started was that "the time is not yet ripe for securing a settlement honourable for our country. . . . Needless to mention that we do not in any way share either your [Sapru's and Jayakar's] view or the Viceroy's that civil disobedience has harmed the country or that . . . it is unconstitutional. English history teems with instances of bloody revolts whose praises Englishmen have sung unstintingly. . . . It therefore ill becomes the Viceroy or any intelligent Englishman to condemn a revolt that is in intention, and that has overwhelmingly remained in execution, peaceful. But we have no desire to quarrel. . . . The wonderful mass response to the movement is, we hold, its sufficient justification."[12] Jawaharlal's florid language reflected much he had learned at Harrow and the intoxication and delight he felt in warfare. "It can be no pleasure to us needlessly to expose the men, women and even children of our country to imprisonment, lathi charges and worse," Gandhi added, for he too signed this letter. "You will therefore believe us when we assure you, and through you the Viceroy, that we would leave no stone unturned to explore any and every channel for an honourable peace."[13]

Motilal was released from jail on September 8, never having recovered even the weak state of health in which he had entered it little more than two months earlier. In those ten weeks he lost fourteen pounds. Three days later he left for Mussoorie with his wife and daughters. "I am sure you will get well soon there," Jawahar wrote optimistically on the eve of his father's departure. "The main thing is to get well. Personally I have little faith in medicines. . . . But I have great faith in nature and in the mountain air of the Himalayas."[14]

On October 11, 1930, Jawaharlal was released from jail, two days before his six-month sentence expired. "I am very much alive and kicking," he announced upon returning to Allahabad. "I hope to do my little bit to hasten the dissolution of the British Empire and take part in its final obsequies." The next day he addressed a large public meeting, at which he insisted, "Today every man has to choose between the two flags, the flag of Indian freedom and the flag of foreign domination. The country has made its choice and has stuck to it in spite of all the frightfulness and methods of barbarism which history has associated in the past with the Huns."[15]

Nehru informed Lord Irwin that "we are in deadly earnest, we have burnt our boats . . . there is no going back for us. . . . [T]he Congress stands for the independence of India and it will fight to the bitter end till it has achieved it." He spoke of Congress today as "the Indian people, including every major and minor community." In a recent speech at Simla the viceroy had criticized some of Nehru's references to violence in his Lahore presidential address, and Jawahar now replied: "It is always interesting to read a sermon on morals from one who does not practise them. If England were invaded by Germany or Russia, would Lord Irwin go about advising the people to refrain from violence against the invader? . . . I stick by every word I said at Lahore. . . . The first phase of the great struggle has come to an end. It has been marked by a national awakening. . . . Now the second stage is beginning, the stage of our laying the foundation of a future, free India. . . . We must be prepared not only not to pay any taxes to the British Government but also to do without any service which they may render to us. While lawyers argue and raise their petty quibbles in London, we in India will fight for the reality, the conquest of power."[16]

A week after delivering that speech, Jawahar was rearrested and brought back to Naini Prison, his fifth term in jail. "The honour of arrest has again been accorded to me," he boldly informed his comrades in a message sent out to them. "I had long hoped to journey to the new places of pilgrimage in this ancient land—to Bombay and Peshawar, Gujarat and Delhi and so many others—and to see with my own eyes what manner of men and women they were who defied the British raj in all its panoply of power. . . . I cannot journey now. But, rest assured, I shall keep my promise and come to you when your heroic sacrifice has borne fruit and made India a free land, worthy of the heroes and heroines that inhabit her. . . . Let this message of freedom be carried to field and factory. . . . Be of good cheer, comrades, for the day of our deliverance approaches."[17] Jawahar now mistakenly believed that victory was in sight and, as Congress president, felt destined soon to preside over free India's republic, or soviet. Yet only after seventeen more years of struggle—much of that time spent behind bars—would he accept the premiership of a much smaller dominion of India. Now, however, Jawaharlal was riding too high even to think of agreeing to dominion status.

"I am charged with sedition and with the spreading of disaffection against the British Government," Nehru told the magistrate who presided over his trial in late October. "Eight and a half years ago I was charged with a similar offence and I stated then that sedition against the present government in India had become the creed of the Indian people, and to preach and practise disaffection against the evil which it represents had become their chief occupation. . . . I have had no other profession, no other business, no other aim than to fight British imperialism and to drive it from India. . . . There can be no compromise between freedom and slavery, and between truth and falsehood. We realise that the price of freedom is blood and suffering—the blood of our own countrymen and the suffering of the noblest in the land—and that price we shall pay in full measure. . . .

"We have no quarrel with the English people, much less with the English worker. Like us he has himself been the victim of imperialism, and it is against this imperialism that we fight. With it there can be no compromise. To this imperialism or to England we owe no allegiance and the flag of England in India is an insult to every Indian. The British Government today is an enemy government for us, a foreign usurping power holding on to India with the help of their army of occupation. My allegiance is to the Indian people only and to no king or foreign government. . . .

"To the Indian people I cannot express my gratitude sufficiently for their confidence and affection. It has been the greatest joy in my life to serve in this glorious struggle and to do my little bit for the cause. I pray that my countrymen and countrywomen will carry on the good fight unceasingly till success crowns their effort and we realise the India of our dreams. Long live free India!"[18]

Jawahar was sentenced to two and a half years but was released early on January 26, 1931, for by then Motilal lay dying at home, and the viceroy feared his son might be too late to say one final farewell to his father. By early December 1930, however, Jawahar was depressed again, for although he had learned in prison of the government's ban of the Allahabad Congress Committee and the Youth League, he was also informed that no protest marches or street demonstrations had followed that "illegal" declaration. He hoped at least that Kamala would march and be arrested, but the old head of Allahabad's Congress Committee advised her to stay home, not only because of her frail health, but also for fear that "her arrest might have a bad effect on father who was ill!" Jawahar wrote in his prison diary. "Why did Kamala agree to this humiliation! It was shameful and I have felt miserable. One wrong act can undo half a year's good work. The whole city is depressed."[19]

The next time she visited him in jail he told his wife how let down he had felt by her reluctance to court prison. She had, of course, done her best, leading a parade of female protesters in Allahabad's midday sun to urge college students to resign and join the civil disobedience movement. When she marched, Kamala wore khaki pajamas and a kurta as well as a white Gandhi cap, looking like a pale boy, her long hair tied in a bun, mostly hidden, shouting as loud as her frail voice would allow, fainting dead away as she passed one local college lined with bored, amused young men, several of whom scoffed at these "freedom ladies." Only one of them, Feroze Gandhi, jumped off the wall to pick up that tiny heroine from the hot pavement and carry her to some shade, where she recovered soon after he gave her some water. He never recovered his own interest in completing college, however, so moved was he by a sudden romantic and ardent devotion for Kamala Nehru. He quit school, joined the Congress, and followed her everywhere, even to Switzerland and Germany, staying at her bedside until the day she died. Then he followed her daughter until she married him, her mother's "son." The pudgy Parsi Feroze Gandhi soon became a constant thorn in his father-in-law's public as well as private life.

On New Year's Day, Kamala marched again and spoke to a large crowd in public, reading aloud her husband's October 12 speech, for which he had been charged with sedition. "Kamala arrested!—a good beginning for the New Year," Jawahar reported cheerfully. "She will be happy now and it is quite possible that she may profit by the rest in prison—Poor Indu alone—What impressions must be produced on a growing child's mind by all these events?" Kamala was sentenced to six months but was released in a few weeks.[20]

Motilal had been driven to Calcutta with his daughters, hoping that several famous doctors there might be able to do something that none of his local specialists found possible. There was no cure, however, for his many ailments. On January 12, 1931, Motilal nonetheless managed to accompany his wife and daughters to his son's prison. "Had interview today with father . . . looking ill and weak," Jawahar recorded in his diary. "I had a shock on seeing him. He told me he was much better than he had been."[21]

In less than a month he would be dead. Still, Jawaharlal liked to pretend that all would soon be well, writing his father after that last visit; "I do hope that you have got the trouble under full control and are marching rapidly to recovery. . . . To worry about anything is a waste of good energy and is profitless. . . . You do not suffer fools gladly, and unhappily there are a fair number of this breed about. . . . Prison life has one sovereign virtue. It teaches one detachment to some extent and the capacity to see things in their proper perspective." Then as a postscript to this letter, he gave his dying father another bit of work to do for him back home. "I learnt . . . that a young lady named Kishori has made Anand Bhawan her residence. She had done so previously . . . and has now reverted to it. I do not know her, but from various reports, which appear to be worthy of credence, I feel that it is not desirable for her to stay in

Anand Bhawan. . . . May I suggest for your consideration—it is, of course, for you to decide—that she should be politely but firmly requested to depart?"[22]

Jinnah, Sapru, Jayakar, and seventy-one other British Indian delegates and princely Indian representatives were still in London at the final sessions of the first round table conference. But without Congress's participation, the talks were rather like trying to stage *Hamlet* without the prince of Denmark. Jinnah proved himself by far the most brilliant spokesman for Muslim India, warning Ramsay MacDonald and the other British officials that unless they swiftly negotiated a settlement to "satisfy the aspirations of India," seventy million Muslims who had until then mostly kept away from the Congress-led revolt might be tempted to join the noncooperation movement."[23] Most Englishmen, like most congressmen, including Nehru, still underestimated Jinnah, however, considering him a mere bargainer, some scorning him as the scion of Bombay "eel-purveyors." He was, nonetheless, a man who meant what he said and usually managed to do precisely what he willed.

A much smaller, far less publicized meeting of the Muslim League was convened in Allahabad in December 1930. Dr. Muhammad Iqbal, Pakistan's greatest Urdu poet, presided over the meeting, poorly attended, for many of its leaders were in London. President Iqbal announced for the first time that the "final destiny" of the Muslims of Punjab, the North-West Frontier Province, Sind, and Baluchistan should be amalgamation "into a single State. Self-government within the British Empire, or without the British Empire, the formation of a consolidated North-West Indian Muslim State appears to me to be the final destiny of the Muslims, at least of North-West India."[24]

Pakistan (land of the pure) was as yet no more than a nebulous dream in Iqbal's mind, but its political shape and ultimate form would evolve over the next decade and a half once Jinnah adopted the idea as Muslim India's best political hope.

On January 19, 1931, Prime Minister Ramsay MacDonald concluded the first round table conference with the startling declaration that until "full responsibility" could be transferred to India "through the new constitution" now under consideration by the Simon Commission in London, "responsibility for the Government of India should be placed upon Legislatures, Central and Provincial, with such provisions as may be necessary to guarantee, during a period of transition, the observance of certain obligations and to meet other special circumstances, and also with such guarantees as are required by minorities to protect their political liberties and rights." Although Defense and External Affairs would still be controlled by the governor-general, other departments would now be transferred to elected members of the Viceroy's Council. Finance, however, would remain "necessarily subject" to ensure the "fulfilment of the obligations incurred" under the authority of the secretary of state for India and maintenance of the "financial stability and credit of India." The governor's Provinces were to be given the "greatest possible measure of self-government," and only minimal "special powers" would be reserved to governors, to preserve tranquillity and maintain the rights of public services and minorities. A more liberal franchise would now be framed to empower many more millions of Indians to elect Indian representatives. The prime minister rightly concluded that "personal contact is the best way of removing those unfortunate differences and misunderstandings which too many people on both sides have been engendering between us in recent years."[25]

For Congress even to think of calling off civil disobedience in response to that declaration, however, Jawaharlal felt would be "suicidal." Even a truce until the April session of Congress could be held (the annual date had been changed at Lahore, as December was too cold), would be "equally dangerous. The only safe course to adopt, therefore, is for the Working Committee to express its opinion boldly and say that the announcement is entirely unsatisfactory."[26] On his deathbed Motilal read MacDonald's declaration more positively, appreciating how much it could mean for India in real terms. He had no more patience with his son's impulsively negative, violently radical reactions to any official gesture of British friendship or conciliation. He told him as much and asked what his alternative was. What would he do?

On January 22 Jawahar wrote back. "My dear Father, I was somewhat confused the other day and could not give a clear answer. . . . [T]he kind of questions that were put to me filled me with anxiety and clouded my brain. . . . Of course I have no business from here to interfere in the decisions outside. . . . A major reason for my doing so is that the main burden of decision must fall on you. . . . I gathered from your note that there were forces in the Working Committee which were for a very weak and temporising attitude. . . . The suggestion that a reference be made to the Congress would, if adopted, have been almost fatal . . . I have always laid stress on the psychological character of our fight—on morale. We have to produce an impression of . . . strength on our own minds, on our people and, thirdly, on our enemy. . . . The day we convince the enemy that we are immovable like the Himalayas and as difficult to crush . . . that day the enemy will crumple up. A really strong-minded man makes the other party realise his strength by his very attitude. He need not shout. . . . Our job is to go on making the country stronger and making the enemy think that we shall never give in except on our terms. The country as a whole is very big. . . ." Motilal barely had the strength—or time left—to read. "Gandhiji and a thousand or so, I am sure, are quite enough to win through in this struggle—provided our own Congress brethren do not rend us. Therefore it does not matter much how many persons get tired or want to rest.. . . I am absolutely confident of the strength of the country. . . . Nothing is more fatal in a campaign than a fluid temperament and outlook. Our meeting to consider the P.M.'s statement is bound to make people think that after all there was . . . something to be considered. . . . Of course any temporising with it, like a reference to the Congress, is abject surrender . . . the government . . . will feel that we . . . are bound to break up under further strain . . . and wait for us to crack. . . . Therefore if the Working Committee has to die through government action, its last message should be 'Don't budge an inch! Carry on!' "[27]

A few days later, on January 26, 1931, Lord Irwin ordered the release of Gandhi, Jawaharlal Nehru, and the rest of the Working Committee. "My Government will impose no conditions on these releases," Viceroy Irwin announced. "Our action has been taken in pursuance of a sincere desire to assist the creation of such peaceful conditions as would enable the Government to implement the undertaking given by the Prime Minister that if civil quiet were proclaimed and assured, the Government would not be backward in response."[28]

"I have come out of jail with an absolutely open mind, unfettered by enmity, unbiased in argument and prepared to study the whole situation from every point of view and discuss the Premier's statement with Sir Tej Bahadur Sapru and other delegates on their return," Gandhi told the Associated Press upon his release. "I have no plan and no policy mapped out."[29] He went to Bombay to celebrate Independence Day there, addressing a huge crowd that welcomed him on Chaupati Beach. The next day he left for Allahabad, where Motilal met with him for the last time, and to chair the meeting of Congress's Working Committee. Motilal had summoned this wise Mahatma to counter his rashly impulsive, angry son. Sapru, Jayakar, and others were returning from London to meet again with Gandhi and the Working Committee of Congress to try to bring Congress to the second round table conference, scheduled for year's end. Gandhi looked forward to meeting them but argued that if after doing so he agreed that there was in the prime minister's declaration "sufficient ground for the Congress to tender co-operation" he still wanted to retain the right to picket foreign cloth shops and for India's "starving millions to manufacture salt."[30]

"I am going soon, Mahatmaji," Motilal told him, when Gandhi reached his bed in Allahabad, "and I shall not be here to see swaraj. But I know you have won it and will soon have it."[31] Motilal faced death as he had lived his life, "gallantly and with a smile," as his elder daughter Vijaya Lakshmi recalled. "Even when speech and breathing had become difficult and he was in great pain, he had a pleasant word for everyone."[32] In early

February his doctors insisted that their only hope of saving him was to take him to Lucknow for special X-ray treatment, unavailable in Allahabad. Although he was driven to Lucknow, it was too late for any "miracle" cure. On February 6, 1931, Motilal Nehru died in the home of his old friend, Varanasi's raja of Kalakankar, the father of independent India's Foreign Minister Raja Dinesh Singh. His body was brought back to Allahabad for cremation at the sacred confluence of the rivers Ganga and Yamuna, worshiped as Hindu mother goddesses, where they were said to meet invisible sister goddess Saraswati, giving that sacred confluence the Hindu name of Triveni (three rivers). "My position is worse than a widow's," Gandhi informed the press that day. "By a faithful life, she can appropriate the merits of her husband. . . . What I have lost through the death of Motilalji is a loss for ever: 'Rock of ages, cleft for me, let me hide myself in thee.' "[33]

"I was dazed all that day, hardly realizing what had happened," Jawaharlal wrote of his father's death. "The swift dash from Lucknow to Allahabad sitting by the body, wrapped in our national flag, and with a big flag flying above—the arrival at Allahabad, and the huge crowds that had gathered for miles to pay homage to his memory. There were some ceremonies at home, and then the last journey to the Ganga with a mighty concourse of people. As evening fell on the river bank on that winter day, the great flames leapt up and consumed that body which had meant so much to us. . . . Gandhiji said a few moving words to the multitude, and then all of us crept silently home."[34]

11

Summit in Delhi

1931

ON FEBRUARY 17, 1931, Lord Irwin invited Mahatma Gandhi to his palatial viceroy's house in Delhi, inaugurating the most remarkable summit in recent Indian history. Sapru & Jayakar had urged Gandhi to "seek an interview" with the viceroy. Jawaharlal, of course, opposed the idea but later wrote that Gandhi was "always willing to go out of his way to meet and discuss anything with his opponents."[1]

The leading members of Congress's Working Committee now moved from Allahabad to Dr. Ansari's house in Delhi, where Gandhi stayed throughout the summit. "A ceaseless stream of people" came daily to call at Dr. Ansari's house now that they knew the viceroy was ready to meet with the Mahatma. "For some years our chief contacts had been with the poor," Jawahar recalled sardonically. "The very prosperous gentlemen who came to visit Gandhiji showed us another side of human nature, and a very adaptable side, for wherever they sensed power and success, they turned to it and welcomed it with the sunshine of their smiles."[2]

"The Viceroy . . . impressed me very well," Gandhi reported, immediately after his first meeting with Irwin. The British Tories, however, were outraged by Lord Irwin's initiative. Winston Churchill found it "nauseating and humiliating" that this "half naked . . . seditious fakir" should have direct access to any viceroy. Irwin now treated Gandhi as a friend; Churchill came to think of him as little better than the devil incarnate.[3]

The second meeting between the viceroy and the Mahatma was briefer than the first and at times acerbic. When Gandhi insisted on eliminating the salt tax, Irwin asked whether he thought "any Government in the world can tolerate disobedience of its laws?" Gandhi replied that "no Government in the world can enforce all its laws." The viceroy felt the same way about altering the land tax and was equally opposed to picketing as a political weapon. Irwin wondered what Gandhi would do if he attended the next round table conference and was unsatisfied with its results. "Must you resume civil disobedience?" the viceroy asked. "I will have to unless we feel that we shall get something from the changes proposed," Gandhi answered. Irwin began to find the "whole discussion embarrassing." He knew, of course, what Churchill was shouting back home about this summit.[4]

"The Viceroy has the virtues and weaknesses of the average Englishman," Jawaharlal told Gandhi that evening as they walked together, for Nehru understood the Mahatma's disappointment.[5] The next day a third invitation came, and Gandhi went to meet Irwin at 2:45 P.M. This time the meeting lasted only half an hour, but Gandhi called it "very satisfactory." The viceroy then cabled Whitehall, reporting on his earlier talks and Gandhi's question as to whether he might raise "the right of secession from the Empire" if he agreed to attend the next round table conference in London. "I said that I presumed he could raise it, . . . I gather he did not attach much importance to the point . . . confirmed by Sapru and others."[6] Gandhi had just met with Sapru and Jayakar, who waited outside the viceroy's office to accost him and ask what had happened inside. Gandhi referred to them both as "the sub-Viceroys."[7] Irwin awaited Secretary of State Wedgewood Benn's reply to his cable about the precise limits of constitutional discussions the British cabinet might accept, hoping to launch his own constitutional conference in Delhi before month's end.

Because he had to confer with his prime minister and other colleagues, the secretary of state took more than a week to reply. On February 27, therefore, Gandhi was called again to confer with Irwin. They met for three hours. "Good night, Mr. Gandhi," the viceroy said as his friend left to march, staff in hand, five miles back to Dr. Ansari's house, "and my prayers go with you."[8] Irwin had urged Gandhi to exert his influence "on the side of peace," to which the Mahatma replied that he "desired peace, but that when in doubt he fell back on 'Lead Kindly Light' and 'One step enough for me.' "[9] They exchanged notes, putting their respective positions in written form. On March 1 Gandhi returned, and Irwin informed him immediately that he could not agree to picketing and boycott. The viceroy then invited in his home secretary to explain why. "After rather a wandering discussion," Gandhi agreed to give up "boycott as a political weapon" and promised there would be "no element of coercion." Next they had a long talk about picketing, and again they restated their positions and moved closer to agreement. The same thing happened on the question of public inquiries into police excess. They also discussed salt, making progress on that point as well. By the end of a long evening they had reached a provisional settlement that cheered Gandhi and most of the Working Committee but deeply depressed Jawaharlal. The next day Gandhi found him "lonely and almost uninterested," so he wrote to young Nehru, "My strength depends upon you. I want your active support . . . criticise, alter, amend, reject."[10]

"I recognize that, in many respects, the terms of settlement are honourable," Jawaharlal wrote to Gandhi that evening, "But I feel that there is . . . a certain limitation of our ideal as laid down at Lahore . . . a limitation of our freedom in regard to defence, external affairs, finances and the public debt." Gandhi returned on the next two afternoons, each of those discussions yielding concessions from the viceroy and modifications of the language, if not the actual terms, of their agreement. Finally, on March 5, 1931, a truce was signed by Gandhi and Irwin at the viceroy's house. Irwin suggested they drink to it and each other's health with tea. Gandhi, however, asked for plain water and lemon for his toast, to which he gleefully added a pinch of the salt he had picked up at Dandi Beach, and carried under his dhoti. It was the first untaxed salt consumed at the viceroy's house.[11]

Gandhi told the press that day that some would be "keenly disappointed" by the truce he and Irwin reached. "Heroic suffering is like the breath of their nostrils. They rejoice in it as in nothing else. . . . [W]hen suffering ceases they feel their occupation gone. . . . Suffering can be both wise and unwise, and when the limit is reached, to prolong it would be not unwise but the height of folly." He was speaking now to Jawaharlal, who reacted so negatively to the truce and draft of the settlement that Gandhi showed him that he had almost resigned from Congress.[12]

"He [Gandhi] came back about 2 A.M., and we were woken up and told that an agreement had been reached," Nehru wrote. "We saw the draft . . . a tremendous shock. I was wholly unprepared for it. I said nothing . . . our leader had committed himself . . . what could we do? Throw him over? Break from him? . . . That might bring some personal satisfaction . . . but it made no difference to the final decision. The Civil Disobedience movement was ended. . . . Was it for this that our people had behaved so gallantly for a year? Were all our brave words and deeds to end in this? . . . [I]n my heart there was a great emptiness as of something precious gone, almost beyond recall. 'This is the way the world ends, Not with a bang, but a whimper.' " Since publication of *The Waste Land* in 1922, Nehru had admired T. S. Eliot's poetry, as his borrowing the last lines of Eliot's "Hollow Men," reveals, showing how depressed and betrayed he felt by the Gandhi–Irwin truce.[13]

Congress had embarked deliberately, on "a career of co-operation," as

Gandhi called it in reporting the truce on March 5. "If Congressmen honourably and fully implement the conditions applicable to them . . . I promise that it will not be long before every one of these political prisoners is discharged."[14] It was a bold and hopeful promise. But if Jawaharlal and his young comrades were so deeply depressed by the peace their leader had won for them, British India's police and the sun-baked "steel-frame" of the Indian Civil Service (I.C.S.) felt not only betrayed but angry at what the viceroy had done to them. They raged and swore they would never surrender a single prisoner to that "seditious little bania-fakir!" So even though Gandhi promised his colleagues and pleaded with the viceroy to save the lives of Bhagat Singh and other Indian terrorist-heroes, he could not obtain their release. *Shaheed* ('Martyr') Bhagat Singh, who had thrown a bomb onto the floor of Delhi's Legislative Assembly from its gallery in 1929, was hanged with two other nationalist "terrorists" in Lahore one day before Congress met again in Karachi in late March 1931.

"The execution of Bhagat Singh and his comrades is a reminder of our inability to protect brave and patriotic young men," an outraged Jawahar announced to the All-India Students Convention in Karachi. "I should like the youth to remember that no living country can accept a settlement which is anything less than complete independence. . . . [T]he struggle will have to be continued. . . . The time for shouting slogans has passed, and you must act, for action alone counts."[15]

Gandhi, however, remained committed to the truce and urged an end to all violence. "We must not put ourselves in the wrong by being angry," the Mahatma insisted. But Kanpur had already exploded, caught in a frenzy of rioting that left 166 dead and 480 injured. Hindus and Muslims looted each other's shops and killed and burned one another while police watched silently, abstaining from any interference merely to restore order.[16]

"We have become so sad and frustrated that we do not know what to do," Jawaharlal cried out as news of the Kanpur tragedy reached him. "Have we all gone mad?"[17] His own frustration was compounded not only by the Gandhi–Irwin truce but also by the Working Committee's choice of Vallabhbhai Patel to serve in Karachi as the Congress's president, with not one voice raised to urge Nehru to continue in that position. He knew that none of those older heads trusted him, fearing he was much too communist. He felt painfully alone again, abandoned by his friends at both ends of the international political spectrum.

Gandhi insisted on going to London as the sole representative of Congress, knowing that if there were two or more Congress delegates, Jawaharlal would have to accompany him. And even though Vallabhbhai might serve as president of Congress for a year, his rural manner—not to speak of his halting English—would hardly prove an asset at the round table conference. Gandhi knew how disaffected Nehru felt but urged him to introduce a Congress resolution endorsing his Settlement with Irwin. "I hesitated," Jawahar remembered. "It went against the grain, and I refused at first, and then this seemed a weak and unsatisfactory position . . . not proper to prevaricate or leave people guessing . . . a few minutes before the resolution was taken up in the open Congress, I decided to sponsor it." Still his heart remained opposed, as his speech on the resolution reflected.[18]

"A year ago we unfurled the flag of complete independence at Lahore, and to achieve that end also fought a battle during the last eleven months. . . . I am now going to move a resolution which may create some doubt in your minds. . . . We took a vow at Lahore that we would not have any relations with the British. When the truce talks were going on in Delhi, our hearts were full of pain. I wondered, and you too would have wondered, whether we were not going against our pledged word. After the glorious year that had passed, should we now agree to such a half-way house? We thought deeply over this and concluded that we would compel the government to give us complete independence, and if we did not get it, we would fight."[19]

It was hardly the sort of support Gandhi desired, but it was the most Jawaharlal could offer, for the mere idea of any truce with the British imperialists was still anathema to him, and the prospect of peace or dominion status intolerable. "We cannot go to the Round Table Conference unless we are able to state our chief aim there . . . [to] place the Lahore resolution before the conference." He thus put Gandhi on notice, and clearly alerted the British to the fact, that he at least would not rest content with half a loaf of freedom or less. He still hoped and liked to believe that he spoke for "Young India," but Gandhi himself had recently reminded him that *Young India* was the name of his own weekly newsletter. Furthermore, throughout India Gandhi remained the most popular hero as well as the Mahatma, and Vallabhbhai was his second at the Karachi Congress and now seemed destined to be his heir, even as Jawahar had once inherited his father's throne and crown. But Motilal Nehru was dead.

To mollify Jawaharlal in Karachi, Gandhi agreed to support a resolution on fundamental rights, which Nehru drafted and introduced, and which committed Congress to a comprehensive program of social and economic reforms, "to end the exploitation of the masses." Congress would only accept a constitution according to Nehru's resolution, which promised "a living wage for industrial workers; limited hours of labour; healthy conditions of work," and "protection against the economic consequences of old age, sickness and unemployment." His resolution also called for an end to "serfdom," the "protection of women workers," the "prohibition" of child labor in factories, a substantial "reduction of land revenue and rent," free primary education, universal adult suffrage, no duty on salt, the prohibition of intoxicating drinks and drugs, and the state control of "key industries and mineral resources," among other things. It was Nehru's first blueprint for the many five-year plans he insisted on introducing when he came to power, as well as for a number of promises to be included in the Republic of India's constitution.[20]

"After the Congress meeting Bhai desperately needed a rest," sister Betty remembered.[21] "I had a little breakdown in health soon after the Delhi Pact," Jawaharlal wrote of what was in fact so severe a depression that he decided to leave India for a sea voyage to Ceylon with Kamala and Indira.[22] "We sleep a lot!" he wrote Nan that April. "I feel better already after two days of sea. I am sure I shall be fit as a fiddle after three or four weeks. All I want is sleep and a different atmosphere."[23] Kamala was very weak, of course, but three days in Ceylon, after their three days at sea seemed, Jawahar thought, to help both of them "a lot." The family holiday lasted a month, and Jawahar returned home feeling better, buoyed by the warmth of his popular reception in Sri Lanka. "Lanka is an enchanted place," he wrote Betty from the ship headed back from Colombo to India. "One night an amazing thing happened to me. I fell off my bed and woke up on the floor with my face all covered with blood. . . . I woke up Kamala, who was in the next room, and as she was washing my head, she fainted away and I had to carry her and look after her. The memory of the anonymous letters I had received in Allahabad [two threatening his life] had come to her and the sight of blood. . . . Hari [Motilal's old personal servant, Harilal, now served Jawaharlal] was then found and washed me, and Kamala later dressed my head. There was just a small cut on the head about 3/4 of an inch long. . . . I must have got it . . . on the hard floor."[24]

While the Nehrus were abroad, Lord Irwin was replaced by Lord Willingdon, a former governor of Bombay. Willingdon was much tougher than Irwin and soon launched a half decade of repression such as India had not known since the turn of the century, when Curzon ruled with an equally iron hand and rod. The Great Depression and growing unemployment at home gave Britain's Conservatives more popular approval for their harder line in India. Lancashire cloth stuffed warehouses in Bombay, Madras, and Calcutta, British woolens rotting in the humid heat or consumed by vermin, thanks to Gandhi's boycott. Economically, 1931 was the worst year in history for India to expect British sympathy for its claims for greater freedom, financial as well as political. In August 1931 the financial crisis in Britain proved so severe that Labour's prime minister was forced to invite more Tories into his national government cabinet, thereby dooming any possibility of Gandhi's winning Purna Swaraj in London.

On the eve of his departure for London it actually seemed as if Gandhi's trip would be aborted. Nothing would have pleased Jawaharlal more, of course, as he stood ready to "resume the fight," writing Fenner Brockway in late August that it "seems highly unlikely" that Gandhi would venture west.[25] Both sides had broken the truce a thousand times, each accusing the other of bad faith. Communal rioting spread, as did terrorist and official violence. Sikhs of the Punjab, as well as Muslims, now wanted their fair share of the political pie about to be divided, they thought, in London. India's princes wanted to keep their sovereignty, which they now enjoyed thanks to Britain's imperial sovereignty over 570 princely states interspersed among eleven provinces of British India and other centrally administered regions and frontier zones. Nonetheless, Gandhi wanted to go, to see for himself how much His Majesty's Government was prepared to offer India. He had not, after all, been back to London since the start of World War I. So he steamed out of Bombay aboard the S.S. *Rajputana* on August 29, 1931. Sarojini Naidu joined his entourage from the ashram, which included Admiral Slade's daughter, Madeleine, whom Gandhi renamed Mirabehn (Sister Mira), and west India's most powerful Hindu mill owner, G. D. Birla, in whose garden in Delhi the Mahatma was assassinated little more than a decade and a half later.

Gandhi was in high spirits and good humor aboard the ship, keeping his ashram routine as much as possible: prayers on deck every morning at 4 A.M. Prayer "saved my life," Gandhi insisted. "Without it, I should have been a lunatic long ago."[26]

"India seems to be a duller and an emptier place without you," Jawaharlal wrote him from Allahabad. Kamala was "ill" with high fever, "probably malarial," "extraordinarily fragile and weak."[27] He still pretended to be unaware that her tubercular fever would prove fatal. Indira was now living in Poona, with the founder-directors of the "Pupils' Own School," where she remained until 1934, when she moved on to Rabindranath Tagore's private school, Santiniketan, in Bengal. Nehru subscribed to *Punch* as well as several other English newspapers to read about Gandhi's adventures in England. His heart and mind were there throughout the round table conference. Although he considered it doomed from the start, he would have loved nothing better than to be there.

"I grow impatient," Nehru wrote Gandhi again before the end of September. "You have hardly been in England a fortnight and I complain of you not having got the British Government . . . to discuss vital matters of principle. . . . How wonderful you are to argue and argue and yet again argue with this motley crowd. I wonder if any purgatory would be more dreadful for me than to carry on in this way. If I had to listen to my dear friend Mohamad Ali Jinnah talking the most unmitigated nonsense about his 14 points for any length of time, I would have to consider the desirability of retiring to the South Sea Islands, where there would be some hope of meeting with some people who were intelligent enough or ignorant enough not to talk of the 14 points."[28] Jinnah was living in Hampstead now and rented chambers at the Inner Temple, whose treasurer, Sir John Simon, was "very glad to have so distinguished a man within our own boundaries."[29] He had emerged as the most effective leader among the Muslims at the round table in London, moreover, and Nehru's undisguised contempt for him was unfortunate, the tragedy of its political magnification and reciprocation later to prove irreversible.

To Dr. Ansari Jawaharlal wrote, "There seems to be too much sugariness in the proceedings at least so far as our side is concerned. That of course is Gandhiji's way and we must not complain. But a little pepper would add to the taste. How I would love to have a nice little chat with the London crowd of Round Tablers."[30] Nehru now spent more time on agrarian problems in the United Provinces, visiting villagers in his district and preparing them for a resumption of the struggle, which by late September he believed imminent. "The truce was but a phase of the fight, a change of tactics," he announced in Delhi. "We will have to continue the struggle till we have won complete freedom for our motherland."[31]

Jaya Prakash Narayan (J.P.) (1902–1979) came to stay with Nehru in Allahabad at this time. The brilliant young socialist had been working as Birla's secretary in Bombay but was unhappy in the employ of that industrial baron, Jawaharlal told Gandhi. J.P. became Nehru's disciple and they remained friends, ideologically attuned and in close touch as well, each an unhappy introvert forced into the dusty fray of politics and labor organizing by public life. Another later equally famous colleague of Nehru's at this time was Lal Bahadur Shastri (1904–1966), who assisted him now as secretary of the United Provinces Congress Committee and remained his political right arm in mobilizing rural areas of his district and of the entire United Provinces. Shastri held many important posts in Nehru's cabinets, succeeding him, but only briefly, as Delhi's premier.[32]

While Gandhi tried without success to convert hostile English and indifferent or antipathetical round table Indians to Congress's radical platform, Jawahar grew too restless to wait for him to return home before launching his own agrarian Satyagraha in the United Provinces. On behalf of his Allahabad District Congress Committee, Nehru notified the chief secretary, K. J. Prasad, that it might be "necessary" to advise tenant farmers in his district to "withhold payment of rent" in mid-October. "I regret greatly that such a step should be contemplated . . . when Gandhiji is attending the Round Table Conference. But the question of payment or non-payment is an urgent issue and cannot await the deliberations of the Round Table Conference." He meant, of course, that he could not wait, even though Gandhi had asked him to before leaving and Vallabhbhai had repeated that request to be more patient and wait until their Mahatma came home before commencing a no-tax campaign and full-scale conflict against the Raj.[33]

"My dear Bapu," Jawaharlal wrote the next day, "I am afraid I will add to your worries. . . . It is really deplorable to what a pass we have reduced the tenants. . . . It is clear that any step the Allahabad district might take would have its effect on adjoining districts which are equally hard hit. The matter may become provincial. . . . I do not know how this will affect your work in London. So far as I can see the Round Table Conference is as dead as a door nail and it does not matter much."[34] Gandhi turned sixty-two that October, as general elections brought more Tories to power in Westminster, but Ramsay MacDonald remained the prime minister of a second national coalition government.

Gandhi had planned a tour of the Continent on his way home but canceled it in November when he received word from both Vallabhbhai and Jawaharlal that the situation had deteriorated so badly in Bengal and the United Provinces and on the frontier that his presence was urgently needed. "I do not want to revive civil disobedience," Gandhi told the round table, still hoping for a permanent truce with the British government and people, whose hearts he had won in great measure and whose friendship he longed to retain, asking his fellow conferees, "For heaven's sake, give me, a frail man sixty-two years gone, a little bit of a chance."[35]

Chief Secretary Prasad, though suffering from "indisposition," phoned Nehru and offered to meet him and explain the government's position on the land tax. Jawaharlal felt even more frustrated, for as he told Vallabhbhai: "Peace we can understand and war we are not unused to. But this is neither peace nor war. It is an impossible situation."[36] By November *goondas* (hooligans) were being hired by the government or the larger taluqdars (landlords) of the United Provinces to "break up Congress meetings," Nehru charged. "They used drums . . . forcibly pushed about people including speakers and the chairman . . . even tried to gag them. . . . Our position is really difficult on account of the Delhi Settlement."[37] Addressing a peasants (*kisans*) conference in Allahabad on November 25, 1931, Nehru insisted, "If we are to die, we shall die after a fight. I do not want a half-way settlement or a half-way fight. . . . I feel that if the fire of satyagraha is fully lit up the kisans' miseries would disappear in that fire and the kisans and the country would then be free."[38] Young J.P. cheered Jawahar with the others at that conference. Forty-four years later when J.P. gave similar advice to peasants and workers in Bihar, Indira, Nehru's daughter, had him thrown into prison.

On November 28, Nehru wrote Willingdon's private secretary, Eric Mieville—later to return as Lord Mountbatten's principal secretary—"I am sorry to inform you that all our efforts to secure an honourable compromise for the unfortunate tenantry have failed . . . and we have been compelled to advise the peasantry in Allahabad district to withhold payment of rent and revenue till relief is obtained."[39] He knew now that he might be arrested at any time. He hoped to welcome Gandhi home in Bombay that December, taking Kamala there for more treatment for her constant fever on December 12. But Gandhi did not reach Bombay until December 28, 1931, so Jawahar returned to Allahabad alone and headed west again on December 26, when his train was stopped. "A Black Maria waited by the railway line . . . I mounted this closed prisoners' van and . . . bumped away to Naini. The Superintendent of Police, an Englishman, . . . looked glum and unhappy. I . . . had spoiled his Christmas. And so to prison! 'Absent thee from felicity a while, And for a season draw thy breath in pain.' "[40]

12

Rigorous Imprisonment

1932–1933

NEHRU WAS SENTENCED to two years' "rigorous imprisonment" on January 4, 1932, for inciting peasants not to pay land taxes. He was also fined Rs. 500, in default of which payment he would have to remain another six months behind bars.

"I should like to say how proud I am of my kisan [peasant] brethren. . . . They have lighted a fire which will spread and burn brightly till it has consumed and reduced to ashes the British domination and exploitation of this country, which has sucked the life-blood out of them . . . and reduced them almost to the level of the beasts of the field."[1]

Gandhi and Vallabhbhai were arrested the morning of Nehru's trial. The Working Committee had met before year's end to pass a resolution calling for the resumption of civil disobedience unless the viceroy agreed to meet with the Mahatma to discuss possible ways of extending the truce. Willingdon was, however, unwilling to entertain the "half-naked fakir" as Irwin had done. The crowds returned to city streets en masse. Police and troops were waiting, firing live ammunition as well as lashing out with sticks and batons. Two protesters were beaten to death that day in Allahabad alone. "So we can settle down for a while!" Jawaharlal wrote in his prison diary.[2]

Nan and Betty organized a huge rally in Allahabad on January 26, 1932, to retake the pledge of Purna Swaraj first articulated in Lahore. Nan was the director of the Allahabad City Congress Committee. The pledge was taken solemnly and the meeting peacefully dispersed. "That night,

Betty . . . and I," Nan recalled, "were arrested and taken to the district jail . . . to the 'female' section . . . inmates of this jail were mostly prostitutes."[3] Nan and Betty were each sentenced to one year's rigorous imprisonment. . . . The Nehru sisters were, of course, Class A prisoners and were sent to the female ward of the Central Prison in Lucknow. Jawaharlal was "glad to hear of their arrest. They have worked hard and deserve rest."[4]

"It was one thing to be a cheerful martyr in the high excitement of being arrested for love of your country," Betty wrote less cheerfully, "and quite another to spend night after night in a filthy cell full of insects crawling about the grim walls. For several nights we could not sleep, lying there imagining that some slimy thing would crawl over our bodies; nor was it all imagination—they did."[5]

In February Jawaharlal was moved to a new prison in Bareilly. "I had a bit of a shock when I entered the ward for 'A' class prisoners. . . . The level was at least six feet lower than the general jail level . . . I suppose I shall get used to the new conditions soon enough. I am likely to be much more cut off from the outside world here . . . I am far from Allahabad and Anand Bhawan . . . a little depressing."[6] Nor had his health been good. He had run a fever for almost a week after his arrest and feared his lungs might be weakening. "The sun is a late comer to our ward and barrack. In the wider world outside, they say it rises before seven. It reaches us after half past eight. . . . For the first time after many years—since Nabha—I was locked in my barrack at night."[7] But summer came early that year. By mid-March, Jawaharlal noted, "I wage ceaseless war against mosquitoes, bugs etc. . . . The sight of a bed bug in my bed galvanized me into activity. . . . The bed had a bath of boiling water with potassium permanganate. Also Flit . . . Flit is sprayed in the barrack at least twice a day and thousands of mosquitoes are slaughtered daily. Still they come. . . . Father has been very much in my thoughts. His picture seems to haunt me, and there is an emptiness in the heart."[8]

He read Andre Maurois' *Byron,* "strangely fascinated" by Byron's life and the "links" he felt thanks to their common Harrow and Trinity Cambridge experiences. "Disgust and admiration," were Nehru's twin reactions to Byron's life, "with disgust preponderating," or so he told his diary.[9] This time prison got on his nerves much more than before. "I have been bored, fed up, angry with almost everything and everybody—with my companions who often get on my nerves, with the country in general for not being as aggressive and active as it might, with Sarojini for having gone and stayed with Fazl-i-Husain in Delhi, with the British Empire . . . above all with myself. . . . I have been getting a substantial rise in temperature daily for ten days . . . So I must give up even my gentle evening walks. As for running this is not to be thought of . . . I must get used to living like an invalid. This idea has obsessed me and oppressed me." Next day, for no particular reason, he felt manic. "Heigh-ho! I feel better to-day—But, oh, the contempt in my mind & heart for the chicken-hearted animal called the Indian moderate—the Sivaswamy Iyer breed! It will not be necessary to punish them for their poltroonery and betrayal. They will give up the ghost themselves when their soul-mates and protectors— the British—have to depart from India. But they are a lowly breed."[10]

By April Jawaharlal's temperature rose higher, and a pain in his side that he had felt a year earlier returned to worry him. "Apart from this there is no other trouble, except a slight feeling of lassitude," he wrote their trusted family physician Dr. B. C. Roy of Calcutta, who had taken care of Kamala's "appendix" surgery. "I have absolutely no cough. My bowels move regularly. My weight has remained steady at about 126 lbs. . . . The superintendent of the jail . . . examined me several times . . . today he suspected a little patch in the right side where I have the pain. He has prescribed iodine . . . and iodex massage. . . . [He] also seemed to think that perhaps something might be wrong with the base of one of my teeth. . . . About a year ago I had all my teeth x-rayed in Bombay and as a result one was pulled out. Since then I have frequently visited the dentist to have an artificial tooth made for the gap . . . I have had no pain . . . I am not worrying in the least about myself but I would like you very much to examine me. My mother and Kamala would feel greatly relieved if you could do so." The "touch of dry pleurisy" he had did not leave him a permanent invalid, but a few days later his mother marched in a parade to celebrate National Week and was knocked down and beaten with lathis.[11]

"After the first shock and alarm I have felt greatly elated at the news from Allahabad," Jawaharlal told his diary, "specially at the part mother took . . . I was thrilled!"[12] Years later in his autobiography, Nehru wrote that "the thought of my frail old mother lying bleeding on the dusty road obsessed me, and I wondered how . . . far would my non-violence have carried me. Not very far, I fear."[13] Betty was "frantic" when she heard of what had happened to their mother. "Had I been there I would not have thought of non-violence at all," she wrote, "it is not in my nature."[14]

By early May, when his mother could visit him, Jawahar found her "full of beans and so proud of having received canes and lathis."[15] Kamala was not well, however, and he urged her to return to Bombay for more antitubercular treatment there. The home member of the United Provinces Governor's Council, the nawab of Chhatari, had visited Nehru in Bareilly and agreed to allow Dr. Roy to see him as well, suggesting it might be best for so important a prisoner's health to transfer him to Dehra Dun's more salubrious climate. "Personally I am not keen on the transfer," he wrote Nan and Betty, advising them to "keep fit . . . like the string of the violin—neither too tight, nor too loose—just right to play the music of life."[16] A few weeks later he was growing impatient to be transferred to the foothills of the Himalayas. "Waiting, waiting to be sent to Dehra! It is annoying to be kept waiting this way."[17] Bareilly was 112 degrees in the shade. A week later he was moved.

"We are settling down in our new quarters," he wrote his diary on Wednesday, June 8, 1932, from Dehra Dun. "The whole jail staff seems to be eager to make us comfortable. . . . All manner of rickety furniture from a second hand shop is coming on hire. . . . It hardly feels like being in prison here."[18]

On June 13 Kamala came to visit. "I was not as pleasant as I might have been during the interview. News of the weakness of this person, the caution of the other . . . irritates and I forget that the fortnightly interview is a precious thing which must not be spoilt by irritation. Poor little girl! She is having a hard time outside. And instead of cheering her up I criticise everything."[19] Seeing Kamala always reminded him of his greatest weakness, his failure, not only as a husband, but also as a man, incapable of standing firm and, saying no to his father and tradition. Yet after all he had done to her, all he had put her through, Kamala now emerged stronger in some respects, certainly in the purity of the chaste and saintly life she had chosen, with her swami's help, than he had, would, or could ever become. She had even found herself an "adopted" son, a most devoted disciple, in Feroze Gandhi, who seemed so obnoxious to him but would clearly do anything for her. She seemed so frail, always coughing. "I found Kamalaji an inspired soul," Swami Sivananda told his Ramakrishna Mission successor, Swami Abhayananda, "she was a fit receptacle for higher spiritual truths and great was my satisfaction in offering such a pure soul at the lotus feet of the Lord."[20] Soon after that the old swami suffered a stroke and died, and Kamala wrote to her new spiritual guide, his successor, "I long to be with God but we are not given eyes to see Him. . . . I have had a persistent cough."[21]

"So you did not know what to write to me," Nehru chided daughter "Indu darling" in mid-June, "and the fortnight went by without any letter. Even mummie did not write. I hope you will make up for it now. . . . [F]ind a little time once a fortnight and tell me of the books you are reading, and what else you are doing in Poona. . . . As I listen to the *loo* blowing, I think of the Harrow song: 'What a tune, Kind June, you are singing, all the Noon!' But the tune is not a very pleasant one in the Indian plains and one gets rather fed up with it. . . . All my love and kisses, Your loving Papu."[22]

Even as he was writing that letter, Kamala and his mother were waiting outside his prison gate for an interview, which was denied to them by the higher authorities for another month. "The little gods at Naini Tal have decreed that I am not to have an interview with mother or Kamala for a month! . . . For ten days mother & Kamala have waited here . . . now they are told to go away. . . . I was put out a little . . . this impertinence and insult to mother and K hurt. I do not think I shall have any more interviews in future. . . . If interviews mean insults for mother & K there will be no more of them." Nehru could be as hot tempered as he was cold blooded. Unfortunately, he had not thought through the implications of what he decided to do to "punish" the prison authorities, cutting off his most important line of communication with his own family and the world outside the bleak concrete wall of his compound.[23]

"I am settling down again to a peaceful existence," he told his diary a month later. "No interview is conducive to peace! . . . [B]ut I have had a few shocks lately . . . bad news from Kamala. She has again been very ill."[24] Two weeks later, "I feel angry again. . . . Govt. have revised their own sentence . . . about stoppage of interviews. Originally it was for a month—Now . . . for three months and correspondence also banned! Heigh-ho! What disgusting meanness. Perhaps this enhanced punishment is due to my writing and making a fuss."[25]

In mid-August, "Kamala has been very ill again. . . . Poor girl! She frets because she is ill just when she ought to be up and doing."[26] He had started teaching Hindi to one of the convict "overseers" in Dehra Dun but confessed, "How hopeless I am as a teacher—utterly impatient and intolerant. I shout at the poor chap and frighten him out of his wits because he makes silly mistakes!"[27]

"Heigh-ho! What a life! The same round day after day, and nothing to distinguish one day from another. An occasional meal missed . . . almost a remarkable event. And books, books—what would one do without them to escape from ennui and depression? Today is Nan's birthday—I thought of her and of the day 32 years ago when she was born. And . . . myself then, ten years of age, and vastly excited at the idea of having got a sister! How important I felt—I remember how some stupid person—one of the doctors in attendance, I think—in telling me of the new sister suggested that I should rejoice at not having a brother to share my inheritance! . . . How well I remember the scene—It was in the old Anand Bhawan and I was in the western verandah of father's old rooms. The church clock in the distance marked a quarter to six in the evening—but I am not sure if it was functioning or not. It often stopped . . . Inheritance—Anand Bhawan—The police sit in Anand Bhawan that was, and is Swaraj Bhawan now. And I sit in Dehra Dun jail and Nan in Lucknow Central Prison. And life rolls on!" Moments later he tried to read the newspaper but found that "I did not see anything." To "my surprise, quite unawares, my eyes were full of tears."[28]

That same mid-August day in Poona's prison, Mahatma Gandhi decided that he would start a "fast unto death" against the British government's Communal Award announcement that separate electorate seats would be reserved in all future councils for India's "depressed classes" or "untouchables." Dr. B. R. Ambedkar, west India's brilliant leader of one untouchable Hindu community, who had studied law at Lincoln's Inn and Columbia University, had attended the round table conferences and spoken so eloquently on behalf of his community that Ramsay MacDonald agreed to add it to the growing list of separately elected communities: Muslims, Sikhs, Christians, Anglo-Indians, and now the depressed classes. Gandhi had warned before leaving London that he would "resist with my life" any such attempt to "vivisect" the body of Hinduism, for he argued that untouchables were as much Hindus as Brahmans. Indeed, he spoke of himself as an untouchable and devoted much of his time and attention in speech as well as writing to that community's impoverished "uplift" and equal access to Hindu temples. In March he had written the new secretary of state for India, Sir Samuel Hoare, to warn him against further "divide and rule" tactics by offering a new batch of specially reserved seats to the depressed classes, but Hoare haughtily commented, "The dogs bark, the caravan moves on."

"Dear Friend," Gandhi now wrote to Prime Minister MacDonald, "There can be no doubt that Sir Samuel Hoare has showed you . . . my letter to him of 11th March on the question of . . . 'depressed' classes. . . . I have read the British Government's decision on the representation of minorities and have slept over it. . . . I have to resist your decision with my life . . . by declaring a perpetual fast unto death from food of any kind save water . . . from the noon of 20th September next, unless the said decision is meanwhile revised. . . . It may be that my judgment is warped and that I am wholly in error in regarding separate electorates for the 'depressed' classes as harmful to them or to Hinduism. . . . In that case my death by fasting will be at once a penance for my error and a lifting of a weight from off those numberless men and women who have childlike faith in my wisdom. Whereas if my judgment is right, as I have little doubt it is, the contemplated step is but the due fulfillment of the scheme of life, which I have tried for more than a quarter of a century."[29]

Ramsay MacDonald wired his response in early September, appealing to Gandhi not to take so drastic a step. Gandhi insisted, however, that "for me this matter is one of pure religion." He wrote again to explain, "In establishment of a separate electorate at all for 'Depressed' Classes I sense the injection of a poison that is calculated to destroy Hinduism and do no good whatsoever to 'Depressed' Classes. . . . I should not be against even over-representation of 'Depressed' Classes. What I am against is their statutory separation, even in a limited form, from Hindu fold."[30] A few days after this exchange, Gandhi's decision to fast was released to the press with the correspondence exchanged between himself and the prime minister.

"What a capacity Bapu has got for giving shocks to people!" Nehru wrote in his prison diary. "A week ago . . . I read about his decision. . . . For two days nearly I was in darkness with no light to show the way out . . . I thought with anguish that I might not see him again. . . . And then I felt annoyed with him for choosing a side issue for his final sacrifice—just a question of electorate. What would be the result on our freedom movement? Would not the larger issues fade into the background? . . . And I felt angry with Bapu at his religious and sentimental approach to a political question. Was he entitled to coerce people in this way? What would happen to this country if the practice spread? And his frequent references to God—God has made him do this—God even indicated the date of the fast. . . . What a terrible example to set! And I thought of Kamala and mother, feeling sure that they would have a terrible shock if Bapu died . . . If Bapu died! What would India be like then? And how will her politics run? . . . [D]espair seized my heart when I thought of it . . . confusion reigned in my head and anger and despair and love for him who was the cause of this upheaval. I hardly knew what to do and I was irritable and short-tempered with everybody. . . . And then—a strange enough thing for me! —I had quite an emotional crisis and I cried and wept. Somehow the tears took away a great deal of my confusion and worry. I felt calmer and the future seemed not so dark. Bapu had a curious knack of doing the right thing at the psychological moment and it might be that his action—impossible to justify as it was from my point of view—might lead to great results. . . . And even if Bapu died; the great fight will go on. So whatever happened one must keep ready and fit for it instead of behaving like a foolish lovesick girl."[31]

"The fast which I am approaching was resolved upon in the name of God for His work . . . at His call," Gandhi told the press on September 16, 1932. "It is intended to sting the Hindu conscience into right religious action . . . to throw the whole of my weight (such as it is) in the scales of justice pure and simple. . . . The separate electorate is merely the last straw. . . . If the Hindu mass mind is not yet prepared to banish untouchability root and branch, it must sacrifice me without the slightest hesitation."[32] Kamala "fainted off immediately" when she heard of Gandhi's decision to fast unto death.[33] Rabindranath Tagore wired from Shantiniketan, "It is worth sacrificing precious life for the sake of India's unity and her social integrity. . . . Our sorrowing hearts will follow your sublime penance with reverence and love."[34]

As Gandhi started to fast on September 20, a conference of the leaders of all communities was convened in Poona to consider the MacDonald award and the Mahatma's fast, both coming in the same week. Gandhi remained in his prison cell, even though the government had offered to release him to confer with Ambedkar and others in Poona during that soul-searching week when Hinduism probed its heart for ways of removing its meanest, most ancient social stigma. "I am an irrepressible optimist," Gandhi told the reporters who filled his prison "temple" cell that first evening. "Unless God has forsaken me, I hope that it will not be a fast unto death." Although not a Brahman by birth, Gandhi's family were high-caste Modh Bania Hindus, yet now he spoke of himself as "untouchable by choice." "I believe that if untouchability is really rooted out, it will not only purge Hinduism of a terrible blot but its repercussions will be world-wide. My fight against untouchability is a fight against the impure in humanity."[35]

On September 22, Dr. Ambedkar came to his cell and bluntly told Gandhi, "I want political power for my community. That is indispensable for our survival. The basis of the agreement therefore should be: I should get what is due to me . . . I should be assured of my compensation." Gandhi softly replied: "You have clarified your position very beautifully . . . I want to serve the untouchables. That is why I am not at all angry with you . . . I will not get angry even if you spit on my face . . . I know that you have drunk deep of the poisoned cup. However, I make a claim which will seem astounding to you. You are born an untouchable but I am an untouchable by adoption. And as a new convert I feel more for the welfare of the community than those who are already there."[36]

At 6 P.M. on September 24, an agreement was reached in Poona between Ambedkar and the other leaders of the depressed classes and the leaders of the Rest of the Hindu community on the exact number of seats, which would hereafter voluntarily be reserved by Congress for the depressed classes in each provincial legislature as well as at the center. Gandhi's youngest son, Devadas, had been sent by his father to meet with the leaders and signed the agreement next day in Bombay, as did many others, starting with Malaviya, Sapru, Jayakar, Ambedkar, and Birla. Krishna Menon came to Poona to interview Gandhi for the India League of London. "The settlement arrived at is to me but the beginning of the work of purification. The agony of the soul is not going to end until every trace of untouchability is gone," Gandhi told him. "I would like finally to assure Britain that so long as life lasts in me, I shall undergo as many fasts as are necessary in order to purify Hinduism of this unbearable taint."[37] At 5:15 P.M. on Monday, September 26, after he had been assured that Great Britain's cabinet would accept the agreement reached among the Hindu leaders, Gandhi broke his fast. "The settlement arrived at is," Gandhi observed, "a generous gesture . . . a meeting of hearts, and my Hindu gratitude is due to Dr. Ambedkar."[38] Ambedkar became Nehru's first minister of justice and chaired the drafting committee for independent India's constitution, which abolished untouchability.

"The fast is broken!" Nehru noted in his prison diary. "I am so glad mother went to Poona."[39] Indira had also gone with Kamala to visit Gandhi the day before he broke his fast. "He looked very weak," Indira recalled. "But he managed to notice that I had put on some weight."[40] "Bapu is an extraordinary man and it is very difficult to understand him," Nehru wrote his daughter the next week. "But then great men are always difficult to understand. . . . It is amazing how he conquers his opponents by his love and sacrifice. By his fast he has changed the face of India and killed untouchability at a blow."[41] Indira first met Tagore in Gandhi's cell and later went to study Manipuri dance and history at his Creative College in Bengal. Indira now "fills my mind," Nehru told his diary, and "I have begun to think of her future. She will be fifteen soon."[42]

A week later he recorded "an extraordinary and annoying experience." He had gone to bed at about 9:15 P.M., and "I had just dozed off when I seemed to have had an uncomfortable feeling which is difficult to describe. I got up and tried to light the hurricane lantern. I lit it and just then I seem to have fainted off with remarkable suddenness. Anyway my next recollection was a taste of grit in the mouth. I was lying on the floor with my mouth on the coir mattress. I got up with some difficulty and managed to lie down on the bed. But for some considerable time I felt queer and light-headed . . . I lay awake for hours—till almost 3 A.M. . . . It was very curious—the whole thing. I suddenly felt as if I was taking gas or chloroform and I went off. One step back would have brought me to the bed. I had no realisation of the coming fall. I simply must have collapsed rather heavily for I hurt my temple pretty badly and there were also bruises on my arm and shins. . . . Why in thunder should I misbehave in this way? Immediately my mind went back to a year and five months ago when I performed the same trick in Colombo, only with worse consequences . . . I thought then that I was overworked and overwrought. But here, in Dehra Dun jail, there is no question of overwork—or strain. I am outwardly fit. I am not even worrying overmuch about any thing. Why, why then should I behave like a hysterical girl?"[43]

Next day he still felt "a little groggy" and "a wee bit light-headed." "Most annoying," for he always prided himself on how healthy he was, exercising to keep fit, standing on his head, and laughing long and hard to keep his "youthful" body in tone, thinking of himself as a yogi. It angered and frustrated him. "That I should have sunk so low! Is it jail or is it too much reading?" he wondered. He was trying to finish the "historical letters to Indu," which he had initially planned to finish by now but confessed, "I have slacked or not been in the mood for writing for weeks and weeks." He had covered world history through the eighteenth century to date. Now he feared that he might be released from jail before his "Glimpses" of the last 132 years were completed, knowing that he found time to write only in prison, never at home or on the political trail.[44]

A few days later on the eve of the Hindu New Year, Diwali, the "festival of lights," a "depressing" letter came from Kamala. "Poor girl! She seems to be unhappy and just a bundle of nerves. I felt that I was partly responsible—My letters to her lately have not been very cheerful or cheering. . . . I must write to her a really nice letter, and I felt sure that this would cheer her up . . . I have also decided to send Kamala a book by post every week or ten days. I have plenty of books. This sending and receiving of books will add to our contacts and will cheer both of us up . . . drive away Kamala's melancholy . . . I must behave in the future—what a pig I am sometimes!" Had his recent fall made him more self-aware, more willing to admit his narcissistic weakness, which had left Kamala so lonely, lost, fearful, and "inadequate" that she had fatally undermined her own powers of resistance to illness? Or was it that with his forty-third birthday less than a month away, Jawahar finally decided it was time to face himself honestly?[45]

"Darling Indu boy," he wrote his daughter on November 1, sending her a book for her birthday. "It is the story of a boyhood spent in the Pampa . . . of the Argentine . . . Hudson, was a great naturalist. . . . *Far Away and Long Ago*—is full of wistfulness and longing for the days that were past. . . . I began to think, also rather wistfully of my own boyhood and especially of the horses I had had and the fine rides . . . How splendid it is to gallop away . . . the cold wind rushing past you! It is long since I have ridden or possessed a horse to ride, and you, my dear, have had no decent chance at all. But if we cannot ride horses we can ride the wind and the whirlwind and they are even swifter, and the excitement is greater! So visions of boyhood came to me and I thought of the old Harrow song: 'Visions of boyhood shall float there before you, echoes of dreamland shall bear you along, twenty and thirty and forty years on.' It is twenty-five years since I left Harrow! And it seems but yesterday. . . . All my love, darling Priyadarshini, Your loving Papu." Priyadarshini was the middle name Nehru gave his daughter, meaning "of gentle visage." Originally used for India's greatest emperor, Ashoka, it is inscribed on many of his Mauryan pillars of peace still standing throughout the subcontinent.[46]

"Kamala has given another fright by having heart attacks & fainting and otherwise misbehaving," Nehru noted in his diary on the eve of his birthday. She soon went to Calcutta, for treatment, and to see her swami guru at the Ramakrishna Mission. Jawahar continued to suffer from toothaches and, as his birthday "gift," had another tooth pulled. He did, however, receive birthday book presents from his Cambridge friend Syed Mahmud and his new beloved Padmaja (Bebee) Naidu. "I felt sure that I would receive something from both of them," he wrote. "How people spoil me!"[47] Next day his entry began "Forty-three!" and ended, "how old I am getting."[48] Three days later it was young Bebee's birthday, and she had "a motor accident! What an extraordinary unlucky girl for accidents! I wired to her and her reply has just come. Very cheerful and all that—she is always cheerful whatever happens—but it is clear that she is far from well."[49] After independence she became his appointed governor of West Bengal and remained unmarried all her life. He now felt "weary and tired and stale . . . all the world going awry, and knaves and scoundrels in the seats of authority shouting and threatening, and good men silent, and fools, oh! so many of them everywhere!"[50]

Jawaharlal's mother, Swarup Rani, had gone with Kamala to Calcutta, and they both stayed near the Ramakrishna Mission, where Swami Abhayananda often met with them. Swarup Rani, the swami remembers, expressed "her anxiety at her daughter-in-law's leading an ascetic life." Kamala "ate very little, was absolutely indifferent to dress, and had given up wearing jewellery entirely." Almost in tears, Swarup Rani used to plead with her daughter-in-law to wear her marriage necklace at least and a few bangles. But Kamala remained adamant. "No, I can't do that! . . . For me it is criminal to walk on the streets exhibiting my jewellery while the people of my country do not have enough food to satisfy their hunger." She went daily to the mission to pray, and often remained there in silence for hours. By December Swarup Rani returned alone to Allahabad, leaving Kamala with her swamis and her prayers.[51]

"My dear Mother, I received your long letter and read all about your trips and your work," Jawahar wrote. "I am concerned about Kamala's health. It is a relief that she is getting the best treatment and I am sure she will be completely restored to health. You did well by asking her mother to join her."[52]

"I felt very fit early this month," Nehru noted in late December 1932. "The cold suited me and I sat for hours in the sun. . . . I could almost imagine my face, as I looked at it in the mirror, growing younger! Vanity! I felt convinced that this was due to the pulling out of the tooth on Nov. 14th . . . and I shall flourish like the green bay tree. So, greedily I consented, and I wrote to Kamala that I was growing in beauty and youth and hoped to look 33 and not 43 when I went out! . . . yet hardly had I written when I began to feel rather poorly. Yesterday as I was walking in the evening my legs became heavy under me and I could hardly lift them. The calves pained as if I had marched for a dozen miles. Pride goeth before a fall! . . . Perhaps it was the cold. . . . Kamala continues ill in Calcutta. At the back of my mind I feel that she is bound to get well. I have a habit—is it merely the fear of facing the truth?—of seeing the bright side of things. Still a shadow seems to hang and I feel a little depressed."[53]

Gandhi was now getting ready to resume his fast. In September he had persuaded a fasting Kerala leader of the untouchables, named Kelappan, to postpone for three months his fast to open to untouchables the Guruvayur Hindu Temple on the Malabar coast. Gandhi appealed to Calicut's Zamorin, who presided over the temple's board, to open its doors to all Hindus. His appeals were, however, rejected. Moreover, the Brahman Shankaracharya, Bombay's leader of Hindu orthodoxy, announced that hundreds of thousands of devout Brahmanic Hindus were by now so opposed to Gandhi's views on free temple entry that they were "prepared to lay down their lives" to prevent it. Gandhi set January 2, 1933, as the day on which he would start fasting again, this time to open the Kerala Temple to all Hindus.[54]

"January 2nd," Nehru noted in his diary. "What will happen? What will happen? I think and wonder and though the faith is strong in me, so many things seem to go awry. What is the reason for the faith? . . . The future is dark. It is difficult to pierce the veil." Gandhi postponed his fast, deciding to undertake that "penance" not simply to open one south Indian Hindu temple to untouchables, but to open every Hindu temple.[55]

In London the third round table conference ended on December 24, 1932, with Hoare promising to remove "every obstacle" to the birth of an Indian federation and dominion as "early as possible." Sind would emerge from Bombay as a separate Muslim-majority province, and Orissa was to be separated from Bengal. Muslims would be assured one-third of all seats at the center, and although Defense would be retained entirely in British hands, much of Finance was to be put under the control of Indian representatives, elected by a still larger segment of the population.

In early January 1933, Nehru confessed to Gandhi as well as himself that he "greatly missed" seeing visitors—"interviews"—which he had insisted on stopping almost seven months earlier, after Kamala and his mother had been kept waiting so many days. "I felt that I had no alternative," he wrote Gandhi on January 5. "I have not yet got rid of my obstinacy—a hereditary failing in me of which you cannot be unaware. . . . [L]argely I have been left to myself, and I have grown a little contemplative, in defiance of heredity and family tradition and personal habit! But that is a thin veneer which I am afraid will rub off at little provocation. How can the Ethiopian change his skin? I have read a lot, and if wisdom could be had in books I would be wise. But wisdom eludes me, and big question marks confront me wherever I look. . . . From all accounts in the papers you are as ever the slave of industry and are over-working yourself even in prison."[56]

To his prison diary he was even more frank in admitting that "the question of interviews has been troubling me. The desire to see some persons and obstinacy! Betty's appeal in a letter saying how much she wanted to see me upset me. . . . I decided to resume interviews even if nothing further done by Govt. . . . Bapu's letter . . . suddenly made me feel that I must end this drifting and resume interviews immediately. . . . I felt light-hearted all day . . . I was all eagerness to have the interview! . . . [N]ow I am full of it. I await Kamala's letter to find out where Indu is. After a day I decide to wait no longer and wire to Kamala. . . . Next day a vague phrase in the paper about Kamala's health put me out and I felt depressed again. . . . It is so cold."[57]

A week later Nan, Betty, and Indu all came to visit him and were permitted to remain nearly two hours, "Nan bustling and full of life and talk as ever—Betty rather quiet and still somewhat childish . . . But Indu is becoming a little woman and remarkably attractive and smart looking. . . . I felt so happy and proud to see her healthy and straight, and growing up apparently without any marked inhibitions." It was a "great event" for Jawaharlal, his first brief reentry into the society of those family members he loved most, and finding his daughter so well, after so wretched a childhood, the ugly duckling emerging healthy, without marked inhibitions. Kamala, of course, could not come, for she was still in Calcutta. After they left, he still "felt excited and light-hearted" for several hours, but then came the inevitable "depression . . . I feel lonely," he confessed. Then "my pain in the right side has returned."[58]

He responded to a card from an old English friend, Evelyn Wood, at this time: "Your card came as a breath of fresh air. . . . It brought a picture in my mind's eye of a pleasant afternoon spent in your flat high up in the clouds . . . far removed from the madding crowd. . . . So you are shaking the dust of India off from your feet. I am not surprised. But the whole world seems to be awry and you cannot escape from it even in the pleasant land of Kent . . . however . . . you can always fall back on hop-picking. . . . Once upon a time long long ago I actually indulged in hop-picking there; it almost seems like a fairy tale now, or a story from another's life."[59] He felt lonely again and distracted. "My mind wanders a lot and I find it a little difficult to concentrate on any reading," he told his diary. "I suddenly catch myself thinking of something quite different and far removed from the subject of the book. . . . I get very angry at the scandalous and shameful way the Liberals and others have agreed to the monstrous abortions which are the outcome of the R.T.C. [round table conference]. . . . But then the question that comes again and again to my mind—how to meet the situation and fight it? Bapu is getting more and more involved in his Temple Entry business. Well, well, why worry? . . . time enough to decide when we are out! . . . no hurry. But oh, these

Liberals and other Round Tablers! What morons they are—or is it merely that they are a lot of old women?"[60]

Gandhi was reaping the whirlwind of Hindu orthodox wrath for his attack against their temples of "purity" that they so zealously guarded against defilement by any untouchable. He had started a new weekly, *Harijan,* the name he used for the depressed or untouchable classes, meaning "child of God (*Hari*)." "The fight against Sanatanists [Hindu orthodoxy] is becoming more and more interesting," Gandhi wrote Jawaharlal in mid-February. "The one good thing is that they have been awakened from long lethargy. The abuses they are hurling at me are wonderfully refreshing. I am all that is bad and corrupt on this earth. But the storm will subside. For I apply the sovereign remedy of *ahimsa,* non-retaliation. The more I ignore the abuses, the fiercer they are becoming. But it is the death dance of the moth round a lamp." Exactly fifteen years later, however, one of those "moths" would extinguish his life.[61]

"I read through the *Harijan* of course with interest," Jawahar replied. "I pity the poor Sanatanists. With their anger and abuses and frantic cursing, they are no match. . . . And yet the power of stupidity and conservatism and entrenched privilege in this world is amazing. Even mahatmas and saints cannot easily upset this combination. . . . Have you read Bernard Shaw's *Saint Joan* . . . ? . . . Of course, if you will stir up stagnant matters there will be [a] bad smell, and the Hindu social structure has stagnated long enough. . . . My own remedy for a stagnant pool would be to drain it completely and let the sun dry and purify the bottom. . . . Kamala has at last left Calcutta. . . . She is at Patna now staying with her brother. . . . By the end of the month she might manage to reach here. She now intends staying in Dehra Dun for two or three months and Indu can join her during her holidays. We are a homeless and a scattered family!"[62]

A few days later it was Holi, the Hindu festival of spring, when villagers spray one another with colored powder and sweets are exchanged, and hugs and kisses and wild dancing and laughter. He heard it all, those "sounds of revelry outside," but then fell asleep in his lonely cell, when his *dhobi* (laundryman) "turned up to pay a *Holi* visit to me! It was a surprise. . . . We embraced and he sprinkled some *gulal* [colored powder] on me and presented me with a plateful of really good sweets. . . . Strange how suddenly a physical longing for the soft things of life comes on me—pleasant surroundings, good friends, agreeable women, interesting conversation. Reading a newspaper yesterday I came across some names of old acquaintances—and their pictures were there—and the old days when I was a student in England came before me—and a certain nostalgia—We live in different worlds and I have not seen some of them for a score of years. The *dhobi* who came to visit me is nearer to me than the princes & princesses of old."[63]

"The idea of writing father's life has occurred to many people," Jawaharlal wrote an old Kashmiri friend, who wanted his help in doing so. "It was also suggested by a dear friend to me that I might utilise my time in prison by writing about father. I found myself wholly incapable of doing so. I had certain personal reasons for this . . . it is an extraordinarily difficult task to write biography well . . . It is not a question of eulogising or praising a person. Anybody can do that. It is most difficult to produce a real picture of an almost living person . . . it involves a true appraisal of complicated currents and under-currents. Do not think I am writing this to discourage you . . . I am merely pointing out to you some of the difficulties of the job which overwhelmed me."[64]

Nehru missed no chance of writing to his daughter, always sending more books to her, as well as offering her advice on any and every subject, perhaps only subconsciously preparing her for the premier role she would play on India's stage, even longer than he would. He sent off a copy of *Saint Joan,* asking if she remembered that they had seen it together in French in Paris, "when a little Russian woman made a charming Jeanne?" He reported that Birju Bhai had brought an Austrian educator to meet with Kamala, asking if they would like to send Indira to "Vienna for schooling! Vienna is a very beautiful place; it is the home of music. The Viennese are delightful people but their country is having a terrible time at present. . . . [A] German lady, who stayed in Anand Bhawan recently, was keen on making arrangements for your education in Germany. Plenty of people seem to be interested in your future education besides your father and mother! Well, perhaps some day, before very long, you will go away to a far country to carry on your education, leaving us rather lonely here."[65]

Next month he wrote to ask if she had received the books he sent, adding, "I do not know what you will do in after life. . . . But whatever you do I have hoped that you will do it well and distinguish yourself in it. I have wanted you to play a worthy part in the world and to be full of life and intelligence and activity. I hope you will do so."[66] A few days later he was comforted by a letter from Indu, "a real intimate letter," he informed his diary.[67] "Papu darlingest," she wrote, "Up till now I had arranged to go for French at six thirty A.M. . . . but now the teacher says that this time does not suit her as she misses her morning walk. So, from next Monday I have to go at 11:30. . . . Now, after having our meal we are free for the whole day, which is very convenient as we all feel very hot and sleepy in the day-time. Of course I have to get up early (I get up at six), but I find it agrees with me. . . . We . . . go to see Bapu [Gandhi] every week on Saturdays. We have just returned from jail. Bapu was very cheerful and pulled my ear so hard that I thought it would come off. It's still a little red. Bapu asked us to bring Tara [Nan's young daughter] too, but she gets bored and prefers staying at home. . . . You asked . . . what science I learn. We do Physics and Chemistry . . . I find them very interesting. . . . I went to see Padmaja day before yesterday. She is better, but is still taking the injections. . . . Padmaja got a wee little tent for me . . . I am thinking of having it dyed. With heaps of love and many many kisses, From your loving, Indu."[68]

"Indu *bien aimee,*" loving Papu replied, "I am glad you are doing Chemistry. Chemistry was one of my special subjects . . . at Cambridge. The other subjects were Geology and Botany. I am afraid I do not remember much that I learnt in those far-off days. And yet, the study of science makes a tremendous difference to a person . . . in life. The help has been chiefly in the training it has given and the outlook of the mind. . . . Almost all our modern life is based on science. . . . Science really means experiment, the finding out of truth by experiment, and not merely accepting facts just because somebody has said so. . . . All my love, sweetheart. I am looking forward so much to seeing you."[69]

Jawaharlal's mother was arrested again on March 31, 1933, at 3 A.M., on her way to Calcutta to attend a Congress session that had been banned by the government. "I am so glad mother was arrested," Nehru wrote when he heard the news. "She must be pleased."[70] It was just a year since the tiny Swarup Rani had been beaten with canes in the road of Allahabad. Four days after her arrest, Swarup Rani was released from Asansol Prison. "You must have been uncomfortable in the jail," Jawaharlal wrote his mother upon her release. "On 22 April Indu's school will close for the holidays. I want her to come to Dehra Dun . . . and stay here with Kamala. Kamala is renting a house. Nan has not yet decided. . . . She will decide after meeting Ranjit. . . . He did well by coming here directly [after his release from Bareilly] and meeting me. . . . The weather here is fine. I am well . . . congratulations!"[71]

Gandhi's "inner voice" returned to awaken him at midnight at the end of April, urging him to fast again, this time for twenty-one days to win temple entry for the untouchables.[72] He resolved to start his fast on May 8 at noon and to continue it through May 29, 1933. He prayed that friends would not press him to postpone or abandon the fast, as he had firmly decided on it to "remove the lid" that hides the truth. Gandhi equated truth with God. "Religion is not familiar ground for me," Jawaharlal wrote Gandhi on the eve of his second fast in this matter. "The Harijan question is bad, very bad, but it seems to me incorrect to say that there is nothing so bad in all the world. . . . All over the world there is the same Harijan question in various forms. . . . Surely it is due to something more than mere ignorance and ill will. To remove these causes or to neutralise their effect appears to be the only way to deal with the roots of the matter. . . . It is hard to be so far from you, and yet it would be harder to be near you. This crowded world is a very lonely place, and you want to make it still lonelier. Life and death matter little. . . . The only thing that matters is the cause that one works for, and if one could be sure that the best service to it is to die for it, then death would seem simpler. I have loved life—the mountains and the sea, the sun and rain and storm and snow, and animals, and books and art, and even human beings—and life has been good to me. But the idea of death has never frightened me. . . . Yet, at close quarters, it is not pleasant to contemplate . . . ever since I had the good fortune to be associated with you. . . . Life became fuller and richer and more worthwhile, and that is a dear and precious memory. . . . And whenever the future happens to be dark, this vision of the past will relieve the gloom and give strength."[73]

Nehru "fretted about these strange methods," like fasting, which he called "magic" ways of doing things, though he knew it was pointless to try to dissuade Gandhi from his "inner voice's" agenda. Yet he feared that "this leads people inevitably to give up troubling their minds for solutions of problems and await for miracles. . . . I have developed a horror of nostrums and the like," he wrote Mahadeva Desai, Gandhi's secretary and cellmate in Poona. He had been receiving several German and French periodicals in prison and for the past six months had been trying to learn German, though "I made very little progress. And now with the bounder Hitler and his Nazis I have begun to hate the language almost!"[74]

Gandhi survived his "great fast," which he ended before the end of May. On launching it, however, he was released from prison and suspended civil disobedience for six weeks, which in itself "shocked" Nehru, and he now confessed to his diary: "I wondered more and more if this was the right method in politics. It is sheer revivalism and clear thinking has not a ghost of a chance against it. All India, or most of it, stares reverently at the Mahatma and expects him to perform miracle after miracle and put an end to untouchability and get Swaraj and so on—and does nothing itself! And Bapu goes on talking of purity and sacrifice. I am afraid I am drifting further and further away from him mentally, in spite of my strong emotional attachment to him. His continual references to God irritate me exceedingly. His political actions are often enough guided by an unerring instinct but he does not encourage others to think. . . . What a tremendous contrast to the dialectics of Lenin & Co.! More and more I feel drawn to their dialectics, more and more I realise the gulf between Bapu & me and I begin to doubt if this way of faith is the right way to train a nation. . . . I cannot understand how he can accept . . . the present social order; how he surrounds himself with men who are the pillars and the beneficiaries of this order. They talk lovingly of 'Bapu' but do not take any risks . . . I want to break with this lot completely . . . to place our ideal crystal clear before the people. But Bapu always talks of compromise and his sweet reasonableness deludes people and befogs their minds. There is trouble ahead so far as I am personally concerned. I shall have to fight a stiff battle between rival loyalties. Perhaps the happiest place for me is the gaol!"[75]

Betty now wrote from her holiday in Mussoorie's posh hotel, where she met all the rajas and ranis who continued to live the "gay life" that Motilal and his rani once lived long ago. "As I read your letter I was suddenly made acutely conscious of a contrast," Jawahar replied. "The gay life! The strange quest for something in the market place . . . not to be found, search as you may, in the gaily-bedecked stalls; the pursuit of a shadow which eludes us for shadows are always elusive; the frantic attempt to possess the beauty of a living rose by grasping it too tightly and crushing it to death. Mussoorie is but a score of miles away from where I sit and write this letter, and the contrast between life at Mussoorie and life in my little high-walled . . . abode is great enough. But greater still is the contrast . . . between . . . the phantom figures that flit on the hilltop and the other phantom that sits and reads and writes and imagines all manner of things. So as I read your letter, reality, if such a thing there is, melted away from me and I fell to musing in a ghostly world of phantoms in which I myself was but a shadow figure. And I saw a procession of Rajas and Ranis . . . dancing away in a veritable *danse macabre* for they danced on a seething mass of hungry and famine-stricken humanity, and their dance led to a sudden precipice over which they toppled, relics of a bygone age, trying bravely to keep up appearances but doomed to inevitable extinction."[76]

He felt so frustrated by Betty's letter and a visit from Birju Bhai and his lovely wife, "Birju Bhabhi," that a few days later "this accumulation of unhappiness and nerves" made him "speak angrily" to the Indian acting superintendent of his jail, Dr. Bhargava, who had always been most kind. "I cursed the whole jail department up and down and called it a sink of iniquity, and a hotbed of bribery and corruption and violence, and said that if I had the power I would smash it into little bits! He listened meekly enough, rather taken aback," Nehru noted, "And Bapu lies slowly recovering from his fast, very weak still. When I think of all this indecision and patent weakness and submission to wrong doing, I feel very angry."[77] Next day, Gandhi's entourage announced a further six-week suspension of civil disobedience, for reasons of health. "And among the mighty ones so deciding was G. D. Birla!" Nehru noted. "Heigh ho! This is a funny world and not an easy place to live in. Perhaps jail is best."[78]

But even that pleasure would soon be denied him, for his mother's health was fast deteriorating, and he was released in late August, a few weeks before his sentence was due to expire. He felt more and more isolated from Congress leaders by now, including Gandhi, who had cabled to the viceroy of "Honourable Peace! What about? With whom & on what basis?" Nehru cried aloud in his solitary cell. "It is amazing how flabby-minded our people have got. . . . I am getting more & more certain that there can be no further political cooperation between Bapu and me. . . . We had better go our different ways." In August 1933 he finished his last letter to Indira, the final chapter of his *Glimpses of World History,* with a poem by Tagore, from that Nobel laureate gurudev's (divine teacher) *Gitanjali:* "Where the mind is without fear and the head is held high; Where knowledge is free; Where the world has not been broken up into fragments by narrow domestic walls; Where words come out from the depth of truth. . . . Into that heaven of freedom, my Father, let my country awake." [79]

On August 23 Nehru was told of his release and driven to the station, where a train took him back to Naini Prison. He was taken out at Allahabad and driven to Anand and Swaraj Bhawan in a police van. Gandhi also was released from prison, on August 16, after having again started a fast, this time to improve conditions inside all jails. His health was so precarious that the government decided on his early release, knowing how dangerous the popular response to his death in prison would be. When Nehru was driven into Allahabad on August 30, 1933, reporters were waiting to interview him.

"I am, of course, dissatisfied with the Liberal ideology and its leaders who need to learn the A.B.C. of political thought and movements in the world today which lay particular stress on the economic issues," he told the press. "In the U.S.A., they have at present, in Mr. Roosevelt, a more intelligent and liberal dictator than the rest. . . . [T]here is a struggle between Fascism and Communism for capital or power in every independent country. . . . [S]tatesmen from all parts of the world are meeting over and over again for . . . peace but their conferences are meeting the same fate as the conferences for Hindu–Muslim unity in India because they are not prepared to sacrifice their individual interests in favour of others. India is not immune from the effects of these international forces." [80]

Nehru's longest interlude behind bars was over, but his political struggle for power had only just begun.

13

Out on India's Left Wing

1933–1934

NEHRU LEFT PRISON in August 1933 convinced that India's destiny was to follow the Russian conception of social and economic change. "The alternative is fascism," he insisted, "the last weapon which vested interests can employ to keep what they have." Twenty months behind bars, much of it spent in virtual isolation—self-imposed for the most part yet no less painful—had embittered him, deepened his hatred for the British Raj, which he viewed not only as India's cruel jailer and enemy but also as a crumbling capitalist society in its last stage of imperialism.[1]

Asked for his reading of the current state of the world, Nehru answered, "The problem is primarily economic . . . [i]n India . . . [it is] agrarian." He expected nothing good to emerge from Britain's white paper, the legislative end product of the Simon Commission and the three round table conferences. The future prosperity of India's masses, Nehru insisted, required a complete "reconstruction of society," starting with the diversion of profits and property from the "haves" to the "have-nots."[2]

After two days, he found the inhibitions and restrictions of freedom "most distressing and suffocating." He had gone to Lucknow from Allahabad but knew he was closely watched wherever he went and resented the Anglo-Indian press criticism. "My whole nature rebels against this compulsion and terrorism of government, and I feel nauseated at the vulgarity, falsehood and sycophancy that accompany them. Politics is a queer game at times, but surely loyalty to the principles one holds to be true and to brave comrades who suffer for them can never be wrong."[3]

Gandhi had suspended all Congress activities relating to civil disobedience, and when Nehru was asked how he felt about that decision, his response was to reaffirm the "creed of independence" adopted at Lahore in 1930. "It is obvious there is a world of difference between the objective, methods and ideology of the Congress and the Round Tablers," he added. "As for the Congress and the British Government, they are poles apart."[4] He had "no regard for men afraid of the present inevitable clash of forces" and quoted Trotsky's "if one is afraid of strife, disorder and revolution, one has chosen the wrong moment to be born."[5]

His sister Betty had chosen the wrong moment to be married perhaps, for their mother was ill in Lucknow, and most of the fabulous fortune in Nehru jewels had been sold over the past decade, though Kamala gave some of what jewelry she had left to her younger sister-in-law. "It was Kamala more than anyone else who made my wedding a gay and happy occasion," Betty recalled. "She had always loved me dearly, more as a daughter than as a sister-in-law." Betty's husband, "Raja" Hutheesing, was a tall, handsome Lincoln's Inn barrister and graduate of St. Catherine's College, Oxford. He was, however, neither a Kashmiri Brahman nor a Hindu but a Jain. He and Betty had dated several times before she so much as told her big brother about the man she intended to marry.[6]

"I hear you are contemplating marriage, my little Beti," Nehru began. "What do you know about the young man?" She said she knew "nothing at all." So he asked what "his people" were like. She did not have the "faintest idea," nor did she know which college at Oxford he had attended. "Absolutely livid, Bhai said, 'This is preposterous!' " He "stalked" out of the room and then went off to Bombay to meet young Raja, who was well versed in Marxist–Leninist literature. When Betty's mother first met him at her bedside in Lucknow he had tears in his eyes and was so gentle and kind that she later told her daughter, "You know you are going to marry a saint." Mahatma Gandhi also approved. So they were married on October 20, 1933, at Anand Bhawan. It was a purely civil wedding, another Nehru first, for neither Jain nor Hindu priests would officiate at so radical a departure from both traditions. "I had wanted to be married in my father's bedroom, but nobody would agree to that," Betty—whose name was now changed to Krishna—recalled, "so the big sitting room was decorated with bright flowers and ribbons, and ashoka leaves were hung on the doorway through which I entered, because that tree is supposed to be very auspicious. When Bhai came upstairs to take me down, he thought I did not look much like a bride, for I had no pearls in my hair. . . . . He took a rose from a vase and stuck it in my jet-black hair and said, 'Well, now you look better.' "[7]

It was one of Jawaharlal's few soft moments since he had emerged from prison. He was usually angry, short tempered, bitter about Congress as well as the British, for he felt let down, ignored for the most part during his long incarceration, his radical program at Lahore all but forgotten between the round tables and Gandhi's prayers. After meeting Raja in Bombay, he visited Gandhi in Poona and, following a long discussion about Congress strategy and tactics, wrote to him, "In our recent conversations you will remember that I laid stress on . . . [a] clearer definition of our national objective. . . . We stand for complete independence. Sometimes a little confusion arises because of vague phraseology." Complete independence, Nehru insisted, included "full control of the army and of foreign relations as well as financial control." Gandhi knew that that was impossible to expect, and Vallabhbhai and most of the other leaders of Congress agreed with Gandhi. Jawaharlal inched further out on his extreme left-wing limb almost alone at this time, but isolation never really bothered him. It only confirmed his feelings of self-righteous indignation, for his mind acquired the tenacity of the British bulldog as it became more Trotskyist in ideology.[8]

"The problem of achieving freedom becomes one of revising vested interests in favour of the masses," Nehru informed Gandhi. "The biggest vested interest in India is that of the British Government, next come the Indian princes, and others. . . . We are all agreed that the Round Table Conference and its various productions are utterly useless to solve even one of India's many problems. As I conceive it, the Round Table Conference was an effort to consolidate the vested interests in India behind the British Government. . . . [I]t was a fascist grouping of vested and possessing interests and fascist methods were adopted in India to suppress the national movement. . . . The problem of Indian freedom cannot be separated from the vital international problems of the world. . . . [W]e must, I feel, range ourselves with the progressive forces of the world."[9]

"India cannot stand in isolation," Gandhi agreed, but he responded to Nehru's arguments that "our progress towards the goal will be in exact proportion to the purity of our means. If we can give an ocular demonstration of our uttermost truthfulness and non-violence, I am convinced that our statement of the national goal cannot long offend the interests which your letter would appear to attack. We know that the princes, the zamindars, and those who depend for their existence upon the exploitation of the masses, would cease to fear and distrust us, if we could but ensure the innocence of our methods. We do not seek to coerce any. We seek to convert them. . . . Now about secret methods. I am as firm as ever that they must be tabooed. I am myself unable to make any exceptions. Secrecy has caused much mischief and if it is not put down with a firm hand, it may ruin the movement."[10]

So the differences between Nehru and Gandhi grew deeper and wider, not only ideologically, but also in tactics and strategy, means and ends, goals and methods of achievement. Jawaharlal was a child of the modern revolution, Marxist–Leninist in theory and Fabian or Trotskyist in practice. Conversely, Gandhi was the child of Hinduism, his revolt reaching back much further in time to the Satya and Ahimsa of the Vedas, relying on religious Truth as its rock, for "Truth was God." Gandhi was only twenty years older, yet centuries—millennia—separated their minds. "It was obvious that we differed considerably in our outlook on life and politics and economics," Nehru wrote of Gandhi in his autobiography. "He was a curious phenomenon—a person of the type of a medieval Catholic saint . . . and at the same time a practical leader with his pulse always on the Indian peasantry. Which way he might turn in a crisis it was difficult to say. . . . He might go the wrong way, according to our thinking, but it would always be a straight way." It was a straight way that Nehru might at times find useful, but one that he would never again wholeheartedly accept as his own or as the "right" way for India.[11]

"Some of us feel," Jawahar pointed out to Gandhi, "that . . . in communicating with each other or sending directions or keeping contacts, a measure of secrecy may be necessary. . . . [S]ecrecy or the avoidance of it . . . cannot be made into a fetish . . . secrecy is certainly involved in the production of printed or duplicated news-sheets and bulletins. These bulletins have often served a useful purpose . . . I do feel . . . it may be desirable for a local or provincial committee . . . to issue bulletins of direction etc. secretly . . . one other small matter which seems to me rather ridiculous. It was right . . . for you to court imprisonment by giving previous intimation of your intention to do so to the authorities. But it seems to me to be perfectly absurd for others . . . to send such notices."[12]

Whenever asked by reporters, Nehru insisted that there was "not the slightest difference between Gandhiji and myself."[13] It was now rumored that Nehru would leave the Congress and form his own party, one with more complete independence and socialism as its goal, but he denied any intention of doing so. Nehru was, after all, angry and impatient but hardly stupid. He knew how important a party the Congress had become and how essential it was for him to retain Gandhi's support if ever he hoped to win support from the masses he always spoke of so warmly, but from whose lifestyle, mentality, and appearance he remained more remote than from his old Harrovian classmates or his "Cantab" friends at Trinity. For all his Marxist learning and Leninist rhetoric he remained a Kashmiri Brahman prince of Anand Bhawan.

That October Nehru wrote a series of articles, published in pamphlet form, entitled "Whither India?" It was his personal economic–political manifesto, the first in a series of pieces he would write to educate his compatriots in the current world situation as he saw it, Indian history as he reread it, and the best way of achieving India's freedom as he understood it. It was the closest he came to "translating" Lenin's pamphleteering brilliance to the Indian scene, and it was published not only in English but in every major Indian language. "Many people ask: What are we to do?" Nehru began. "Right action cannot come out of nothing; it must be preceded by thought. . . . What exactly do we want? And why do we want it? I write with diffidence because I have long been cut off from the nationalist press." Few who read on would believe his modest disclaimer. "On the gaily-decked official stage of India or England phantom figures come and go, posing for a while as great statesmen; Round Tablers flit about like pale shadows . . . engaged in pitiful and interminable talk which interests few and affects an even smaller number. Their main concern is how to save the vested interests of various classes . . . their main diversion, apart from feasting, is self-praise. Others, blissfully ignorant of all that has happened in the last half a century, still talk the jargon of the Victorian Age. . . . Yet others hide vested interest under cover of communalism or even nationalism. And then there is the vague but passionate nationalism . . . an idealism, a mysticism, . . . something of the nature of religious revivalism. Essentially all these are middle class phenomena." But then Nehru offered his answer: "Our politics must either be those of magic or of science . . . I have no faith in or use for the ways of magic and religion . . . I can only consider the question on scientific grounds." [14]

His "scientific" analysis was based on the primacy of economic interests that lead to class struggle. "Nothing is more absurd than to imagine that all the interests in the nation can be fitted in without injury to any. At every step some have to be sacrificed for others . . . . We cannot escape having to answer the question, now or later: for the freedom of which class or classes in India are we especially striving . . . ? Do we place the masses, the peasantry and the workers first, or some other small class at the head of our list? . . . [A]nd when a conflict arises whose side must we take? To say that we shall not answer that question now is itself an answer and taking of sides, for it means that we stand by the existing order, the *status quo*." [15] He then launched into a Marxist–Leninist review of world history, a brief summary of the long study he had done over the past year and a half in prison, concluding that a new system was required because the old capitalist system had "outlived its day." By now positioning himself ideologically as the leader of the socialist "vanguard of the masses," Nehru rallied many former comrades who had given up hope for him, fearing that he had sold out to the Mahatma and his Tata and Birla bosses, His "Whither India" resurrected him as the first among India's nationalist-comrades.

"In India . . . we find a struggle today between the old nationalist ideology and the new economic ideology. Most of us have grown up under the nationalist tradition and it is hard to give up the mental habits of a lifetime. And yet we realize that this outlook is inadequate . . . the nineteenth century cannot solve the problems of the twentieth, much less can the seventh century or earlier ages do so."[16] This, of course, is an allusion to Gandhi's methods of fasting or praying for change. "We have got into an extraordinary habit of thinking of freedom in terms of paper constitutions. Nothing could be more absurd than this lawyer's mentality which ignores life and the vital economic issues. . . . Even the halt and the lame go slowly forward; not so the lawyer who is convinced, like the fanatic in religion, that truth can only lie in the past." Here he attacks both Jinnah and Gandhi at once. He clearly rejects his own legal training, rightly perhaps regarding himself as never having been a lawyer: more the political activist, now the pamphleteer. "India's struggle today is part of the great struggle . . . all over the world for the emancipation of the oppressed. Essentially, this is an economic struggle, with hunger and want as its driving forces, although it puts on nationalist and other dresses." And what "dress" did Jawahar now wear? Surely "hunger and want" had never been driving forces for any Nehru. The answer may lie in the words of George Bernard Shaw, whom Nehru quotes at the end of this political manifesto.

"Whither India? Surely to the great human goal of social and economic equality, to the ending of all exploitation of nation by nation and class by class, to national freedom within the framework of an international cooperative socialist world federation. This is not such an empty idealist dream as some people imagine. It is within the range of the practical politics of today and the near future. We may not have it within our grasp but those with vision can see it emerging on the horizon. And even if there be delay in the realisation of our goal, what does it matter if our steps march in the right direction and our eyes look steadily in front. For in the pursuit itself of a mighty purpose there is joy and happiness and a measure of achievement. As Bernard Shaw has said: 'This is the true joy in life, the being used for a purpose recognised by yourself as a mighty one; the being thoroughly worn out before you are thrown on the scrap heap; the being a force of nature, instead of a feverish, selfish little clod of ailments and grievances, complaining that the world will not devote itself to making you happy.'"[17]

Nehru needed a "mighty purpose." At first only Norwegian mountain peaks and Himalayan glaciers seemed to satisfy that insatiable thirst deep within him. Later it was Gandhi and prison cells, but now those had failed. Gandhi was too archaic a magician, and magic never worked on Nehru. And lingering too long in prison had nearly turned him into a "selfish little clod" full of "ailments." So now it would be socialism, scientific and national, leading the army of India's masses up the summit of Purna Swaraj. Surely there he would find the "true joy" in life, would he not?

No sooner was "Whither India?" widely published, however, than the critics emerged, from the Left as well as the Right. The "left criticism which accepted the main line of thought," Nehru informed the Indian press, "said that it did not go far enough, and the right criticism which attacked the very premises of my argument . . . rejected with anger my conclusions. On both sides the personal element was brought in and my seeming contradictions and weaknesses were pointed out. . . . Personalities . . . should not intrude themselves when world problems and world forces are analysed and a meaning is sought to be drawn from them. It is therefore desirable that my many failings and deficiencies might be forgotten for a while. . . . Personally I am not conscious of any glaring inconsistencies in my ideas or activities during the last thirteen years or so. . . . I have grown mentally during this period and many a vague idea has taken shape and many a doubt has been removed. It is also true that as an active politician, having to face day to day problems, I have sometimes had to make compromises with life. . . . But even so I am not aware of any betrayal of the ideal that drew me on or the principles I held."[18]

That November he turned forty-four. Indu wrote her father from school to wish him "very many happy returns of the day. . . . All I can give is my love. And this has always been yours and always will be. Often I do and say things which I ought not to. And sometimes you get angry. Of course it is my fault—but will you please forgive and forget? . . . How old are you now?"[19] "Indu sweetheart," he replied. "When one gets so old as I am birthdays are not quite so welcome as they used to be. But a birthday letter from you is always very welcome and your few lines made me happy. . . . My age is becoming a delicate subject, not fit for public reference. I was born on the 14th November, 1889—so you can calculate it. The date is a hundred years after the storming of the Bastille . . . to help you remember it."[20]

Gandhi had just received a proposal from Sindhi Saraladevi Chowdharani for Indira to marry her son at this time, which he passed on to Jawaharlal. "I have told her Indu is left free to do as she chooses and that she is not likely to entertain any marriage proposal as she is still studying. . . . I am forwarding the letter to you. If Indu was at all prepared to consider . . . I do regard Dipak to be a good match." Jawaharlal never even mentioned the offer to his daughter, responding negatively to Gandhi, but may later have regretted that he did so, considering how he felt about the man she would choose for herself.[21]

Since his release Nehru had been anxious to meet with all other members of the Working Committee, of which he still remained general secretary. No meeting could, however, be arranged before December. Several members of the committee resigned as soon as they heard from him, others were ill, and some were reluctant to risk arrest, since Congress and its various committees were illegal organizations, according to viceregal Ordinance. Nehru went to Juddulpore on December 5 to meet with Maulana Abul Kalam Azad and to give an address. "So long as the bread and unemployment problems of the millions of poverty-stricken peasants and educated youth remain unsolved, the freedom fight cannot be stopped even by Mahatma Gandhi. Although the continuous ordinance regime has paralysed the fight for freedom . . . it has failed to kill the Congress, which still breathes. . . . [T]he flag will fly as high as ever." Three days later, Kamala's disciple, twenty-one-year old Feroze Gandhi, was arrested in Allahabad for encouraging people not to pay their rent and taxes. He was sentenced to six months of rigorous imprisonment, his first prison term. Sixteen-year-old Indira knew Feroze, of course, but only as a brash young man obsessively devoted to her mother. His arrest, however, gave him heroic stature in her eyes, almost as great as that of her father.[22]

On December 10, 1934, Jawaharlal went to Delhi, trying to convince Working Committee colleagues to meet openly with him there to "clarify" the Congress's policy. Dr. Ansari had returned from his treatment abroad but was still reluctant to risk prison, and Gandhi was preoccupied with a Harijan tour he launched after his own prison release. Gandhi did come to Delhi but remained only briefly at a "meeting of Congress workers" as it was called by the press, for the viceroy's home member, Sir Harry Haig, had just issued an ordinance banning any All-India Congress Committee meeting from being held. "As for Sir Harry Haig's ban," Jawaharlal told his Congress friends in Delhi, "I am immensely pleased. . . . The assertion of the government that it has wiped out the national movement and crushed the spirit of the people is only a proof . . . that government is trying to cover its moral defeat by mere words. Real freedom cannot be won until real power passes into the hands of the people. . . . Do your duty to these struggling peasants, serve them to your best. Those who cannot carry on this mission can carry on the campaign against untouchability but I am personally interested at present only in the political and economic struggle of India. . . . The political struggle must always have the first place in your heart." At that point, Gandhi walked out, and most of the others followed him, off to his next Harijan meeting.[23]

Jawaharlal thus learned before year's end that few of his Congress elders were ready to follow his call to revolution, but many younger radical socialists, J.P. among them, were eager to follow if he would agree to organize a new socialist party of India. He rightly recognized, however, that such a course would leave him only to preside over so minuscule a club of India's leftist elite that they could never capture the masses they claimed to represent. But he wanted power, the power he had known briefly in Lahore. Without Motilal's commanding presence, however, Jawahar seemed much smaller to his colleagues on the Working Committee, and Gandhi knew how negatively he felt about Harijan work, so he wisely left him to sort things out for himself. Nor were any of the British lords who ruled over Simla and Delhi eager to ask Nehru's opinion, for they knew precisely what he thought, having read and discarded "Whither India?"

Nehru's mother rallied to attend Betty's wedding, and although she suffered her first stroke in 1935, she survived as an invalid until 1938. Kamala, however, was less strongly committed to life, finding no joy in it and praying only for "release" *(moksha)* from the painful bonds of karma. She insisted now on going back to her swami in Calcutta. Jawaharlal went with her, for he tried to maintain some public pretense of their marriage as more than the phantom formality it had become. Their last intimate link was Indira, and her education was the one subject that could at times engage both their minds, as it did on this visit to Bengal, for they planned to enroll her by the fall of 1934 in Tagore's Santiniketan.

"Darling Indu-boy," Nehru wrote to his daughter that January. "We had three and a half very strenuous days in Calcutta and now we are at Santiniketan . . . welcomed by the Poet and all the students and staff. . . . We have been put up in the Poet's own house. . . . Art is of course the strong point of this place and everything has the artistic touch. The electric lights have all gone out and I am writing this by the light of a hurricane lantern. . . . Mummie is asleep. . . . We shall be travelling for another three days before we reach Allahabad. . . . Our next halt will be Patna. We are staying there for the day to see Rajendra Babu [Prasad] and confer with him about relief work for the earthquake areas. There has been terrible destruction and loss of life. . . . We felt the earthquake distinctly in Allahabad on the 15th. I was standing at the time in the verandah. . . . Suddenly I started wobbling and when I discovered that it was an earthquake I was interested and greatly amused. Nothing much happened in Allahabad but in Bihar . . . Whole cities have been destroyed."[24]

That Bihar earthquake of 1934 was the most devastating natural disaster in recent Indian history. More than ten thousand lives were lost, with hundreds of thousands of people in north Bihar left homeless. Small towns were reduced to rubble by the quake that struck shortly after 2 P.M. and was felt over a thousand miles away. Smoke and dust filled the sky, bringing darkness to an area of thirty thousand square miles, uprooting a thousand miles of railroad track and all but paralyzing commerce and communication across north India's major Gangetic plain artery. Millions of Indians believed that God had spoken, and the world was coming to an end. Rajendra Prasad, who had been in prison, was released by Bihar's government to help organize earthquake relief. Gandhi was touring south India at this time and called "this earthquake . . . a divine chastisement sent by God for our sins. . . . For me there is a vital connection between the Bihar calamity and the untouchability campaign. . . . Let this Bihar calamity be a reminder to us that, whilst we have still a few more breaths left, we should purify ourselves of the taint of untouchability and approach our Maker with clean hearts."[25]

Nehru used the earthquake to preach a different message: "Nature is often pitiless and cruel. It plays its pranks regardless of what happens to the millions of mites that crawl on the surface of the earth. We feel helpless and bow to it. . . . But there are other earthquakes, human earthquakes, which are not caused by unthinking nature but by thinking men. Human masses, when their lot becomes unbearable, rise up and break up the order that enslaves them." Communal riots in East Bengal around the port of Chittagong led to a crackdown by the British, who ordered a complete curfew of all "Hindu men" in the region under the age of twenty-five, first for a week and then for a month. Nehru compared such "wholesale orders affecting the entire population" with the Inquisition. While in Calcutta, he also attacked a recent speech by the British commissioner of Midnapur, as revealing "to what depths of vulgarity and bluster even a seemingly powerful government can sink when it has lost all moral hold on the people it governs." For those statements he was returned to jail the following month, charged with sedition.[26]

Nehru visited Bihar's Muzaffarpur, in the heart of the earthquake disaster region, reporting to Allahabad's press that the most urgent needs were for food, clothing, blankets, shelter against rain, medical relief, and money. The crowded camps were filthy, generating fears of epidemics. Water was scarce, since most of the wells in the region had been choked with sand. "Amazing" false rumors spread that the Ganges River was "drying up." Nehru issued several further appeals for funds and urged young men to "help in clearing the debris," inquiring caustically, "Does the enthusiasm of our young men run to this kind of physical labour, or is it merely a clerkly enthusiasm which looks to a desk?"[27] In early February he toured north Bihar's rural "earthquake areas" and the "once rich and flourishing city" of Monghyr. "I will never forget," Nehru reported, "the actual sight of mile after mile of desolate sand-covered land, full of cracks and fissures . . . all that I had heard and read had not prepared me for the fearsome sight . . . the heap of utter ruin and desolation."[28]

Jawaharlal's shock at what he witnessed led to his fearless criticism of the government's slow response to the disaster. Many other Indians "expressed their admiration and gratitude" to officialdom for what was done several weeks after the quake hit, and thanks to those published words of praise, Sir Samuel Hoare paid further "tribute" from Whitehall to the "magnificent services" of the British Raj. "But . . . vital facts cannot be ignored," Nehru pointed out, "The government we have has a measure of efficiency . . . in collecting taxes, in repressing political activity, in running its bureaucratic machine. . . . It takes action swiftly and often ruthlessly when its interests are at stake. It passes new laws overnight to strengthen its position. . . . But . . . [w]as it efficient in the days immediately following the earthquake? . . . Large numbers of living persons were lying under the debris suffering untold agonies. . . . What steps did the government take to rescue them in these vital early days? What has it done since in regard to this? . . . Government has not attempted to remove the debris from any considerable number of houses. Some roads have certainly been cleared . . . but the houses with their dead and half-dead remained. . . . Could it [the government] not declare a state of emergency in the affected areas and pass special measures to deal with the new conditions swiftly as other governments have done?" That was Nehru at his best, forthright, constructively critical, and eager to help those who needed help.[29]

Such critical honesty, however, hardly endeared Nehru to any officials, Indian or British, who knew how right he was in all that he said and wrote about their sluggish response to the earthquake. That evening as the sun set over Allahabad, a police car drove up to Anand Bhawan, and the superintendent saluted him with a warrant for his arrest. "I have been expecting you for many days now," Nehru told the policeman in Hindi, soon to be India's other national language. "I went to mother but . . . mother was breaking down. I consoled her and she pulled herself up a little—brave little woman. Then packing. Suddenly Kamala broke down . . . I drafted telegrams to Indu, Bapu & Rajendra Babu. . . . We started about 6:30 P.M."[30] To Indira, he wired, "Am going back to my other home for a while all my love and good wishes cheerio."[31] He was not, however, taken to Naini Prison in Allahabad but transported by rail to Calcutta's Lalbazar Prison, and then moved again to Alipur Prison where he was sentenced to two years' imprisonment a few days later. "This is a big prison. Pop. over 2000," Nehru noted in his diary. "The Supt. is a Punjabi—some Singh [Major M. A. Singh]. And so begins my seventh term."[32]

14

Whither Nehru?

1934–1935

NEHRU ASSURED Bengal's magistrate's court in mid-February 1934 that for "many long years" his activities had, indeed, been "seditious," "if by sedition is meant the desire to achieve the independence of India and to put an end to foreign domination. . . . Individuals sometimes misbehave in this imperfect world of ours. . . . But it is a terrible thing when an organized government begins to behave like an excited mob; when brutal and vengeful and uncivilized behaviour becomes the normal temper of a government . . . [I]n India today . . . pricked by a guilty conscience, the government tries to suppress all criticism, all freedom of thought and speech and press and public assembly . . . to perpetuate the greatest sin against any people, to injure human dignity, to crush the dignity of the Indian people . . . we cannot and will not submit to this national degradation and humiliation of a great people. . . . I trust that though my voice may be silenced for a long while . . . we shall remain unyielding and uncompromising emblems of the spirit of the Indian people, which will . . . fight its way to full freedom."[1]

Nehru's voice thus became India's cry for freedom, its most eloquent defender of those human rights that we have come to think of as universal but that were then still limited in application, if not appeal. His fearless assertion of those rights inspired generations of young Indians, his daughter among them. From prison he wrote urging her to visit Kashmir with Nan and Ranjit and their children after her exams. "Kashmir is a place well worth visiting and as you know it is our old homeland. . . . Long,

long ago we left it and since then the whole of India has been our home. But the little corner of India which is Kashmir draws us still both by its beauty and its old associations. We have not been there for seventeen years or more and you have not been there at all. . . . I should have liked mummie and you both to go there but I do not know if she will agree." Kamala went nowhere now but to her doctors and her swami. Indira did, however, go to Kashmir and soon developed almost as romantic an attachment to the old Nehru homeland as her father had.[2]

Gandhi continued his antiuntouchability tour of the south, and wherever he went he repeated his claim that the Bihar earthquake was "God's chastisement" of India for the "sin of untouchability." Rabindranath Tagore responded in "painful surprise" to the Mahatma's accusations concerning "God's vengeance." Tagore, like Nehru, had more faith in science and natural laws but knew that Gandhi's "unscientific" arguments were "accepted" by a "large section of our countrymen." "We do not know the laws of God," Gandhi insisted. "I believe literally that not a leaf moves but by His will." When Patna's Dr. Rajendra Prasad asked Gandhi to come to Bihar in March to assist him there, the Mahatma cut short his Harijan tour and went to Patna, which he toured in the company of the Quaker Agatha Harrison (1885–1954), who had first visited India in 1929 on a royal commission on labor. Agatha remained an influential advocate of Indian freedom throughout her life, one of Gandhi's most devoted British disciples and a friend of Krishna Menon, Nehru, and, later, Indira.[3]

In early April 1934, Drs. Ansari and B. C. Roy came with Bhulabhai Desai, soon to be speaker of the assembly, to Patna to meet with Gandhi, urging him to call off the civil disobedience campaign entirely and to support their intention of reviving the Swaraj Party to contest forthcoming elections to the assembly under recently announced reforms that had emerged from the round table conferences. Gandhi agreed to suspend the nationwide Satyagraha campaign but would himself remain "the sole representative of civil resistance in action." One of his reasons for ending the campaign, he announced, was having recently learned "about a valued companion of long standing," meaning Nehru, who had focused on "his private studies" in prison rather than on spinning cotton. "The friend said he had thought that I was aware of his weaknesses. I was blind. Blindness in a leader is unpardonable." During their last meeting in Poona, Nehru admitted that he had stopped spinning, considering it more important to devote himself to writing *Glimpses of World History,* published in 1934.[4]

When Jawaharlal learned of Gandhi's decision and his announcement of the reasons for it, he resolved in his prison cell in Bengal that he would thereafter go his own way, wherever that might lead, marching with Gandhi only if he felt it politically advantageous. "I read in the weekly *Statesman* today Bapu's statement about the withdrawal of C.D.," Nehru noted on April 13, ". . . it bowled me over. I have spent an unhappy day feeling rather lost. I read & re-read it. It marks an epoch not only in our freedom struggle but in my personal life. After 15 years I go my way, perhaps a solitary way leading not far. . . . [M]eanwhile there is prison and its lonely existence."[5]

In his autobiography, soon to be written in that same prison cell, Nehru quoted Lenin on this point: "Any and every political crisis is useful because it brings to light what was hidden, reveals the actual forces in politics; it exposes lies and deceptive phrases and fictions; it demonstrates comprehensively the facts, and forces on the people the understanding of what is the reality." Abandoned by Gandhi, Nehru turned to his other guru in revolution, Lenin, whose cold brilliance he had long admired, at least since seeing his dead face and mummified body inside the Kremlin seven years earlier. Reflecting now on the impact of Gandhi's "retreat" on Congress, Nehru wrote, "Probably some of its weaker elements might drop out . . . no loss. And when the time came for . . . a reversion to so-called constitutional and legal methods, the advanced and really active wing of the Congress would utilise even these methods from the larger point of view of our final objective."[6]

Hereafter Nehru considered himself leader of the "advanced and really active wing" of his party. Like Lenin, he would never hesitate to take one or even "two steps backward" in order ultimately to inch himself and his party forward toward the "final objective," which he always kept at the top or back of his brain, where each tactical or strategic change was dictated "from the larger point of view." As for Gandhi's statement about the personal reason for his withdrawal of civil disobedience, Nehru wrote: "The imperfection or fault, if such it was, of the 'friend' was a very trivial affair. I confess that I have often been guilty of it and I am wholly unrepentant. But even if it was a serious matter, was a vast national movement involving scores of thousands directly and millions indirectly to be thrown out of gear because an individual had erred? This seemed to me a monstrous proposition and an immoral one. I cannot presume to speak of what is and what is not Satyagraha, but in my own little way I have endeavoured to follow certain standards of conduct, and all those standards were shocked and upset by this statement of Gandhiji's. I knew that Gandhiji usually acts on instinct . . . and very often that instinct is right. He has . . . a wonderful knack . . . of sensing the mass mind. . . . The reasons which he afterwards adduces to justify his action are usually afterthoughts and seldom carry one vary far. A leader or a man of action in a crisis almost always acts subconsciously and then thinks of the reasons for his action." Nehru himself was a master of that art of subsequent self-justification.[7]

Nehru's attack against Gandhi in Alipur Prison reveals the depth and intensity of self-repression he exercised whenever he pretended—before or in future years—to follow in Gandhi's footsteps. "Why should we be tossed hither and thither for, what seemed to me, metaphysical and mystical reasons in which I was not interested? Was it conceivable to have any political movement on this basis? I had willingly accepted the moral aspect of Satyagraha. . . . That basic aspect appealed to me and it seemed to raise politics to a higher and nobler level. I was prepared to agree that the end does not justify all kinds of means. But this new development . . . held forth some possibilities which frightened me . . . and oppressed me tremendously. And then finally the advice he gave to Congressmen was that 'they must learn the art and beauty of self-denial and voluntary poverty. They must engage themselves in nation-building activities, . . . through personal hand-spinning and hand-weaving, the spread of communal unity of hearts by irreproachable personal conduct . . . the banishing of untouchability in every shape or form . . . the spread of total abstinence from intoxicating drinks and drugs . . . cultivating personal purity. . . .' This was the political programme that we were to follow. A vast distance seemed to separate him from me. With a stab of pain I felt that the chords of allegiance that had bound me to him for many years had snapped."[8]

Yet Nehru feared that without Gandhi's support he could never achieve power over Congress, and without Congress behind him he could never rule India. Although he admired and secretly identified more closely now with Lenin than with Gandhi, he knew that Mahatma Gandhi, not he, was for India and Congress more nearly what Lenin had been to Russia and the Bolshevik Party. "Suddenly I felt very lonely in that cell. . . . Life seemed to be a dreary affair, a very wilderness of desolation. Of the many hard lessons that I had learnt, the hardest and the most painful now faced me: that it is not possible in any vital matter to rely on any one. One must journey through life alone; to rely on others is to invite heartbreak." Gandhi had abandoned him, he knew, because he had broken his Satyagraha pledge of spinning daily in prison and then was fool enough to admit it, to confess his "weakness." He long thought of Gandhi not only as his "little father" but also as his father-confessor, to whom he could always unburden himself. This time he knew better and would never forget that lesson: Trust no one, confess to no one, rely on no one, for "to rely on others is to invite heartbreak."[9]

The longer he thought about it, the more certain Nehru became of how "hopelessly in the wrong" Gandhi was, in so many matters. "What, after all, was he aiming at? In spite of the closest association with him for many years I am not clear in my own mind about his objective. I doubt if he is clear himself. One step enough for me, he says. . . . Look after the means and the end will take care of itself, he is never tired of repeating. Be good in your personal individual lives and all else will follow. That is not a political or scientific attitude. . . . Personally I dislike the praise of poverty and suffering. I do not think they are at all desirable. . . . Nor do I appreciate the ascetic life as a social ideal. . . . Nor do I appreciate in the least the idealisation of the 'simple peasant life.' I have almost a horror of it. . . . We cannot stop the river of change or cut ourselves adrift from it. . . . Gandhiji is always thinking in terms of personal salvation and of sin, while most of us have society's welfare uppermost in our minds. I find it difficult to grasp the idea of sin, and perhaps it is because of this that I cannot appreciate Gandhiji's general outlook . . . [including] Gandhiji's attitude to sex. . . . For him 'any union is a crime when the desire for progeny is absent,' and 'the adoption of artificial methods must result in imbecility and nervous prostration.' . . . Personally, I find this attitude unnatural and shocking, and if he is right, then I am a criminal on the verge of imbecility and nervous prostration."[10]

Nehru was shifted in May from Calcutta's Alipur Prison to the cooler Dehra Dun. "Quite a decent place . . . [but] I am not going to have many of the privileges that I have previously had. . . . No walking outside. . . . No keeping of more books than the regulation 6 at a time. . . . Also no smoking as at Alipur. The effect of all this is to irritate me greatly and some of this irritation flows out to the Supt. I feel a little sorry for him afterwards, but why be sorry for worms."[11] He had started smoking cigarettes again during his last long prison term. Soon he "felt poorly" and "feverish." But it was much cooler than in Alipur, and before month's end Ranjit stopped by to visit on his way to Kashmir. Nehru learned from Ranjit that Kamala had decided not to return to Dehra to be closer to him, and "I felt angry with her for giving up the idea of a Dehra stay. Had a heavy and depressing afternoon."[12]

Nor did he feel better in June, rather "very depressed & poorly . . . Kamala is again in the grip of her old disease . . . Indu does not write to me and I get very angry." He was "full of irritation" against both Kamala and Indu but then realized that Kamala's illness was hardly her fault and so felt guilty again. "What is the poor girl to do? But this means that I shall be lonelier here than ever. And my mind goes back to all the petty grievances I have had against her in the past and dwells on them and this increases my ill-humour."[13]

Now that Gandhi had called off civil disobedience, the A.I.C.C. met in Patna in mid-May, and the Mahatma moved a resolution permitting congressmen to enter councils again. A parliamentary board was appointed under Dr. Ansari and Malaviya to choose Congress candidates for the forthcoming elections in October. Jinnah had, moreover, returned to Bombay from London and was viewed as the most prestigious leader of the Muslims. He now indicated his willingness to work with Ansari and

Bhulabhai Desai and other congressmen in the assembly in Delhi to achieve a true national opposition, trying to revive the Hindu–Muslim pact he had drafted with Motilal and passed through Congress and the Muslim League at Lucknow almost twenty years earlier. Jinnah was elected unopposed by Bombay's Muslims and soon emerged as leader of the Independent Party in the assembly.

Indu wrote her father from Kashmir: "I love this place . . . and I'm glad I came. The sweet [children] are ever so sweet but very dirty. Puphi [Auntie] and I often make plans that years later we are coming back here . . . to open a 'Scrubbing Washing Home' for children!" Her Poona school mistress had recommended that she stay in a special "cottage" in Santiniketan, "as the boarding will not be a convenient place at all for me, specially the food. . . . I could take a servant from Allahabad who would cook for me & also do some of the other work. . . . Do you think I should do this?"[14] Nehru replied, "I do not agree at all. I dislike very much the idea of your keeping apart from the 'common herd' and requiring all manner of special attention, just as the Prince of Wales does. . . . This seems to me to savour of vulgarity and snobbery. It is a bad beginning to make in any place. . . . Wherever we go we must keep on a level with our surroundings. . . . My fear is that you will be too much looked after there, not too little. That can't be helped because you happen to belong to a notorious family. . . . Do you know what I had to put up with at Harrow? We never had a full meal at school unless we bought it for ourselves. As junior boys we had to wait on the seniors as fags, get their food, clean their places up, sometimes clean their boots, carry messages for them etc. and be continually sworn at by them and sometimes beaten. So I think you had better go to Santiniketan without making any special arrangements." He trained his "Indu boy" well for the premier job she would inherit soon after his death.[15]

On June 20 he noted to his diary he had learned that "the Working Committee—is of opinion that confiscation and class war are contrary to the Congress creed of non-violence." His temper flared: "To hell with the Working Committee—passing pious and fatuous resolutions on subjects it does not understand—or perhaps understands too well!"[16] He got so angry he sprained his back "in my stretching exercises" a day or two later. "Ailments in gaol are not at all pleasant," he decided. "They make the loneliness greater."[17] Next day, the pain "continues. How old I am getting! My hair is turning grey at a great pace, it is already mostly grey and I have developed quite a distinguished (?) elderly look. The world passes on and goes ahead, while I . . . sit silent and inactive and grow older & older. Life stops in jail in many ways but age creeps on. My eyes grow dull from staring at the walls. They are beginning to sag—like father's."[18]

He grew more impatient, more frustrated. "The future, the future—

what of the future? It looks dark and I feel weary. There is little 'pep' in me. . . . The days drag through slowly . . . and I find it difficult to do any solid work."[19] Indira, however, cheered him with her letters from Santiniketan in July, when she started there. "Everything is so artistic & beautiful & wild. . . . The only difficulty here will be the language. Bengali is a very sweet & nice language. . . . Gurudev [Tagore] suggested that I do some gardening in my free time. He seems to be very fond of it and was complaining that so few girls take it up."[20]

Nehru responded, "I knew you would like the place. . . . Some people . . . are so used to the official universities that anything new strikes them as undesirable; or they don't like the Poet. . . . Curiously enough it is the Bengalis themselves who criticise it most. Bengalis, in spite of being advanced and cultured in many ways, are very conservative folk. . . . My own feeling . . . is that it is a good place with one thing lacking. It does not give quite an up to date education for the modern strenuous life. It concentrates too much on the artistic side . . . [A]rt and culture without anything else are apt to make us rather helpless persons in the present-day world. . . . No person can call himself educated today unless he or she knows something of science and economics and technology." Therefore, at her father's urging, Indira studied chemistry, as well as history, English, Hindi, painting, and dancing. Her year spent at Santiniketan was hardly wasted, though after leaving it in April 1935 her life was much more painful than either she or her father imagined it would be.[21]

By July Kamala was bedridden. Nan came with their frail mother to visit Nehru, and "my spirits went up," though their news of Kamala's deteriorating condition quickly made him "angry." Officialdom feared that Kamala might die while Nehru was still behind bars, so he was briefly released to visit his wife "on parole" in August. Crowds gathered round Anand Bhawan as soon as he returned home; word of his release spreading with electrifying speed throughout Allahabad. His life had become the stuff of legends, epic tales, national dreams.

"I am writing this letter to you at midnight," Nehru wrote Gandhi two days after he got home. "All day there have been crowds of people coming. . . . When I heard that you had called off the C.D. movement . . . I had a sudden and intense feeling, that something broke inside me, a bond . . . snapped. I felt . . . absolutely alone, left high and dry on a desert island. . . . The keenness of my feelings . . . amounted almost to physical pain." Jawaharlal knew how best to chastise him, for by now Gandhi had suffered so many spiritual defeats in his dealings with the other members of the Congress Working Committee—most of whom were primarily interested in running for office—that he was ready to quit Congress entirely. His vocation, after all, had never been purely political. And Gandhi did, in fact, resign from Congress that October and settle down to carry on his constructive program of spinning, weaving, and writing in *Harijan* to change the heart and restore the lost soul of Hinduism in his newly founded Central Indian Ashram of Sevagram, near Wardha.[22]

"The leading figures of the Congress suddenly became those people who had obstructed us, held us back, kept aloof from the struggle and even co-operated with the opposite party in the time of our direst need," Nehru continued in his powerful Leninist "What is to be Done?" letter from parole to the world-weary Mahatma. Just as Nehru understood that it was Gandhi who held the heart of India's peasant "masses," Gandhi recognized that it was Jawaharlal—more than any other member of Congress's Working Committee—who had the vision, brains, and courage to seize power and govern India. "And so the flag of Indian Freedom was entrusted with all pomp and circumstance to those who had actually hauled it down at the height of our national struggle at the bidding of the enemy. . . . And what of the ideals they set forth before . . . the nation? A pitiful hotch-potch, avoiding real issues . . . expressing a tender solicitude for every vested interest, bowing down to many a declared enemy of freedom. . . . The Congress from top to bottom is a caucus and opportunism triumphs."[23]

"It is the leaders and their policy that shape the activities of the followers," Nehru continued, reminding Gandhi of much he had learned from him. "It is neither fair nor just to throw blame on the followers. Every language has some saying about the workman blaming his tools. The [Working] committee had deliberately encouraged vagueness in the definition of our ideals and objectives and this is bound to lead . . . to the emergence of the demagogue and the reactionary. . . . I feel that the time is overdue for the Congress to think clearly on social and economic issues. . . . A strange way of dealing with the subject of socialism is to use the word, which has a clearly defined meaning in the English language, in a totally different sense." Now Nehru played the Cambridge tutor, and Gandhi was humbled, soon asking everyone he met for books on socialism, trying to find better dictionaries than the inadequate one kept in his ashram.[24]

Adding insult to the injury, Nehru ended, "Probably I have written in a confused and scrappy way for my brain is tired. But still it will convey some picture of my mind. The last few months have been very painful ones for me . . . I have felt sometimes that in the modern world . . . it is oft preferred to break some people's hearts rather than touch others' pockets."[25]

Gandhi responded immediately, thanking Nehru for "your passionate and touching letter. . . . I understand your deep sorrow. You were quite right in giving full and free expression to your feelings. But . . . [l]et me assure you that you have not lost a comrade in me. I am the same as you knew me in 1917 and after. I have the same passion . . . for the common good. I want complete independence for the country. . . . And every resolution that has pained you has been framed with that end in view. I must take full responsibility for the resolutions and the whole conception surrounding them. But I fancy that I have a knack for knowing need of the time. And the resolutions are a response thereto. Of course, here comes in the difference of our emphasis on the methods or the means which to me are just as important as the goal, and in a sense more important. . . . Greatest consideration has been paid to the socialists, some of whom I know so intimately. Do I not know their sacrifice? But I have found them as a body to be in a hurry. Why should they not be? Only if I cannot march as quick, I must ask them to halt and take me along with them. That is literally my attitude. I have looked up the dictionary meaning of socialism. It takes me no further than where I was before."[26]

J.P. and Narendra Deva had formally founded the Socialist Party of India in May 1934, which held its first All-India Conference under Deva's presidency, attracting close to a hundred young enthusiasts, most of whom viewed Nehru as their true leader, no matter who might preside over their meetings. Gandhi, who indeed had "a knack for knowing," realized the potential power of such a party if Jawaharlal ever should opt to take overt, rather than retaining covert, command of its idealistic and enthusiastic cadre of young men, including some of the best and brightest of nationalist India's educated minds. He remembered what Motilal had said about the impatience of the next generation and their frustrations at being told to wait ever longer, and he knew from Jawahar's frank letter how angry he was, how betrayed he felt from behind prison bars at the opportunism of congressmen who had never served a day in prison but were eager to reap the fruits of freedom and elective office. He decided, therefore, that perhaps the best way to keep Nehru in Congress, rather than losing him to a splinter party, was to step down himself, as he wrote Vallabhbhai in early September. "I have come to the conclusion that the best interests of the Congress and the nation will be served by my completely severing all . . . connection with the Congress. . . . This does not mean that I cease to take any interest. . . . But I feel that my remaining in it any longer is likely to do more harm than good. I miss at this juncture the association and advice of Jawaharlal who is bound to be the rightful helmsman of the organization in the near future. I have, therefore, kept before me his great spirit . . . [H]is reason would endorse the step I have taken. And since a great organization cannot be governed by affections but by cold reason, it is better for me to retire."[27]

Personally Gandhi was much closer to Vallabhbhai, his Gujarati "brother," than to Jawaharlal, but Gandhi was always realist enough to know that even though Sardar Patel was a rock and a man capable of leading any peasant march or standing firm before any army, whether of police or foreign power, his mind remained provincial. Nehru's mind was not only national but also international. "My presence more and more estranges the intelligentsia from the Congress," Gandhi continued his letter of resignation. "I feel that my policies fail to convince their reason. . . . [T]here is the growing group of socialists. Jawaharlal is their undisputed leader. I know pretty well what he wants and stands for. He claims to examine everything in a scientific spirit. He is courage personified. He has many years of service in front of him. He has an indomitable faith in his mission. The socialist group represents his views more or less. . . . That group is bound to grow . . . I have fundamental differences with them. But I would not . . . I may not interfere with the free expression of those ideas, however distasteful some of them may be to me. . . . For me to dominate the Congress in spite of these fundamental differences is almost a species of violence which I must refrain from. . . . I leave [Congress] only to serve it better in thought, word and deed. I do not leave in anger or in a huff. . . . Everything will go well, if we are true to ourselves . . . living in complete detachment, I hope I shall come closer to the Congress. Congressmen will then accept my services without being embarrassed or oppressed."[28]

For millions of Indians, Gandhi's withdrawal from the political fray and the coming struggle for power only confirmed his saintly nature. Who but a mahatma, after all, would want nothing for himself? What better preparation for *moksha* (release) than to live in complete detachment? Yet Gandhi never closed a door to any friend or congressman willing to "accept my services." And so for the next half decade and more, the Working Committee members and every Congress president would wend his way to Wardha's dusty, remote village ashram, where the Mahatma sat spinning, sleeping, or fasting, to seek his saintly advice or blessing or merely to sit cross-legged and look at his naked torso and bowed bald head long enough to find the "answer" to countless questions that troubled and perplexed most mortals. After Gandhi stopped going to Congress meetings, its leaders all came to him, seeking the peace of heart and mind that always rewarded those who made that arduous journey to India's geographic center, where only the music of a spinning wheel would be heard by the countless visitors from afar. As Sarojini Naidu later quipped, "Who knows how many millions of rupees it costs India to keep Gandhi in poverty?" Nehru, too, journeyed to Wardha when free to do so, but the bond that once tied him to Gandhi had snapped, and the distance dividing their mental worlds thereafter remained far greater than the mileage between them. Having learned that "hardest lesson" so late in life, Nehru could never return to sit at Bapu's feet as once he had done. His years of political discipleship were over.

In October, Kamala was taken to Bhowali Sanatarium high in the foothills of the Himalayas, in the United Provinces' Almora district. As if her tubercular condition were not bad enough, she now also suffered "a severe heart attack." Nehru was moved, therefore, to neighboring Almora Prison, from which he could easily be brought to visit his bedridden wife, whose fever now rarely fell below 103 degrees. He was forty-five that November 14, but received no flowers or gifts and had no visitors. "One of the blankest birthdays I have had," he observed morosely. "I waited in the morning. Nothing came—Afternoon also blank. Curiously enough not even the jailer, who comes almost daily, turned up. . . . Late in the afternoon . . . [f]lowers & fruits arrived from Bhowali & a doctor's report which was not good."[29] Padmaja Naidu and Mahmud sent books, but they reached him a week late. "One of the books that Padmaja sent was a beautiful one about [Anna] Pavlova with scores of fine pictures. In my student days in England I had often seen her and I was very fond of her dancing. . . . The book brought back the old days very vividly to me . . . how happy I was."[30] He hoped for possible early release, but Hoare responded negatively to a question in the House of Commons about that possibility. "So that's it!" he told his diary. "I am not to be discharged. Strange how on reading this my prospect of the future changed. Immediately I began to think of the many books I can read in prison and the other things I can do. The outside world receded into the distance."[31] The next day, however, he felt "very depressed." Dismal December in the hills. Rain and mist. He sat wrapped up inside his cell, shivering. "And yesterday I acknowledged the onset of age by having a hot water bottle in my bed. . . . Heigh ho! And so the year is almost past."[32] He was allowed to spend New Year's Eve with Kamala. "We talked and talked about the past, present & future and I think we succeeded in unravelling many a knot. . . . And it really surprised me how attached we are to each other." Nehru told his diary at the start of the New Year (1935) that "I learnt that Bapu is likely to return to prison before very long. . . . It will be a good thing. . . . Today I feel light-hearted." Misery loves company. Nehru's informants were, however, wrong.[33]

Nan soon wired him of their mother's first stroke. She was "unconscious . . . and then came the thought that I had seen the last of mother. . . . How she loved me and was wrapped up in me. She would ask my advice in the most trivial things. . . . How she had suffered because of my long absences in prison and how she had bravely faced . . . a police lathi charge. She did this more for her love of father and me than for any principle. . . . I collapsed and wept. . . . Poor little mother—what a tortured life she had led for many years—all her children in prison. . . . And now the end had come suddenly. Perhaps it was as well that it was sudden and perhaps painless." But the end did not come for Swarup Rani

Nehru for another three painful years, not until a third massive stroke finally released her in 1938.[34]

Faithful Feroze Gandhi remained in Bhowali at Kamala's bedside; only he and her nurse stayed there all day long. Lonely, isolated Indira soon fell in love with him, almost as much as he had loved her poor mother. When Jawaharlal much later learned of their secret love, his rage proved almost as passionate. The prison superintendent asked Nehru if he wished to be shifted to Bombay, to be closer to his mother. "I told him that certainly I would like to be by her side. . . . And if I go what of Kamala? She has few companions now—indeed only Feroze & the nurse. . . . If I go to Bombay how long will I be kept there? . . . Am I going to be treated like a shuttlecock . . . ? What a life has fallen to my lot! Indu is full of dissatisfaction at the present college course. . . . Wants to chuck it up. . . . How cold it is. Last night at Bhowali it froze hard."[35]

A week later, "Bad news of Kamala. She seems to be going down and down. . . . The only treatment they have is failing . . . hopelessly."[36] Another week dragged by, and this time after visiting Kamala, he felt "terribly miserable." Kamala seemed "reserved," remote. He tried to ignore the "psychological change" in her and talked to her for more than an hour, reading some of what he had written, reciting a few poems, "and then she told me of her desire to give away to charity part of the proceeds of the sale of her jewellery. I was a little surprised as she had always looked upon this more or less as belonging to Indu. But of course I did not mind it in the least, only I criticised the proposed objects of her charity." The final irony was that *his* wife's treasure in jewels would go to the Ramakrishna Mission and its swamis. "I was taken aback. I had known for long that she had been religiously inclined . . . but it was all rather vague. It seemed to me very far from religion or search for God, whatever that may be, and much more a type of hysteria. . . . I seemed to be losing her—she was slipping away and I resented this and felt miserable. Many of our little tiffs . . . were due to this background of conflict. Even when I was out of jail . . . this shadow seemed to lie between us. . . . Naturally I told her that she had perfect freedom to think & act as she wanted to. I said so but I did not feel like it."How wretched he felt, how sorry for himself. "In politics I was an unhappy, lonely figure, and now even my home life was ending for me. Loneliness everywhere. Nothing to hold on to, no life-boats or planks to clutch while I struggled with the rising waters."[37] Kamala was dying, but it was himself that Nehru felt sorry for.

Then a few days later Indira wrote and cheered him up. "The snow was very pleasant to see and feel . . . I thought so much of you and of Switzerland. . . . And then Geneva with our flat there . . . and the Ecole Internationale . . . and Montreaux with its picture-like beauty . . . and Chamonix and Mont Blanc . . . and Bex; and the Establissement Stephani at Montana. How small you were then and how you have grown since."[38] What had triggered that memory was Indu's letter recalling her first sight of snow, "all wrapped up in a blanket with my nose flattened against a hotel window pane in Lausanne. . . . That was when I had bronchitis at the Hotel Mirabeau, where a French lady took a fancy to me and sent me flowers & chocolates and a card that I still have, on Easter day."[39]

"Languages are also desirable and languages are tricky things after a certain age," Nehru advised his heir and successor. "Of the Indian languages I understand you are taking up Bengali and Hindi. French of course you are taking and presumably English literature. Are you doing German also? . . . If you want to take it, do so by all means. . . . I look upon all these studies of yours as preparatory to specialization, probably in Europe . . . I do not yet know what your own desires are in regard to this. . . . In the ordinary course you will remain at Santiniketan till the summer of 1936. If possible I should like to go with you and mummie then to Europe but it is quite impossible to make any plans so far ahead." They all would, in fact, be together in Europe for the first part of 1936.[40]

Nehru worked at training Indira to care for her body as well as her mind. "I want you not merely to keep well but to be aggressively fit and, as far as possible, to make yourself impervious to disease. What a terrible waste of energy is illness. . . . A little care now might make such a lot of difference later on. Many people take tonics but really the best thing, especially in youth, to build up a strong body is cod liver oil. Horrible stuff you will say and I entirely agree with you. But there is a very good substitute—halibut liver oil. This is as bad but the great advantage is that one has only to take two or three drops. . . . Even I have started taking them regularly! I would be glad if you followed your father's example in this respect at least. Ask your doctor there and tell him to get it for you from Calcutta. . . . It is called Crooke's Colossal Halibut Liver Oil. I am sure it will . . . make you strong and capable of resisting any disease."[41]

Next night, "I had a strange dream," he confided to his diary. "We were living in the old Anand Bhawan . . . and Kamala was unwell. Something happened—I have forgotten already what this was—and Kamala decided to leave me! Indeed she went off. I am not clear if she went alone or with someone. I followed her to the Allahabad station but I do not seem to have met with success there. What a curious mix-up is one's mind! . . . Well, I suppose another year will go by somehow. . . . What could I do if I was out? Nothing much. I would sit quietly at Bhowali. . . . Meanwhile Sastri grows lyrical & hysterical over the virtues of the British connection, & Sapru chants the praises of King George & blesses the jubilee. It takes all sorts to make this world!"[42]

Jinnah had met with Congress President Rajendra Prasad in Delhi in January, their "heart-to-heart" talks aimed at resolving the continued communal deadlock. No agreement was reached, however, for Congress refused to accept Jinnah's demand for more reserved seats for Muslims than their percentage in the population appeared to warrant, and Madan Mohan Malaviya and other Hindu Mahasabha communalists prevented Rajendra Babu from budging throughout that Congress–League summit, which might still have averted the disaster of partition. In February, Jinnah rose in the new assembly in Delhi to propose accepting Ramsay MacDonald's Communal Award "until a substitute is agreed upon between the communities concerned. Now, it may be that our Hindu friends are not satisfied . . . but at the same time I can also tell the House that my Muslim friends are not satisfied with it either. . . . But why do I accept it? . . . I accept it because we have done everything that we could so far to come to a settlement . . . [W]hether I like it or whether I do not like it, I accept it, because unless I accept . . . no scheme of Constitution is possible . . . [T]his is a question of minorities and it is a political issue . . . [W]e must face this question as a political problem; we must solve it and not evade it."[43]

Nehru believed that the Hindu–Muslim problem was first of all stirred up by British imperial policy of divide and rule and that it was mostly an economic, rather than a political, issue. Nor did he really think that barrister Jinnah was much of a Muslim or had half the Muslim support enjoyed by Congress, especially in the United Provinces, where some of Nehru's best friends and allies were Muslims, or in the North-West Frontier Province, where he and Gandhi had so assiduously courted and won the undying support of Khan Abdul Ghaffar Khan and his Pathan army of "red shirts" *(khudai khidmatgars)* and his Cambridge-educated brother, Dr. Khan Saheb.

Indira's warm response to her father's advice about her curriculum elicited a much longer, more detailed "tutorial" on literature from him in late February. "Most of Plato's books are very interesting and thought provoking. Try one of them—say the *Republic*—and see how you like it. The old Greek plays are also fascinating. Some of them are so powerful they make one shiver almost. Sophocles, Euripides, Aeschylus for tragedies—Aristophanes for delightful comedies . . . talking of plays—have you read *Shakuntala?* Not of course in Kalidas's original [Sanskrit] but in translation. It is worth reading. Shakespeare also makes fascinating reading if one takes to him for pure pleasure and not for examination stunts. . . . [H]is sonnets are extraordinarily beautiful. I do not know if you are keen on poetry . . . you have been studying Walt Whitman and Browning—excellent persons but not poets after my heart. Modern poetry is . . . some good, some totally incomprehensible to me. The most lyrical and musical of modern English poets is Walter de la Mare. . . . When we were together . . . last August you told me that you would read Tolstoy's

War and Peace. Did you read it? . . . Another great novel by Tolstoy is *Anna Karenina.* Have you read much of Thackeray or Dickens? . . . I used to enjoy them in the old days. You should certainly read Thackeray's *Vanity Fair.* . . . I have always been interested in utopias and books peeping into the future. William Morris's *News from Nowhere* was an early favourite of mine. Then there is Samuel Butler's *Erewhon,* and . . . H. G. Wells's *Men Like Gods.* Bernard Shaw you have read a little. Read more of him. Almost all his plays are worth it and his prefaces. . . . A favourite author of mine is Bertrand Russell . . . eminently sensible. I think you will like him . . . within a few days you will get the 2nd volume of *Glimpses* and you might honour it with a perusal. . . . Perhaps, if I am provoked, I might write another book for you!

"Why does one read books? To instruct oneself, amuse oneself, train one's mind. . . . Ultimately it is to understand life with its thousand facets and to learn how to live life. Our individual experiences are so narrow and limited, if we were to rely on them we would also remain narrow and limited. But books give us the experiences and thoughts of . . . often the wisest of their generation, and lift us out of our narrow ruts . . . as we go up the mountain sides fresh vistas come into view, our vision extends further and further. . . . We are not overwhelmed by our petty and often transient loves and hates and we see them for what they are . . . hardly noticeable ripples on the immense ocean of life. For all of us it is worthwhile to develop this larger vision for it enables us to see life whole and to live it well. But for those who cherish the thought of rising above the common herd of unthinking humanity and playing a brave part in life's journey, this vision and sense of proportion are essential to keep us on the right path and steady us when storms and heavy winds bear down on us." Even at seventeen, Indira knew that she was destined to rise above "the common herd of unthinking humanity," even as she had known since she identified with Joan of Arc that hers would be "a brave part" in the drama of life's journey, following her beloved Papu up India's mountain to its highest peak of power.[44]

Back from a visit to Kamala a week later Nehru felt less sure of himself. "How curious is life! I seem to be learning its mysterious ways still. . . . It is overwhelming at times and how it mocks! I have returned from Bhowali full of strange thoughts and feelings which I hardly dare put down in black & white," he confessed. "There is a sense of emptiness & loneliness and yet there is also some relief. . . . At least I feel I know what the trouble is which caused this tension. Understanding is often painful but it is better than a blind search for something which eludes one. And yet I do not know—so many fresh mysteries crop up. So many dark alleys leading I know not where. What can one do except to keep calm and preserve an untroubled exterior and wait for life's playful fantasies to develop? . . . [O]ne can read Blake:

> Never seek to tell thy love,
> Love that never told can be;
> For the gentle wind does move
> Silently, invisibly."[45]

His plan now was to send Kamala with Dr. Madan Atal to Europe, and he had reason to believe he would be released before his term expired to follow her there. "Govt. might be informed, informally and through the 'friends' who delight in such tasks," he confided to his diary on March 5, 1935. Those "friends" were Sapru and his colleagues. "What was to be done? There was me—and Indu—and mother, weak and ailing and liable at any moment to a relapse . . . if I am let out I can hardly remain behind . . . I want to go of course—I want very much to be with her and yet . . . perhaps it would be well if I kept away from her for a while . . . yet I don't know. I feel at sea in this strange world. . . . Heigh-ho!

> How could I dream that I
> With my two hands should touch the sky?"[46]

A week later there was an Associated Press report, he noted, that "the U.P. Government are prepared to release me if I want to accompany Kamala to Europe. Ifs & buts. I don't want to go out on any such condition—but what am I to do? I want to go to Europe—out of India—for a while for my own sake and even apart from Kamala. But there is mother. . . . Then Indu—what of her? Another thought worries me—the political situation . . . so many of my colleagues are in trouble, in jail and outside, and I am deserting them. I don't suppose I could do anything, but one's presence . . . is something after all. Ordinarily I would have liked to be out for a few months and to meet friends and colleagues. . . . In October or thereabouts K and I and Indu might have sailed. . . . Meanwhile I . . . feel much fitter than I have done for long."[47] The mere thought of returning to Europe proved medicinal.

Nehru felt less like a deserter at the prospect of traveling abroad when he reflected on what Gandhi had recently done, resigning from Congress to retreat to his ashram, where "he is writing long articles on the relative merits of cow's milk and buffalo's milk!" Nehru noted. "And of course the injury done by eating mill-ground rice and sugar. These are the great planks of village uplift."[48] Now that he had made his plan of "escape," however, it was not easy to get Kamala to agree, for she had no strength left and knew how futile it was to travel so far from home. She had hated it there the last time they went to Switzerland and Germany, so many strange foreign faces and cold places. There was no cure for what ailed poor Kamala. "What a child K is! That irritates me often enough. . . . How my moods change when I think of her. How much she means to me and yet how little she fits in or tries to fit in with my ideas. That is really the irritating part, that she does not try, . . . I have been feeling somewhat unhappy for the last two or three days," he confessed on March 19. "No letter from Indu for a long time. How she forgets me! . . . When will I go out?"[49]

Indira had begun to enjoy her life at Santiniketan much more and was "not frightfully keen on going"[50] to Europe, as she wrote Darling Papu in response to his letter telling her of the possibility. "Of course if I can be of any help to Mummie *en route* or at the sanitorium, I had better go." "I feel clearly now that you should accompany her," Nehru insisted in his next letter. "You will be of the greatest help." He alerted her in early April to be prepared to "pack up your tooth-brush and march off!"[51] A week later Dr. Atal came to his prison, and all the details of the passage to Europe by sea for Kamala and Indu were arranged. Then Nehru wrote to inform Tagore "that it has been decided that my wife must proceed to Europe for further treatment. . . . As I am incapacitated from accompanying her, it has become all the more necessary that Indira should go with her. . . . This new development distresses me . . . but I see no way to avoid it. . . . Her [Indira's] own testimony, and that is important enough, is clear . . . she has been very happy at Santiniketan and has no desire whatever to leave it. I was looking forward to her remaining there for a longer period . . . I rejoiced that I had been fortunate enough to choose Santiniketan for her education at this stage of her life."[52]

Kamala's and Indira's passage was confirmed for May 23, 1935. The cabin "is not at all good," Nehru confessed. "This is troubling me as it will mean a painful voyage. . . . Another matter that worries—will K be strong enough to walk a little—to the ship, from cabin to deck? So far she cannot walk at all. It will be bad if she has to be carried about. . . . Various activities—making plans, arrangements, meeting family people after nearly 6 months—fill the day and kept me a bit exhilarated. A change from the jail routine." He also noted a recent Hindu–Muslim communal tragedy in Firozabad that had left eleven dead and another thirty-one injured in the ensuing riot. "What a disgusting, savage people we are. Politics, progress, socialism, communism, science—where are they before this black religious savagery?"[53]

Two weeks later he read of the A.I.C.C. meeting, berating the "spirit of terrible caution" that characterized its discussions. "Rajendra Babu has disappointed me greatly," Nehru confessed. "But what can he do with his

Working Committee? Rajagopalachari ["C.R." of Madras, the only Indian later to serve as governor-general] is probably the ablest of the lot and he has gone terribly to the right. Sarojini, Nariman etc. were always of the cautious right except when Sarojini's poetical fervour made her say fine nothings. . . . Even Bapu—he is either a noncooperator or a full-blooded cooperator. The fire or a sofa. . . . He can think only in extremes—either extreme eroticism or asceticism. Was it not Aldous Huxley who said that the ascetic was the counterpart of Don Juan? . . . Meanwhile Vallabhbhai, that great peasant leader, has presided over a provincial peasants conference at Allahabad, under the august auspices of Purushottamdas Tandon, and no doubt he preached love and goodwill towards those pillars of society, the landlords. What does Tandon stand for? I don't suppose he knows himself. Singular how an intelligent person like him can be so lost in cloudland. . . . Religion is really more materialistic than science."[54]

Nehru kept waiting for word of his early release, for Sapru knew what he wanted, but the viceroy was Lord Willingdon, and he was in no rush to set "Comrade Nehru" free, sick wife or no wife. "Felt weary and stale and rather disgusted with everything," Nehru confessed. "Again there is a vague rumour that I might be let out to go with Kamala . . . only hints thrown out by newspaper men. And yet even a rumour of this kind, which I do not believe in, excites me a little and automatically I begin to plan. I want to be out of this damned hole but still I had rather not be beholden to govt. for their grace."[55]

Indira bid a tearful farewell to Gurudev Tagore and her classmates at Santiniketan that blistering April, and in early May she reached Bhowali to meet her father there, as well as Kamala's cousin, Dr. Madan Atal, and the ever-faithful Feroze, all at Mummie's bedside. Indu's own frustrations and feelings of sadness at having left her lovely school in midcourse were compounded by the gloom she felt at seeing her dying mother again, still coughing up blood on the eve of her last futile journey. The visit proved so painful that Indira had to cut it short to go to Allahabad, insisting she had to pack there and prepare herself for the sea voyage and her higher education in Europe that her father explained to her.

Sensitive as he was to her many moods, he wrote on May 6, "As . . . you must be speeding to Allahabad . . . in all likelihood this will be my last letter to you before you sail with mummie. . . . How far you will be from me then, moving away rapidly across the Arabian Sea. . . . And I wonder when and where I shall see you again. We create these distances in our minds. . . . The world grows narrower daily and you in Switzerland will only be four or five days flight from Allahabad. And yet . . . Switzerland does seem a long way off. . . . You will have to shift for yourself a little more. . . . Home is good but it has a tendency to narrow one. . . . It does not prepare one sufficiently for wider contacts and interests and when one goes out into the bigger world, it is apt to hurt. Other people do not take so much interest in us and we are apt to resent this. . . . The sooner we get used to cooperation with others the better fitted we are for the ways of the world.

"I want you to leave India in a happy and expectant frame of mind. Do not worry at all about me. I am all right. . . . The mind cannot be enchained and I have developed the habit of undertaking great journeys mentally. . . . Parents are a curious phenomenon. They seem to live their lives again in their children. I have many wider interests in life which sometimes envelop me . . . but still I am not free from that preoccupation of parenthood and I am vastly interested in your growth and preparation for life. . . . Parents . . . have a tendency to mould their children after their own fashion and to impress them with their own ideas. To some extent I suppose this is inevitable, and yet . . . I have tried, with what success I cannot say, not to force my ideas and pattern of life on you. I want you to grow and develop after your own fashion and only so can you fulfil your life purpose. Inevitably you will carry through life certain hereditary habits and ideas . . . and I am conceited enough to think that your hereditary background is rather good. But the foreground must be your own creation. . . . Right education must be an all-round development of the human being, a harmonising of our internal conflicts and a capacity to cooperate with others. . . . As I grow older and perhaps wiser I attach more and more importance to real education. . . . I think a proper intellectual training is essential to do any job efficiently. But far more important is . . . the internal harmony, the capacity for cooperation, the strength to be true to what one considers to be right, the absence of fear. If one attains this internal freedom and fearlessness it is difficult for the world, harsh as it is, to suppress one. One may not be happy in the narrow sense of the word for those who are sensitive can seldom be crudely happy, but the loss is not great. . . . It is a little foolish of me to . . . burden your mind when it should be as free of burdens as possible. At your time of life you should grow in happiness for otherwise your youth would be darkened with care and worry. I want you to be happy in your youth for so I renew my own youth and participate in your joy. I do not want you to be a quarrelsome and disgruntled specimen of humanity!

"You cannot help carrying the burden of your family with you, not so much in Europe but very much so in India. As it happens, your family has attained a great deal of prominence in the Indian world. . . . I am proud of my father and the example of his life has often inspired me and strengthened me . . . I believe he was a really great man. If your grandfather's example strengthens and inspires you in any way that is your good fortune. If your feelings towards your father or mother also help you in that way, well and good. But your grandfather and father and mother . . . have many failings also like all human beings. The public mind, however, especially in India, has a habit of idealizing and dehumanizing the persons it likes and this is apt to irritate, in particular those who are supposed to live up to these imaginary standards. The family and one's forebears thus become a nuisance and a burden. I do not want you to feel this way about us!"[56]

Indira had bridled at the sudden changes imposed on her life and routine by the "family responsibilities" with which she had been harnessed by her father overnight. She was, after all, only seventeen and had just begun to enjoy Assamese dancing at Santiniketan and loved music and art and all the joys of carefree schoolgirl life. But Nehru knew that for his daughter to inherit the power he hoped one day to pass on to her, she now had to leave India and study, as he had done, in Europe. "There is a terrible lot of vulgarity in the world and we see it everywhere in India," he wrote her. "And when I talk of vulgarity I do not refer to the poor . . . for they do not try to pose and appear to be something other than they are. It is our middle class that is often vulgar. It has no artistic standards and it has got rather lost between Eastern and Western culture. It is hardly to blame for . . . [p]olitical circumstances have largely made us what we are and then there is our narrow domestic life. And so when we go out into the world we are often making false gestures which jar on the sensitive. I confess that I find this very painful. . . . My pen runs away with me. I have little to say about your studies in Suisse. We have discussed them already and you will fix up with Mlle. Hemmerlin. . . . I think you had better consult Mlle. Rolland . . . I think you had better take, besides languages, history and economics. If you join the university at Lausanne or Geneva you will have to choose some fixed course. I don't think you will have any difficulty about opening a bank account in Switzerland . . . I had none in Cambridge although I was a minor . . . about your age . . . I do hope you will be able to take with you volume 2 of the *Glimpses*. . . . Please take three copies. . . . Also have a copy sent to me."[57]

On May 15, 1935, Kamala was carried by Sherpa porters in her bed from Bhowali's hilltop down to the all-weather road, where a truck waited to drive her, Dr. Atal, and Indira to Kathgodam, where they caught the train for Bombay. Nehru had walked down to the road with them, "bade goodby to them," and then headed back to Almora jail. "I took the high road and she took the low. When will we meet again? And where? . . . I have the flowers at least for company and they are coming out bravely. . . . Mother was present at the parting at Bhowali. . . . She was terribly affected. Poor brave old lady."[58] Two weeks later he received Indira's cable from Aden and in early June wrote her again, after having a second cable from Vienna:

"So you have arrived at the end of your sea journey . . . I have followed the course of the *Conte Rosso* since it bore you away from Bombay, and have tried to find vicarious joy in the voyage. Did not the dark water of the Indian Ocean sparkle with phosphorescence in the night? And then Aden, drab and dull, with the white glare of the hot sun. . . . Do you remember Djibuti near by? That was bare enough and dreary, but it had some romance for it is the gateway to Abyssinia. . . . The only thing worth seeing was sometimes the magnificent body of a negro, like an ebony statue.

"From Aden I accompanied you through the Red Sea . . . [and] the Suez Canal with its sandy stretches on either side and an occasional camel going by. . . . It was pleasant to enter the cool, blue Mediterranean with all its memories of early civilizations. The Isles of Greece—'where burning Sappho loved and sung' (Byron's lines come back to me from the old days in Harrow when I had to learn them by heart)—and Ithaca, home of Ulysses . . . with its bare rock standing out of the sea. . . . Up the Adriatic, a troubled sea with rival nations on either side glaring at each other. And Venice, queen of that sea, which once 'held the glorious East in fee.' But you could only have a fleeting glimpse of her for you were bound for Trieste. . . . So we travelled together, you and mummie and I, but now I do not know . . . I have to stay in Almora jail . . . this has been a record summer all over India. . . . Even in Almora it has been 100°F in the shade, and in Allahabad and Lucknow 115° and 116°. . . . It is June. 'What a tune, kind June, you are playing all the noon,' says the Harrow song. But June is not kind here. . . . But all this grousing about the heat seems trivial and out of place in view of the appalling catastrophe that has overwhelmed Quetta. . . . On the last day or night of May a terrific earthquake laid a fine city and numerous towns and villages low, and where Quetta stood is now a heap of ruins, a wilderness of brick and plaster . . . covering thousands of human bodies. . . . The Quetta earthquake is evidently on a vaster scale even than the Bihar one and estimates of the dead alone exceed fifty thousand."[59]

"My flowers are fighting bravely for life but their thirst is seldom quenched," Nehru continued in his June letter to Indira. "Perhaps *en route* you looked through the second volume of *Glimpses*. It is a fearsome object, enough to frighten the bravest. Yet I hope that the inside is not so bad or heavy. . . . I should like to know what parts of it interested you. . . . Probably the letters dealing with economics and financial affairs you found dull. . . . My own private belief is that some of my last letters dealing with the world financial situation are rather good. . . . They are very elementary of course and yet, strange as it may sound, most of our big politicians and the like are hopelessly ignorant of the subject. If you read through those two volumes you will have a better knowledge of world history and affairs and the economic and financial basis of the modern world than many a well-known politician. . . . Nothing like blowing one's own trumpet!"[60]

A few days later Betty and Raja came to visit, reporting that Ranjit was "very ill." He and Nan had moved into Anand Bhawan with their children, but now suddenly it seemed as if the old curse had returned to wreak its vengeance on poor Ranjit, whose health remained fragile for the next seven years, until his early death from pneumonia and pleurisy. "What an arid waste is Indian humanity with few cases of intellect & character—very few," Nehru complained to his diary. "I feel depressed when I read the paper daily and notice the vulgarity all round. I cling to a hope, based on no sufficient reason, that great things can be done in India and rapidly. But doubts arise—with this material! And happenings all over the world are also depressing. Soviet Russia seems to be changing for the worse. The fine idealism that moved her is no longer apparent."[61] Then came a bundle of books from Padmaja to cheer him, although on the same day he learned of a dreadful automobile accident in which Nan was thrown out of her car on Fort Road and had a "narrow escape from death. . . . Ranjit ill—mother ill and weak—Kamala ill and far away—Indu also inaccessible—and I sitting in prison helpless and incapable of helping others!"[62]

From Vienna, Kamala was taken to Berlin, where she was operated on immediately. Part of her lung was removed, yet the operation only increased her suffering and extended her life for little more than half a year. From Berlin they went back to Badenweiler's sanatorium in the Black Forest, where they had taken her last time. Subhas Bose welcomed them to Vienna and escorted them to Berlin. Bose was in Europe at this time, studying fascist doctrine and the methods of Mussolini and Hitler, which he later hoped to integrate into his own scheme of "integral" National Socialism for India after he returned home.

"So you are all in Badenweiler now and likely to remain there for two or three months more," Nehru wrote Indira in early July. "Not far is the Rhineland and the banks of the Rhine must be covered with the vine. . . . You were with us, were you not, when we steamed up the Rhine . . . passing its great rocks with frowning castles seated atop of them, and legends of long ago. . . . And so we passed from Cologne to Mainz and thence to the old delightful city of Heidelberg. . . . I am very fond of Heidelberg and several times I have been there, once with Dadu [Motilal] so long ago as 1909 to pay a visit to Shridhar Chacha [Uncle Shridhar] who studied there and lived in a pension run by a professor over eighty years old. That professor's one consuming passion was hatred for England. . . . So you are in a beautiful land and . . . I hope you will take some advantage of your position and have an occasional excursion. I am sure you will give pleasure to Mummie if you do so, rather than if you remain with her all the time."[63]

Nehru long feared that Indira might come down with her mother's morbid disease, as he wrote to Dr. Atal at this time. "It is curious how nothing radically wrong has ever been found with Indu and yet she does not prosper. I remember, when we were last in Switzerland, I had her thoroughly overhauled by children's specialists and other doctors and they were all rather surprised at finding nothing wrong with her. And yet obviously she was below par. . . . I am not worried about her health. I think she will get over these infantile weaknesses. . . . There is one aspect, however, which doctors do not often pay attention to, and that is the psychological. I am convinced that to give psychological satisfaction is far more necessary than . . . tonics. This, in my opinion, requires a life of mental and physical activity in a pleasant atmosphere with companions and fellow students." Nehru was quite right, but he had just removed her from that very atmosphere. "I do not mean to imply that Indu is at all morbid or given that way. I am happy to know that her outlook is fairly sane and healthy but she has just a tendency to introspection which is not to be encouraged at her age," he cautioned Madan Atal.[64]

To "Darling Indu" Papu wrote the same day: "Madan Bhai . . . told me that you were not exactly flourishing like the green bay tree. . . . I do not particularly fancy your hob-nobbing too closely with the tribe of doctors. . . . Of course under certain circumstances one must go to them and even swallow medicine, or get cut up. Surgery attracts me rather more than medicine. . . . More and more I feel that health comes from inside rather than outside, from the observance of simple rules of life and activity. . . . The body must be looked after. . . . But to make of it an invalid and to think and speak continually of its pains and troubles is not only a most distressing habit (alas, so prevalent in our country!) but is calculated to make its condition worse. I sometimes feel that speaking about disease and illness, except in the case of necessity, should be forbidden by law."[65]

Indira only became more preoccupied with her ailments, however, convinced that like her mother, she was dying, driven to distraction and depression by fear and the frustrations of finding herself daily facing death and hearing Kamala's prayers for "release" from her misery.

"I have always been a bit of a student, trying to learn to understand, but largely this effort was intellectual," her father wrote on and on from his own enforced isolation in a cell ten thousand miles away. "There was also the emotional element . . . learning from crowds, the appreciation of mass psychology. Latterly I have felt drawn more and more towards nature—to plants and animals. Maybe it is a relief and an escape from human folly, human cowardice and human knavery! . . . Are you not tired of me, Indu-boy, writing all this stuff to you and wearying you even at this greater distance with my wordy chatter?"[66]

The next month Kamala's condition deteriorated so rapidly that Dr. Atal cabled Nehru, urging him to come quickly if he ever hoped to see his wife again. The viceroy was also cabled on August 31, 1935, from Whitehall, advising the release of his most famous prisoner. Agatha Harrison knew the secretary of state well and had told him to "let Nehru go" or face "disastrous" consequences should his wife die without one last meeting. On Monday, September 2, at 8 P.M., Nehru was informed by Joint Magistrate Vira of Almora, who had just received his message from the chief secretary of the United Provinces, that "the Governor-General in Council have decided to allow Pt. [Pandit] Jawaharlal Nehru to proceed at once to Germany to enable him to join his wife, and for this purpose have suspended his sentence under section 401 of the Criminal Procedure Code."[67]

A day later Nehru reached Allahabad and a week later flew out of India, headed west.

15

European Refresher

1935–1936

"IN EUROPE, I am told, I can go where I like," Nehru noted on the eve of leaving Almora jail. "Here, before I go off, govt. would not like me to indulge in public political activity . . . a kind of gentleman's agreement." Being a gentleman, of course, he agreed to the viceroy's stipulation.[1]

"The Governor-General in Council trusted to my honour not to make any political speeches during this period," Nehru informed the press that awaited his sudden early release with many questions. "I might add that the burden on my honour is not a heavy one. After over nineteen months of seclusion it would be extraordinary vanity and folly on my part to rush suddenly to the platform and presume to give advice on public questions to my colleagues and others."[2] He had, however, spent most of his time during those months doing precisely that in the brilliant and blistering autobiography he had just completed, the only typed copy of which went with him to Europe a week later. "It would have been an impertinence on my part and unfair both to myself and my colleagues, to come to any decision without the fullest consultation with them," the diplomatic gentleman Nehru added, excusing himself to hurry home to Anand Bhawan, where Nan and her daughters awaited him with floral garlands, hugs, and kisses.

That very night, however, he had to leave Allahabad by train for Delhi. "New Delhi was a blaze of illumination," he noted, amazed at how much had been built, how beautiful it all looked, how busy, happy, and healthy people seemed, even without his having been there to lead and advise and care for them, all these months and years. Then he went south to Jodhpur, in princely Rajasthan's desert, with its airport. "We . . . were taken to the state hotel . . . a very up to date affair," Nehru admitted. "A card was handed to each one of us informing us that we would be called at 3.30 A.M., *chhota hazri* [breakfast], luggage to be put out at 4 A.M., departure from hotel at 4.15 A.M., plane starts at 4.30 A.M.," Nehru recorded in his travel diary. (Indian airlines' schedules have changed little since then. The cool predawn remains the best time for take-offs, just as midnight or later is still the time most favored for landings.) "I managed to shave, have a good hot bath—there was running hot water and an English bath—and to get ready at the scheduled time. . . . And so we were off in the starlight."[3]

"I found Kamala very much changed for the worse," he wrote Gurudev Tagore shortly after reaching her sanitorium in Badenweiler. "She was suffering from an acute inflammation of the pleura and was running very high temperature. . . . The inflammation seems to be no better and pus has to be taken out every few days by aspirations. . . . The acute condition has almost become semi-chronic. . . . It is difficult to know how matters will shape themselves. She has made a gallant fight of it."[4]

She had, indeed. Poor Kamala prayed daily for release from her torture all the months she had lain coughing in Bhowali, where she wrote her swami in March, "I am fully confident that I shall be with God and that it will not be long before I am liberated from the snare of worldly life. I feel I have nobody but God and my way to Him is being cleared by circumstances. . . . What are the rules regarding staying at Mayavatti? How does one get there? . . . The president and secretary of the Banaras Mission are here. Somebody has donated land in Almora for TB patients. They came to see it. They return tomorrow. Let us see when God sends for me." That had been her fondest wish, to remain in the mission's TB hospice in India, to die there as soon as possible, in relative peace. She had no faith in German surgeons, but her husband had no faith in God.[5]

"I have been at peace during my illness," Kamala wrote her swami on the eve of her operation in Berlin. "God is helping me. I have left everything in His hands. I am happy in whatever He does. . . . I have passed on my burdens to Him and therefore feel serene. . . . Belief in God gives one much comfort and one feels free of worries. But people do not always realise this. They say there is no God, because if there were God an innocent person would not be suffering and going through the poverty one sees around. It is not easy to make others understand this."[6]

The day after Nehru reached Badenweiler, Agatha Harrison came to greet him there and soon reported to Gandhi: "I met Indira . . . what a pathetic figure—though young in years—old beyond her years in experience and suffering. . . . Jawaharlal Nehru arrived about 1.30 that night. . . . We met early the next day. . . . He was standing in the sun talking to Dr. Atal, Subhas Bose and Indira. . . . What touched and moved me most was to watch this Father and daughter together. Indira was holding tight to his arm, every now and then rubbing her head against his shoulder and some of the 'years' that I noticed the day before—seemed to have slipped away, and she was a different person. . . . The setting in which your loved friends now are—is very beautiful."[7]

Agatha had started London's India Conciliation Group in 1931, which included not only Quakers like herself and C. F. Andrews, but also Liberal and Labour members of Parliament and professors and intellectuals from the London School of Economics, Oxford, and Cambridge, all of whom were eager to meet with and listen to Nehru. Krishna Menon was also a friend of Agatha's and worked closely with her in arranging Nehru's forthcoming gala receptions in London. Krishna Menon was busy lobbying his friend Allen Lane of the Bodley Head to publish Nehru's autobiography. He would have a very busy few months ahead of him in Europe, Jawahar knew, for he also wanted to arrange for Indira's enrollment in Oxford before he had to return to India.

"My wife's condition continues to be grave and it is difficult to say how matters will shape themselves in the future," he replied to Horace Alexander, another Quaker follower of Gandhi's, who also was anxious to arrange several talks for him in Birmingham. "Certainly I propose to visit England if my wife's health permits it and I hope to have my daughter, Indira, with me when I go there. I imagine that in any event this visit is not likely . . . before the middle of October. . . . [I]f it is at all possible, we would very gladly avail ourselves of your kind invitation to visit Birmingham and stay with you."[8]

One of Nehru's passionate, adoring and beautiful admirers, the younger daughter of India's third most powerful industrial textile family, Bharati Sarabhai, was studying at this time in Oxford. She wrote to invite him to Oxford. "My dear Bharati," Nehru replied from Badenweiler, "It is a little difficult for me to think of you as anything but the little girl you were, full of enthusiasm—sometimes misplaced perhaps—and so I do not find it very easy to write to you. I do not know how you have grown . . . it is four years since we met . . . for you years of physical and mental and emotional growth. What has Oxford made of you? I am told you have liked Oxford and been happy there. That is as it should be. But that does not convey much, except to me a vague nostalgia, almost, shall I say, envy of youth and its glory. You were on the threshold of life in those old days, you tell me. Do you imagine that you have passed it now and tasted of life in its fullness? Linger awhile on the threshold, do not hurry. I shall see you soon . . . I intend paying a short visit to England. Indira will accompany me. We expect to reach London on the 29th October afternoon and probably we shall spend a week or so there. A day at least will be spent at Oxford . . . I want to find out if it is possible and desirable for Indu to go there next year. I have been vaguely considering this possibility." [9]

Bharati's enthusiastic response was so infectious that Jawahar replied at once, "I feel eager to enter into the spirit of the game and join you in the 'wild plans' you are hatching. How I would love to! But I cannot start earlier than the 28th, and after all what does it matter? Any day can become *Diwali* and New Year's Day if we know how to make it so. . . . But there is another ordeal which I have to face. I am told that there is—horror of horrors!—a 'Nehru Reception Committee' in London and I am to be exhibited and carried about from place to place and generally made pretty miserable. And then I must meet and hold solemn converse with many earnest and excellent persons and hollow-eyed politicians (my own kind) and behave in a dignified and leader-like manner. The 'wild plans' seem rather far off, like so many things that we desire. . . . [S]o, little woman, I must refashion the picture of you in my mind. . . . Must I also teach myself how to behave as 'heroes' are supposed to, so that I might impress your friends? That is a tiring prospect and, in any case, is not likely to be a successful attempt for I am only a bogus hero. The heroes of our childhood days stand unmasked when we grow up. . . . We shall stay at Mount Royal [near Marble Arch] in London. We reach there on the 29th afternoon. Love." [10]

Bharati was at Victoria Station waiting with her kum-kum powder to adorn Nehru and with garlands of roses to welcome him, and also to accompany him wherever he went and spoke during his tour of London, Oxford, and Cambridge. For many years after, whenever possible, she marched at his side or stood in the front rank of any audience he addressed, later writing poetry for him. Like several others, she imagined and even believed that someday he would ask her to marry him.

"Bharati dear," he wrote the day he returned to Badenweiler after their whirlwind month. "I returned to a huge pile of letters and . . . my recent memories, powerful as they were, grew dim, and I entered another world—a world of conflict and unhappiness and sordid manoeuvering and gallant endeavour wasted and helpless impotence and doubt and indecision and mutual recrimination—and yet, through all this misery . . . I began to feel a pull and a call which ever grew stronger. It was the call of India, whatever this may be. And I wondered at this pull and how strongly it influenced me, even though spiritually I often felt a stranger in my own country . . . I went to London in a strange frame of mind. Purposely I made myself receptive and tried to remove the veil which covers what has been and to look back to what I had been in the old days. The charm worked to some extent, helped by others no doubt, and behind my political talks in London, Oxford and Cambridge, my mind was wandering and exploring strange and half-forgotten by-ways of my past being. The charm worked, to my own pleasant surprise, but no charm could get over the bars of my own temperament or remove the steel barriers that my contacts with life had provided. . . . We are strange mixtures, are we not?, of the soft lotus bloom and the mud in which it grows, of putty and iron, of earth and sky. You have got Sappho all wrong. So far as I remember the lines run thus:

> The moon has set and the Pleiades,
> It is the middle of the night and time passes
> And I lie awake and alone."[11]

Once again the old struggle for the Congress crown had begun in earnest, the "sordid maneuvering" as Nehru called it. But he wanted that crown even more now, though he feared that his endeavor would be wasted, leaving him again with feelings of "helpless impotence and doubt and indecision." No sooner did he inhabit one of his two worlds than he yearned for the other. How lovely, how tempting, London always seemed, from India. But now that he was there, lionized, feted, he began "to feel a pull and a call," the "call of India, whatever this may be." He was uncertain of everything. He closed his letter to Bharati by advising her to read in Walter de la Mare's *All That's Past* the poem that starts "Very old are we men";

> Our dreams are tales Told in dim Eden
> By Eve's nightingales;
> We wake and whisper awhile,
> But the day gone by,
> Silence and sleep like fields
> Of amaranth lie.[12]

Again she responded overnight, as did he, as soon as her passionate, letter arrived: "Bharati dear, To write is difficult, you say, to speak equally so. How limited are words and how they hide us from ourselves. Sometimes we seek refuge in their triteness, at other times we lose ourselves in their muse, and ever the reality eludes us . . . for reality included the pit of hell as well as heaven. We avoid it in spite of our brave talk and to escape from it we build up illusions and fanciful castles in the wide expanse of our imagination. . . . Does a person ever understand another, or does he understand himself? Individuals meet and pass each other with eyes closed, unknowing and unknown, . . . drab exteriors covering a deep mystery. Sometimes there is a flash of understanding, a strange revelation, and then darkness. . . . And sometimes . . . as we gaze into the eyes of our dearest and nearest we find that a stranger is looking at us.

"How to bridge the gap between the dreams that fill us and the reality that is so different? How to do it? How? It seems easy when one is young. . . . In our pride of youth we care little for walls and obstacles. . . . And when you pass on to a fuller and more complex life, carry something of the breath of youth with you . . . try to widen your humanity. Only then can we endeavour to bridge the gap. At best that bridge is a ramshackle affair, liable to collapse at the first touch of storm and tempest. But the effort is always worthwhile; it brings its own reward. . . . It was good for me to see your bright face and to sense your enthusiasms. Some of the weariness and the fret, that is our common lot, left me for a while and I kept company again with thoughts of long ago . . . ideas, old and new, a strange medley, and I wondered what I was and what you were. . . . Where am I going for the Christmas vacation? I have no such vacations . . . I suppose I shall remain here in Badenweiler, unless, it is a very remote chance, I take Kamala to Switzerland before Christmas. In the latter event I shall be with her probably round about Lausanne. . . . I would of course love to see you—need I say it?"[13] So it remained, at best "a ramshackle affair," but in his own way he did, indeed, love seeing her again. And again. Yet he slept now with Kamala's amaranth silk "sari under his pillow," for he loved the feel of that silk as well.[14]

"My dear Jawaharlal," Gandhi now wrote him from Wardha. "Your letter about the wearing of the next year's crown was delightful. I was glad to have your consent. I am sure that it would solve many difficulties and it is the rightest thing that could have happened for the country. Your presidentship at Lahore was totally different from what it would be at Lucknow. In my opinion it was comparatively plain sailing at Lahore. . . . It won't be so in any respect at Lucknow. But those circumstances I cannot imagine anybody better able to cope with than you. May God give you all the strength to shoulder the burden."[15] Now that Gandhi had left Congress, he still managed to decide, before the Working Committee ever made up its mind, who its next president would be, year after year. By entering into the gentleman's agreement he had reached with the viceroy before leaving prison early, Jawahar had proved again how flexible a mind his was, that he could be trusted to toe the mark, holding his tongue when the situation called for restraint and diplomatic behavior rather than rabble-rousing. Agatha's reports from London sufficed to convince Gandhi that Jawahar was still loyal to him, for Nehru knew that Gandhi had many friends who kept him well informed. So Nehru wrote to Agatha in September, soon after she had left him in Badenweiler to return to London:

"Your suggestion that I should 'talk with some of the people with special responsibility for India' during my visit to England has set me thinking. It was almost a new idea for me and so far I had not thought of it as a possibility. . . . I was and am looking forward to my visit to England as an entirely private affair with nothing in the nature of public engagements. . . . I feel reluctant to come out of my shell. . . . You must remember that during the last four eventful years I have been mostly in prison and no one in prison can ever be in real touch with developments outside. . . . One lives in a world apart, viewing external happenings as through a glass, darkly, with phantom and almost unreal figures moving hither and thither. . . . I want of course to meet as many friends in England as possible and to get into personal touch with many whom I have desired to meet for long. I am also prepared for frank talks about Indian affairs. . . . But the light does dazzle one, after a long darkness. . . . I think it is always good to meet people on the other side. Personal contacts do not . . . solve hard problems but they do help in bringing a . . . human element in an otherwise impersonal and inhuman atmosphere. So ordinarily I would welcome such contacts. But just at present my mind rebels against any such step and it is very difficult for me to explain just why this is so. . . . I do not consciously think that I have any grievance against particular individuals, but I do find a certain hard core inside me against the system that functions in India. I suppose most of us have been hardened during these last few years and the iron has entered our souls. We may not bark as we used to do and as Sir Samuel Hoare so chastely put it, but silent dogs have also their feelings. We expected oppression and cruelty . . . but it is harder to bear them when they are accompanied by ostentatious vulgarity and a sickening hypocrisy. . . .

"What can I say to anybody in authority? I represent nobody and I am completely out of touch with recent happenings. Yet, strangely enough, I am still the General Secretary of the Congress. The only person who represents India more than anyone has done or can do is Gandhi. I may differ from him in a multitude of things but that is a matter between him and me. . . . So far as I am concerned he is India in a peculiar measure and he is the undoubted leader of my country. If anybody wants to know what India wants, let him go to Gandhi. But the British Government has tried deliberately to insult him and ignore him. . . . But if Gandhi is to be ignored and insulted where do we come in? Are we to raise ourselves by stepping over his body? Many will prefer the earth with him to the height without him. They have their own conception of loyalty. I do not know what Gandhi or the Congress will do in India, nor have I any clear idea yet of what I am likely to do on my return. . . . But if anyone wants to know what nationalist India stands for, as a Congressman, I can only refer him to Gandhi." This letter sufficed to win him back the crown.[16]

Nehru himself had not, however, abandoned any of his socialist theories or dreams and, in a letter from Badenweiler that November, admitted, "In many ways I have far more in common with English and other non-Indian socialists than I have with non-socialists in India. I am quite convinced that any real solution of our Indian problems is vitally connected with the world solution of present-day problems. It has become difficult for me to think of India apart from the world . . . I am quite sure that there are individual English socialists . . . who can think of India on real socialist lines."[17] Harold Laski was one such English socialist, to whom he was introduced in London by Krishna Menon, as "one of us." Feroze Gandhi had by now come to Germany to be near Kamala and her daughter during this last watch. Nehru sent him off to London, armed with letters of introduction to Agatha and Krishna Menon, and soon Feroze enrolled in the London School of Economics, where he became another in the growing circle of young Indians to sit at Professor Laski's feet.

Nehru worked day and night revising his books, which he had resolved to leave in Krishna Menon's hands for final editing and publication. Krishna Menon and his friend Allen Lane had started the Penquin paperback house, but the year before Nehru's return to London, the mercurial and unstable Krishna Menon had broken down entirely. He was, in fact, still in St. Pancras Hospital almost to the eve of his friend's return, but now he snapped back and poured enough tea into his emaciated body to revitalize it in many more creative, if no less acerbic, ways. Nehru wrote Congress President Rajendra Prasad from Badenweiler that "in London I met with . . . various India groups. . . . There is the Conciliation Group [Carl Heath, Agatha Harrison, and so on] . . . good people, mostly Quakers, who believe in bringing about contacts between prominent Indians and . . . the big noises. I like these people, but. . . . To me it is wholly immaterial what the personal virtues or failings of an Irwin, a Willingdon, or a Linlithgow might be. We have to deal with the policy of British imperialism. . . . Then there is the 'Friends of India' group. Also good . . . but generally ineffective. . . . A third group is the 'India League.' This has become connected with the left wing of Labour and has some prominent men in it, like Harold Laski. Because of this it is definitely socialistic in outlook. Of the three it is the only really political organisation. The man who runs it is V. K. Krishna Menon whom perhaps you know. . . . He is very able and energetic and is highly thought of in intellectual, journalistic and left-wing Labour circles. . . . I was very favourably impressed by him. Unfortunately he had been very ill and has spent the last six months in the hospital."[18]

"My dear Menon," Jawaharlal wrote him a week later, "I have sent you this morning the remaining part of *In and out of Prison*. . . . I would like the book to appear as early as possible. It has a certain political significance for India at present. . . . The feeling has been growing upon me that I must go to India for the Congress . . . quite apart from the question of the presidentship. I would, from many points of view, prefer not being president. However, I propose to do nothing in the matter and to await the course of events. But I must attend the next Congress session, and it is desirable that I reach India about a fortnight before the session begins. This means that I must sail from Europe—I cannot afford another air journey—about the 10th of February. This means that I cannot go to England in February. . . . At the latest therefore I can visit England [again] in the last week of January, and for not much more than a week. During such a short visit, I doubt if it will be desirable to have any public functions, workers' meetings etc., which you suggested. Besides they do not seem to fit in with the role I imagine I am playing just at present. . . . [U]nder different circumstances, I would gladly indulge in this kind of activity. But pure propaganda does not seem to me to be indicated from now."[19]

It was getting harder for Nehru to remember from day to night to day exactly which role he was supposed to be playing. He was no longer even sure of which "role I imagine I am playing just at present," for the same day he wrote to Krishna Menon he also wrote to Lord Lothian. The eleventh marquis of Lothian (1882–1940) had been parliamentary undersecretary of state for India in 1931–1932 and chaired the Indian Franchise Committee during all the round tables, and Agatha was anxious for Nehru to meet him. "Dear Lord Lothian, I have been looking forward to meeting you greatly. . . . It is always a pleasure to meet people who open out new avenues of thought and help one to see a little more than the tiny corner of the world which is the average person's mental bent. . . . [T]he attempt must be made to cultivate friendly contacts for without them the world would be a drearier place than it is. . . . I should have liked to meet you. . . . I like the beautiful houses and countryside of England and your superlative description of Blicking attracts me, but it is really the man that I want to see, not the house he owns."[20]

Lord Lothian was so favorably impressed with Nehru's letter that when he next spoke to Agatha and she told him that Nehru was planning to return to London at the end of January 1936, he immediately said, "I will delay my sailing to the States if he can come earlier." Agatha was very excited by this gesture and wrote a confidential memo to Nehru about it, insisting that "Lord Lothian's friendly gesture has done something important."[21] Lothian, who had edited *The Round Table* for many years and was later Britain's ambassador to Washington during the last year of his life, was one of England's leading intellectual politicians and became something of a bridge between Nehru and the Tory establishment, including the next viceroy, Lord Linlithgow.

"Dear Lord Lothian," Jawaharlal next wrote him from Badenweiler in January, "I have read your long letter more than once, as well as your article in the *Twentieth Century* . . . on subjects which interest and affect us all so deeply. . . . I entirely agree with you that we are in the midst of one of the most creative and changing epochs in human history. It does seem that we have reached the end of an era and are on the threshold of another. I also agree that the two ideals which are moving most intelligent and sensitive persons are: the ending of the present anarchy of sovereign states . . . and the creation of a world order; and the socialistic ideal, aiming at 'a system whereby the earth and its fruits will be exploited for the benefit of all members of the community in proportion to the services they render to it and not according to the accident of property ownership.' The League of Nations, you say, represents the former ideal. . . . In actual practice, however, it hardly functions that way. . . . The League today does not look beyond the present capitalistic system. . . . It is essentially based on the *status quo* and its chief function is to preserve it. . . . It is right and proper that the League should condemn Italian aggression in Abyssinia and try to curb it, but . . . [i]t does seem rather illogical to condemn Italian bombing in East Africa and maintain a dignified silence about British bombing on the north west frontier of India. . . . The second ideal, of socialism, indeed includes the first, and it may be said that real world order and peace will only come when socialism is realised on a world scale. . . . The main question for us to consider is how to create an environment and circumstances under which these deeper changes can take place. . . . Under present conditions the environment is against us and instead of lessening our mutual hatreds and selfishness and acquisitiveness, which lead to conflict, actually encouraging all these evil traits. It is true that in spite of this grave disadvantage some progress is made and some of us, at least, begin to challenge our old habits and opinions. But the progress is very slow."[22]

For Lord Lothian this response from Nehru was most heartening. Merely for this former fierce advocate of class struggle and Marxist dogma to write that "the main question for us to consider" was how to reshape the "environment" that made for hatred, greed, war, and the like and further to find him agreeing that in spite of the twin monsters of Marxist–Leninist theory, capitalism and imperialism, some progress was being made, was of monumental importance. For it meant that Nehru would play the game when the court was finally cleared of its rubble of horrors and hate. Both Agatha from her side and Gandhi from his knew Nehru in some ways perhaps better than he knew himself, and so they could recognize that Lothian and Nehru essentially agreed. Each of them knew that this world was changing much faster than any but a handful of their most prescient colleagues realized.

"It is true, as you state it, that the capitalist system has not created international anarchy; it merely succeeded to it," Jawaharlal told Lothian.

"It may be that Marx overstates the case for the materialist or economic interpretation of history. Perhaps he did so for the simple reason that it had been largely ignored. . . . It may be that a great deal of private initiative is left; in some matters, cultural etc., it must be left. But in all that counts, in a material sense, nationalisation of the instruments of production and distribution seems to be inevitable.

"Coming to Britain and India . . . It is perfectly true that in politics, as in most other things, we cannot start with a clean slate. . . . We have to take things as they are, whether we like them or not, and to reconcile our idealism with them. But we must move in the right direction. . . . Personally I am perfectly prepared to accept political democracy only in the hope that this will lead to social democracy."[23]

Nehru returned to London on January 26, 1936. At Victoria Station a reporter asked him, "May we congratulate you?" He was surprised. "I do not know why. . . . If you think my wife is better you are mistaken." "No, Panditji," was the reply, "I congratulate you on your being elected President of the Congress. We have this news in today's newspapers." "Oh, I see," he said, sounding rather sheepish, "but that is news to me. . . . I have no official intimation from the Congress Committee. . . . From what I know of it, the election could not have taken place by now and I think . . . congratulations are a bit premature."[24]

Krishna Menon managed his schedule during the second visit. He took Nehru to the Chinese exhibition at the Victoria and Albert Museum and to Allen Lane's for lunch. Lane read Nehru's autobiography during that week and urged the Bodley Head to publish it in April. It was an immediate success, reprinted four times by July 1936. Krishna Menon deposited Nehru's sterling royalties in his account at Lloyd's, which later proved useful for Indira's education and other private expenses in England.

Nehru addressed a packed, cheering Labour Party meeting in London's Caxton Hall, chaired by Ellen Wilkinson. He was introduced effusively by Harold Laski but insisted: "I assure Professor Laski that there are thousands of men and women in India, whose names you may not have heard . . . who deserve to be compared with men like Tom Mooney, Dimitroff and Ernst Thaelmann. . . . [M]y name . . . might have achieved notoriety, but there are hundreds and thousands in India who can be mentioned with those great men and not me. . . . Many people imagine that the Indian problem is settled. . . . But obviously . . . it is impossible for imperialist England and nationalist India to agree on anything. . . . The Congress is fundamentally a joint front against British imperialism . . . I do not think there can be any healthy basis of cooperation between imperialist Britain and subject India. But, in spite of all the years of horrors that we have undergone, I may tell you that there is an amazing amount of good-will; the bitterness is not deep-rooted amongst the people. Our bitterness is against the system that governs us and not against the people of Britain."[25]

Next day Nehru went to a Friends' meeting of Agatha's India Conciliation Group, chaired by Carl Heath.[26] He was invited to address them again each time he returned to London, and Carl Heath, as well as Agatha and other members of the board, became his champions among their friends in Parliament. Stafford Cripps and Harold Laski lobbied Labour on Nehru's behalf, as India's best and wisest spokesman and the future leader of his nation-to-be. This visit to London thus proved more fruitful for Nehru than the previous one had been. His name was now known in many influential intellectual circles, and soon his book would go to press. Then all the most advanced, intelligent people of England would begin reading it and learn to view the problems of India through Nehruvian lenses. Five years later an abridged version of his autobiography, retitled *Toward Freedom,* was published by John Day in New York. Americans then would also learn more about India and about Nehru, whose name was thereafter coupled with Mahatma Gandhi's as the bravest, most brilliant spokesmen for India's national Congress and its struggle for freedom.

On the eve of his departure from London, to return to Badenweiler and the dying Kamala, Nehru was interviewed once more. He urged Britain to "teach Indians to think in terms of bread and butter and not in terms of temples and mosques and you will find that the communal question has receded. . . . Soon I shall be going back, entangled again, in spite of myself, in Indian politics. . . . You get caught in the coils."[27]

Kamala's condition had deteriorated so much that soon after he saw her on his return to Germany he decided to remove her from Sanitorium Waldeck in the Black Forest to the brighter heights of the Sanitorium Sylvana in Switzerland, where she died before February ended. He had planned to fly home from Marseilles on February 29 and hoped to leave Lausanne on February 28, for it hardly seemed possible that she could last beyond that. Only two days earlier, he responded to a letter from her swami that "she had been keeping fairly well for over two weeks . . . no rise in temperature and it seemed that she was getting over her pleural trouble. But . . . yesterday her temperature shot up again. Her body, after the terrible long fight it has put up with, seems to have exhausted all its strength. . . . It cannot cope with little ills. . . . One never knows what she may not be capable of even now but, ordinarily speaking, there is no hope."[28] Indira had come back from her school at Bex a few days before. She was not, however, with her father at her mother's bedside when Kamala died at 5 A.M. on the morning of Friday, February 28, 1936. Dr. Atal was there, and they had her cremated immediately. Nehru carried his wife's ashes home with him in a small urn, immersing most of them in the sacred confluence of the Ganga and Yamuna Rivers at Allahabad after he flew home. Nehru kept some of Kamala's ashes in his bedroom, later asking his daughter to immerse them with his own.

He changed his flight for one leaving Marseilles in early March, stopping first in Rome, where Mussolini wanted to see him. Nehru found the suggestion "very embarrassing," as he told Krishna Menon, since "I fear that any such meeting will give rise to all manner of misconceptions."[29] But he was interested. He found much about Mussolini's socialism "fascinating," and Subhas Bose was eager for them to meet. "I was curious to know what kind of a man the Duce was," he admitted but claims to have declined to accept the invitation hand delivered to him by Mussolini's attaché aboard his long-grounded plane in Rome. "Today on rising from Rome we had a fine view of the city and of St. Peter's," he wrote Indira from Cairo on March 8, 1936. "We were flying high . . . over 14,000 ft. . . . not very much below the height of Mont Blanc . . . it was difficult to see out. . . . The Mediterranean was of a wonderful . . . deep blue, sometimes turning to emerald green near the coastline. . . . Our plane is a beauty. . . . It bears the name *Perkoetet,* which in Dutch means . . . wood pigeon. . . . We had five passengers in all. . . . On arrival at the Cairo airport I found Fouad Bey waiting for me. He was the same as ever and embraced and kissed me." Nehru again enjoyed Cairo's posh Heliopolis Hotel, "reeking of luxury and jazz and dancing . . . and crowds of waiters and . . . page-boys dashing about. I take advantage of the lovely bathrooms by having two baths during my short stay."[30]

His next stop was in Gaza, then on to Baghdad on that long ride home, where he spent the night in the "very mediocre affair" River Front Hotel. "We covered the 600 miles from Gaza to Baghdad in three and a quarter hours. . . . We start very early. . . . Tomorrow's run is a very long one. It will take us to the heart of India. We pass Karachi at tea-time and reach Jodhpur in the evening. There we spend the night and another two hours . . . will bring us to Allahabad."[31]

From Baghdad he cabled Krishna Menon, informing him that he wanted to dedicate his autobiography "To KAMALA, who is no more." He shed no tears for her, however, knowing that death was her only solace, the release she had prayed for so many long years.

"So you return leaving Kamala for ever in Europe," Gandhi wrote him the day Jawaharlal spent in Baghdad, not knowing that he was carrying the urn with her ashes back to Anand Bhawan.[32]

16

Reclaiming the Crown

1936–1937

NEHRU RETURNED HOME to find his mother "sad-looking and silent . . . terribly shrivelled up. The shock had been very great for her." People came to pay their respects, bowing to him and Kamala's ashes, set atop a small mountain of flowers, taken in silent procession through the city for three hours and across the wide stretch of sand to the merging of the river goddess Ganga and her sisters Yamuna and the invisible Sarasvati. Nehru entered that most holy confluence of Hindu waters wearing only his hand-spun dhoti and the Brahmanic thread with which he had been initiated into the faith he so often disavowed and claimed to disbelieve until his own death, yet whose rituals he nonetheless followed. He emptied Kamala's ashes into the water while the Brahmans in attendance chanted their mantras.[1]

Soon to reclaim Congress's crown, Nehru understood how mistrustful some of his older colleagues were of his radical, mercurial mind. He was not insensitive to their fears of him, as this genuflection to tradition and Brahmanic ritual proved. "I know there is a certain difference between your outlook and that of men like Vallabhbhai . . . and myself," Rajendra Prasad had written him. "But I believe . . . it will still be possible for all of us to continue to work together. You are undoubtedly dissatisfied with the present condition of things. . . . But the difficulties are inherent in the situation and . . . it is not possible to force the pace or cause any wholesale change. In all big struggles . . . however much we may chafe and fume we have to lie low and work and wait for better times. We are passing through one of such crises. But I see no reason to be disheartened. The spirit of freedom is not crushed. . . . In any case you have certainly a free hand to shape things as you would like and to appoint any Working Committee of your choice. . . . It has been wrongly and unfairly assumed that the Working Committee has been thinking of nothing except offices under the New Constitution."[2]

India's British rulers knew very well when finally they completed their new Government of India Act in 1935—the combined legislative end product of all three round table conferences as well as Simon's commission and the prime minister's Communal Award: tens of thousands of pages of documents, filling dozens of heavy official white papers and bluebooks—that Congress could not easily ignore it. The ball was now in their court, after all, and if they opted for a boycott of the new act, as they had done initially with its earlier incarnation in 1920, events would simply pass them by. Rajendra Prasad—whose mind was much closer to Gandhi's and Vallabhbhai's than Nehru's was—knew the danger to India and Congress of rejecting the new act. "So far as I can judge no one wants to accept offices for their own sake," he explained to Nehru. "No one wants to work the constitution as the Government would like it to be worked. The questions for us are . . . What are we to do with this Constitution? Are we to ignore it altogether and go our way? Is it possible to do so? Are we to capture it and use it as we would like to use it . . . ? Are we to fight it from within or from without and in what way? It is really a question of laying down a positive programme . . . in the light of circumstances as they exist . . . we have to consider and decide the question irrespective of everything except the good of the country."[3]

Subhas Bose was the only young leader of Congress who remained Nehru's potential rival for the presidency. Bose's militancy and Bengali brilliance made him as much a hero among India's youth as Jawaharlal was. Bose was C. R. Das's foremost disciple, moreover, though not his son, and Gandhi never developed the same intimacy with Das that he had with Motilal, for in 1920 Das had initially opposed his rise to power in Congress. Indeed, Gandhi never liked Netaji (Leader) Bose, who always favored armed revolt over Satyagraha and looked ideologically to Hitler and Mussolini as his external role models. "My dear Jawahar," Subhas wrote from Austria that early March, "Since leaving you I have been thinking . . . I have definitely decided to give you my full support. Among the front rank leaders of today—you are the only one to whom we can look up to [sic] for leading the Congress in a progressive direction. . . . [Y]our position is unique and I think that even Mahatma Gandhi will be more accommodating towards you than towards anybody else."[4]

Gurudev Tagore added his support, speaking at an early March memorial service at his school to honor Kamala Nehru: "Her husband Jawaharlal has his undoubted right to the throne of Young India. His is a majestic character. Unflinching is his patient determination and indomitable his courage, but what raises him far above his fellows is his unwavering adherence to moral integrity and intellectual honesty. He has kept unusually high the standard of purity in the midst of political turmoils, where deceptions of all kinds, including that of one's self, run rampant. He has never fought shy of truth when it was dangerous, nor made alliance with falsehood when it would be convenient."[5]

On the eve of the Lucknow Congress, Nehru was asked for a "message" to youth that could be printed in the Lucknow University *Union Journal.* "The hunger for messages and the like in India fills me with wonder," he responded impatiently. "Whence does it come? Why does it continue in spite of a surfeit of pious sentiments repeatedly expressed? . . . [T]his is the way of age when the blood runs cold and petrification of mind and body sets in. What of students and young men and women? Are they satisfied with this ineffective substitute for activity? Look at the world around us, pulsating with the fever of change. Study it, understand it, fit into it, and play your part in it—or else collect messages and wax fat on them."[6]

A few days later he left for Delhi to meet with Gandhi, staying at the Harijan colony, where Gandhi continued his constructive work, seeking to reform Hindu society. Nehru arrived in Delhi on March 17, and his talks with Gandhi began that night and continued for more than a week. "I am still carrying on here but as the days go by I get more and more tired," he wrote Indira from Delhi. "Probably I shall be here for another three days & then I return to Allahabad. . . . Padmaja is here, or rather in New Delhi. She has grown much thinner and is very weak and frail." One of his few diversions in Delhi that hard week was to call on his beloved Padmaja; they always cheered up each other.[7]

Nehru returned to Allahabad on March 26, finding some three hundred letters awaiting him at Anand Bhawan, three from Indira, to whom he wrote: "I go to Lucknow to attend the opening ceremony of the Swadeshi exhibition by Bapu. I return to Allahabad and have just under a week to write my presidential address. Bapu is also coming here and so I am not going to have an easy time . . . then the Congress. There is a devil of a lot of trouble and much pulling different ways and the strain of all this is great."[8] Four days later he wrote her again, venting his frustrations on her and sharing his concerns about Congress. "Perhaps you know that the British Govt. has informed Subhas Bose that if he returns to India he is liable to be arrested and sent to prison. This news upset me and made me very angry. On arrival here from Delhi I had a letter from Subhas asking me what he should do. A difficult question for me. . . . All my inclination was that he should come back in spite of the Govert.'s intimation to him.

And yet it is not easy to send another to prison. So at first I cabled to him to postpone his departure. But on further consideration and after consulting colleagues I cabled again suggesting to him to return immediately. I have thus made myself responsible for his return to prison."[9] Bose did return and was arrested, for "inciting terrorism" in Bengal, but on his release in 1938, he was elected Congress president, just as Nehru feared he might be. Bose was even reelected president in 1939, but Gandhi and Nehru then decided he needed deflating, so they swiftly arranged to remove the crown from his swelled head.

"Darling Papu, I long to come to you to help you in some way so that you may get a few moments of rest," Indira replied from Bex. "But what can I do from here or even if I were in India? And if I were there beside you it would be more painful to see your dear face so tired and your eyes closing with weariness and I so close by but unable to prevent its being so or to help in any way. Please try and snatch a little rest every free moment."[10]

At the heart of the struggle between Nehru and Gandhi and Prasad's Working Committee was the question of whether or not Congress should accept high office—"ministries"—under the viceroy, thus advancing the step-by-step process toward full dominion status, as envisaged in the Government of India Act of 1935. Most Congress leaders favored accepting office, taking that half-loaf of responsibility, sharing the burdens and opportunities of leading India toward its ultimate goal of freedom. The older Congress leaders understood that time was required to bring India's social, economic, and political status to the point of preparation for complete independence. "At the time of the Lucknow Congress, acceptance of ministries was a dead certainty as far as the Congress was concerned," Nehru informed Krishna Menon six months later. "Today it is a very doubtful proposition." He felt proud of himself, for he had done something even Motilal could not do; he had argued with the Mahatma for almost a week, and he had won! Or at least they had exhausted each other enough to put off the next round in that battle for another year.[11]

The crowds gathered at Lucknow at the end of March for the huge Swadeshi exhibit, and all the Congress leaders came for the April meetings of the Subjects Committee. Some of Jawaharlal's socialist comrades feared that he had been coopted by Gandhi and the right wing, that the presidency was enough perhaps to make him "sell out," as their most extreme members, like Ram Manohar Lohia, later put it. Even J.P. was worried, but Nehru nominated him and Narendra Deva, as well as another brilliant young Maharashtrian socialist, Achyut Patwardhan, to be on his Working Committee. Nehru had to threaten to resign on the eve of his presidential address in order to persuade Gandhi to force Vallabhbhai and Prasad to bow to his wishes. Yet he could not win everything. The right wingers still outnumbered the leftists, at least in the All-India Congress Committee, which continued to dictate Congress policy as well as the composition of its board of Working Committee directors. Nehru and his socialist comrades hoped to change all that by proposing the "collective affiliation" of workers' and peasants' organizations, which had sprung up all around India since Nehru's return from Moscow nine years earlier, but the old guard defeated that motion in the Subjects Committee, just as it rejected several other radical resolutions that Nehru favored but would not abandon his crown for. The night before his presidential address, however, rising in the Subjects Committee, he did speak on the issue of accepting office.

"I must break my silence," he said. "Ordinarily it is not right and proper for members of the Working Committee to say anything against a Working Committee resolution, but that Committee itself is, as you know . . . practically ended . . . a new Committee will take its place. . . . I find there is a definite difference of outlook. . . . I feel that the difference is between . . . the reformist and . . . the revolutionary. If we wish the country to advance towards independence, if we wish the country not to be disillusioned, then we must think many times before we take any steps which increase this reformist mentality. . . . [A]cceptance of office tends to reformism . . . those hopes must be dashed to the ground. Are we going to think in terms of a revolution, or are we going to support reformism, and cut up Congress flags and make boys and girls in schools sing our songs? Is this going to bring independence to this country? A postponement of the question means hesitancy and indecision. . . . But . . . we cannot afford to split up, and break up the Congress."[12] Thus, postponement it would have to be, for Nehru knew that any vote on rejecting office would have gone down to certain defeat.

"Comrades, after many years I face you again from this tribune—many weary years of strife and turmoil and common suffering. . . . [I]t is good for me to see this great host of old comrades and friends . . . I am heartened and strengthened by you, though even in this great gathering I feel a little lonely. Many a dear comrade and friend has left us. . . . But . . . [w]e cannot rest, for rest is betrayal . . . of the cause we have espoused and the pledge we have taken . . . of the millions who never rest. I am aweary and I have come back like a tired child yearning for solace in the bosom of our common mother, India." It was a long speech, his recapitulation of the past sixteen years of struggle. "Only by constant self-questioning, individual and national, can we keep on the right path," Nehru argued. "Real failure comes only when we forget our ideals and objectives and principles . . . therefore, let us look into ourselves . . . without pity or prejudice. . . . We dare not delude ourselves." He spoke of the changes in Europe and Asia since World War I, praising the Soviet

Union, where "in marked contrast with the rest of the world, astonishing progress was made in every direction."[13]

"Where do we stand, then, we who labour for a free India?" He insisted that Indian nationalism could find no "common ground" with British imperialism. Nehru deplored the "depths of vulgarity" to which British rule had descended. He called it a "clear and definite fascist mentality," noting the lack of "civil liberties" in India, the suppression of the press and literature, and the banning of "hundreds of organizations" while keeping thousands of people in prison "without trial." He also attacked most Indian leaders, not just the "handful" who were "closely allied to British imperialism," but virtually all "middle class leadership," which he termed "a distracted leadership, looking in two directions at the same time."[14]

"Nothing less was to be expected of Jawaharlal," Gandhi wrote Agatha, responding to her anxiety about what she had read of Nehru's "rabble-rousing" address in London's newspapers. "His address is a confession of his faith. . . . Of course the majority represent my view . . . Jawaharlal's way is not my way. I accept his ideal about land etc. But I do not accept practically any of his methods. I would strain every nerve to prevent a class war. So would he, I expect. But he does not believe it to be possible to avoid it. I believe it to be perfectly possible if my method is accepted. But though Jawaharlal is extreme in his presentation of his methods, he is sober in action."[15]

"I am convinced that the only key to the solution of the world's problem and of India's problems lies in socialism," Nehru insisted from his presidential platform in Lucknow. "That involves vast and revolutionary changes in our political and social structure, the ending of vested interests in land and industry, as well as the feudal and autocratic Indian states system . . . the ending of private property, except in a restricted sense, and the replacement of the present profit system by a higher ideal of cooperative desires. . . . Socialism is thus for me not merely an economic doctrine which I favour; it is a vital creed which I hold with all my head and heart . . . I should like the Congress to become a socialist organisation. . . . But I realise that the majority . . . may not be prepared to go thus far."[16]

As for the new Government of India Act, Nehru called it "a charter of slavery." He argued that Congress's attitude toward such a charter could only be one of "uncompromising hostility" and a "constant endeavour" to end it. "How can we do this? . . . I think that, under the circumstances, we have no choice but to contest the elections to the new provincial legislatures. . . . We should seek election on the basis of a detailed political and economic programme, with our demand for a constituent assembly in the forefront. I am convinced that the only solution of our political and communal problems will come through such an assembly, provided it is elected on an adult franchise and a mass basis. That assembly will not come into existence till at least a semi-revolutionary situation has been created in this country."[17]

A month later in Bombay Nehru called for an "anti-imperialist united front" led by Congress. He had been told that Bombay's wealthy merchants were dissatisfied and agitated over his Lucknow address, but he tried to assuage such fears, saying he meant no harm to anyone. He warned, however, that unless existing poverty in the country was removed voluntarily, that problem would have to be solved by socialism.[18]

While in Bombay that May, Nehru addressed a women's group, responding to many complaints he had heard about not choosing a woman for his Working Committee of thirteen members, Congress's "shadow cabinet." Sarojini Naidu had expected to be named at Lucknow, but Nehru noted now that "several of the things that I advocated were not accepted. Possibly a more self-respecting man in my position would have tendered his resignation. . . . But I thought that I should continue as President. I paid the price for my views. . . . Hence the question of choosing the Working Committee personnel was not an easy job." The *Hindu* and other press reports of his statement indicated that Gandhi had "dictated" to Nehru the retention of many of the more conservative members of his Committee.[19]

"My dear Jawaharlal," Gandhi wrote him immediately. "The exclusion of women was entirely your own act. Indeed nobody else had even thought it possible to exclude a woman from the cabinet."[20] Nehru's reply "does not give me satisfaction," Gandhi continued. "If you had shown the slightest desire to have a woman on the committee, there would have been no difficulty whatsoever. . . . But so far as I remember you yourself had difficulty in choosing a substitute for Sarojini Devi [Naidu] and you were anxious to omit her. You even went so far as to say that you did not believe in the tradition or convention of always having a woman and a certain number of Mussalmans on the cabinet. Therefore so far as the exclusion of a woman is concerned, I think it was your own unfettered discretion. No other member would have had the desire or the courage to break the convention. . . . As to the other members . . . your statement which your letter confirms has given much pain to Rajen Babu [Prasad], C.R. and Vallabhbhai. They feel and I agree with them they have tried to act honourably and with perfect loyalty towards you as a colleague. Your statement makes you out to be the injured party."[21]

A month later a majority of seven members of Nehru's Working Committee wrote to tender their resignations. "We feel that the preaching and emphasising of socialism particularly at this stage by the President and other socialist members of the Working Committee while the Congress has not adopted it is prejudicial to the best interests of the country and to the success of the national struggle for freedom which we all hold to be the first and paramount concern of the country."[22]

That letter, written by Rajendra Prasad, was also signed by Vallabhbhai Patel, C. Rajagopalachari, and J. B. Kripalani, and three less well known Working Committee members, all important contributors to Congress's organization and its coffers. The letter itself was drafted and signed in Wardha. Gandhi read every word of it, nodding his head in approval before it was sealed, to be handed to Nehru, who had just finished an exhausting tour of Punjab, which "laid me low," he confessed to Indira, losing his "voice." He wrote in the train on the day of his colleagues' revolt: "I am on my way to Wardha for the Working Committee meeting. I shall be there for two or three days and then I intend going to Bombay for a day for some business—also to see Padmaja after her operation."[23]

Nehru read the mass resignation letter next morning and well understood that without those pillars of support, the Congress as it then existed must collapse. Nehru knew the resignations could prove a fatal blow to all his hopes and aspirations. Much as he felt angered and humiliated by these people's threats and betrayal of him—for he took it as nothing less than that—Nehru knew he had to swallow his pride and keep the rebels from resigning. And of course he knew that only one person could bring them round again to toe the mark.

"My dear Jawaharlalji," Rajendra Prasad wrote him on July 1, "Since we parted yesterday we have had a long conversation with Mahatmaji and a prolonged consultation among ourselves. We understand that you have felt much hurt by the course of action taken by us and particularly the tone of our letter has caused you much pain. It was never our intention either to embarrass you or to hurt you." So they all agreed to withdraw their resignations. They hoped, nonetheless, to "make . . . clear in this private communication" what still troubled them, for Gandhi had badgered them to withdraw, insisting that Jawahar had been hurt by their attack but that if they were more open with him, he would see the wisdom of their concerns and come around, without anger. So Rajendra Babu tried doing that. "We have felt that . . . [t]here is a regular continuous campaign against us treating us as persons whose time is over, who represent and stand for ideas that are worn out and that have no present value, who are only obstructing progress of the country and who deserve to be cast out of the position which they undeservedly hold. The very ideals, methods of work and tactics which we have learnt in the company with Gandhiji forbid any scramble for power in any organisation and we have felt that a great injustice has been and is being done to us by others, and we are not receiving the protection we are entitled to from you as our colleague and as our President. . . . This hurts us. . . . We are sorry for having hurt your feelings and I only hope that this letter will help to smooth matters.

. . . [T]his letter is meant personally for you and not intended in any way to form part of the official record."[24]

Nehru never forgot that letter nor the previous one it was designed to remove from his mind as well as from any official record. He had to take another step or two back now, but he also was better informed about just whom he had to worry most. Leaving Wardha for Bombay, where he went to see his loving Padmaja, Nehru added a postscript to the letter he took with him: "As usual I have had a very tiring and distressing time with all-day debate and argument, my colleagues in the Working Committee are greatly irritated with me as they think that my talking about socialism puts them in a false position . . . so they have almost come to the conclusion that we should part company. This may result in my resigning . . . I am not sure yet. . . . It is a depressing business."[25]

Back in Allahabad a few days later he wrote to Gandhi, "Ever since I left Wardha I have been feeling weak in body and troubled in mind. Partly this is no doubt due to . . . a chill which has aggravated my throat trouble. But partly also it is due to other causes which touch the mind and the spirit directly. Since my return from Europe, I have found that meetings of the Working Committee exhaust me greatly; they have a devitalizing effect on me and I have almost the feeling of being older in years after every fresh experience. . . . It is an unhealthy experience and it comes in the way of effective work.

"I was told, when I returned from Europe, that the country was demoralised and hence we had to go slow. My own little experience during the past four months has not confirmed this impression. Indeed I have found a bubbling vitality wherever I have gone and I have been surprised at the public response. . . . This public response has naturally heartened me and filled me with fresh energy. But this energy seems to ooze out of me at every meeting of the Working Committee and I return feeling very much like a discharged battery. The reaction has been greatest on this occasion. . . . But it was not about my physical or mental condition that I wished to write you. . . . I do not wish to act in a hurry. . . . But even before my own mind is decided I want to tell you which way I am looking. I am grateful to you for all the trouble you took in smoothing over matters and in helping to avoid a crisis. . . . And yet, where are we now and what does the future hold for us? I read again Rajendra Babu's letter . . . and his formidable indictment of me. . . . My own impression before Lucknow, and to some extent even at Lucknow, was that it should not be difficult for all of us to pull together this year. It is evident now that I was mistaken. . . . Perhaps the fault may lie with me; I am not aware of it; but one can seldom see the beam in one's own eye. The fact remains, and today there is no loyalty of the spirit which binds our group together. . . . [A]fter last week's incidents I am beginning to . . . think that the right thing for us to do will be to put the matter . . . before the A.I.C.C. at its next meeting and take its direction. . . . Presumably the result of this will be that I shall retire and a more homogeneous Committee will be formed."[26]

"I have just received your letter," Gandhi replied. "The letter of withdrawal does not bear the meaning you put upon it. . . . It was sent to you after I had seen it. . . . In any case I am firmly of opinion that during the remainder of the year, all wrangling should cease and no resignations should take place. A.I.C.C. will be paralysed and powerless to deal with the crisis. It will be torn between two emotions. It would be most unfair to spring upon it a crisis, in the name of democracy, which it has never been called upon to face. You are exaggerating the implications of the letter. I must not argue. But I would urge you to consider the situation calmly and not succumb to it in a moment of depression so unworthy of you."[27]

Nehru accepted Gandhi's advice but knew now that he had approved of both those treacherous letters that Rajendra Babu dared write to him. So he carried on, touring, campaigning for Congress wherever he went, and everywhere he went crowds gathered and cheers went up, rhythmic chants of "Jawaharlal Nehru-ki-jai!" (Victory to Jawaharlal Nehru!). He wrote Indira from a train running north from Sind's Hyderabad to Punjab's Multan. "Terrible programmes they draw up for me everywhere, as if I was a machine. It is going to be worse in the Punjab . . . I am keeping well, however, in spite of all this and my throat is no worse if no better. Travelling brings little rest, for at every station there is a crowd and much shouting of slogans, and big baskets of fruits and sweets, and of course garlands & flowers. These symbols of affection and good will are very welcome but they are burdens and often a nuisance. . . . [C]askets and addresses accumulate and I have now two packing cases full . . . all probably to go to the Allahabad municipal museum, and later I hope to the national museum we shall establish at Swaraj Bhawan."[28] More than a decade before independence, he was already planning the national museum he would build after becoming prime minister, saving the ornate caskets presented to honor his presence amongst them by local Congress workers at every whistle stop on his tour of the subcontinent, each carrying its elaborate message, like the prayer inside a Buddhist prayer wheel.

"Yesterday I was in Hyderabad . . . a day of continuous activity and yet I liked many of the functions, especially those connected with the children and girls. . . . The procession in Hyderabad was unique of its kind. Across the narrow bazaar, where the rich shops were situated, hung all manner of articles . . . silks and caps & hats & curtains . . . & even false beards! . . . Two or three days ago I had a longish camel ride—the first I have had . . . I liked it and found it fairly comfortable. . . . It is a question of swinging your body with the camel's motion and a person used to horse-riding should have no difficulty. . . . But I did not like the smell of the camel. . . . In a remote Sind village . . . I saw that one of the gates . . . was named 'Indira Gate.' I thought of your heritage of storm and trouble and how, whether you liked it or not, you could not rid yourself of it.

"None of us . . . can have an easy time or freedom from storm & trouble. But to some of us fall a greater share . . . and it is your lot, because of your family and other reasons, to have to bear this heavier burden. May you be ready for it when the time comes and accept it willingly and take joy in it," he exhorted his heir to the Crown of India. "Those who seek to do higher things must also face bigger difficulties. . . . I remembered your admiration for Jeanne d'Arc many years ago. Does that endure still? . . . And now our train is approaching Multan. . . . Processions, speeches, interviews, arguments without end—it is a tiring business."[29]

"The Punjab is an astonishing province," he wrote in the aftermath of his three-day tour. "Hundreds of thousands of brave faces met me, eager eyes full of hope, ears listening to the call of political and social freedom, and strong arms that were quivering for action. The sight of them made me forget . . . weariness . . . and in response to that call of youth my own blood tingled. . . . I can never forget this tour of . . . Punjab and the unbounded love and affection. . . . I shall venture to hope that this fine and intoxicating enthusiasm will not end in froth but will be harnessed and directed to right ends; that the Punjab will remember that the real problems which face it, as well as the rest of the country, are those of poverty and unemployment and national freedom; that all others are minor issues. . . . Especially I hope that the Punjab will extricate itself from the morass of communalism and pay no heed to the false and misleading counsels of narrow-minded communalists. . . . I hope the Punjab will stand solidly with the Congress, for only the Congress has the will and the capacity to face these tasks."[30]

As soon as he returned to Allahabad, Nehru wrote Agatha, from whom three letters awaited him, explaining his delay. "My usual programme began with the dawn and carried me on till midnight. I have completely lost count of the number of places I visited and the meetings I addressed. If I travelled by train almost every station meant a disturbance, if by car every few miles our car was held up by a crowd . . . I think I deserve some kind of a prize or medal for endurance."[31]

Three days later he issued a statement to the press in Allahabad: "The elections are yet far off; half a year. . . . But their long shadow darkens the horizon and hoarse and strident voices assail our ears . . . our middle class intelligentsia talk of little else. Yet as I wandered in Sind and the Punjab, this tumult and shouting seemed . . . unreal, the talk of candidates and pacts and manoeuvres and intrigues ruffled the surface only. Underneath this surface I sensed the strange currents, I heard a deep rumbling. Why did these vast crowds, especially in the rural areas, gather together or wait long hours by the roadside? Not surely to hear a person who had gained notoriety, or just to pay their homage to the Congress. There was a deeper urge, a hunger that gnawed and required satisfaction. . . . It was an extraordinary experience to see these scores of thousands of Punjab peasants. They were not exuberant. . . . [O]utward signs of enthusiasm . . . [were] lacking. There they sat quietly and stolidly but behind that quietness there was commotion and underneath that peasant stolidity there were reserves of power and a deep unrest. As I watched them and tried to look within them I thought of a volcano which has long seemed extinct but which shakes again with inner fire, and of the sea which begins to darken before a storm."[32]

In August, Nehru was invited to address the All-India Students Conference, over which Jinnah, who had just been elected permanent president of the Muslim League and remained the leading independent member of the Legislative Assembly, presided. "I have come here on your bidding," Nehru began softly with mock modesty, "but I must confess that I am not as eminent as our president, Mr. M. A. Jinnah, who is a distinguished leader. . . . First of all, may I thank you, Mr. President, for your kind words about me?" For Jinnah had been more than polite, graciously hoping to prove to Nehru by this invitation and his generous introduction of the President of Congress that it was still possible to resurrect the pact between Congress and the league, which he and an older Nehru had drafted together exactly twenty years ago here in this very same city of Lucknow.[33]

This was still Jinnah's fondest dream, but Nehru felt nothing but contempt for him whom he considered a Muslim "collaborator" with British imperialists. "May I remember an occasion about a quarter of a century ago, when I first met you?" Nehru reminded him vaguely. "It was in Europe—I am not sure whether it was in Paris or somewhere else, but it was in Europe, and at that time I was a student at Cambridge while you had already achieved distinction in the political service of the country. So far as most of us are concerned we always try our best to tackle the problems with which we are concerned in the best way we can," he added, so subtle a slight that one less sensitive than Jinnah might have missed it. But Nehru well knew of whom he was talking and felt quite clever in reminding Jinnah, as well as this audience of bright university students, that he was the Cambridge man and much closer in mind and heart, as well as age, to this student body than one who had achieved his political distinction at least a generation earlier.

"As many of you know, it is one of my failings, of which I am constantly reminded," Nehru continued, "that I am always fighting shy of the immediate problems which face us . . . I confess to that failing. With your leave I will talk for a while about various big questions which do not perhaps immediately affect you, but which I think have a vital consequence for each one of us." Then he lectured them on the Spanish civil war and its implications for all of Europe and Asia. He urged that "as Indian citizens" it should be "your business to understand all this," insisting, however, that "on account of vested interests that exist in India," these problems could neither be solved properly nor well understood. Then he turned to India's "communal question," which he called "a nuisance."[34]

Nehru knew, of course, how that word hit Jinnah, for Jinnah had hoped perhaps that after this conference Nehru might sit down with him and quietly reconsider the thorniest of all internal problems that threatened permanently to divide India and would soon partition the entire subcontinent. "Why is it a nuisance? It is a nuisance for many reasons, but it is a nuisance chiefly, I think, because it diverts your attention from the real problems of the country. It is bad in itself because it makes us petty-minded, but ultimately it is utterly bad because it hides from our view the really big problems that affect our country and the people of the world at large, and I want you as students especially to try to understand these . . . problems . . . and try to solve them."[35]

By now Jinnah must have wished he had never invited Nehru to speak, for not only had he been called a "nuisance" once, but that hateful word was repeated four times. "We are apt too much to think in terms of law and legal circumstances . . . and safeguards and compromises—as if half a dozen prominent individuals and leaders can by meeting together solve the vast problems," Nehru went on. "The vast majority of the eminent statesmen who have attended these conferences have really tried to solve the big problems . . . although they wanted to solve them for their selfish reasons. . . . If you ask me why they have not been solved, my answer is clear. I have no doubt—because these problems cannot be solved within the fabric of the structure of the present-day government." That elicited prolonged cheers from his audience and yet a grimmer visage from Jinnah, for "safeguards" "compromises" were two of the words he had often used in the long difficult negotiations he and other lawyers and eminent statesmen engaged in at every round table conference and assembly meeting he attended.[36]

"Now, you cheer me, but I suspect your cheering," Nehru continued in his best Cambridge debating style. "I suspect it, I tell you, because it is a kind of emotional display of sympathy. Well and good. Emotions are good, but . . . [t]hese problems cannot, I think, be solved by British capitalism." He then gave them a lesson in Marxist dialectic, as well as early

Indian history and British imperial rule. He reverted to the communal question before concluding his speech, pointedly insisting that "you cannot consider the strength of a group or community by counting of heads. . . . The real thing is that if there was the question of numbers we thirty five crores [350 million] of people would not have become a slave country . . . the fault lies with us. . . . It is because British people are united and we are not. So do not bother about percentages. It does not matter how many people there are in this group or in that group." Jinnah said not a word to him when he finished.[37]

"I am off again soon on my 'hurricane' tours," Nehru wrote Agatha early in September. "It is an exhausting business and yet I find a strange relief in it from the politics of committees and individuals. . . . I suppose the enthusiasm and the crowds cheer me up. I have just learnt by telephone that Gandhiji is seriously ill with malaria. . . . Malaria by itself is not uncommon or dangerous. But in his state of health it is a disconcerting business. He has been removed from his village hut to the hospital at Wardha. I am afraid I do not appreciate his living in village huts."[38]

Gandhi had recently been asked by two foreign visitors how he "felt" about Nehru. "We are not estranged from each other," Gandhi insisted. "There are obvious differences in outlook, but in spite of them our affection has not diminished . . . I have never had even the suspicion that Jawaharlal's policy has ruined any part of my work. . . . His enunciation of scientific socialism does not jar on me. I have been living the life since 1906 that he would have all India to live. To say that he favours Russian communism is a travesty of truth. He says it is good for Russia, but. . . . [a]s for India, he has said plainly that the methods to be adopted . . . would have to answer India's needs."[39] Nonetheless, Gandhi remained troubled and, shortly before coming down with malaria, argued at bitter length with Nehru. "Our conversation of yesterday," Gandhi wrote on August 28, "has set me thinking. Why is it that with all the will in the world I cannot understand what is so obvious to you? I am not, so far as I know, suffering from intellectual decay. Should you not then set your heart on at least making me understand what you are after? . . . Yesterday's talk throws no light on what you are after. . . . You know what I mean."[40]

Nehru wrote Indira from Anand Bhawan in early September. "I have been to Cawnpore and back . . . monster processions and innumerable meetings. Day after tomorrow I am off again to other parts of the U.P."[41] "This touring business is becoming more and more difficult for me," Nehru told the press. "Crowds become vaster and vaster, and the most carefully made plans go to pieces because of the pressure of innumerable human beings. All this enthusiasm is exhilarating, one feels intoxicated by it."[42]

"We had had a heavy day full of meetings and processions," Nehru wrote that evening. "Night has fallen . . . I could hardly keep awake. Suddenly we had to pull up, for right across the road sat a crowd of men and women. . . . They came to us and when they had satisfied themselves as to who we were, they told us that they had been waiting there since the afternoon. . . . And then we had *Bharat Mata ki jai* ['Victory to Mother India'], and other slogans. 'What was all this about,' I asked them, 'this . . . *Bharat Mata ki jai?*' . . . They looked at me and then at one another and seemed to feel a little uncomfortable at my questioning. . . . The Congress worker in charge of that area was feeling unhappy. . . . If the Congress people told them to shout, why they would do so, loudly. . . . Still I persisted in my questioning and then one person, greatly daring, said that *Mata* referred to . . . the earth. . . . I told them that *Bharat* was Hindustan, how this vast land stretched from Kashmir . . . to Lanka. . . . How all over this great land they would find millions of peasants like themselves, with the same problems to face . . . crushing poverty and misery."[43]

Nehru's preelection campaign tours for Congress thus proved to be tutorials at the roadside as well as more formal lectures to enormous crowds, turned out by local Congress workers in every district and village, or bused and trucked to towns for such meetings from the surrounding villages. *Tamashas*—"big parties"—they were called, with free sweets and tea or clean water, free transport, and colorful Congress flags flying everywhere, and *burra sahibs*—"big people"—giving speeches. It was always exciting to be there. But "What are you after?" even Gandhi had by now begun to ask him, for most of those to whom Nehru spoke could not vote. Yet he kept stirring them up, telling them to think about their "crushing poverty and misery," of which he now kept reminding them.

"Sir," Nehru wrote the editor of Allahabad's *Leader* that September, "Napoleon, in spite of his genius, does not happen to be a hero of mine, and a modest person, as you rightly characterize me, can hardly have Napoleonic ambitions."[44] A week later Gandhi chided him for appearing to believe it seemed at times that he alone had power to turn back the tides of poverty and misery. "That is why we have made you the King Canute," Gandhi twitted Jawaharlal, "so that you may do it better than others."[45] A Maharashtrian woman, Prema Kantak, who helped Gandhi organize his Women's National Volunteer Corps, was frightened by Nehru's stress on socialism and its impact on women. "I am surprised to learn of the controversies . . . in Maharashtra over the question of women volunteering," Nehru replied to her. "What has Russia or socialism got to do with this? I am a socialist and I want to spread socialistic ideas but . . . what have questions relating to marriage and sex relations got to do with volunteering or even with socialism?"[46] Solicitor Chimanlal Shah of Bombay was worried more about Nehru's repeated emphasis on a constituent assembly, which was not mentioned in the new act or white paper. "It is the obvious culmination of our struggle for independence and for a democratic constitution," Nehru informed him. "I lay stress on adult franchise so that the mass elements in India may make their weight felt and thus divert attention to mass problems—poverty, unemployment, the land question, industry etc. . . . When we have to face these mass problems . . . inevitably we shall have to think more and more on socialist lines."[47]

"This election business here is getting on my nerves," he confessed to Krishna Menon. "It is curious, or perhaps it is the normal state of affairs, that elections seem to bring out all the wrong things in a man."[48] He had been trying to lure Krishna Menon back to India. He needed someone he could trust completely, other than Indira, who had also moved to London now, living in a third- story apartment on Fairfax Road, near Swiss Cottage, while she prepared for Oxford. But Krishna Menon was too attached to London to return to Bombay. "I suggest to you," Nehru wrote him that October, to "take charge of the civil liberties organisation here. . . . Don't be put off by the abruptness of the proposal. There are many reasons . . . if you gave a push to the Civil Liberties Union in India . . . it would go a long way. Apart from this I think it would be a good thing for you to come to India . . . and meet people. You cannot do effective work . . . unless you renew contacts in India. . . . Then again I should like you to come here so that I might discuss many things with you. . . . The India League work will of course suffer. It may even stop, but that cannot be helped. . . . About the finances. The Civil Liberties Union, at the present moment, lives on air and has no resources . . . [but] [i]t will not be difficult for us to get some money . . . paying a wholetime secretary . . . who will be the principal executive as the honorary president [Tagore] and the head of the Council [Sarojini Naidu] cannot give much time or energy to it. . . . I shall also try to arrange for your steamer fare to India . . . I am writing this practically off my own bat, though . . . I am informing Sarojini . . . and I shall mention the fact to Tagore. . . . I am quite sure, however, that all of us will welcome your taking charge of the civil liberties business. We want a few intelligent men about. There is no superfluity of them. . . . Think over all this and send me a brief cable . . . you can place some reliance in me. I cannot let you down."[49]

Nehru had just returned from the south, where less than a week earlier he had entered one of Hinduism's oldest and greatest temples in Madurai. "A strange way to see a place," he wrote Indira, "with a mass of seething humanity! . . . I was taken to the inner sanctum, the holy of holies, the shrine of Menakshi [the mother goddess], an incarnation of Parvati [Shiva's consort]. A great honour. I was struck by the curious way of the Hindu faith which refuses to part with any born within its fold. Here I am accused of irreligion and yet treated as one who is a devout follower. A present of a silk scarf was also given to me—a special gift of the goddess. I realised in this vast temple . . . the great psychological influence of these religious edifices . . . when the priests were triumphant and ruled over the minds of men . . . not many people . . . can resist. . . . I have never seen so many roses in my life. . . . Every where there are enormous garlands of roses, fat, heavy things, each containing a thousand to two thousand roses. In the course of the day I might get 200 of such garlands! . . . overwhelming. It seems such a pity to waste them. . . . The car gets filled up and then I distribute them to girls and women and children by the roadside."[50]

Indira had begun to feel overburdened by letters and numerous engagements in London, and so her father tried to teach her how to cope with such popularity and power. "I have learnt by long practice how to deal with them," he wrote his successor. "Some—the useless ones [letters]—I destroy, many others I pass on to someone else to answer with brief notes on the margin. Still a large number remain and I try to deal with them personally . . . rapidly enough. . . . Gladstone used to say that most letters answer themselves after some time if you simply leave them alone. . . . [O]ne must choose worthwhile persons to see and correspond with, or else one is snowed under by the wrong sort of people."[51]

Agatha had arranged for Indira to go to the Badminton School in Bristol, where she would be able to learn enough Latin to enter Oxford. It was cold and constantly raining there, but Indira joined the Left Book Club and had enough reading to keep her busy while sneezing indoors, reporting to Papu darling "news of Queen Mary's cold . . . in the *Times!* Meanwhile the rebels got nearer & nearer Madrid. . . . Fascism seems to be spreading almost like flames, while the various Labour parties fold their arms & vote for non-intervention. . . . The atmosphere at school is terribly anti-Fascist & very pacifist. . . . But on the whole, imperialism seems to be inherent in the bones of the girls. . . . They worship the King, admire [Stanley] Baldwin & although [Anthony] Eden's popularity is waning off, he is still considered by some as the last word in cherubic innocence!"[52]

Nehru had just been to Calcutta to visit Tagore, whose strength was waning. "We talked of you and he told me what a good influence you had been. . . . And I fell to wondering what influences had shaped you—how good they were or otherwise—and how you were reacting to your present environment. It is eight months since I saw you, my dear, and I wonder how you are growing and changing. Perhaps in another six months' time we might meet again." He wanted her to come home for the summer, and was also hoping to escape abroad with her.[53]

By late November Nehru's younger supporters argued that he alone should continue to lead Congress through next year's elections, for his campaigning proved uniquely popular. Rajendra Prasad and most of the more conservative or moderate leaders preferred Vallabhbhai, however, and Gandhi remained diplomatically uncommitted. "The idea did not attract me," Nehru modestly told the press, "for I do not believe in the same person functioning again and again in one office. . . . The burden that a Congress President has to carry is no light one and his lot is not enviable." That "burden" sounded Kiplingesque. He had pressed other comrades, notably Khan Abdul Ghaffar Khan, who had declined. So "I felt that I could not myself adopt a wholly negative attitude," for "in a way I represented a link between various sets of ideas," therefore, "should . . . the choice of my countrymen fall on me, I dare not say no to it." After reading Nehru's statement, Gandhi advised Vallabhbhai to back off, and a week later he announced his decision "to plump for Pandit Jawaharlal as . . . the best person to represent the nation and guide . . . in the right channels the different forces that are at work in the country."[54]

Congress met that Christmas in the central Indian village of Faizpur, the first of its village sessions, reflecting its new mass base of peasant support and rural appeal. An estimated 200,000 people "poured into Faizpur and to feed them itself was a terrible problem," Nehru reported to Indira after his ordeal was over. "So I hurried through the business and . . . then I stole a day for Ajanta, which was not far. . . . How beautiful are the painted Bodhisattvas and the women of Ajanta!"[55]

On Christmas day the A.I.C.C. met, and Nehru told them that no fewer than fifteen hundred candidates had been selected to represent Congress in the ensuing elections by the Central Board, led by Sardar Vallabhbhai Patel. Some prominent Congressmen had resigned, with others threatening to resign after candidates, known to be incompetent or corrupt, had been chosen to represent their localities. "No organisation can grow unless its final decision is respected by the rank and file," Nehru told them. "I want you to carry the message back from here that threats of resignations are no good. Who dare run away from the Congress? . . . He cannot do so and yet carry the Congress name with him. . . . The election is only one of the things that the Congress has to do. Even if the heavens fall the Congress will go on." The next day he addressed the entire Congress, a sea of Gandhi-capped faces staring up at him.[56]

"Comrades, Eight and a half months ago I addressed you from this tribune, and now, at your bidding, I am here again . . . grateful to you for this repeated expression of your confidence, deeply sensible of the love and affection that have accompanied it, somewhat overburdened . . . fearful of this responsibility. Men and women, who have to carry the burden of responsible positions in the world today, have a heavy and unenviable task and many are unable to cope with it."[57]

Nehru was thinking of Edward VIII, who had abdicated just two weeks earlier to marry Wallace Simpson. "Personally, my sympathies are with the King," Indira had written him from Bristol, sending the news. "Although I do agree with most of the newspapers which have long articles about the duties of a king-emperor . . . large crowds gathered outside the Palace, shouting that they were with him . . . perhaps this is the beginning of the end of the monarchy in England."[58] "I must say I like many of their qualities," her father replied, referring to the English, "—most of all their restraint. It impresses one. It is an aristocratic quality. . . . The human element in the drama was powerful and when kings behave as simple humans, people are gripped by the story. On the whole Edward came out rather well. Not as a very great person, but at any rate as one who refused to behave as an automaton and who could decide for himself in spite of all the pressure that was brought to bear on him. I listened to his farewell speech on the wireless."[59]

"Soon after the last Congress I had to nominate the Working Committee and I included in this our comrade, Subhas Chandra Bose. But you know how he was snatched away from us on arrival at Bombay and ever since then he has been kept in internment despite failing health. . . . Helplessly we watch this crushing of our men and women, but this helplessness in the present steels our resolve to end this intolerable condition. . . . We are all engrossed in India at present in the provincial elections . . . and yet I would ask you . . . to take heed of the terrible and fascinating drama of the world. Our destinies are linked up with it." Then Nehru spoke again of the tragic civil war in Spain, the "rape of Abyssinia," and the impotence of the League of Nations to avert such international disasters. He accused India's rulers of "inclining more and more towards the fascist powers. . . . But we 'Move with new desires. For where we used to build and love/Is no man's land, and only ghosts can live/Between two fires.' What are these new desires? The wish to put an end to this mad world system which breeds war and conflict and which crushes millions; to abolish poverty and unemployment and release the energies of vast numbers of people . . . for the progress and betterment of humanity; to build where today we destroy."[60]

Then he outlined the policy of Congress, insisting that it "stands today for full democracy in India" and "not for socialism." But the "logic of events" will "lead to socialism[,] for that seems to me the only remedy for India's ills," Nehru added. He thanked Krishna Menon, who had "ably" represented Congress at a "world peace congress" held in Brussels in September, to which Nehru had been invited, sending Krishna Menon instead. "I trust that the Congress will associate itself fully with the permanent peace organisation that is being built up and assist with all its strength in this great task. . . . The League of Nations has fallen very low. . . . We must work for a real . . . League of Peoples." He said the new Government of India Act "stares at us offensively," calling it now a "new charter of bondage which has been imposed upon us despite our utter rejection of it." Congress had "entered into this election contest," he explained, simply to "combat the Act and seek to end it. . . . We are not going to the legislatures to pursue the path of constitutionalism or a barren reformism. . . . The elections must be used to rally the masses to the Congress standard." Congress should "have nothing to do with office and ministry. Any deviation from this . . . would inevitably mean a kind of partnership with British imperialism in the exploitation of the Indian people." He was loudly cheered for his principled stand.[61]

"Imperial Delhi stands as the visible symbol of British power, with all its pomp and circumstance and vulgar ostentation and wasteful extravagance; and within a few miles of it are the mud huts of India's starving peasantry, out of whose meagre earnings these great palaces have been built. . . . And the new Act and constitution have come to us to preserve and perpetuate these contrasts, to make India safe for autocracy and imperialist exploitation."[62]

Nehru closed the Faizpur Congress by thanking all his comrades and colleagues for their warm outpouring of love. "You brothers and sisters, all of you love me and tolerate me," Jawaharlal told them. "A few months ago somebody said that I am the spoilt child of the Congress and will wreck the Congress and the country. I had thought that this conclusion may be wrong, for I have never wrecked anybody, but he is right in thinking that I am the spoilt child of the Congress. By your love and showers of praise you add to my pride. You lift me high because you talk so highly about me. That is not good. We should not shower too much praise on each other. You do not consider how we or my predecessors reached this high pedestal of glory. We are small men."[63]

But then he may have noted the puzzled look in those hitherto happy faces around him and confusion in the eyes of those staring up at him. So he changed his frank confession of human weakness to a tone and words more appropriate to the upbeat conclusion of this annual session. "But chance had it that we joined the huge task and came under the banner of a great leader . . . Gandhiji came to our country. That was our good luck. But not even the biggest leader can accomplish this stupendous task. In this struggle we all went forward. Our strength was the strength which our army gave us . . . the army of workers which has contributed to our greatness . . . soldiers on the same battlefield. . . . Our relationship cannot break. I take leave of you."[64]

"The flame that we lighted at Faizpur," Nehru told the press next day, "will shed its radiance not only in the rural and urban areas of Maharashtra but all over India."[65]

17

Provincial Powers

1937–1938

"I LIVE IN a kind of moving cyclone—in trains and motors and vast gatherings," Papu wrote Darling Indu from his campaign train in mid-January 1937. "It is a race between my completing my programme or breaking down in health. . . . I shall win the race I hope. . . . The aeroplane is going to be called in to save time. . . . I think of you often and my love goes to you."[1] To Krishna Menon he had earlier written, "I am pretty tough but by the end of February I might have to retire to a nursing home for a while."[2] Yet he loved the dust and heat of electoral battle.

"For seven days I sped like an arrow from the bow from place to place in Bihar carrying the message of the Congress," he told the press. "Everywhere I found [an] enthusiastic response to the message of the Congress, everywhere love and goodwill beyond measure. . . . Men and women of Bihar, dear comrades in a great and glorious enterprise, I wish you good fortune."[3] He sensed victory in the air. Congress would win in most of British India's eleven provinces, in greatest measure because of him. Not himself alone, of course, for legions of people supported him. But as more and more peasants turned out to cheer, the costs of transporting and feeding them rose so dramatically that he warned, "It must be clearly announced that the Congress cannot . . . carry voters by lorry or motor, nor do we provide food as others do at the polling booths. The cry from us everywhere must be, 'On Foot to the Polling Booths.' "[4]

Thirty-five million Indians, 10 percent of the population, were eligible to vote that February, and more than half of them went to the polls. On the eve of balloting, Nehru's message to the country was strangely redolent of a World War I marching song: "Vote for the Congress and pack up your knapsack for the march to Swaraj."[5] "The villagers came out in thousands on foot, on cycles, and in bullock carts," Nan, who ran for an assembly seat in United Provinces and won, recalled. "All were dressed in their holiday best, the bullocks wearing bright garlands of marigolds. . . . Carts, jeeps, and bicycles were festooned with the Congress flag."[6] Congress won easily in six of the eleven provinces and dominated two more, collecting a total of 716 seats out of little more than 1,500 in all provincial assemblies. The second most powerful party, the Muslim League, managed only to win 109 seats, failing to gain absolute control over a single provincial government. "Away with reaction and the enemies of freedom," Jawahar shouted. "Line up! Line up! And let us all march together to Swaraj. Who dares to ignore this call?"[7]

"We have now to face this ministry business which the A.I.C.C. will have to decide within a month," Nehru wrote Krishna Menon from Allahabad in late February, enclosing a copy of the letter he was mailing the same day to Stafford Cripps.[8] "My dear Cripps, I have long wished to congratulate you on the joint front of left-wing elements [his Socialist League, Independent Labour Party, and British communists] in Britain that you have succeeded in bringing about. . . . You must have heard of the Congress victory at elections here."[9] He then told Cripps of the thousands of meetings he had addressed, coming into "direct contact" with "ten million people." He also informed him that Congress had reached beyond its electorate to hundreds of thousands of nonvoters as well and that "only a microscopic handful at the top, fearful of social changes, might be said to be against us." He admitted that the "Muslim masses are more apathetic" but insisted that "even these Muslim masses are getting out of the rut of communalism and are thinking along economic lines. . . . Congress is supreme today. . . . Even the Muslim masses look up to it for relief."[10] Thanks to Krishna Menon, Nehru knew that Cripps would play an ever more powerful role in England's future decisions on India. He therefore was eager to counter Jinnah's impact on British thinking, for in early January, when Nehru announced that the British government and the Congress were now the "only two parties" in India, Jinnah had reminded him of a "third party," the Muslims.

"Mr. Jinnah, it seems to me, has said something which is surely communalism raised to the nth power," Nehru argued. "He objects to the Congress interfering with Muslim affairs . . . and calls upon the Congress to let Muslims alone. . . . Who are the Muslims? Apparently only those who follow Mr. Jinnah and the Muslim League. . . . We have a new test of orthodoxy. What exactly Mr. Jinnah would like us, of the Congress, to do with the large numbers of Muslims in the Congress, I do not know. Would he like us to ask them to resign and go on bended knee to him? And what shall I say to the great crowds of Muslim peasants and workers who come to listen to me? . . . [W]ith all deference to Mr. Jinnah, may I suggest that such ideas are medieval and out of date? . . . [T]o stress religion in matters political and economic is obscurantism. . . . To encourage a communal consideration of political and economic problems is to encourage reaction and go back to the Middle Ages. . . . Thus, in the final analysis, there are only two forces in India today—British imperialism and the Congress representing Indian nationalism. . . . May I suggest to Mr. Jinnah that I come into greater touch with the Muslim masses than most of the members of the Muslim League? I know more about their hunger and poverty and misery than those who talk in terms of percentages and seats in the councils and places in the state services."[11]

It was hardly the sort of "suggestion" Jinnah would accept without challenge or ever forget. Nehru would have been wiser to ignore Jinnah's boast, as Gandhi had so long ignored communal Hindu attacks against his Harijan policy. Instead, he urged his comrades on all provincial Congress committees to increase Congress's contacts with the Muslim masses, importuning them "to concentrate on enrolling Muslim members of the Congress." This "mass contact" campaign for enrolling Muslims not only failed but backfired, for it challenged the Muslim League to launch a far more successful campaign of its own.[12]

President Nehru also continued strongly to oppose Congress's acceptance of provincial office, but most members of his Working Committee favored taking provincial responsibility, as did Gandhi, who drafted a compromise formula that Nehru was willing to accept, allowing Congress representatives and ministers to take office in those provinces whose governors promised not to use their special veto powers provided under the 1935 act. Lord Zetland, Britain's new Tory secretary of state for India, refused, however, even to consider waiving the emergency powers enjoyed by his governors; hence deadlock returned to Anglo–Indian politics, and for several more months all the Congress-majority provinces continued to be ruled autocratically. Nehru returned from the Congress meeting in Delhi feeling "tired and rather empty," as he confessed to Krishna. "What I want badly is rest to refresh a jaded mind." He was eagerly awaiting Indira's return and planned to take her with him to Southeast Asia. Before she reached Allahabad, however, he felt so weary and depressed that his usually robust health broke down.[13]

"Why should you become ill?" Gandhi asked him in early April. "Having become ill, why will you not give yourself rest? . . . [A]bout your grouse . . . everything I say and even perhaps do jars on you. . . . You must bear with me till my understanding becomes clear or your fears are dispelled. . . . Is there anything at the back of your mind that I do not understand?"[14] It was not the last time Gandhi wondered about the strange thoughts at the back of Nehru's mind. British governors had invited non-Congress members to form interim ministries in the Congress-majority provinces. "This autonomy is still-born," Gandhi wrote to Agatha that same day. "But the teachers of the world teach us to pray when human effort proves vain. I believe in them and therefore do not lose hope but am praying. Jawaharlal is on a sick-bed."[15] Prayer never helped Nehru. He had promised Tagore to open his Chinese Hall at Santiniketan that month but now wrote to beg off, feeling not only sick but also tired and sorry for himself, frustrated at the Muslim League's sudden strong showing in several by-election campaigns, now that Jinnah had resolved to prove to Nehru and Congress that his was, indeed, a party to be reckoned with. Even Subhas Bose, released from prison, became a nuisance to Nehru, as well as his foremost rival for the next Congress presidency.

"I am still more or less bed-ridden," Nehru wrote Subhas on April 16, the eve of Indira's return home. "It seems to me that the best approach to the Muslim masses should be directly through local Congress committees. . . . But the real point is that you should have an effective agrarian programme . . . it will draw the Muslim masses. . . . I think that flagrant cases of indiscipline should be dealt with. But too much disciplinary action sometimes creates the impression of an inquisition." Bose's own dictatorial quest for power was starting to assert itself over his Bengal Provincial Congress Committee, and soon it would extend nationwide. Nehru knew, however, that no one was more to blame for Bose's meteoric rise to Congress power than himself.[16]

The joy of Indira's arrival, however, brought Nehru back to good health and happiness, at least while she was there to cheer him with her news of London, Bristol, and Oxford and all her plans. But she said nothing about Feroze, nor did either of them mention how often they had been in touch and the plans they had made before she flew home, to meet him in Paris on her way back to Oxford. He asked her again there, "on the steps of Montmartre," she later recalled, to marry him. "That's when I finally said yes. . . . But we didn't tell anyone." Young Indu proved almost as good at keeping that secret from her darling Papu as he had always been at hiding the things that mattered most to him.[17]

In May 1937 Nehru and his daughter left for a working holiday tour of Burma, Malaya, and Singapore. The crowds who turned out to greet them and wave flags of welcome cheered him up almost as much as being with Indira who was holding tight to his arm and leaning her head on his shoulder. "I have brought to the people of Burma the goodwill and affection of the people of India," he told them in Rangoon. He talked about the recent elections and the current deadlock between Congress and the government, which appointed "phantom minority ministries" instead of popular Congress ones. "I want Indians to spread all over the world, but I do not want them to . . . exploit the people of those countries," Nehru told the prosperous Indian community in Rangoon. "I would appeal to Indians living in Burma to help their Burmese brethren in their fight for freedom. I would also appeal to the Burmese to help India in her struggle for independence."[18]

On the eve of his departure from Burma, Nehru thanked the "men and women of Burma" for their warm welcome and "the affection that you have showered on me in such abundant measure. . . . This fair land is pleasant and beautiful, but . . . more delightful are the people of this country, their bright young faces, their women with the laughter in their eyes. . . . I came to you tired and weary in spirit, but your joyous enthusiasm removed that weariness. . . . Your eyes told me . . . of the comradeship that is ours."[19]

On route he wrote to Krishna: "After two strenuous weeks in Burma, Indira and I are going to Penang in Malaya. . . . I have received two of your letters. . . . About the Trotsky business, there is no question of my associating myself with it. It is true that I was asked to give my name to an appeal for an enquiry into the Moscow trial. I have no intention of doing so. It is true that when I was in Europe I expressed a measure of sympathy for Trotsky because of what I considered unnecessarily harsh persecution. But recent events have not advanced him in my opinion. I must confess, however, that all these recent trials in Russia still remain partly a mystery to me." Krishna Menon toed Moscow's Communist line more faithfully than Nehru did, warning him about the "widespread and overt conspiracy against the U.S.S.R." launched by Leon Trotsky.[20]

"My Burma visit proved to be very tiring. I covered a great part of the country, mostly by seaplane. . . . Politically and industrially they are behind India. . . . Another fact that might interest you is this: oil and petrol which are produced in Burma in large quantities are sold at a higher price in Burma than in England or in India. . . . A more flagrant example of capitalist monopoly it is difficult to conceive. The Burma Oil Company is of course all powerful." Nehru and Krishna Menon stayed in close and continuous contact whenever possible. Nehru never lost confidence in his British-based comrade, even though Allen Lane's publishing house had failed, owing him some £800 of royalties on his autobiography.[21]

On May 27 Nehru addressed the Indian Ladies Union of Singapore, preferring to call them women, rather than "ladies," since he considered the latter "rather vulgar." It was a brief address but a strongly opinionated one. "I do not," he told them, "like the woman who is like the flower in the field . . . which just blooms and does nothing else. I do not like weaklings. I like physically fit men and women who can face mobs if necessary, and . . . control a mob, if necessary, single-handed." That was the sort of fearless courage he had always had and admired in others. "I just cannot appreciate the idea of women covering themselves with horrid jewels, like prisoners' chains. . . . I tell you, persons covered with brilliant chains and glittering diamonds really repulse me. I may however say that women should not neglect themselves. They should take some care of . . . their looks. We appreciate it. We look for self-reliance, character and beauty in women, but do not want them to be like dolls."[22] Indira was "slightly hurt" in a motor accident in Singapore, and Nehru was still worried about her health, finding her "keeping rather poorly."[23] She recovered completely as soon as her rendezvous with Feroze in Paris began.

Nehru returned home to find that Viceroy Lord Linlithgow spoke in a "softer" tone than did his secretary of state about Congress and its demands for taking office. Lord Zetland had adopted "a minatory and pompous attitude," Nehru said, but the viceroy's speech to his assembly was pitched to "a somewhat lower key."[24] C.R. and others on the Working Committee were eager to take office. That July Nehru bumped down the dirt road in pouring rain from Wardha to Bapu's ashram at Segaon in the back of a bullock cart wedged in next to Maulana Azad. The Working Committee now decided that "it will not be easy for the Governors to use their special powers" and hence resolved "that Congressmen be permitted to accept office where they may be invited thereto."[25]

Nehru's own sister Nan was one of those who now became a minister in the United Provinces, sworn in by the British governor at a solemn ceremony in Lucknow, on which the monsoon broke with its usual downpour. Nonetheless, "the pledge was taken with solemnity," Minister Vijaya Lakshmi Pandit recalled, but when her turn came to vow her "loyalty" to the "King Emperor of India," Nehru's sister "could hardly speak." The governor, Sir Harry Haig, "whom I had known since I was a young girl and he the Commissioner of Allahabad" later asked at his reception "if I was feeling well. 'Thank you, I'm well,' I answered. 'It's just that the King is stuck in my throat.' 'Well, you must wash him down, then,' said the Englishman with a smile.' " So she swallowed a sip of the soft drink passed around at the reception. "Bhai, who was traveling at that time, had sent me a telegram. . . . 'Remember the Chinese philosopher with four sons. The first was . . . a poet. The second . . . learned the arts. The third went into the army. The fourth was the despair of his famous father. . . . [His] intellect [was] limited so he might do well as a cabinet minister." As minister for local self-government and medical and public health in the United Provinces, Nan got along very well with her I.C.S. colleagues, English and Indian alike. She even enjoyed the work and found it challenging. "There was much to be done and a lot more to be undone," she wrote. "It was interesting work."[26]

Jawaharlal, however, became more depressed as he watched so many of his colleagues move into new jobs as junior partners in the firm of British India. Like fish thrown back into water they got on well with the British officials, and seemed to relish their jobs, enjoying good salaries, fine housing, free transport, and bearers carrying red-taped boxes. Nehru viewed it all as only a hopeless snare the British set for them, into which they all rushed. And it was *his* fault. *His* Victory had done this. His killing campaign, covering 50,000 miles in a few months, rallying tens of millions to the Congress cause! Then they had all rushed back to Bapu, seeking his guidance, his solution to the deadlock, and suddenly it was broken, knots untied, sweetness and light, all round. He had stormed the Bastille so that conservative old C.R. could rule in Madras, and Pandit Vallabh Pant could run the United Provinces with the help of his sister Nan!

"My darling one," he wrote Indira from Lucknow in September. "I got involved in the Cawnpore labour situation, which is very grave, and in many other local and provincial troubles. . . . I felt tired. I was not keen on going back to Allahabad and to Anand Bhawan empty and rather desolate. . . . You must be in France now—how far you are from me! I am afraid I shall not come soon to you, for I am tied up by the strong ropes of circumstance and responsibility." He wanted to visit her in England but could not hope to escape again until perhaps next year. "I am tired of this life but I must carry on till February next when the Congress meets at Haripura. . . . Already people are beginning to talk of my continuing as President for another year—an absurd proposition to which I cannot agree. But somehow I have come to occupy a curious position in Indian politics and people are afraid that another person as President may not be able to hold the balance between various forces."[27]

Yet for all the frustration he felt and all the weight of dull administrative routine he hated, part of him loved being president and craved more power. Gandhi's coterie began to fear Nehru and to wonder about his sullen, brooding ambition, his long silences, or that almost too self-satisfied look that at times suffused his handsome face. "Bebee dear," he wrote Padmaja Naidu on October 5. "This morning I returned from Lucknow. As an after-dinner exercise I have written the enclosed essay. Do what you like with it. I have not got it typed here as I did not want to show it to anybody. . . . I would rather you did not tell others that I wrote it for this would take the bloom off."[28]

Nehru called his piece "The Rashtrapati," meaning president, and chose as his pseudonym Chanakya, the name of an ancient Mauryan emperor's Brahman adviser and chief minister, who wrote the Machiavellian Hindu text on realpolitik, *Arthashastra.* "Rashtrapati Jawaharlal ki Jai," it began (victory to President Jawaharlal). "The Rashtrapati looked up as he passed swiftly through the waiting crowds, his hands went up and were joined together in salute, and his pale hard face was lit up by a smile . . . a warm personal smile. . . . The smile passed away and again the face became stern and sad, impassive in the midst of the emotion that it had roused in the multitude. Almost it seemed that the smile and the gesture accompanying it had little reality behind them; they were just tricks of the trade to gain the goodwill of the crowds whose darling he had become. Was it so?

"Watch him again . . . tens of thousands of persons surround his car and cheer him in an ecstacy of abandonment. He stands on the seat of the car, balancing himself rather well, straight and seemingly tall, like a god, serene and unmoved by the seething multitude. Suddenly there is that smile again, or even a merry laugh. . . . He is godlike no longer but a human being claiming kinship and comradeship with the thousands who surround him, and the crowd feels happy and friendly and takes him to its heart. But the smile is gone and the pale stern face is there again.

"Is all this natural or the carefully thought-out trickery of the public man? Perhaps it is both. . . . The most effective pose is one in which there seems to be least of posing, and Jawaharlal has learnt well to act without the paint and powder of the actor. With his seeming carelessness and insouciance, he performs on the public stage with consummate artistry. Whither is this going to lead him and the country? What is he aiming at with all his apparent want of aim? What lies behind that mask of his, what desires, what will to power, what insatiate longings?"[29] Nehru's ability to observe and describe one side of himself with such analytic detachment was almost enough to cure his tortured mind. "These questions . . . have a vital significance for us, for he is bound up with the present in India, and probably the future, and he has the power in him to do great good to India or great injury. . . .

"For nearly two years now he has been president of the Congress and some people imagine that he is just a camp-follower in the Working Committee . . . kept in check by others . . . yet steadily and persistently he goes on increasing his personal prestige and influence both with the masses and with all manner of groups. . . . With an energy that is astonishing at his age, he has rushed about across this vast land of India . . . like some triumphant Caesar passing by, leaving a trail of glory and a legend behind him. Is all this for him just a passing fancy which amuses him, or some deep design, or the play of some force which he himself does not know? Is it his will to power . . . that is driving him from crowd to crowd. . . . What if the fancy turn? Men like Jawaharlal, with all their capacity for great and good work, are unsafe in democracy. He calls himself a democrat and a socialist, and no doubt he does so in all earnestness, but every psychologist knows that the mind is ultimately a slave to the heart and logic can always be made to fit in with the desires and irrepressible urges of a person. A little twist and Jawaharlal might turn a dictator sweeping aside the paraphernalia of a slow-moving democracy. He might still use the language and slogans of democracy and socialism, but we all know how fascism has fattened on this language and then cast it away as useless lumber."[30]

Painfully conscious of his own ambition and its potential danger to India, Nehru here tried to warn his comrades, but his conclusion frightened him too much to leave it unchallenged, so he quickly added that "Jawaharlal is certainly not a fascist. . . . He is far too much of an aristocrat for the crudity and vulgarity of fascism." His "face," his "voice" could never belong to a "fascist" temperament. Yet in probing his own split personality and masked motivation in this extraordinary attempt to analyze the darker side of himself, Nehru was so honest that he refused to absolve his hidden half so lightly: "One wonders as one hears it or sees that sensitive face what lies behind . . . what thoughts and desires, what strange complexes and repressions, what passions suppressed and turned to energy, what longings which he dare not acknowledge even to himself . . . his mind wanders away to strange fields and fancies, and he forgets for a moment his companion and holds inaudible converse with the creatures of his brain. Does he think of the human contacts he has missed in his life's journey, hard and tempestuous as it has been; does he long for them? Or does he dream of the future of his fashioning and of the conflicts and triumphs that he would fain have? He must know well that there is no resting by the way in the path he has chosen. . . . As [T. E.] Lawrence said to the Arabs: 'There could be no rest-houses for revolt, no dividend of joy paid out.' Joy may not be for him, but something greater . . . if fate and fortune are kind—the fulfillment of a life purpose."[31] Like Lawrence, Nehru was driven, and his "secret passions" yearned for more than mere power, so there could be no rest, no revelation of those "strange complexes and repressions" that drove him.

Thus again Nehru felt obliged to repeat, "Jawaharlal cannot become a fascist. And yet he has all the makings of a dictator in him—vast popularity, a strong will . . . energy, pride . . . intolerance of others and certain contempt for the weak and the inefficient. His flashes of temper are well known and even when they are controlled, the curling of the lips betray him. His overmastering desire to get things done, to sweep away what he dislikes and build anew, will hardly brook for long the slow processes of democracy. . . . In normal times he would be just an efficient and successful executive, but in this revolutionary epoch, Caesarism is always at the door, and is it not possible that Jawaharlal might fancy himself as a Caesar?

"Therein lies danger for Jawaharlal and for India. For it is not through Caesarism that India will attain freedom, and though she may prosper a little under a benevolent and efficient despotism, she will remain stunted and the day of the emancipation of her people will be delayed."[32] Here, in fact, his prescient intellect perceived what the greatest weakness would be in his own era of "benevolent despotism," still a decade away. For he knew better than most of his critics—even those who hated and feared his socialism—just what his fatal flaws of pride and fear of discovery could do to the nation he had resolved to lead. He understood just how dangerous the part of his mind that he never completely mastered or fathomed could be to himself, to India, to Indira, and to her progeny as well. Yet he could not back off entirely. Part of him wanted to, and that part warned him to, but his love of battle was too great, his passion for victory too strong.

But he now felt "tired and stale" and so ended his pseudonymous piece by arguing that "he will progressively deteriorate if he continues as President. He cannot rest, for he who rides a tiger cannot dismount. But we can at least prevent him from going astray and from mental deterioration under too heavy burdens and responsibilities. We have a right to expect good work from him in the future. Let us not spoil that and spoil him by too much adulation and praise. His conceit is already formidable. It must be checked. We want no Caesars."[33] A few weeks later when he lost the Congress presidency to Bose, it would be easy for Nehru to reveal that he was "Chanakya," thereby "proving" that he never really wanted the job. His term as president would expire early in 1938, but he never dismounted the tiger he rode, for that tiger was inside his own heart and mind and would never let him rest.

"My dear, what am I write to you?" he replied to Padmaja. She had typed the strange piece for publication, expressing her own admiration of it and amazement at what he had written, asking if he truly meant it. And if so, did he realize what it said about him? "How am I to answer your questions? . . . [Y]our complaint is . . . no doubt justified. But it is justified only because you . . . imagine me as something other than I am, something nobler perhaps, more mysterious, more complicated. Have I not warned you against that fatal error? We expect too much from people and then are disappointed. Many things are wrong with me, not only my theoretical knowledge of psychology . . . itself a vain delusion. I do not possess it and yet I have somehow managed to impress you with it. The first thing that you must remember about me is that I have a knack of imposing on people . . . and I produce in their minds exaggerated notions about myself. . . . Even in examinations I usually did far better than others who were my superiors in knowledge. In life I have done the same because consciously or unconsciously, or both, I always try to create an impression. I succeed often enough. And so I have succeeded in impressing you with my theoretical knowledge of psychology when as a matter of fact

I hardly know anything about it. But you are perfectly right in saying that I have little sense of intuitive perception. Conceited and self-centred people seldom have it.

"Again you are wholly right in saying that I have been a failure in my individual relationships—curiously enough there have been singularly few such relationships in my life. Perhaps I felt my weakness, or was afraid of interfering with my public activities. To find a reason is rather silly. Such things just happen because one is made that way. My most successful relationships are of a casual variety. I suppose the reason for this failure is my incapacity to give. You mention Bapu, but I am quite sure that I have not given him anything that was valuable or worthwhile. I took much from him; what little I gave was not to him as an individual but to him as an abstraction. . . . I have been and am one of those who take from individuals . . . if I give at all, it is to the group. . . . Your questions put me on enquiry and the result . . . was not a pleasing one. When the sanctuaries are empty, what is there to reveal? Out of that barrenness and poverty, what is there to give?"[34]

For loving him as she did, Padmaja asked for more than Jawahar was ready or willing to give any woman. His "sanctuaries" were not empty but, rather, too full of demons and other creatures too strange to "reveal." His mind was hardly bare and impoverished, but so seething full of treasure in dreams and memories, plans for the world and future schemes for himself and Indu and India that he dared give none of them away. Then, as was his way, he added, "I do not think I am a secretive individual; I am probably franker about myself than the average man. And there are very few happenings in my life that I would take the trouble to hide. And if I hide them, it is because they are trivial and commonplace and my conceit wants a nobler background. Legend has almost invested me with that background. . . . I have a measure of restraint but that too is the product of a long loneliness. If thoughts pass through my mind, as they do in abundance, I let them go through and fade away. Some stick or come back again and again. Is it worthwhile giving them the clothing and shelter of words? And how can words imprison the vague fancies of an uncharted mind."[35]

"How I wish you would take things easy for a time at least!"[36] Gandhi wrote him that October, knowing perhaps better than anyone but Nehru himself just how driven and restless Jawaharlal was. On the eve of his forty-eighth birthday, Sarojini Naidu also wrote him, from "The Mahatma's Camp" in Calcutta, that "the Little Man [Gandhi] is sitting unconcernedly eating spinach and boiled marrow while the world ebbs and flows about him breaking into waves of Bengali, Gujarati, English and Hindi . . . He is really ill . . . the most lonely and tragic figure of his time. . . . India's man of destiny on the edge of his own doom. . . . To you the other man of destiny I am sending a birthday greeting . . . I have been watching you these two years with a most poignant sense of your suffering and loneliness, knowing that it cannot be otherwise. What shall I wish you for the coming year? Happiness? Peace? Triumph? All these things that men hold supremely dear are but secondary things to you. . . . I will wish you, my dear . . . unflinching faith and unfaltering courage in your *via cruces* that all must tread who seek freedom and hold it more precious than life . . . not personal freedom but the deliverance of a nation from bondage. Walk steadfastly along that steep and perilous path. . . . Remember Liberty is the ultimate crown of all your sacrifice . . . but you will not walk alone."[37]

"Darling Papu," Indu wrote from Somerville College in Oxford that November. "I am getting on fine. But I am afraid I am acquiring some of your bad habits. Yesterday I had a terribly full day. I had to go out in the morning—then at two there was a Labour study group . . . at seven I was having supper with 'The Darb' [headmistress of Somerville, Miss Darbyshire] and at eight thirty I had to go to a Majlis meeting—for Krishna Menon was speaking. After that I had coffee with Krishna and got home at eleven fifteen P.M.—just in time not to be locked out! My essay on the Evolution of Parliament had to be read at a class at ten A.M. this morning. . . . Well, I read until twelve forty-five and then wrote until three fifteen A.M. It was a job getting up this morning. . . . However I got Very Good for the essay! . . . The best of luck & please do try to rest once in a while."[38]

His own days were hardly so exciting. "I rejoice in your letters which tell me of your life full of activity and work and joy," he wrote back to her, telling how they recalled "innumerable pictures to my mind, a crowd of memories and visions of days gone by, and the sense of emptiness in this silent deserted house goes from me. For otherwise . . . Anand Bhawan is sometimes quite oppressive and reminds me of my days in prison. . . . I enclose an article [his piece] which has appeared in the current number of the *Modern Review*. It is well written and will interest you. I think it is very largely true. Don't you?"[39]

Indira wrote nothing in response about his pseudonymous article for several months, finally prompting him to confess in January 1938, "I wrote that article! It gave me some amusement and the idea of watching other people's reactions to it was also entertaining." He sounded sorry that "nobody found out," so he decided, as he told her, to take "a number of persons into the secret."[40] Indira was hardly surprised. "In fact I had almost guessed it and wanted to ask you," she replied, explaining, "The style of it and the English [are] so typically yours. But it was the viewpoint—that special way of looking at things and people—that first gave me the idea of the true author. . . . I was sure that whosoever had written it was not an Indian—or else he was of a type that I have never come across. So you see you cannot keep such secrets from me!"[41] A surprising, though rather insightful, conclusion for his own daughter to reach. Indira had by now become remarkably English herself, closing a letter to "Darling Papu" with "I have to rush off to a Latin Tute now—so cheerio, and do keep well & fit & fresh."[42]

In late November Nehru wrote a "personal" letter to the United Provinces prime minister, Govind Pant, to complain that "if I may put it in technical language, the Congress Ministries are tending to become counter-revolutionary. This is of course not a conscious development but when a choice has to be made, the inclination is in this direction." Now that the "time for going forward comes," Nehru argued, "we show a marked tendency to go back." He would not remain president of the Congress much longer, however, as Pant and all the other senior members of the Working Committee knew. By the end of his letter, even Jawaharlal seems to have sensed that the real problem was not in Pant's ministry but in his own head. "It may be that I have got the wrong perspective, but I can only think and act according to my own lights."[43]

Nehru's mother had her third and final stroke in January 1938. Both his sisters, Nan and Betty, had returned to Anand Bhawan with their children for a few days of family conviviality, the largest family party in years. Minister Nan was going back to Lucknow on the night train, and they all were in her dressing room when Swarup Rani collapsed. "I took her gently to her bedroom," Nehru wrote Indira on January 14. "She tried to walk but was not very successful. We put her in bed and soon she was wholly unconscious. . . . At four forty-five [A.M.] it was all over. That was exactly the time seven years ago when Dadu ["Grandfather" Motilal] died."[44]

On the eve of the Haripura Congress in February 1938, Nehru wrote his own report as the retiring president of the A.I.C.C., warning his comrades: "We cannot shirk the responsibility that has been cast on us. . . . [W]e have to be clear in our own minds . . . that we shall be no parties to imperialist war, and if British imperialism seeks to drag us into it, we shall resist the attempt. . . . [W]e will not tolerate the use of Indian troops for the purposes of British imperialism. . . . [W]e must be equally opposed to any increase in armaments in India. The army in India is not a national army. . . . To increase its strength or effectiveness is to strengthen imperialism and we can be no parties to this."[45]

A few days later, Subhas Bose reaffirmed Nehru's position in his presidential address to a crowd estimated at more than a quarter million, who squatted in fields beyond the giant bamboo-enclosed town built overnight at Haripura village. "As Lenin pointed out long ago," Netaji Bose told them, "reaction in Great Britain is strengthened and fed by the enslavement of a number of nations."[46] While Congress moved to the left, in its platform rhetoric at least, the Muslim League, now consolidated firmly under Jinnah's direction, assumed a more powerfully self-assertive stand. At its annual session in Lucknow a few months earlier, President Jinnah told the more than five thousand Muslim delegates who cheered him, "The present leadership of the Congress . . . has been responsible for alienating the Musalmans of India more and more, by pursuing a policy which is exclusively Hindu; and since they have formed Governments in the six provinces where they are in a majority, they have . . . shown, more and more, that the Musalmans cannot expect any justice or fair play at their hands."[47]

Now Jinnah called on his Muslim followers to "organize yourselves, establish your solidarity and . . . [e]quip yourselves as trained and disciplined soldiers. . . . Work loyally, honestly and for the cause of your people and your country. . . . There are forces which may bully you, tyrannize over you and intimidate you. . . . But it is by going through this crucible of the fire of persecution which may be levelled against you . . . that a nation will emerge, worthy of its past glory and history. . . . Eighty millions of Musalmans in India have nothing to fear. They have their destiny in their hands. . . . There is a magic power in your own hands." A decade later, under his leadership, Pakistan was born.[48]

Soon after that, Nehru wrote Jinnah to ask him "what the points of difference" between the Muslim League and Congress were. But Jinnah believed such written requests to be disingenuous and replied, "Do you think that this matter can be discussed, much less solved, by and through correspondence?"[49] Nehru knew, of course, that there was little if any hope of his ever agreeing with Jinnah on this most basic issue covering the entire spectrum of Hindu–Muslim conflicts in India. Nor did he wish to spend his last weeks or months as Congress president locked in mortal political conflict with one of those counterrevolutionary Indians he disliked more than any aristocratic Englishman. He had urged Jinnah instead to discuss this matter with Maulana Abul Kalam Azad, "one of our most respected leaders . . . better fitted to explain the Congress viewpoint in regard to the minorities problem or any other matter."[50] But Jinnah contemned Azad as nothing but a "showcase Muslim," a Congress "stooge." Later, he refused even so much as to shake hands with Azad, when the latter, as president of the Congress, was invited by the viceroy to attend summit meetings on this matter at Simla.

A few days after the Haripura Congress ended, Nehru wandered alone to the banks of the Tapti River by that now deserted village. Walking "to the edge of the flowing water," Jawahar "felt a little sad," he recalled, "and the desire that I had long nursed . . . to go away to some far-off place, became strong and possessed me. It was not physical tiredness, but a weariness of the mind which hungered for change and refreshment. Political life was an exhausting business and I had had enough of it for a while. . . . [D]istaste for this daily round grew, and while I answered questions and spoke as amiably as I could to comrades and friends, my mind was elsewhere. It was wandering over the mountains of the north with their deep valleys and snowy peaks, and precipices. . . . It panted for escape from the troubles and problems that encompassed us. . . . I was going to have my way, to pander to my secret and long-cherished desire. How could I trouble myself with ministries coming or going . . . when the door of escape lay open before me?"[51] He planned to get away for two weeks in March 1938 in the hills above Almora, to Khali, where Ranjit and Nan had bought land enough for a flower "farm" that Ranjit enjoyed developing, where Jawahar could lie bare-bodied in the sun, "imbibing warmth and energy," as he wrote Indu. Bharati was coming to join him there the next day.[52] Her older, unmarried sister Mridula was her chaperone and also needed rest, having led eight hundred women volunteers who had done much to make the Haripura Congress a great success.

Nehru stopped for a day in Allahabad on his escape to the northern hills. He found communal conflicts brewing there, not far from his home, and of course, "I grew irritated and angry with myself. Was I going to be thwarted and prevented from going to the mountains because fools and bigots wanted to create communal trouble? I reasoned with myself and said that nothing much could happen, the situation would improve. . . . So I argued with and deluded myself, possessed by the desire to go away and escape. Like a coward I crept away when my work lay in Allahabad. But soon I had forgotten Allahabad and its troubles and even the problems of India receded into some corner of my brain. The intoxication of the mountain air filled me as we climbed up the winding road. . . . Day succeeded day and I drank deep of the mountain air and took my fill of the sight of the snows . . . the world's ills seemed far away and unreal. . . . In the early morning I lay bare-bodied in the open and the gentle-eyed sun of the mountains took me into his warm embrace." And Bharati was there as well.[53]

Hitler's army had just locked Austria in a different sort of embrace. Neville Chamberlain continued to try doing business as usual with him, but Anthony Eden at least had the courage to quit that cabinet of appeasers. Indira was alarmed to learn that her father had accepted an invitation from Lord Lothian to spend a weekend at his estate during his forthcoming visit to England. "Krishna thinks of him—Lothian . . . a very prominent member of the 'Cliveden Set' . . . commonly known as 'Hitler's friends in Britain' . . . a thorough Fascist. . . . Your staying with him would amount to . . . spending a weekend with Hitler himself or with Mussolini. It would create a terrifically bad impression on all people in this country who are even slightly 'left' & who sympathise with India & the Congress. So please do think it over. . . . It is not only I or Krishna Menon who feel this way but every student—Indian or English—who believes in an independent & socialist India, and the whole Left element."[54]

Her father responded, "I am not surprised at your feeling strongly about Lothian. I feel more or less the same way. . . . But still after careful consideration I decided to accept his invitation. . . . I happen to be something more than a prominent leader of a group or party. I have a special position in India and a certain international status. I have to function as such whatever my personal likes or dislikes might be. . . . I happen to know something about my work and I am not unacquainted with international affairs. . . . And I am quite clear in my own mind that I cannot say no to Lothian. . . . If I am so weak as to be influenced by him then I am not much good anyway. . . . It is quite possible that Linlithgow might want to meet me in England. I am not keen on seeing him but if he expresses a wish to see me I shall not refuse."[55] Nehru's holiday in the hills had refreshed him, soothed his frazzled nerves and calmed his troubled spirit. He was now ready for a longer but much less relaxing trip abroad, for Western Europe in that summer of 1938 was hardly a Himalayan paradise.

18

Western Thunder

1938–1939

NEHRU TRAVELED first class from Bombay to Naples and Genoa on the Lloyd Triestino *Biancamano,* which left port on June 2, 1938. "It has been a pleasant voyage," he wrote on entering the Mediterranean. "The Indian Ocean was close and sticky . . . [T]he Red Sea was cooler . . . now . . . it is chilly."[1] A lovely young radical Parsi friend, Miss Bee Batlivala, who had been staying at Anand Bhawan[2] with him for some time before they sailed from Bombay, accompanied Nehru to Europe, acting as "my secretary during the voyage and has helped me greatly."[3] Bee accompanied him to Spain and thence on to London as well.

Krishna Menon was waiting in Genoa to welcome Jawahar and Bee on the morning of June 14. They all flew from Italy to Marseilles and spent "the whole day without food or drink," Nehru complained, "in search for various visas and endorsements for Spain. . . . At last we got them all and returned worn out. . . . We leave at four thirty A.M. . . . for Barcelona. . . . Bee . . . wanted to go to Spain and is accompanying Krishna and me . . . but I must not write much now. A bath is immediately indicated and then a brief three hours' sleep."[4]

"I was astonished at the normality of Barcelona in spite of air raids and all manner of dangers. The trams were running, the shops open, the theatres and cinemas filled," he reported to the *Manchester Guardian.* "I had a room on the sixth floor of a Barcelona hotel. Every night I heard an air-raid alarm and the sound of anti-aircraft guns, and from my seat on the balcony I watched a bombardment. . . . I was greatly impressed with the spirit and steadfast air of officers and men whom I met. . . . I think that psychologically Franco is weak, that if the struggle is prolonged . . . he would only remain in Spain as long as the foreign troops are there. Every Spaniard who spoke to me is bitter against the British Government, and especially against Mr. Chamberlain. . . . Recent events have intensified the determination of the Congress that India should be independent."[5]

After three days in Spain, Jawahar, Bee, and Krishna Menon flew to Paris, where Indira had come fresh from her exams at Oxford to meet him. He spoke over Paris Radio on June 20, "glad of the opportunity to say a few words to our friends in France. . . . War hangs over all of us. . . . We in India stand for world peace and collective security and that is why we have gladly associated ourselves with the International Peace Campaign."[6] The next night he spoke in the Hall of Nations and said that although he had come to Europe "on a private visit . . . [w]herever I am, I am a bit of India, and my mind is naturally full of problems of India and her fight for independence."[7]

Three days and nights in Paris, caught up in the "whirlpool that is Europe" left Nehru quite tired as he headed toward London. Now that the most intense part of his trip was about to begin, "I began to deteriorate physically," he wrote Bharati, trying to explain why he had not even acknowledged her letter to him in more than a month, "and some doctors insisted on my taking injections and ultra-violet rays and the like." A decade earlier he had consulted English doctors about his baldness, but now he feared he was losing more than his hair and hoped those vital injections and potent rays might restore his physically deteriorating powers. Bharati had thought, after their brief but beautiful interlude in the Himalayas, that he felt about her as she did about him, sending him *Vasantsena*, her passionate intimate memoir of their romance. "It is astonishingly personal," he wrote in surprise. "I doubt if I could have written anything quite so personal. But then I am a practising politician and have to be careful."[8]

"After seeing British policy at work in Spain . . . India is more than ever determined to free herself from British domination," Nehru informed the press upon his arrival at Victoria Station on June 23, 1938. "Of course, it is always hard for a layman to make an expert judgment, but . . . I do not see how the Republican Spain can lose."[9] At a reception in the House of Commons that evening, which Krishna Menon had arranged, he cautioned his friends that the Indian question was much bigger than most Englishmen seemed to think and had "enormous potentialities for good or evil."[10]

Lothian wrote next day from Blicking Hall, "I'm glad to see that you have arrived safely in England . . . I am looking forward to entertaining you here for the week end of July 9th. . . . I hope to have Lady Astor, who will interest and amuse you, General Ironside who is one of the best soldiers in England and will be able to give you a conspectus of the military and general state of the world, which you may not be able to get from others, Mr. Thomas Jones, who was Baldwin's most intimate adviser when he was Prime Minister. . . . My main object is to give you a quiet week end in beautiful surroundings."[11]

In his later note to the Working Committee of Congress, reporting on his European trip, Nehru told them, "I met a host of people of all shades and views. . . . Some I met in country houses during week-end visits and had repeated opportunities of talking to them. These talks covered a wide variety of subjects . . . Indian defence forces (which I discussed with a very high military officer with intimate knowledge of Indian and world conditions), and international affairs. . . . The dominant impression that one gathers is that the whole question of India is looked upon from the point of view of what India might do at a time of grave international crisis, such as war. . . . Crudely put, India had a tremendous nuisance value. Because of this, India cannot be ignored."[12]

Nehru's meetings with the Clivedon set thus gave him an exaggerated estimate of the importance of Congress's potential power to undermine Britain's military capability in dealing with Germany and Italy. He also now doubly underestimated Jinnah's importance, hearing less about communal matters from Englishmen who understood how negatively he felt about religious conflicts. But given the inordinately high proportion of Muslims in the British Indian army, Nehru should have realized that in time of war the political stock of any great leader, who claimed to speak for all Muslims was bound to rise rapidly in value in Whitehall as well as in Delhi–Simla.

Nehru met with Lord Linlithgow, who had come home for summer leave, and described their talk as "long, frank and friendly." Linlithgow asked if Nehru would "agree" to an offer of dominion status, and "I said that in spite of the background of hostility . . . we had every desire to maintain friendly contacts to our mutual advantage. We recognised further that the dominion conception was a dynamic and changing one."[13] A week or so enjoying England's stimulating climate of social and diplomatic hospitality was enough to remove most of Nehru's antipathy and mistrust of Great Britain and its lords and ladies. He still argued, of course, for "very radical economic reforms and changes," but Linlithgow showed no surprise and expressed no disagreement with that or, indeed, with anything Nehru said to him. The bland viceroy was a brilliant listener; Nehru was even more brilliant at talking.

Nehru did not see Neville Chamberlain or Winston Churchill in London but had a brief talk with Lord Zetland, which was neither as long nor as friendly as his meeting with the viceroy. His weekend with Cripps, however, brought him into close contact and a basic meeting of minds with the leaders of the Labour Party, who in the next decade continued to confer with him and generally to support his position. Clement Attlee, Aneurin Bevan, Hugh Dalton, R. H. S. Crossman, and Herbert Morrison all were invited, as were Harold Laski and Krishna to Cripps's Nehru weekend. "Labourites and leftists generally . . . accepted my contentions almost in their entirety and were convinced that India had the whip hand if only we would use it," Nehru gleefully informed his colleagues on the Working Committee. "They . . . hoped that pressure from India might influence British foreign policy, which they detested." [14]

Nehru's socialism and Congress's aspirations for self-determination were, of course, much closer to the British Labour Party's policies than to either the Tory or the Liberal positions. Nehru personally, however, often felt less comfortable with most of the leaders of Labour than he did with the Tories, including old Harrovian Winston Churchill. He liked Laski, however, and developed a warm intellectual appreciation of Cripps's brilliance, though Nehru later blamed Cripps for the failure of his mission to India in 1942. At this time Nehru told his Congress colleagues that Labour had "no power" in Baldwin's government, and although he enjoyed his weekend, he really expected nothing substantial from Labour in the way of immediate political support for Congress's struggle. Krishna and Cripps were more optimistic, however, as was Laski, who planned to take Congress's appeal for help in its struggle for freedom to the Labour Party's next annual seaside conference.

Clement Attlee, who led the Labour Party opposition in Parliament, was more cautious and somewhat more conservative than Cripps, as were Bevan, Dalton, and Morrison. Hugh Dalton, however, had attended Cambridge when Jawaharlal did, and as president of the Board of Trade from 1942 to 1947 and chancellor of the exchequer under Attlee from 1945 to 1947, he played an important role in expediting the final transfer of power and treaty of secession. What none of them anticipated or discussed, however, was India's partition, although the previous year's precedent of what had happened in Ireland, where an independent Republic of Eire emerged, leaving Northern Ireland still bound to England, was certainly fresh in all their memories.

"The Indian states offered some difficulties," Nehru reported to his colleagues back home. "It has been suggested that the door should be left open for them to enter the constituent assembly. . . . We should rely on the compulsion of events which is bound to be considerable, and . . . a large number of them would join, if the British Government's attitude was clear and they could get no help from it. . . . I felt sure that the creation of the constituent assembly would give rise to such a power in India that no one would be able to withstand it." This proved to be true for most of the princely states, but the two largest, Kashmir and Hyderabad, refused to join India until they were forced to do so.[15]

"In regard to the public debt . . . [i]t is proposed that we pay this not in cash but in purchase of British goods. . . . That is to say that in paying the debt we get our money's worth in machinery and other goods. . . . Our developing industries will require plenty of heavy goods and machinery and we shall, in any event, have to import them. . . . I pointed out that in the event of a peaceful, speedy and friendly transition to independence, India would not be unwilling to assume some burdens in the shape of pensions and other charges. The cost of conflict would be greater than these." Nehru thus proved himself remarkably flexible, willing to untie some of the toughest knots of empire that had seemed impossible less than a month before but now loosened when the best British and Indian brains focused on them in an atmosphere of goodwill and mutual interest.[16]

"I might mention that the attitude of these Labour leaders was very different this time from what it was two and a half years ago," Nehru reported. "I should like to know the reactions of the Working Committee to this line of approach. I shall meet the Labour leaders again before I return to India and I want to tell them then, informally of course, what the Congress attitude is. . . . There should be a pact of mutual non-aggression between Britain and India. . . . A trade agreement would obviously be necessary." In less than a month Nehru, Cripps, and Attlee all but drafted an agreement that eluded final settlement between India and England for most of the next decade. The Congress's general secretary, Jivan B. Kripalani, wrote Nehru early in September to say that Mahatma Gandhi and the entire Working Committee, including President Bose, were "highly satisfied" with those "talks" and his report.[17]

Krishna also arranged a public reception for Nehru in Caxton Hall in late June, where Harold Laski, and the dean of Canterbury, Cripps, and the Indian communist R. Palme Dutt all spoke. Paul Robeson sang. "Indignation is felt here at the bombing of Barcelona and Canton," Nehru told that cheering audience. "I want you to realise that there is no difference in principle between the bombing of Barcelona and Canton and of the north west frontier . . . England is responsible."[18] His English comrades cheered Nehru much louder than his peasant audiences in Punjab had. That July he addressed an Anti-Imperialist League rally that Krishna organized in Trafalgar Square, wearing for the first time since he had arrived in Europe a dark Indian sherwani and black Gandhi cap, rather than his Saville Row suits or Harris tweed jackets and ties. He looked as handsome in Indian clothing as he did in his tighter-fitting Western garb, and Krishna insisted that for the mass audience in Trafalgar Square the symbolism of his dress was as significant as the short address he gave. Moreover, his choice of north Indian Punjabi dress was tailored to identify him with

Muslim India, as it was much the same costume that Jinnah had worn when he presided over the previous year's meeting of the Muslim League in Lucknow.

From London, Nehru took Indira back to France with him, where they both rested for a week at the seaside in Houlgate. From there he wrote his long note to Congress and a letter to Bharati. They then went from Paris to Munich, where he addressed Indian students in Germany and met with a number of German officers, eager to talk to him.

Nehru and his daughter flew to Prague. By then, Indira had started to break down. Whether it was the pace of his visit or her acute sensitivity to the breakdown of Europe and world order that triggered her collapse, or the combination of both and her desire to be with Feroze, she found it virtually impossible to sleep and kept losing what little weight she had. Nan joined them in Prague, and they met Jan Masaryk on the eve of Germany's march into the Sudetenland, less than a month before Chamberlain's shameful capitulation to Hitler.

"I have no doubt in my mind that war can be prevented if the British Government adopted a strong attitude in favour of Czechoslovakia," Nehru wrote Jivan Kripalani from Prague, "but that is exactly what it is not doing . . . astonishing how the British Government goes on encouraging Hitler and his gang." He feared "the complete domination of Hitler right up to the Black Sea" but rightly predicted that "if war comes, it is unlikely to remain confined. . . . It will spread and thus all manner of terrible things will happen. Personally, I think that it is bound to end in Hitler's defeat." His visit to Prague thus armed Nehru with acute awareness of the impending tragedy of Czechoslovakia and the imminence of war in the West. It helped change his earlier position, removing him further from Gandhi's "well-meaning 100% pacifism."[19]

"I write this article as I sit in a great city in Central Europe," Nehru wrote from Budapest on August 21. "I wonder what the fate of this city will be, as of many another city in Europe, when war comes. . . . Yet war need not come, perhaps it will not come. . . . But sometimes one feels as if we were all in the grip of tragedy, after the Greek fashion, which moves inevitably . . . to a predestined catastrophe."[20]

In Budapest Indira was diagnosed as having pneumonia, "and Bhai and I had an anxious time," Nan recalled.[21] "I think I caught a chill," Indira later wrote of that start of her nervous breakdown. "I felt very tired but I didn't realize I was ill. Everybody remarked that I looked terrible and I was very thin. . . . From the hotel I went to [a] hospital."[22] On August 30, 1938, Nehru wrote Kripalani from Budapest, "I have stayed on so long . . . owing to Indira's illness. She is much better now and we are therefore going to London by air on September 1st. Vijayalakshmi [Nan] is accompanying us. Europe is a volcano on the point of bursting . . .

solely due to Hitler's determination to dominate Europe by crushing Czechoslovakia and thus reaching Rumania with her oil fields and wheat. Nevertheless, the minority question is full of interest and warning for us and we might learn something from it. . . . It is obvious that a war once started will shake and paralyze the world and affect India deeply. Our attitude and the actions we might take will have the most vital consequences for India's future . . . [T]he Working Committee should meet and give full thought to the situation."[23]

That September they flew back to London, and Indira was hospitalized in Brentford, Middlesex, where she remained till she felt strong enough to return home to India before year's end. Krishna had organized the Peace and Empire Congress, scheduled to start in Glasgow on September 24, to which Nehru was asked to speak again from the same platform as would the archbishop of Canterbury. He decided, however, to let Krishna represent him there. Kripalani wrote Nehru a week later, urging him on behalf of Vallabhbhai and other members of the Working Committee to cut his trip abroad short and hurry home in time to attend the next meeting in Delhi, when Congress's response to the possibility of war would be considered. Jinnah had already stolen a march on Congress by leading his Muslim League members in the Central Assembly to support the government's "anti-recruitment bill" that would penalize any Indian issuing "propaganda against recruitment." Nehru had hoped to return to Moscow before going home and wanted, if possible, to go south through Soviet Central Asia, visiting Samarkand and Tashkent en route to India. He was, however, unable to get the requisite Soviet visa in time.[24]

After reaching London Nehru wrote his most depressing note to the Working Committee. The gap between Congress and the British government was much greater than he had previously believed. "The old spirit continues with hardly any change," he now judged, blaming it partly on the "astoundingly reactionary and smug" character of Britain's government, which he considered "quite foolish enough to take some steps which would lead to conflict." He concluded that "the essential thing" was that "India cannot and must not be dragged into war by the decision of the British Government."[25]

Nehru wrote Indira from London in mid-September: "I am writing this after midnight. The morning papers are out already telling us of Chamberlain's flight to Hitler. His first flight! Well I shall stick to my flight to Geneva. . . . Look after yourself and get strong & well soon. Only the fit in mind & body can do much and there is big work ahead."[26] He decided to wait in Switzerland for his Soviet visa. Indira replied from Brentford Hospital on the same day: "I am more or less the same . . . feeling terribly tired all day. In the afternoon I was weighed—85 lbs. . . . I am going to miss you no end."[27]

Nehru now planned to return home by the end of October. From Geneva he reported on September 20: "Depression reigns supreme. The League assembly is sitting, but who cares for it? Geneva does not count, the League is dead. Prague counts and London, Paris, Moscow and, of course, the mountain retreat of Hitler. The Palace of the League looks like a mausoleum built to honour the dead body of peace and collective security."[28] Two days later he wrote home from Paris. " 'We have been abandoned, betrayed,' cried a vast multitude of the Czechoslovakian people in their agony . . . their cabinet . . . had decided to accept the Anglo-French ultimatum, and a hundred thousand citizens of . . . Prague poured out into its streets . . . tasting the dregs of humiliation." Nehru viewed the "great betrayal" by Chamberlain from the vantage point of India's "long experience of promises broken and betrayals by the British Government. Yet it is well that this new experience has come to us also, lest we forget. None so poor today as would care to have the friendship of England or France, for open enemies are safer and better than dangerous friends who betray. . . . Let India cut herself away from this connection with Britain which makes her a sharer in dishonour and betrayal. We must rely on *ourselves alone.*" That was the meaning of Sinn Fein, as he had learned long ago.[29]

Back in London, Nehru told the press, "It is pitiful and absurd for the Indian princes to shout out their loyalty and promise to fight for democracy," as the maharaja of Bikaner had done a few days earlier. "It is scandalous that they should crush their own subjects denying democracy, yet talk tall what they will do abroad." He also attacked Sir Feroze Khan Noon, India's high commissioner to London at this time, for having told Reuters that since all seven Congress provincial ministries were "already cooperating with Britain" and had "sworn allegiance to the Crown," it was "reasonable to assume" that all would support Britain in the event of war. "Whom he speaks for I don't know," Nehru commented sarcastically. Noon was later to speak for Pakistan, first as foreign minister and then briefly as prime minister.[30]

Nehru appreciated President Franklin D. Roosevelt's "dignified, moving appeal" to Hitler, in which Roosevelt urged him "for the sake of humanity everywhere" not to break off negotiations for "a peaceful, fair, and constructive settlement of the questions at issue. . . . It is a question of the fate of the world, today and tomorrow." "Even the printed word of President Roosevelt shows that there is a man behind it. . . . Is Hitler absolutely mad that he should risk . . . plunging into war? Does he not know that defeat and disaster will certainly be his lot in a world war; that many of his own people will turn against him?"[31]

Nehru watched from the gallery at Westminster as Chamberlain spoke that night in late September 1938 when thunder rumbled over every capital of Western Europe and the world hung suspended by a madman's threat. "The Prime Minister begins . . . there is no nobility in his countenance. He looks like a business man. . . . Somehow I feel (or is it my imagination?) that the man was not big enough for the task he undertook. . . . He is excited and proud about his personal intervention, his talks with Hitler, the part he is playing in world affairs. . . . A Palmerston or a Gladstone or a Disraeli would have risen to the occasion . . . so would Churchill in a different way. . . . But there was neither warmth nor depth of intellect in what Mr. Chamberlain said . . . he was not a man of destiny."[32]

Nehru's report of this drama reflects his own passionate conviction by now that he, unlike Chamberlain, was a "man of destiny." Churchill, of course, was the other Harrow man of destiny in Westminster that dark night. "There was no talk of high principles, of freedom, of democracy, of human right and justice, of international law and morality, of the barbarity of the way of the sword, of the sickening lies and vulgarity of the high priests of Nazism, of the unparalleled coercion of minorities in Germany, of refusal to submit to blackmail and bullying," Nehru lamented, writing the speech he would have delivered while the world waited and watched, and the sword of world war hung so precariously over Big Ben's tower. "There was no mention of President Roosevelt and his striking messages . . . no mention of Russia, although Russia is intimately concerned with the fate of Czechoslovakia. And what of Czechoslovakia herself? . . . [N]ot a word about the unparalleled sacrifices of her people, of their astonishing restraint and dignity in the face of intolerable provocation, of their holding aloft the banner of democracy. It was an astonishing and significant omission, deliberately made. . . . He was going to Munich tomorrow. . . . And as a great favour Hitler had made a striking concession—he would defer mobilisation for twenty-four hours! . . . Was there going to be another betrayal again, the final murder of that nation? This sinister gathering . . . at Munich, was it the prelude to the . . . pact of fascism-cum-imperialism to isolate Russia, to end Spain finally and to crush all progressive elements?"[33]

Govind Pant, as chief minister of United Provinces, wired Nehru at this time, asking him to try to find European experts to help India develop industrially. But Nehru was hardly anxious to ask or suggest anyone for that sort of early, desperately needed planning, which might well have given the United Provinces a head start in economic development. "I fear the schemes of the United Provinces Government in regard to industrial and rural development are lacking in any vision or plan," Jawaharlal replied, throwing cold water on the launching of any early plan. His mind was preoccupied with diplomacy.[34]

Nehru now was thinking of ways to link India to its modernist Muslim neighbors. On his way to Europe he had briefly met with Egypt's ex-prime minister Nahas Pasha, who led the popular Wafd Party. Nahas invited him to attend the Wafdist congress in Cairo on November 23, and Nehru wrote from London to Maulana Azad, who was Congress's leading "nationalist Muslim," urging him to go as one of two Congress representatives. "I think it highly important that we should send . . . two, a Hindu and a Muslim." He had heard in London that Jinnah planned to attend that Cairo meeting on November 7, hence was eager to send Azad to Nahas Pasha's meeting as proof of the Congress's pro-Egyptian, pro-Palestinian policy.[35]

Agatha asked Nehru to speak at Friends House a few days later on the current world crisis, and he did so with passion and power. "If the Treaty of Versailles was bad the concord of Munich is a million times worse, and nothing but evil can come out of it," he thundered. "What had been done at Munich would be remembered for all time all over the world . . . in India particularly, to the shame and dishonour of France and Britain . . . I do not like war but I would have a thousand wars before lowering my head to that evil thing called fascism. . . . It is essential that India should be independent, and direct her own foreign policy completely free from that of Britain. Only in that way can she play her part in maintaining moral standards. That is what self-determination—a word prostituted in these last few days—really means for India." Such words would later haunt him in Kashmir.[36]

Nehru sailed home on November 10. Indira was well enough to sail with him, but once he had gone, Krishna became unwell, not surprising given his sleepless routine and diet of tea and tomatoes. They reached Bombay on November 17, 1938, and Nehru informed the press corps that awaited him: "My personal impression of this [British] government is that it has succeeded quite remarkably in tying itself up into a knot, and in a mad frenzy it goes on tying itself into more and more knots."[37] The only hopeful sign he noted was "a remarkable change" in the British Labour Party's outlook. "Labour now admits that the only solution of the Indian problem is complete self-determination." Despite Churchill's conservatism, Nehru called him the "ablest politician in England today."[38]

Nehru was warmly welcomed in Bombay. Bharati was waiting with her garland of roses, and he was soon engaged in speaking about the Spanish civil war, and appealed for food grain for Spain, as he wrote to Krishna from Wardha. "We concentrate on Bombay, as usual, in such matters. The bourgeoisie of Bombay helps us far more than the rest of the country put together. . . . I expect the equivalent of Rs. 50,000/–will be forthcoming from Bombay. I am fortunate enough, in spite of my leftist views and tendencies, to have a measure of popularity with the merchant class in Bombay." For all his leftist tendencies Nehru now worried more about the international public relations value of press reports that impoverished and starving India was sending some of its food to Spain than he did about using those substantial funds for Bombay's own underfed millions.[39]

Jaya Prakash Narayan had not been able to come to Bombay to welcome Nehru home but wrote a few days later from Calicut on the Malabar Coast, where he was "undergoing a special Ayurvedic treatment for my sciatica. I feel improved though not cured," J.P. reported. "I hope that having been in the midst of tremendous happenings you have not forgotten the small affair of the Socialist Book Club of which I wrote you. We have been able to make some progress . . . and with the help of Subhas Bose we were able to raise about Rs. 3000/–for it at Calcutta. . . . In the letter you wrote from Europe you expressed your inability to join the Club as a Foundation Member till you had occasion to know more about it. You had also expressed your reluctance to identify yourself with any Group. . . . Your refusal to join . . . would be a great blow to us. . . . We are, I think, not unjustified in expecting that, . . . you will, as a Socialist, at least help us in doing well the little we may undertake. . . . In your letter you had said that politics in India had fallen into a rut. In your absence they have only gone deeper . . . converting the Congress . . . into a handmaid of Indian vested interests. A vulgarisation of Gandhism makes this transition easy and gives this new Congress the requisite demagogic armour. . . . We are faced today with the real danger of Indian industry being made a synonym for Indian nationalism. . . . The Socialist movement, as you know, has placed in the foreground the programme of labour and peasant organisation. . . . You have on innumerable occasions made your position clear about this programme. But I feel the time has come when you should go further and take a hand in moulding and developing it."[40]

J.P.'s cry long continued to challenge Nehru's once loudly proclaimed principles. Now that Nehru felt he was becoming a man of destiny rather than India's radical outsider, he was much more cautious about joining anything, even a book club.

To British officialdom in India, however, Nehru remained an "extreme Red," whether a Trotskyite or Stalinist was totally irrelevant to the United Provinces' governor Sir Harry Haig or Viceroy Lord Linlithgow. Haig met with Minister Madame Pandit as soon as possible after she reached home with her brother and his daughter. "I always find her very frank," Governor Haig reported in his secret note to the viceroy on his meeting with Nan, whom he called "a moderate-minded person. She is thus a valuable link, as through her brother she is in the closest touch with extremist opinion and policy. She said to me . . . the public expected her to be 'red' and she often finds it very difficult to live up to this expectation."[41] A week later Haig noted in another secret message to his viceroy that Nehru's return to the United Provinces had "proved an encouragement to the left wing" and that communal relations were getting worse throughout the province.[42]

Tagore pressed Nehru to come visit him at Santiniketan as soon as possible, as he was "anxious about Indira's state of health" but also very eager to get Jawaharlal's support for Subhas Bose's reelection as the president of Congress. "My province is clever but morally untrained and supercilious in her attitude towards her neighbours," Tagore wrote of troubled Bengal that November, "she breaks into violent hysteric fits when least crossed in her whims. I know her weakness but I cannot maintain my detachment of mind and passively acquiesce in her doom of perdition." He felt that Bose was the best man to keep Bengal together.[43]

Nehru agreed to Bose's request that he chair Congress's Planning Committee, a subject destined to remain very high on Nehru's own agenda after he became prime minister. He first convinced Congress and later India to adopt the Soviet Union's model of five-year plans as the blueprint for its economic development, thus emphasizing state planning with all its bureaucratic incompetence when encouraging private initiatives, competition, and greater freedom for entrepreneurial growth might well have moved India's economy ahead more quickly.

"Soon after my return from Europe in November I was asked about the Congress presidentship," Nehru claimed. "Would I agree to accept office again? I had not given a moment's thought to the matter and was not particularly interested. . . . Some time later I had occasion to discuss this matter with Gandhiji. I gave it as my decided opinion that Maulana Abul Kalam Azad would be the right choice. . . . He could carry on the old tradition of the Congress and yet not in any narrow or sectarian way . . . the ideal emblem of united working which I sought, especially at this critical juncture . . . Subhas Bose was thinking in terms of reelection. I did not like the idea . . . I disliked it for the same reason as I disliked my own election. . . . [H]e and I could serve our cause much better without the burden of the presidentship."[44]

Nehru sent Indira off with Betty to Raja's farm in Almora's hills in December. "I am afraid you have not had as much rest in Allahabad as we had hoped," he wrote his still frail, always tired, daughter. "You realise . . . you have undertaken a biggish job—to build up your health on an unshakeable foundation. This is not easy unless tackled in a business-like way . . . getting tired easily must be conquered. . . . (1) Take your temperature morning & evening. (2) Three hours' rest in the afternoon. . . . (3) If you feel tired increase your rest. . . . You had better take a new supply of your tonics . . . I am convinced that you will prosper there. Take care not to catch chills. . . . I shall be here," he wrote her from Wardha, "till the 15th. Then Bombay."[45]

Despite his good medical advice to Indu, Jawaharlal himself got sick, probably because he felt obliged to sit all day with the Congress Working Committee. "I have not been keeping as well as I ought to be," he admitted to Indira, "some kind of a chill or cold, or perhaps a mild dose of flu, . . . pursued me from Wardha . . . [where] I sat for long weary hours in committee, with aching limbs and fevered brow."[46] Maulana Azad had refused the "crown of thorns" that Gandhi and Nehru tried to press upon him, and Jawahar would not make it a contest against Subhas, who coveted it for a second time. So Gandhi pressed his devoted disciple, Congress stalwart Pattabhi Sitaramayya, to run. The struggle for power proved to be a bitter contest, the first campaign for A.I.C.C. votes, which Subhas won by a narrow margin of a little more than two hundred votes out of almost three thousand.

"I hope it is well with you," Nehru wrote Indira that December of his own discontent. "I plan to visit when I can . . . but the web of life encircles me and makes me prisoner. . . . The Working Committee meets again in January . . . in remote Bardoli. . . . From now to the middle of January my days are terribly full. Three days in Allahabad . . . with an important conference of Congress Muslims . . . many guests in Anand Bhawan. On the 25th night I go to Lucknow . . . on the 27th to Fyzabad [Ayodhya] for the Provincial Conference over which I am presiding. . . . I am not looking forward to this visit. It will mean a great strain."[47]

With close to a million and a half four-anna members of Congress, the United Provinces had the largest Provincial Congress Committee as it was the largest, most populous province. Ayodhya, the venue of the 1938 meeting, was the epic capital of King Rama, Hinduism's royal incarnation of Lord Vishnu, whose mythical reign was viewed as ancient India's Golden Age of perfect polity, Ram Rajya. Many Congress delegates now viewed Nehru as modern India's reincarnation of Rama, the ideal hero-prince who sacrificed comfort and the luxury of his palace to venture into the demon forests beyond the borders of Ayodhya, waging battle against the demon-king Ravana, ultimately destroying him in single combat.

The political bosses of the United Provinces commanded more votes than all the other provinces and thus long remained India's central government leaders. Ayodhya, with its hundreds of temples to Rama and his wife, Sita, was a perennial magnet for Hindu pilgrims. But just beyond Ayodhya's limits was the Muslim city of Fyzabad, most of whose working population were impoverished Muslim laborers. Hindu–Muslim riots often started there, sparked by irate Hindu attacks on Muslim butchers leading sacred cows to slaughter, or by raucous music that accompanied a Hindu wedding party as it wended its way past a Muslim mosque, where silent worshipers considered such noises to be insults to Allah. There were countless causes for the communal conflicts that proliferated, many of them economic, others reflecting doctrinal religious differences, now intensified by the political competition between the Congress and the Muslim League.

In Lucknow and Cawnpore and other major cities of the United Provinces the Muslim minority populace of many millions felt increasingly aggrieved during this interlude of Congress's provincial rule. The Congress flag flew everywhere, and the Congress anthem, "Bande Mataram" (Hail to Thee, Mother) was sung each morning by schoolchildren and at every public function. Hindi rather than Urdu was taught in elementary schools, and most Congress appointees favored their own caste brethren in ways familiar to all democratic governments, in which patronage, nepotism, or lobbying often appear to favor friends and relatives of the party in power, who then appear to ignore the just claims of minorities or actively discriminate against some of them. Jinnah and his Muslim League lieutenants had compiled several lengthy reports of Muslim grievances against various Congress appointees and several of the Congress ministries, including that of United Provinces, during their year in office. Most of the Muslim members of Congress advised Nehru to mollify Jinnah, some of them reiterating earlier suggestions to form a United Provinces coalition government with the league. But Nehru rejected such advice. He considered political coalitions reactionary in India's current climate, and he strongly opposed any softer policy toward the Muslim League. He denied, moreover, that any discrimination against Muslims had taken place, insisting in his presidential address at year's end in Ayodhya that the United Provinces Congress ministry was always perfectly fair.[48]

"Nehru is either utterly ignorant of what is going on in his own province or he has lost all sense of fairness and justice when he characterises the charges against the Congress Governments as baseless," Jinnah replied.[49] "Mr. Jinnah says that I am ignorant of what is taking place in my own province," Nehru countered. "May I request him to have the courtesy to inform me of the atrocities committed in U.P. by the Congress government. . . . May I also suggest to Mr. Jinnah, who is an eminent lawyer, that one-sided charges made would have to be proved before they are to be believed?"[50] Jinnah then called Nehru's attention to a report prepared by a Muslim League committee led by the raja of Pirpur, but Nehru could find no copy of it. Gandhi was more worried about Jinnah's complaints than Nehru was, for as he wrote that December, referring to Hindu–Muslim unity as news of Maulana Shaukat Ali's sudden death reached him, "No other unity is worth having. And without that unity there is no real freedom for India."[51] Nehru believed that freedom would bring Hindu–Muslim unity through radical economic change. But he grossly underestimated the destructive powers of communal conflict even as he overestimated the reforming potential of his five-year plans.

Nehru still hoped in January 1939 that either he would be urged so strongly to take back the Crown of Congress that he might be persuaded to do so or that Maulana Azad might agree to wear it, thus symbolically answering most of the league's charges and complaints. For what better proof could there be of how fairly Congress treated Muslims than to elect one as its president? But Azad finally withdrew his name from any such contest, after a long private meeting with Subhas Bose in Santiniketan in January. Nehru knew then that Subhas was determined to stand and fight against Gandhi's conservative old guard for what he now considered a socialist left-wing policy for Congress. Subhas hoped that Nehru would help him, as Tagore was doing. But Tagore had no political ambitions of his own, and for all his universal creative genius, Tagore was Bengali; hence Subhas Bose and his brother Sarat seemed to him incapable of doing any wrong. To Nehru, however, Bose had suddenly become the one obstacle to his own rise as India's man of destiny. Gandhi was too old and wanted no office. None of the others was popular enough, not Vallabhbhai, or C.R., or Rajendra Prasad. They all would be happy to serve in his cabinet if he wanted them. But Subhas Bose really considered himself India's *netaji*—supreme leader, a South Asian Hitler, Mussolini, and Stalin—or as he preferred to call himself, a "socialist."

"As I told you," Jawahar wrote Bose, "your contested election has done some good and some harm. . . . Obviously, it is not good enough for any one of us to get into a huff because matters have not shaped as we wished them to. We have to give our best to the cause whatever happens." He wrote that part to himself. "The first thing we have to do is to understand each other's viewpoints as fully as possible. . . . Before I can determine my own course of action I must have some notion of what you want the Congress to be . . . I am entirely at sea about this . . . I do not know who you consider a Leftist and who a Rightist. . . . Strong language and a capacity to criticise and attack the old Congress leadership is not a test of Leftism. . . . You will remember that I sent long reports from Europe to you. . . . You sent me . . . not even an acknowledgement. . . . In effect you have functioned more as a speaker than as a directing president. . . . Your desire to have a [Muslim League] coalition Ministry in Bengal seems hardly to fit in with your protest against a drift towards constitutionalism . . . it would be considered a Rightist step. . . . Then there is foreign policy. . . . I do not know yet exactly what policy you envisage. . . . Public affairs involve principles and policies. They also involve an understanding of each other and faith in the *bona fides* of colleagues. . . . What am I to do with the finest principles if I do not have confidence in the person concerned? . . . [B]ehind the political problems, there are psychological problems, and these are always more difficult. . . . I do not expect you to answer this letter immediately . . . but I would like you to send me an acknowledgement."[52]

Bose felt betrayed by Nehru and soon found that he had no support on the Working Committee, twelve of whose senior members resigned after consulting Gandhi, who did not even bother to attend the Tripuri Congress, for he had clearly made it known that Sitaramayya was his candidate. Gandhi took Sitaramayya's defeat as "my own." Subhas was not surprised at the resignation of Vallabhbhai and his colleagues, as he never wanted them in the first place. Jawahar was the one he wanted, plus his own brother, Sarat, and J.P., and Narendra Deva—all of the Congress socialists. Bose wanted a left-wing Working Committee to take control not simply of the Congress but of the country. He wanted action, and he had been urging Nehru to join forces with him and call on the nation to rise up and revolt against the hard-pressed British Raj, calling for complete *swaraj* immediately. Bose was ready to launch a massive noncooperation movement that would have crippled not only India's commerce and machinery but also all of Great Britain on the eve of its greatest conflict in history.

"Ever since I came out of internment in 1937," Subhas wrote Nehru, "I have looked upon you as politically an elder brother and leader and have often sought your advice. When you came back from Europe last year, I went to Allahabad to ask you what lead you would give us . . . you put me off by saying that you would consult Gandhiji and then let me know. When we met at Wardha after you had seen Gandhiji, you did not tell me anything. . . . Twelve members resigned. They wrote a straightforward letter. . . . But your statement—how shall I describe it? I shall . . . simply say that it was unworthy of you. . . . When a crisis comes, you often do not succeed in making up your mind . . . you appear as if you are riding two horses. . . . I may tell you that since the Presidential election, you have done more to lower me in the estimation of the public than all the twelve ex-members of the Working Committee put together. Of course if I am such a villain, it is . . . your duty to expose me. . . . But perhaps it will strike you that the devil who has been re-elected President in spite of the opposition of the biggest leaders including yourself . . . must have some saving grace. He must have rendered some service to the cause of the country during his year of Presidentship. . . . [Y]ou wanted me to define exactly in writing what I meant by . . . Left and Right. I should have thought that you were the last person to ask such a question. . . . You have charged me further with not clarifying my policy. . . . In my humble opinion, considering the situation in India and abroad, the one problem—the one duty—before us is to force the issue of Swaraj with the British Government. . . . You have told me more than once that the idea of an ultimatum does not appeal to you. . . . I fail to understand what policy you have with regard to our internal politics. . . . Now, what is your foreign policy, pray? Frothy sentiments and pious platitudes. . . . For some time past I have been urging on everybody . . . that we must utilise the international situation to India's advantage . . . but I could make no impression on you or on Mahatmaji, though a large section of the Indian public approved of my stand. . . . Another accusation you made . . . was that I adopted an entirely passive attitude in the Working Committee. . . . Would it be wrong to say that usually you monopolised most of the time of the Working Committee? . . . To be brutally frank, you sometimes behaved in the Working Committee as a spoilt child and often lost your temper. . . . [W]hat results did you achieve? You would generally hold forth for hours together and then succumb at the end. Sardar Patel . . . had a clever technique for dealing with you . . . let you talk and talk and . . . ultimately finish up by asking you to draft *their* resolution. Once you were allowed to draft the resolution, you would feel happy. . . . Rarely have I found you sticking to your point till the last. . . . As a doctrinaire politician you have decided once for all that a Coalition Ministry is a Rightist move. . . . What is the use of your sitting in Allahabad and uttering words of wisdom which have no relation to reality? . . . Regarding Bengal, I am afraid you know practically nothing. During two years of your presidentship you never cared to tour the province. . . . [W]e should have . . . a Coalition Ministry. . . . I should now invite you to clarify your policy . . . I should also like to know what you are—Socialist or Leftist or Centrist or Rightist or Gandhist or something else?"[53]

"The sun was setting as I trudged back, with Kripalani for my companion, along the dusty road from Segaon [Sevagram] to Wardha," Nehru recalled, after the Working Committee that resigned out from under Bose had left Gandhi's ashram. "Loneliness gripped me in that empty plain and the lengthening shadows seemed ominous. I was walking away not from Segaon but from something bigger, more vital, that had been part of me these many years. . . . [The] newspapers say that I have resigned from the Working Committee. That is not quite correct and yet it is correct enough. . . . The reasons that impelled me to act as I did differed in many ways from those that moved my colleagues. . . . I felt an overwhelming desire to be out of committees and to function as I wanted to, without let or hindrance."[54]

This break would remove Nehru from his closest comrades on the Left, not only from Subhas Bose, but also from J.P. and all of India's leading socialists, who would never again truly trust him. For as Subhas so poignantly asked, Who was he? How could he continue forever to ride "two horses?" Or more than two? Left, Center, Right! Or was he "something else?" "My dear Krishna," Nehru wrote the only man who came close to filling the vacant slot in his heart and life. "It is not an easy matter to explain the various developments briefly and I cannot say that I am particularly happy at events. I had a bad time . . . because I did not fit in with any so-called group. Behind the words, right and left, there were many other factors at play, the most notable being the emergence into prominence of certain adventurist individuals who wanted to exploit the left to their own advantage. Nothing could have been worse for the left than to be closely associated with these individuals. . . . Many people in the left realised that they were drifting into an unholy alliance with people who had no policy. . . . Leftist circles, after a feeling of triumph due to the presidential election, feel somewhat disheartened by the course events took."[55]

Indira was now headed to Feroze in London. "Darling Papu," she wrote in mid-April. "I did hate leaving India & you at this time. It has left a strange sort of emptiness inside me. But here I am and there is no turning back. . . . It wouldn't be so bad if I did not keep seeing your face—so sad, with something more than just sadness. Darling, don't be so defeatist—no one can defeat you except yourself. You are so much above all the pettiness that is invading Indian politics. It is distressing to watch it taking hold—but you mustn't let it make any real difference. . . . I miss you terribly. . . . With all my love—darling mine."[56]

"You give me a lot of good advice, my dear to keep smiling," Nehru replied from his empty home. "But it is a little difficult. I suppose age is telling upon me and I am losing my resilience." In less than seven months he would be fifty.[57]

19

War Again

1939–1940

"IT IS A difficult and terrible business to carry on a war," Nehru wrote Indira that September 1939. "I have been going backwards and forwards between Allahabad and Lucknow and tomorrow I am going to Delhi to see the Viceroy. What will come of all this I do not know. . . . I have lost that keen incentive that gave me vitality and drove me to action. . . . Perhaps it is age. . . . It is just a sense of weariness and futility that has been stealing over me these three years or more. . . . Probably I am not a big enough man for the job that fate has thrust upon me."[1]

Nehru was in China when World War II started. Madame Chiang Kai-shek's invitation proved irresistible, even though Krishna warned him not to get caught that far away when the war began. But he was in Chungking on September 3 and was flown out in "a fine Douglas passenger plane" that Marshal Chiang ordered specially for him, going from Hong Kong to Burma. He was then driven to Mandalay, caught a train to Rangoon, and flew to Calcutta. Then another train took him to Wardha, where the new Working Committee had been meeting all week. "We . . . are not out to bargain," Nehru told the press before leaving Rangoon. "We do not approach the problem with a view to taking advantage of Britain's difficulties. This war is going to change the face of things. The old order is dead and cannot be revived. . . . If England stands for self-determination the proof of that should be India. . . . I should like India to play her full part and throw all her resources into the struggle for a new order."[2]

Without consulting any of the leaders inside or outside of the assembly, Lord Linlithgow proclaimed India at war. He then invited them to Simla, and Gandhi arrived there on September 5, when he met with the viceroy. "I told His Excellency that my sympathies were with England and France from the purely humanitarian standpoint," Gandhi reported. "I also told him that I could not contemplate without being stirred to the very depth the destruction of London. . . . I have become disconsolate. In the secret of my heart, I am in perpetual quarrel with God that He should allow such things to go on. My non-violence seems almost impotent."[3] Jinnah had met with the viceroy a day earlier and urged him to "turn out" all Congress provincial ministries, also indicating that he now believed that the "only ultimate political solution" for India "lay in partition."[4]

Nehru reached Wardha on September 11, and remained there through September 14, drafting and redrafting the resolution, which Congress's Working Committee adopted as its immediate response to the declaration of war. "The Congress has repeatedly declared its entire disapproval of the ideology and practice of fascism and Nazism. . . . The Working Committee must therefore unhesitatingly condemn the latest aggression of the Nazi Government in Germany against Poland. . . . The Congress has further laid down that the issue of war and peace for India must be decided by the Indian people, and no outside authority can impose this decision upon them, nor can the Indian people permit their resources to be exploited for imperialist ends. . . . Cooperation must be between equals by mutual consent. . . . India cannot associate herself in a war said to be for democratic freedom when that very freedom is denied to her. . . . A free, democratic India will gladly associate herself with other free nations. . . . She will work for the establishment of a real world order based on freedom and democracy. . . . [T]he Committee cannot associate themselves or offer any cooperation in a war which is conducted on imperialist lines. . . . [T]he Committee therefore invite the British Government to declare in unequivocal terms what their war aims are in regard to democracy and imperialism and the new world order . . . in particular, how these aims are going to apply to India."[5]

Subhas Bose had resigned from the Congress presidency, and Rajendra Prasad was elected acting president by the Working Committee, led by Vallabhbhai and Azad, with the warm support of Gandhi and Nehru. The brothers Bose never returned to Congress, soon forming their own party, the Forward Bloc, which attracted many Bengalis and was outlawed by the British. Subhas Bose was arrested but later fled, first to Germany and then to Japan, where he raised his own Indian National Army before the war ended. Nehru now chaired a war emergency subcommittee of the Working Committee, consisting of Vallabhbhai, Azad, and himself. Linlithgow did not respond to the Congress resolution, nor did Secretary of

State Zetland indicate any intention of shifting Whitehall's gears in favor of the Working Committee's request for an "unequivocal" declaration of "war aims" for India, designed to end both "imperialism and fascism." "My own impression is," Linlithgow wired Zetland on September 18, "we have carefully avoided any suggestion that we are in the war to further or to defend democracy." [6]

"Elemental forces sweep the world, disdaining the scheming of those who, from their seats of authority, had sought to stem them," Nehru wrote two days later. "Men and women become playthings of destiny and are drawn into the seething whirlpool of war. Whither do we all go . . . what part will India play? . . . India can no longer consent to be treated as a part of an empire. . . . Whether in peace or war she must function as a free nation." [7]

Jinnah had no intention of joining Congress, however, nor did any of India's princes hesitate to express loyal support for Great Britain in this hour of its greatest need. Indian troops were sent west again, and India's wheat was mobilized along with its Muslim and Sikh military forces to help fuel the war. "We have taken a dignified stand," Nehru told a skeptical audience in Lucknow on September 25. "We have not taken a final step and our final step depends upon . . . Britain. There is no bargaining spirit. Bargaining will not be compatible with our professions of freedom." [8] The next day Gandhi was invited back to Simla by the viceroy. "I fear the position is not improving," Jawaharlal wrote Krishna from Lucknow. "It is likely that the British Government will try to play off the Congress against the Muslim League and the princes. That kind of thing is just what will irritate Congress people." [9] Linlithgow and Zetland understood perfectly. Zetland, in fact, had reprimanded Congress from the House of Lords for the ill-chosen timing of its reiteration of political claims.

"If Lord Zetland's speech represents the mind of the people . . . there can be no compromise between the Congress and the British Government now or a thousand years hence," Nehru snapped back. "Lord Zetland has not profited by or learnt anything from events during the past twenty years." [10]

Nehru cabled his views to London's *News Chronicle:* "India can take no part in defending imperialism but she will join in a struggle for freedom. . . . This is no small offer that India makes for it means the ending of a hundred years of hostility between India and England, a great turning point in world history and the real beginning of the new order we fight for. Only a free and equal India can cooperate. . . . Autocratic and ordinance rule will alienate public sympathy and lead to conflict." [11] He had just been to Delhi, and "I had a long interview with the Viceroy and I also met . . . my old friend Mr. Jinnah. The position is entirely uncertain," he reported to Indira.[12] Jinnah was friendly, but Linlithgow told Nehru "that

I moved too much in the air." Responding to the viceroy, Nehru admitted, "Probably you are right. But it is often possible to get a better view of the lie of the land from the heights than from the valleys. And I have wandered sufficiently on the solid earth of India. . . . It was a pleasure to meet you for a second time . . . with no trace of unfriendliness, and realising the difficulties which encompass us and which compel us to pursue different paths."[13] "The viceroy was not hopeful of meeting our demands," Nehru reported to Krishna. "The Muslim League is unfortunately misbehaving to an extraordinary extent." Nehru believed that Jinnah wanted "to take advantage of the situation" and was "exceedingly backward and reactionary and has opinions and thinks in terms of twenty-five years ago . . . I think he has overshot the mark."[14]

Nehru was in basic disagreement with Gandhi as well at this time, although Gandhi completely supported the Working Committee's resolution on the war. "What is worrying Gandhiji very greatly is another aspect. . . . He does not want to see India becoming a military nation and he does not want to associate with anything in the nature of militarisation. His attitude in this matter is not even accepted . . . by some of his closest colleagues," Jawahar noted, meaning himself.[15] The A.I.C.C. now met in Wardha, and Gandhi stepped back into his shadow of silence while Nehru took charge, assuring his colleagues that the Working Committee's resolution "will have a deep effect on the country. . . . The issues before us are of a momentous nature. . . . The international situation is by no means clear . . . be prepared for everything . . . Mahatma Gandhi is firmly convinced that the principle of nonviolence should be applied in the present case also, but I do not see how we can ward off an armed attack by nonviolence."[16]

Cripps wrote to him from London the next day. "I saw Zetland and tried to impress upon him the seriousness of the position. I also put forward . . . suggestions along the line of those which I now understand from Krishna you have approved. . . . We have, I think, succeeded in getting a very good measure of publicity for Congress action. . . . I have taken the opportunity of introducing the 'democracy and freedom' argument as illustrated by our attitude to India, so that I am sure the Cabinet is fully alive to the implications though I am not so certain that they are yet awakened to the realities of the actual situation. . . . The addition of Winston Churchill has not added to the friends of Indian freedom, though . . . I am quite convinced that for the good of the British as well as the Indian people Congress should now stand as firm as a rock upon its demands. Naturally I do not refer to any details upon which I know you are only too ready to compromise once the reality of freedom and democracy is conceded in action."[17]

Nehru overestimated Cripps's and Krishna's power as much as he underestimated Jinnah's strength and resolve. Both errors of judgment hurt India more than his own prospects of ultimately emerging as its man of destiny, to preside over a much diminished Indian republic. Nehru now learned from a Muslim friend, Nandan Saran, in Delhi that Laiquat Ali Khan, Jinnah's leading lieutenant in reorganizing the Muslim League and destined to serve as Pakistan's first prime minister, had called to say how much he "sincerely desired the settlement of the Hindu–Muslim question. . . . If only our leaders would rise to the Great Occasion we could successfully exploit this great opportunity to win our freedom. . . . I [Saran] enquired why the conversation so happily begun between you and Mr. Jinnah could not be continued to the logical end. He expressed a little surprise and said that since Mr. Jinnah had fully acquainted you with his views . . . the next move lay with you. . . . Jinnah would stay on in Delhi . . . a while. If anything is to be done in this connection, it should be done before the Government makes any pronouncement." Nandan Saran offered to do whatever Nehru advised him to expedite another meeting with Jinnah, going himself to see Jinnah a few days later.[18]

"He [Jinnah] received me [Saran] with the utmost courtesy and warmth. . . . He seemed to be in a particularly good mood and humour. At the outset he asked me to warn you [Nehru] against lies and gossip. . . . [H]e said that he had affection for you coupled with high regard for your character and integrity. . . . [A]bout the Hindu–Muslim Problem . . . the next move lay with you. . . . It is a tragedy that the matter could not be settled in a friendly spirit, he said! He went on to say that . . . we were very much closer than we thought. . . . He would be here at least up to the 22nd . . . I venture to submit that now is the time to forge an understanding. Mr. Jinnah is in the proper mood."[19]

Nehru wrote Jinnah the next day, saying he would be glad to meet again. "If I had time now I would have come up to Delhi," he wrote, suddenly too busy and still considering Jinnah so backward and reactionary that he was in no rush to go slightly out of his way to meet with India's most powerful Muslim leader at this time, almost half a year before Jinnah's Muslim League irrevocably committed itself to "Pakistan." So Nehru returned instead to Allahabad to plan Congress's revolution if the British cabinet failed to accept his terms, by immediately proclaiming the promise of *purna swaraj* for India as one of its wartime aims. A golden opportunity for Hindu–Muslim reconciliation was thus wasted, ignored.

Linlithgow's response to the Congress's resolution came a few days later, leaving Nehru so livid that he did not even wait to speak with Gandhi before composing his own answer. "We would hesitate ordinarily to comment in haste," Nehru wrote, for Linlithgow had said that same day in mid-October that he was "authorised by His Majesty's Government to say that at the end of the war," they would be very willing to reexamine the Government of India Act of 1935, to frame whatever modifications might be deemed desirable by all the parties.

"Only the blind can imagine that the present-day world of empires and colonies and dependencies will survive the holocaust of war. The world of yesterday is dead, the world of today is dying and not all the king's horses nor all the king's men will be able to keep it alive. . . . Every man and woman of intelligence has some realisation of these profound changes . . . in this era of war and revolution. But not so the British Government. . . . They live in Whitehall and New Delhi apart from humbler, though perhaps more intelligent folk, and they neither see nor remember. . . . We asked for independence and are promised a consultation at the end of the war. . . . We know now, beyond a peradventure, that Britain clings to her imperialism and fights to preserve it. . . . What of the British Labour Party now and all those radicals and lovers of freedom in England who talk so eloquently of the brave new world that is coming? What of America, that great land of democracy, to which imperialist England looks for support and sustenance during the war? Does Britain think that the people of the United States will pour their gold and commodities to make the world safe for British imperialism? . . . The issues are clear and so are we in our minds. This is not a matter for Congressmen only but for all of us, whether we belong to the Muslim League or Hindu Mahasabha or Sikh League. . . . For India's honour and India's freedom are involved. . . . The Viceroy has told us to think of the unity of India. His Excellency's reminder was not necessary. But even the unity of India cannot be purchased at the cost of India's freedom. We want no union of slaves in bondage. . . . Meanwhile we may have to go into the wilderness again."[20]

Jawaharlal's young United Provinces volunteers and comrades had begun painting signs and issuing orders for civil disobedience, which was soon to include tearing up railway tracks and blowing up bridges as well as British soldiers, now that he was ready to launch the revolt, with or without the approval of Gandhi and his old guard. Nehru cabled London the next day: "British Government's reply to the Congress invitation for a declaration of war aims is a negation of democracy and freedom. . . . We cannot believe that the British people accept the imperialist aim. . . . The great democracy, America, has not extended her sympathy for this. . . . To this India can only say that our answer is an emphatic no."[21] A day later he wrote another editorial in his *National Herald* newspaper, reminding readers of the civil disobedience that had followed the national pledge taken a decade earlier on the banks of the Ravi River. "Who could crush or extinguish that bright flame of freedom that warmed our hearts and illumined our minds?"[22] Lord Zetland spoke that day in London,

Jawaharlal ("Master Joe") Nehru in Harrow Cadet Corps Uniform, 1906. *(Hulton Deutsch Collection)*

Above: Motilal Nehru, c. 1930. *(By permission of The British Library)*

Right, above: Jawaharlal Nehru, 1940. *(By permission of The British Library)*

Right, below: Indira and "Papu" Nehru, holding his grandson Rajiv, 1945. *(With permission of the Library Committee of the Religious Society of Friends in Britain/Agatha Harrison Papers)*

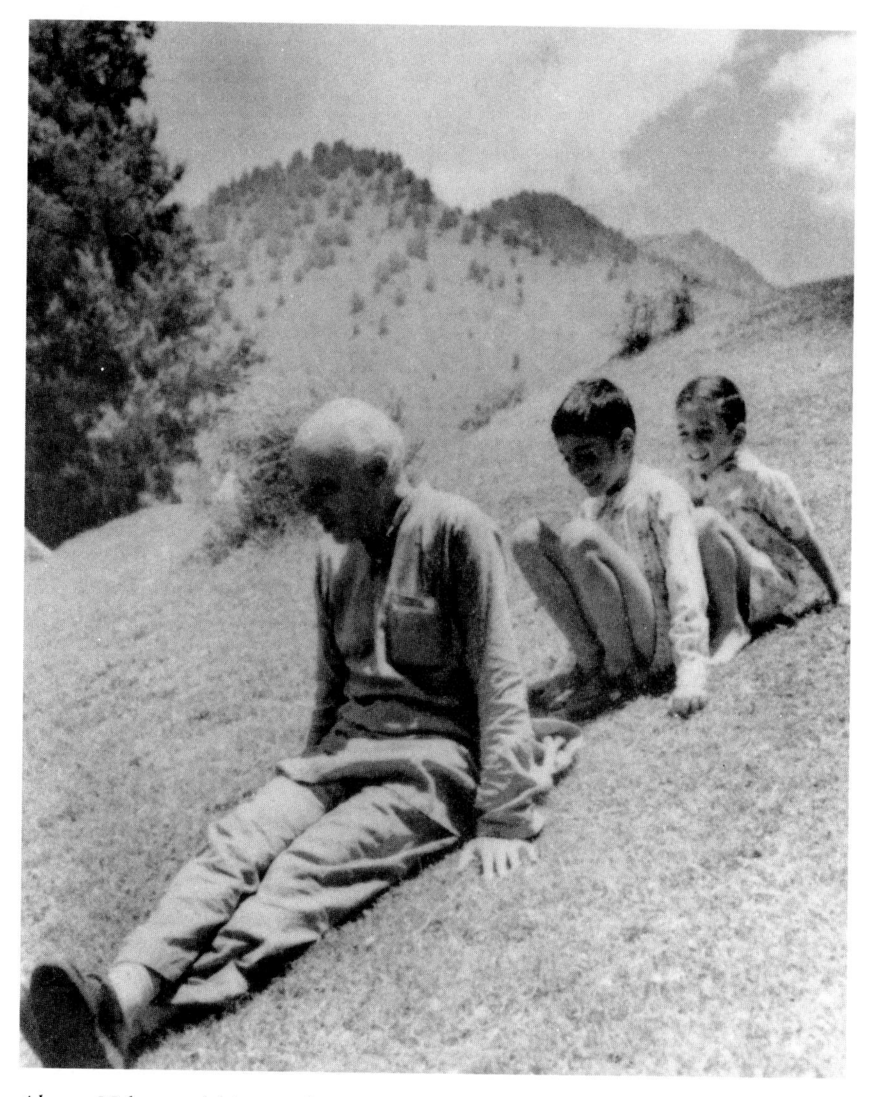

Above: Nehru and his grandsons, Rajiv and Sanjay, sliding off a mountain near Kashmir, 1951. *(With permission of the Library Committee of the Religious Society of Friends in Britain/Agatha Harrison Papers)*

Left: Pandit Nehru in New Delhi, April 1946. *(By permission of The British Library)*

Above: Prime Minister Nehru and Edwina Lady Mountbatten at a reception in London, February 11, 1955. *(Hulton Deutsch Collection)*

Left, above: Prime Minister Nehru received at London's Northolt Airport by Krishna Menon, January 1951. *(Hulton Deutsch Collection)*

Left, below: Prime Minister Nehru and High Commissioner Krishna leaving 10 Downing Street, London, February 1, 1955. *(Hulton Deutsch Collection)*

Prime Minister Nehru and Jacqueline Kennedy in the White House Rose Garden, Washington, D.C., November 10, 1961. *(Hulton Deutsch Collection)*

hailing Mahatma Gandhi as the "most outstanding" leader of India, never mentioning Nehru's name.

"The Marquis of Zetland has spoken soft and soothing words and bestowed praise on . . . Mahatma Gandhi. Perhaps he did not know when he spoke in the House of Lords of what Gandhiji had said about the viceroy's statement," Nehru angrily responded. "Some days ago Lord Zetland had referred in terms of patronising condescension to the Congress statement, and we had heard the old voice of British imperialism, hard and metallic to the ear. Foolishly many of us had imagined that Lord Zetland was of the old guard and did not represent the spirit of . . . the British people. Vainly we had expected more wisdom from others. But we were mistaken. The old imperialist guard holds the citadel in England and in India and its . . . voice is that of Lord Zetland." Zetland had also noted that until Hindus and Muslims resolved their own disagreements among themselves, it was futile for Great Britain to try to impose a constitution. "It is true that there are individuals, hangers-on of imperialism or the possessors of special rights created by British imperialism, who fear a change," Nehru noted. "Is India's progress to be measured by the capacity to walk of the halt or the lame? Are we to wait till we have converted the feudal princes? . . . Lord Zetland and his colleagues have conceived a new interpretation of democracy. Before anything is done, everyone must agree, or else the British . . . will sit here comfortably and carry off the spoils."[23]

Jinnah, however, welcomed the viceroy's response as well as Lord Zetland's speech, both of which assured him and his league that no new constitution would be drafted by Great Britain, turning India over to Congress alone. If Congress refused to reach agreement with him, Jinnah was now reassured that Britain would never abandon its Muslim friends. The Muslim League was soon to announce its historic Pakistan resolution in Lahore. Nehru cabled Krishna on October 25: "All Congress circles consider that the only course is a complete rejection of the viceroy's proposals. Gandhiji's condemnation is particularly drastic. Other groups have also condemned it except for some unimportant individuals. Muslim League attitude is uncertain but younger Leaguers are in favour of rejection. Conflict seems inevitable and is apparently desired by the Government of India."[24]

"The world is knit into one unit by science but it is divided by two forces. These forces sweep the entire humanity," Nehru told a cheering youthful audience in Bombay. "One force is the exploiting clique. The other is the exploited fraternity. India belongs to the latter. . . . India and Indians are struggling to be free. The curtain is about to be rung down on one act and we are looking ahead with eager expectancy to see what the next act has to reveal. A funny show is going on in Bombay and elsewhere. That show will close down tomorrow. The Congress ministers will be leaving their posts. . . . Today there is a conflagration in foreign lands and we are feeling the heat of it here. . . . Now India is going to show that hers is also the same old soul, athirst for liberty, and the same old roar. . . . Congress has never relied on other people's kindness; it relies on its own strength." It was his unilateral call for civil revolt.[25]

"Though your affection and regard for me remain undiminished differences in outlook between us are becoming most marked," the seventy-year-old Gandhi wrote to Nehru that day in late October. "Perhaps this is the most critical period in our history. I hold very strong views on the most important questions which occupy our attention. I know you too hold strong views on them but different from mine. Your mode of expression is different from mine. I am not sure that I carry the other members [of the Working Committee] with me. . . . I cannot move about. I cannot come in direct touch with the masses, not even with the Congress workers. I feel that I must not lead if I cannot carry you all with me. There should be no divided counsels among the members of the W.C. . . . [Y]ou should take full charge and lead the country, leaving me free to voice my opinion. . . . [I]f you all thought that I should observe complete silence, I should. . . . If you think it worth while you should come and discuss. . . . Love, Bapu."[26]

Nehru was hardly more ready to listen to the old Gandhi, however, than he was to reach any agreement with the almost equally old Jinnah. But a week later Gandhi wrote again to him, having learned from Kripalani about the "great ferment and preparation for C.D. [civil disobedience]" going on in the United Provinces. "He told me too that anonymous placards had been circulated asking people to cut wires and tear up rails. My own opinion is that there is at present no atmosphere for C.D. If people take the law into their own hands I must give up command of C.D. movement. . . . It was this that I had intended to discuss with you. But it was not to be. At this critical time in our history there should be no misunderstanding between us and if possible there should be one mind." Gandhi still signed his letter "Love" but knew that he and Nehru would never again be of one mind. The Working Committee agreed, however, almost unanimously, with Jawaharlal's analysis of the current situation and the tough response he urged to Linlithgow and Zetland. The war raged on, and soon India's own mini-war would begin. Following Nehru's lead, the Working Committee called on all Congress ministries to resign from all ministerial offices they held in their joint provincial venture with the British Raj. So act 1 was over, but Nehru's war of national liberation, act 2 had not quite begun.[27]

"One lives in a state of tension, not knowing what will happen,"

Nehru wrote Indira in early November, "and meanwhile rushing about in railway trains—Allahabad to Wardha—to Bombay—to Lucknow—to Delhi—to Allahabad—to Lucknow again. . . . The Congress Governments are over and the ministers have reverted to humbler roles. . . . A cable . . . sent by Agatha . . . informed me that you were going to Switzerland. . . . I presume you are going to Leysin. That is a safe and quiet corner . . . to Dr. Rollier's place [sanatorium]." Indira's health had not recovered sufficiently to allow her to return to Oxford. Xrays indicated spots in her left lung, and Les Frênes, Dr. Rollier's sanatorium in Leysin, became her home for the next eight months.[28]

"Whitehall and India Office are under a delusion if they think that the Congress does not mean what it says," Nehru told Krishna that November. "Speaking for the U.P., I can tell you that it is becoming increasingly difficult to hold our organisation in check. We are prepared for any contingency. Gandhiji is continually laying stress on discipline and nonviolence and says that he will not tolerate any civil disobedience unless this is formally decided upon. . . . I think it is recognised all round that a major struggle is inevitable."[29] Gandhi had just met with Jinnah, driving in his green Packard with Rajendra Prasad to the viceroy in Delhi, where all four of them met without Nehru but resolved none of their differences. Jinnah could hardly believe his eyes when he read that Congress decided to resign from every provincial office and would accept none of the higher offices the British were prepared to offer at the center, on the viceroy's Executive Council. On December 2, Jinnah issued a proclamation on behalf of the Muslim League, announcing that Friday, December 22, 1939, would be celebrated by Muslims throughout British India as a "Day of Deliverance and thanksgiving as a mark of relief that the Congress regime has at last ceased to function."[30]

"I cannot wish you the conventional 'good gift,' " Sarojini Naidu wrote "My beloved Jawahar" on his fiftieth birthday. "Sorrow, suffering, sacrifice, anguish, strife. . . . Yes, these are the predestined gifts of life for you. You will transmute them somehow into the very substance of ecstasy and victory—and freedom. . . . You are a man of destiny born to be alone in the midst of crowds, deeply loved and but little understood. . . . May your questing spirit find its goal and realise itself with splendour and beauty. This is the benediction of your poet sister and fellow seeker."[31]

Nehru's old Oxford friend, the liberal historian Edward Thompson, had long been trying to convince him that it might be wiser to take full advantage of the British offers to cooperate with Congress, rather than toss them aside and rush back into the wilderness of revolt at every rebuff from a viceroy or secretary of state. "I still cling to my old foolish dream that we might see the British Empire pass into a wider grouping of equal nations, a United States of the World," Thompson wrote to Nehru, trying to keep him from hurling himself again over the edge of that precipice. "Listen! You are not in the habit of listening to older and wiser men, are you? I remember Sarojini's stories of the way you used to behave on the President's platform. 'Oh, he used to *rush* down into the audience and seize some poor fellow who had contradicted him, and cuff him and kick him all round the *pandal* [Congress tent] until everyone was terrified for the poor man!' Did you pick up these habits at Harrow? However, I am too far away for you to cuff and kick me ('You *irritate* me!!'), so I suggest that you *listen*. . . . I have been slaving night *and* day, getting my Rhodes Report done. . . . Two matters (1) The Rhodes Trust wants to spend some money on India. . . . I am thinking of a Fund that would make grants to Indians who were doing really valuable research in economic lines—kinds of fruit or sugarcane. . . . Now shake your ears, and listen again. . . . The biggest honour . . . this ancient University can offer any man is to invite him to be the Rhodes Memorial Lecturer. . . . [H]e meets all our leading politicians. . . . He comes in the Summer term, and gives at his own choice 3 or 4 or 5 lectures. . . . The Lecturer must be a foreigner . . . *always* of the very front rank; we have had Einstein, Smuts, Flexner. . . . It is a very big chance for a statesman. . . . Between ourselves the Rhodes Trustees are going to invite a man called Jawaharlal Nehru. It is not in our power . . . to declare India an independent nation. But we can . . . pick out a man and say, We want you to allow us the honour to say publicly before the whole English-speaking world that we recognise you as a man in the very first rank . . . YOU MUST NOT . . . [here Nehru cut out of the published letter Thompson's strongest admonition]. Please think it over. This question of India's status is tied up with the whole perplexing problem of this mad world that is now being broken to pieces, and must be remade some time soon. It would give you an opportunity to do some statesmanlike work on a great scale. You could put before us some creative picture from your own mind and experience. . . . [F]rom your very exceptional international experience, you could shake our isolation and insularity, you could put your own people on the world's map (you once said to me that you were a good ambassador and you are), and in the delightful English summer you could see our young men (the finest in the world) . . . give innumerable dons who have never seen an Indian . . . an entirely new idea of India . . . meet all our political leaders, for they all come to Oxford. You could do the job which Tagore began to do, and then failed to do by publishing so much wishwash. The job is entirely non-political; the India Office does not thrust its nose within a hundred miles of it. . . . If the job is offered you, take it, for it comes to you by . . . COMMAND from your destiny and your demon. . . . If you do not take the job, then (between ourselves) it will be Sapru, who will do the job brilliantly . . . but will fortify our politicians in a worse die-hardness than they have now. I shelved Sapru (he might come later) for you. By the way, you would be able to see Indira. You would be away from India only about six weeks. . . . [T]hink as kindly of me as you can. I mean well, and an Englishman cannot help being stupid. . . . Please do not despair of my country. Some of us are doing our best for India, and despite our faults we are better than some other nations."[32]

"Dear Edward . . . Whether the British Empire will pass into a wider grouping of equal nations, a united states of the world, I do not know. But if your old dream materialises I shall be most happy. It would be a very good thing. . . . Having shaken my ears . . . I appreciate all you say about the honour and distinction and all of us are greedy for both. But, though you may not believe me, I am a rather modest person and I hesitate to venture into new fields. Anyhow, I shall keep my mind open and see how things shape themselves. . . . For various reasons I should very much like to go to England and possibly to America. I always feel that I can be of more use to India outside India. The feeling that I do not quite fit in here pursues me and depresses me." But the "rush of events," of war and revolution, prevented Edward Thompson and his colleagues at Oxford from luring Master Joe back to his other world, the second home that part of Nehru always loved best.[33] It was too late for Panditji to retreat. The tiger raced off with him.

20

On the Eve of Revolt

1940–1941

"I WOULD HAVE to repudiate all my past, my nationalism and my self-respect if I were to resume the talks with Mr. Jinnah in the face of his appeal to the Muslims to observe a 'day of deliverance,' " Nehru announced angrily in response to Jinnah's proclamation that December. "I emphasise that the allegations made against the Congress ministries are baseless. The whole campaign is a malicious one." Nehru's emotional reaction made it more difficult for Congress to negotiate with the Muslim League.[1]

Stafford Cripps, who was soon to join the British war cabinet as lord privy seal, flew to India in mid-December 1939 and met with Jinnah and his leading lieutenants of the Muslim League, as well as with Nehru and Gandhi and the Working Committee of Congress, and with Ambedkar and several princely state prime ministers *(dewans)*. "Stafford Cripps has had a busy time since he came to India," Nehru reported to Krishna late that December. "It is a confusing situation though basically it is simple enough. The communal stunt is being played for all it is worth and Jinnah and Ambedkar have said things and behaved in a manner which passes all bounds of decency." Ambedkar and other "untouchable" leaders strongly endorsed Jinnah's call for a day of deliverance, hailing the end of Congress ministries. For Nehru and Congress, such professions of relief at the end of "Hindu-Raj," as Jinnah called Congress's provincial governments, "produces bitterness and difficulty." Nehru blamed the viceroy's initiative, of which "Jinnah has taken full advantage."[2]

"If Jinnah claims to represent the Muslims, why is he afraid of an election by adult franchise for the constituent assembly," Nehru argued. "The fact of the matter is, he does not trust the Muslim masses. What he is after now is to have . . . Muslim League ministers in every cabinet." Jinnah now called for a royal commission to inquire into "Congress atrocities," and Nehru told Krishna that in the unlikely event of such a commission's being appointed, Congress would refuse to accept it or appear before it.[3]

Gandhi's approach to Jinnah was devoid of bitterness. "I silently joined . . . in the wish 'Long Live Quaid-e-Azam Jinnah,' " Gandhi wrote in mid-January 1940. "The Quaid-e-Azam is an old comrade. What does it matter that today we do not see eye to eye in some matters? That can make no difference in my goodwill towards him. . . . And now he commands further congratulations on forming pacts with parties who are opposed to the Congress policies and politics. He is thus lifting the Muslim League out of the communal rut and giving it a national character. . . . I observe that the Justice Party [of Madras] and Dr. Ambedkar's party have already joined Jinnah Saheb. . . . Jinnah Saheb is giving the word 'minority' a new and good content. . . . This may any day convert itself into a majority by commending itself to the electorate. . . . If the Quaid-e-Azam can bring about the combination, not only I but the whole of India will shout with one acclamation: 'Long Live Quaid-e-Azam Jinnah.' For he will have brought about permanent and living unity for which I am sure the whole nation is thirsting."[4]

That January Nehru raced off on one of his "old-style" tours of the United Provinces, hoping to counter the Muslim League's influence by making contact with the Muslim "masses." Though "our peasant masses depress me with their poverty," he found their enthusiasm uplifting, a partial antidote to "the listless and argumentative folk of the cities." The tour was primarily political, but it was also to counter depression, for he was feeling older and finding himself more prone to the annoying pains of even "slight injury," as he confessed to Indu. "We age. I cannot judge about myself but I am often surprised to see others around me showing signs of age. Nan's hair is almost all grey now. Betty is very matronly."[5] The next day he was off to Wardha for a Working Committee meeting, which he found even more "tiring" than his village tour. He felt bored again. The All-India Women's Conference met at Anand Bhawan in late January, and although Nehru returned to welcome Padmaja, Sarojini, Rajkumari Amrit Kaur, and the other leading women of Congress to that important meeting, it was hardly his meeting. "Suddenly I have begun to feel as if I had nothing special to do," he confessed. "I do not like playing the spectator's role and yet there is no other obvious job to be done." For not only had he refused to join the viceroy's Executive Council, but he also had insisted on the resignation of all Congress ministries in the provinces. Thus, there was really no work—either administrative or political—to occupy his time and easily distracted attention. Boredom had always been his demon. He longed to leave for Europe again and told Indu that "perhaps it is the desire to see you that makes me so restless" but quickly added, "There is no escape for me for the present and in the near future."[6]

At Wardha, Gandhi had asked Nehru about Russia's foreign policy, for to eyes unclouded by Marxist–Leninist–Stalinist dogma, the Soviet Union seemed almost as bad as Germany at this moment of Moscow–Berlin friendship and the invasion of tiny Finland. Nehru's reply had been "rather vague," but he wrote Gandhi and tried to explain: "Russia has acted very wrongly in regard to Finland. . . . But what concerns us even more is the fact that behind the Anglo-French-German war, what is really happening is a consolidation of the imperialist and Fascist power to fight Russia. It is clearer now than it was even before that the war is a purely imperialist venture on both sides."[7] Gandhi, however, agreed to meet with Linlithgow in early February. Nehru had not been invited but wrote to inform Gandhi on the eve of his meeting with the viceroy that "it is exceedingly difficult for me to think of going to Delhi during the next two weeks. I am fully occupied all this time." His schedule had suddenly become chock full of meetings in Lucknow, Allahabad, Bombay, and Gorakhpur.[8]

Nehru nonetheless wrote at length to Gandhi, to advise him that "there is not the slightest ground for hope that the British Government will accept our position." He wanted to be sure that Gandhi knew every negative thing he had learned about what was said or done anywhere in London during the past few months, of which Krishna Menon always kept Nehru well informed. "I wonder if your attention has been drawn to a recent social function in London," he asked, at which Lord Zetland presided over a meeting of the Royal Central Asian Society, whose ostensible object was to establish a Muslim cultural center in London, but the real object of which was to exploit Muslim sentiment in India. He argued that Britain was playing the same old game, as the viceroy's calling in "a procession of people headed by you to interview" seemed to Nehru clearly to prove. "An atmosphere of approaching compromise pervades the country," he warned Gandhi, insisting that there was no ground for it. "It is enervating and depressing because it does not come out of strength but, in the case of many individuals, from the excessive desire to avoid conflict at all costs." Not that he preferred conflict, of course, but "I fear that the impression is widely prevalent in England as well as in India that we are going in no event to have any conflict and therefore we are going to accept such terms as we can get."[9]

That same day, February 4, 1940, he wrote Indira to report that all the women had gone and Anand Bhawan was quiet again. "Tonight I am going to Lucknow . . . then to Bombay. Up and down I travel to attend various . . . meetings. . . . But my mind is elsewhere and my heart still further away. Doubts creep into my mind if all this is worthwhile or not. Still I go on. . . . [R]esponsibilities grow in spite of me and the coils of fate bind me. Not that I have any desire to shirk them but the future is so uncertain that the urge to activity lessens. Activity to what end? I wish often enough that I could retire for a while in some remote faraway place where I could read and write and there is no engagement, no visitors, no letters. . . . Sometimes a foolish desire to be ill and to lie in bed—but the illness should not be serious!—comes to me. How pleasant it must be, I think, to laze and not be forced to do this and that when the mind is unwilling and rebellious." Soon he would return to jail, but first he planned a trip to Kashmir.[10]

Sheikh Mohammad Abdullah, the leader of Kashmir's Muslim Conference, a nationalist party he had founded in the princely state of Jammu and Kashmir, was staying with Nehru at this time. The thirty-five-year-old sheikh, later hailed as the "Lion of Kashmir," was an ardent nationalist of progressive mentality. Nehru liked him very much, and they remained friends, at least until Prime Minister Nehru felt "obliged" to put Sheikh Abdullah, whom he had elevated to run Kashmir for a while, behind bars, fearing his outspoken demands for Kashmiri self-determination and his charismatic appeal to most of Kashmir's predominantly Muslim populace. Nehru now took Abdullah with him to Bombay for the States Peoples' Conference meeting in early February. Nehru presided over the conference, and Sheikh Abdullah was his deputy. Now he promised Abdullah to visit the state he had not set foot in since he had almost dropped to his death there on his honeymoon, twenty-four years earlier. Before the month's end Nehru attended a Kashmiri wedding, the first he had witnessed since his own. "Children whom I knew long ago were grown-up boys and . . . young women. . . . They looked an attractive crowd," he wrote Indu. "I am not depressed, *Cara mia,* although the world, as well as my own particular world, goes all awry. . . . My mind goes back to a year ago. Were we not at Almora, then? . . . I broke off a cactus cone . . . and here it is lying in front of me now, all dried up. But I keep it and often handle it for it reminds me of . . . you."[11]

He also wrote Bharati that last week in February: "Your letter has been with me for nearly a month and I have kept it back to answer at leisure. . . . You say that when we talk we go round and round. It is true. But that is not because life is round. Life is inconclusive and its problems unending but it is very pointed, painfully so . . . it is only a vague, pleasurable or painful or both, notion in our minds. . . . Yet to conceive of life one must get out of the self, forget it and think in other terms . . .

so we go round and round entangled in self and no-self. . . . I am sending you a copy of a picture I took of you. It is one of my successful efforts." Although Bharati remained romantically devoted to him, Nehru had long since lost interest.[12]

The Working Committee met in Patna in early March 1940 to confirm Gandhi's choice of Maulana Azad as the next president of the Congress. Had he been asked, of course, or pressed, Jawaharlal himself would have agreed to shoulder that heavy burden again. Other than his old friend Syed Mahmud, however, at whose home he stayed in Patna, none of Congress's Working Committee really wanted Nehru back at their helm. Most of them feared his radical ideas and found his mercurial mind and impulsive passions disconcerting. Congress's crowning of a Muslim, moreover, in the very month when the Muslim League met in Lahore to articulate its "Pakistan" demand for a separate Muslim nation, seemed positively inspired. "I . . . felt that in the crisis of the war, it was my duty to serve in any capacity to which I was called," Maulana Azad wrote. "When Gandhiji again requested me to become Congress President, I readily agreed."[13]

The March 1940 session of Congress met in remote Ramgarh, a village on the plateau of Chota Nagpur, to which Congress's faithful ventured in their thousands and resolved that "Great Britain is carrying on the War fundamentally for imperialist ends and for the preservation and strengthening of her Empire."[14] After his meeting with the viceroy, Gandhi continued to hope that another national struggle might be avoided. Linlithgow had emphasized Britain's "earnest desire that India should attain Dominion Status at the earliest possible moment."[15] He promised to do everything within his power to "shorten the transitional period" and was prepared to expand his Executive Council immediately by adding members of Congress as well as "other parties" if Gandhi and Jinnah would only agree to support the war effort. Defense and Finance would remain in British hands, but even in those departments, "transitional" steps could be taken. Gandhi knew, however, that he could never convince Nehru or any of his younger supporters to agree to so tepid an offer. "But nothing has been lost by our meeting," Gandhi wrote of it. "Non-violence requires great patience. . . . If India is not to be a co-sharer in the exploitation of the Africans and the degradation of our own countrymen in the Dominions, she must have her own independent status. . . . What is to be done? Declaration of civil resistance? Not yet. . . . I ascribe sincerity to Lord Linlithgow. He is doing his best to understand us. . . . With all his traditions he cannot be made to jump to our position. He cannot be hustled. . . . And we must not despise our opponent or belittle his strength. . . . Our duty is . . . putting our own house in order. . . . We may not resort to civil resistance out of our impatience or to cover our shortcomings."[16] That was Gandhi's way of telling Nehru to stay calm and as civil and patient as possible. Gandhi recognized that the no-longer-so-young Jawahar was chafing at the bit, eager to revolt. Just a week before Congress met, the younger J.P. had urged Indian steelworkers in Jamshedpur to "stop supporting the Imperialist War machine." He was then arrested for sedition. Several of Bose's Forward Bloc activists had also taken to throwing bombs as well as brickbats at British troops; they too were put behind bars.

Nehru promised Gandhi to wait a bit longer, and President Azad brought him his address in Urdu, appealing to Panditji, Congress's foremost scribe, to put his message into flawless English. "I am no good at this job," Papu complained to Indu on the eve of Ramgarh. "The whole address even has not reached me yet . . . coming in driblets. Yesterday I spent about eight hours over it, wrestling with Maulana's . . . flowery Urdu. . . . Maulana is a curious type. . . . He reminds me very forcibly of eighteenth-century Rationalists. . . . That does not mean that he is reactionary but he is out of touch with many modern developments. . . . If he had had the chance to learn well one or more European languages . . . he would have been a very remarkable person." Yet Nehru later appointed Azad minister of education in his first cabinet.[17]

The resolution of the Ramgarh Congress and Azad's address thus reflected Nehru's own thinking and his militant anti-British position, more radical than that of Gandhi or most of the Working Committee. Congress reaffirmed its commitment to complete independence, and called for a constituent assembly as soon as possible to decide on India's own constitution. Britain's stress on the prior need for Hindu–Muslim agreements was seen as further evidence of divide and rule. "It has become clear now that a struggle is inevitable," Nehru argued at Ramgarh, "but how and when will have to be decided for us by . . . Mahatma Gandhi. We should be ready for everything." Militant socialists accused him again of "selling out," but he charged them with "misguided enthusiasm," arguing that "such enthusiasts are counterrevolutionaries and rebels. Our object should be to get the entire army moving and not a few headstrong people who can be described as adventurers. They are no better than terrorists."[18]

That very week in Lahore, Muslim League President Jinnah insisted that "the Musalmans are not a minority. The Musalmans are a nation. . . . The problem in India is not of an intercommunal but manifestly of an international character, and it must be treated as such. . . . If the British Government are really in earnest and sincere to secure the peace and happiness of the people of this Subcontinent, the only course open to us all is to call the major nations separate homelands, by dividing India into 'autonomous national States.' "[19] "The whole problem has taken a new complexion," Nehru responded after reading Jinnah's address and the Lahore League's "Pakistan" resolution. "The knot that is before us is incapable of being untied by settlement; it needs cutting open. . . . Without mincing words, I want to say that we will have nothing to do with this mad scheme. . . . [T]he League leadership today, with its contempt for the country and its hymn of hatred . . . has at last unmasked itself. . . . [O]ur goal is clear and we will march on our path. A struggle is inevitable."[20]

He went to Bombay to chair another meeting of the Planning Committee of Congress, "intent on planning . . . of everything; thinking or trying to think of an ordered sane world, a new order as they call it," Nehru wrote his daughter. "And way across the seas war rages and spreads and casts its shadow. . . . Even Bombay . . . has semi-darkened streets. I read and write and answer letters normally but civil disobedience looms ahead. . . . Yet the mind wanders . . . I think of you so often. . . . Bombay wearies me . . . I am in no mood for meeting large numbers of people and, here in Betty's flat, there are always people. . . . It is long since I heard from you."[21]

That May Germany marched through Holland and Belgium, and Churchill replaced Chamberlain as the prime minister of Great Britain's tottering war cabinet while Congress hovered on the brink of its longest and harshest civil disobedience campaign. Indira remained in her Swiss Alpine sanatorium at Leysin, which proved to be, as her father hoped, "about as quiet a place as you could find anywhere in Europe."[22]

Nehru flew off to Kashmir in mid-May for a week's holiday, "to freshen the picture I have in mind." In Kashmir he felt "excited and moved." As he wrote to Indu: "I find Kashmir exhilarating and I have a sense of coming back to my own—it is curious how race memories persist, or perhaps it is all imagination."[23] Kashmir and her people were "surpassingly lovely," and whether from inherited memories or by virtue of his imagination, Nehru's love affair with that poor Himalayan state and its lovely vale blossomed and indeed grew stronger over the rest of his life. "What an enchanting land with an air that vitalises!" he wrote of Kashmir. "I am convinced that it is superior to Switzerland and, I would say, healthier. . . . I wish I myself could go there for two months every year—one month trekking towards Ladakh or Baltistan or the various glaciers and upper valleys, and one month's rest and intensive reading and writing also in a higher valley. That would be an ideal life."[24]

On June 11 Italy entered the war, and three days later German troops marched into Paris. The Congress Working Committee met three days after that in Wardha and was "deeply moved by the tragic events that have taken place in Europe in startling succession." Nonetheless, the Working Committee argued that "the war in Europe, resulting from a desire for imperialist domination over other peoples and countries, and a suicidal race in armaments . . . has demonstrated the inefficiency of organised violence . . . for the defence of national freedom and the liberties of peoples."[25] Congress remained adamant in withholding support from Britain's war effort, but Mahatma Gandhi was not ready to launch a Satyagraha campaign. Gandhi favored total nonviolence, urging the abolition of armed forces in India for all time. Nehru, Azad, and most of the others on the Working Committee would never be ready to disarm. "Congress governments have used the police," Nehru argued. "The state has certain duties and obligations to its people and it should endeavour to discharge them peacefully. . . . But . . . it cannot abdicate its functions." As prime minister, however, he never hesitated to use force, internally or externally. Nehru insisted, however, that "the bonds that unite us to Gandhiji will remain as strong as ever."[26]

Gandhi truly believed that the best way to counter Hitlerism was for Great Britain to disarm and welcome the German invaders as Vichy France had done. "I think the French statesmen have shown rare courage in bowing to the inevitable," Gandhi argued. The viceroy invited Gandhi to meet him in Simla again on June 29, seeking his advice as to how best to win the support of Congress. Gandhi told Lord Linlithgow that the best thing for the British to do was to "fight Nazism without arms . . . invite Herr Hitler and Signor Mussolini to take what they want of . . . your possessions. Let them take possession of your beautiful island. . . . If these gentlemen choose to occupy your homes, you will vacate them. If they do not give you free passage out, you will allow yourself, man, woman and child, to be slaughtered."[27] Maulana Azad recalled that "Lord Linlithgow was taken aback by what he regarded as an extraordinary suggestion. It was normally his practice to ring the bell for an ADC to come and take Gandhiji to his car. On this occasion he was so surprised that he neither rang the bell nor said goodbye. . . . Gandhiji reported this incident to me with his characteristic humour."[28] Two days later Gandhi issued his longer appeal "To Every Briton," reiterating at greater length what he had told the dumbstruck viceroy, adding: "This process or this method . . . non-violent non-co-operation, is not without considerable success in its use in India. . . . Your representatives in India may deny my claim. If they do, I shall feel sorry for them." More Englishmen now felt sorry for Gandhi, convinced that he was mad. He would never again be taken seriously by any viceroy or Secretary of State, including Mountbatten. Unfortunately.[29]

The Congress Working Committee met again in Delhi, without Gandhi, in early July. It resolved this time to call on Great Britain to issue an immediate declaration of complete independence for India, with the formation of a provisional national government at the center, which would then be followed by Congress's cooperation in the war effort. "This would be the preliminary and transitional stage towards the establishment of a new popular constitution framed by a constituent assembly," Nehru told the press on July 8, 1940.[30] Sporadic violence broke out now in several parts of the United Provinces and Bengal, and more young Indians were arrested throughout early July. "*Quatorze Juillet,* the Fourteenth of July," Nehru wrote for his *National Herald* editorial. "The people of Paris storm the Bastille and . . . set free the prisoners. But the fall of the Bastille was a symbol and a portent of much else. . . . The Revolution came and the Terror . . . and Napoleon. . . . [T]he Third Republic of France was born . . . now the Third Republic is no more. . . . France's ruling class has sold her glorious heritage . . . for not even the proverbial mess of pottage. Defeat in battle brings sorrow and humiliation; yet it may be borne if the spirit lives. But what shall we say when the spirit dies and the soul of a country surrenders?"[31]

The A.I.C.C. met in Poona at the end of July 1940. Gandhi was not present. He had in fact completely withdrawn now from Congress meetings and Congress policy, devoting himself entirely to his ashram work and the preaching of nonviolence. "Gandhiji has not parted company from us," Nehru told the A.I.C.C. "But he has, for the moment, left us to shift for ourselves." He then urged Congress to remain "strong" and prepare for "struggle and travail and sorrow." He continued, however, to speak softly, decrying "creating chaos for chaos' sake."[32]

"Some people have thought fit to criticise me for my earlier statement that I do not want to embarrass the British Government," Nehru told his Gandhi-capped Congress colleagues. "I have spoken with the dignity and responsibility of a statesman who has the privilege to represent a great organisation. We cannot think in terms of petty things like embarrassing Britain. Our aim is . . . more noble. The Communist cause is becoming a weak cause. . . . I am proud to regard myself as a socialist, but that does not mean that we should do anything to make ourselves ineffective," Nehru argued. "I can continue to sympathise with the British people in their hour of trial, but that has nothing to do with British imperialism which I want to see humbled." From the vantage point of his own two-world consciousness Nehru thus emerged at this time of world conflict and confusion among Congress leaders as the man best suited to rally his party to a middle path while convincing the wisest of British statesmen that there was at least one popular Indian who knew their language and could be relied on to do the right thing. Jinnah, of course, was another such leader, but he could rally only the Muslim quarter of India's more than 300 million, a powerfully important quarter during the war, indeed, yet still a minority.[33]

At this time, Rajendra Prasad, Abdul Ghaffar Khan, and several of Gandhi's followers on the Congress Working Committee submitted their resignations to President Azad. "I was deeply hurt to receive this letter," Maulana Azad recalled. "I immediately wrote . . . [to] request them to continue as members of the Working Committee."[34] Ghaffar Khan, "Frontier Gandhi," refused to return, but Prasad and the others did, albeit reluctantly. "Gandhiji's withdrawal has created a new and difficult situation," Jawaharlal cabled Krishna. "Present demand is the unequivocal declaration of Indian independence, immediate establishment of provisional national government at the centre, full powers to responsible elected members of Central Assembly and the formation of a constituent assembly later . . . no expectation of satisfactory response . . . no question of Congress weakening its present position. In all probability we will be going ahead."[35] "Ahead" meant, of course, with the last great Indian rebellion that would remain neither civil nor nonviolent but was to become as violent a revolt as unarmed Indians could launch against the British Raj. Before the war ended, disruptions of every sort of road and rail transport, sabotage of military cantonments, the blowing up of British troops in cinemas as well as barracks, and every violent and nonviolent style of struggle devised in the century of resistance and revolt that had begun in India in 1857, emerged to plague the British Raj. Krishna Menon knew all about the plan, and his job was to try to convince Cripps and Laski and anyone else with sympathy, power, or influence in London that before all hell broke loose throughout India it would be wiser, cheaper, and saner to reach agreement with Nehru and Congress. But Krishna Menon's friends did not include Prime Minister Churchill or his new secretary of state for India, Leo S. Amery. Gandhi also knew of Nehru's militant plan for the liberation of India, which was why he had withdrawn completely from Congress and from any role in leading that sort of bloody revolt.

"The hounds of war are yet distant from India, but the drums of war are heard here also and their sound grows louder," Nehru wrote for his *National Herald* in early August. "We in India have escaped thus far the physical impact of forces, but we cannot escape this conflict. . . . The world is fluid and seeks new levels before it solidifies afresh. . . . Nothing is more astonishing than the ineptitude of the British ruling class. . . . Peril and disaster do not teach them any lesson, the nearness of catastrophe has no meaning for minds that are closed and incapable of understanding. The Old Brigade continues as of old, the old gang holds on, imagining that the Colonel Blimps of the empire will triumph. *Quem deus vult perdere prius dementat.* . . . If any there were who thought that in this hour of grave peril, British imperialism would hand over power to the Indian people, they must realise how mistaken they were. . . . *Theirs not to reason why, theirs not to make reply, theirs but to do and die.* . . . Our course is set and we shall run it. There is no stopping, no resting-place, for us till we have created the free India of our dreams."[36]

Two days later Nehru wrote Indira from Cawnpore: "The situation here is developing rapidly. . . . One must keep ready for emergencies.

. . . Today I came here . . . especially to see the Congress volunteers. . . . A certain Government of India order about [banning] volunteering . . . produced a curious situation and this visit . . . today became an all-India event. There was a procession, a march past and speech by me to a huge crowd of about 50,000. For an hour during the procession I stood at the back of the car. Then I stood saluting while nearly 1000 volunteers marched past, and then I held forth for an hour and forty minutes! How long-winded I have become! . . . Perhaps I shall have real rest soon."[37]

He knew that he would soon be back in prison, for his speeches were not only long but also filled with the contempt he felt for the Raj and its current rulers. He dressed smartly for volunteer rallies, wearing a short-sleeved khaki uniform and shorts with a Gandhi cap, taking the salute with his baton, as a field marshal should.

Having given up on Gandhi, Linlithgow invited Azad to Simla that August, hoping perhaps that the Maulana might agree to join his cabinet, but Azad knew as well as Nehru how far the plan had progressed and dared not even accept the viceroy's invitation to tea. "The students of the country . . . should be prepared to undertake the burden in an age of conflict when the fittest alone will survive," Nehru told a students' meeting in Cawnpore the day after he took their salute, reviewing his troops. "Students in European countries are fighting for their country at this critical juncture, and I ask you to bear this in mind as it is your duty to work for the good of the country. India will play an important role in the affairs of the world by becoming a united nation. . . . Only those persons who follow international events and understand them correctly can control the destiny of a nation."[38] Nehru's vision flew ahead, and his mind soared to thoughts of a future Asian federation, what he later called the "Third World," over which he hoped one day to preside. "I am thinking of India's freedom, but at the same time I am thinking of a world federation of nations," he told a Benaras audience a few days later. "The world is passing through a struggle and revolution and the people of India cannot keep themselves out."[39]

Hundreds of Nehru's young followers had by now been arrested in the United Provinces. "The time for our trial approaches, and the zero hour for us, as for others, is near," Nehru told the press on August 21. "The cause we have loved and have been privileged to serve, beckons to us again. . . . Sir Maurice Hallett, the Governor of the U.P., said recently that 'those who are not with us are against us' . . . then let me assure him that we are not with him. We owe allegiance to India alone, and we serve the cause of her freedom. . . . [S]oon . . . I shall join my comrades of the U.P. My place is with them and my heart longs to be by their side."[40]

Linlithgow and Hallett knew, as well as Gandhi and Azad did, just how eager Nehru was to be taken prisoner. Virtually every socialist worth his salt was behind bars. He almost felt ashamed, walking free to say whatever came to mind, with not a police lathi or bayonet so much as pointed at his balding head. It was quite frustrating, humiliating. "It is a queer world we live in," he wrote Darling Indu from Wardha on August 20.[41] The Working Committee moved back to visit with Gandhi, hoping to convince him of the need for launching at least an individual Satyagraha campaign, as a symbolic token of Congress resistance. Secretary of State Amery told the House of Commons on August 14, 1940, that "India's future house of freedom has room for many mansions."[42] He had the Muslims and princes primarily in mind and still thought, as did Linlithgow, in terms of a unitary federation of India, one that might somehow accommodate all interests and minorities, without resort to fragmentation or partition. But Nehru wanted none of it. "The Viceroy's statement and Mr. Amery's speech have made two points crystal clear," Nehru told the press in Bombay. "The British Government have no intention whatever of giving up any real power to the Indian people. . . . The glimpse of the future that is shown to us . . . dim and distant . . . does not promise us any kind of freedom. . . . [T]he white man must wearily continue to shoulder his burden. The whole thing is fantastic and absurd and has not even the merit of decent phraseology about it." The latter point most offended his Harrow–Cambridge sensibilities.[43]

"The spark that has been lit in my province is sure to spread throughout India and annihilate many of us and will certainly put an end to the British rule in this country," Nehru told a huge Bombay audience on September 1. Only Bombay remained quiet and mostly calm, its stately homes atop Malabar Hill, one of which was occupied by his sister Betty, another by Jinnah, unruffled by the growing chaos of resistance that spread throughout several United Provinces districts around Allahabad and Kanpur. "I am glad that the stalemate has ended. . . . [T]his has given us peace of mind. The only question . . . is how to advance and achieve independence for the teeming millions of India . . . the opportunity has arrived for you and me to make history by relying on our own strength. None of us can remain aloof from this struggle."[44]

He could hardly wait for his arrest. "So far we have restrained ourselves but from now on we shall shout at the top of our voices. . . . We would rather die than participate in such a war."[45]

Yet another month passed, and he remained free. He raced about as much as he could, trying to say things that would exhaust the phlegmatic viceroy's patience. "Kisan brethren!" he addressed the peasants in the eastern United Provinces' Gorakhpur in early October. "These days the biggest problem before us is. . . . We do not want foreign rule. We are now ruled by the British . . . even if the collector or any other officer is an Indian the ultimate authority is the British. . . . The viceroy draws Rs.

20,000 as his pay and lives in a palatial building. Where does this pay . . . come from? . . . [F]rom your pockets. . . . These people live upon your own earnings and then rule over you. . . . [A]ll our wealth goes to foreign countries. . . . We must change the very system. . . . We want the reins of government in the hands of the public. . . . We have to establish such a government and put an end to the British capitalistic form."[46]

Gandhi finally agreed to launch an "individual" Satyagraha campaign, choosing Vinoba Bhave, his saintly ashram disciple, as the first embodiment of national struggle. Vinobaji, "India's walking saint," was as yet unknown to the world outside Gandhi's circle.[47] He had been a faithful disciple of the Mahatma's since 1916 and remained true to Gandhi's teachings all his life, initiating several important movements of village land reform long after Gandhi was assassinated. "He believes in communal unity with the same passion that I have," Gandhi explained in announcing his choice of Vinoba as India's first Satyagrahi. "In order to know the best mind of Islam, he gave one year to the study of the Koran . . . learnt Arabic. . . . Vinoba believes in the necessity of the political independence of India. He is an accurate student of history. But he believes that real independence of the villagers is impossible without the constructive programme of which khadi is the centre. He believes that the spinning wheel is the most suitable outward symbol of non-violence which has become an integral part of his life. . . . He has never been in the limelight on the political platform. With many co-workers he believes that silent constructive work with civil disobedience in the background is far more effective than the already heavily crowded political platform."[48]

In his choice of Vinoba, as in his explanation of just why he chose him, Gandhi was pointedly reminding Nehru of how wide the gap between them had grown in this last decade. It was a blow to Nehru's ego not to be chosen first, for Gandhi knew exactly how eager he was. Yet Gandhi feared that Nehru's passion for communal unity was much less ardent than his own or Vinobaji's. Nor was he sure of just how accurate a student of history Jawaharlal was, although he knew how Nehru prided himself on his historical knowledge and brilliance. The old man well understood, moreover, how little Nehru believed in the powers of spinning or weaving, finding the limelight of every political platform far more seductive. And how heavily crowded those political platforms had become of late! The Working Committee had gone back to Wardha to meet with Gandhi before this announcement in mid-October. Congress President Azad agreed with Gandhi's choice of Vinoba rather than Jawaharlal, although Azad's concerns did not focus on Nehru's indifference to spinning and nonviolence but, rather, on his inability to appreciate the complexity of Hindu–Muslim communal conflicts, which he continued to view in simplistic economic terms.

For more than a week after Vinoba was chosen and arrested, Nehru continued his tour of United Provinces villages, desperately trying to fill his time in a frenzy of speaking and moving about, to keep his mind from focusing on what he knew Gandhi thought of him now and what he suspected Azad also felt. "The shades of evening were falling when Jawaharlal Nehru arrived," he wrote of his tour for the *National Herald.* "A vast concourse of people had gathered to hear him. . . . He spoke of Swaraj . . . of the mighty revolution that was taking place all over the world; of the war in Europe and how India was dragged into it; of the satyagraha started at Gandhiji's instance by Vinoba Bhave; of the next step that would follow; of the vast responsibility of each one of us at this tremendous crisis. Be ready and disciplined! Organise yourself, hold to nonviolence, put an end to all internal squabbles . . . face the future with unity, strength and confidence. . . . Nehru asked the audience to remain seated till he had left and to make a narrow passageway for him to go through . . . not a single person got up. Right through that mighty gathering he marched. . . . The stars were shining brightly as Nehru motored away."[49]

He returned to Anand Bhawan to find a letter from Gandhi waiting, asking him "whether you can see anything to commend itself to you in all I am writing and doing. . . . My present conception requires those who believe in the plan."[50] Nehru cabled immediately, "Agree generally." Then he wrote to explain: "My mind is by no means clear as to what the future is going to be. More and more I think about the problem in India in its larger world perspective . . . from the political point of view. I think that your plan has great possibilities and therefore I would like it to go ahead and to help it. . . . But . . . I feel a little confused. . . . I think the horror of war is bound to lead to a strong reaction. But this reaction will only come when some way out, which guarantees individual and national freedom, is found. Without this guarantee nonviolence becomes associated in people's minds with submission to slavery and force." After reading this letter, Gandhi agreed to name Nehru as his second choice to wage individual Satyagraha against the government of British India.[51]

While waiting alone at Anand Bhawan for Gandhi's decision to reach him, Nehru wrote to his Darling Indu in Switzerland. "You are seldom out of my mind and sometimes I have a feeling that you are very near me. The door between your room and mine is always open and I walk in and out of it frequently for no apparent reason. Your room is as it was. I do not like to have changes made or even an alteration of the furniture. In my own room you look at me from every side. You will be amused to learn how many pictures of you I have round about me . . . twenty-six of them—all different ones of course—in my sitting room and dressing room! From babyhood upwards you sprawl or sit or stand and the past comes up before me and becomes more real than the present. . . . What a world we live in! It is a nightmare. And yet even the most sensitive grow used to its daily horrors. I read about the bombing of London and then remember that Chungking has had this kind of thing for years—and still carries on. . . . I did not feel the slightest bit nervous as I watched the bombing of Barcelona . . . I was fascinated by the sight. Indeed I resented being pushed into a dug-out . . . one must face risk in the course of doing something worthwhile. . . . Let us get our minds and bodies in fit condition. . . . You ask for news. . . . There is plenty of course but then there is also an army of censors to get through. But we seem on the eve of big events. Indeed they have begun already."[52]

Two days later Gandhi summoned him to Wardha and gave him his approval to march forth in publicly announced defiance of the Raj. On October 31, 1940, Nehru was arrested at Chheoki and driven to the house of Allahabad's superintendent of police, then sent by car to Gorakhpur, where he was placed on trial for speeches he had earlier made in that district. "My trial, within the jail premises, was a curious *Alice in Wonderland* affair. Of course it was all formal and proper. . . . But an air of utter unreality hung over it and somehow it was difficult to take it seriously. . . . India is so unreal today—a Mickey Mouse affair or Snow White. Or perhaps the world itself is behaving like that. What a mad lot of people we all are in this mad world!"[53]

Nehru was found guilty of sedition for three speeches he had made earlier in the month and was sentenced to one year and four months of rigorous imprisonment on each count, the three to run consecutively. "Four years seems a long time to look forward to," he wrote Indira from Gorakhpur District Jail. "And yet in this world of shock and change, it makes little difference what period is fixed for a sentence. For my part I might as well be here as elsewhere. For the last five days my tired mind and body have been clamouring for rest and I have slept more than I have done for many months."[54]

21

War Behind Bars

1941–1942

"AFTER A GAP of many years, five and a quarter to be accurate, I begin writing a diary again. My . . . jail productions," Nehru wrote in his cell. "I have been two weeks in prison. . . . Today is my birthday . . . a terrifying thought, this creeping on of age. Fifty-one. . . . In China it entitles one to all the respect that is paid to age. I can't get used to the idea that I am getting old, and others help me to delude myself. . . . True, I . . . am still full of vitality . . . I think I am almost indifferent to death. There are no strong personal bonds which would hold me back if death beckoned." [1]

Vallabhbhai was also arrested that November, and before year's end eleven members of the Congress Working Committee were behind bars. On New Year's Eve Azad was sentenced to eighteen months. Gandhi remained free, announcing that perhaps it was because the government "consider[s] me a fool. In so doing, they only follow many others who think that Congressmen fool me and that the latter's non-violence is but a cloak for hiding their violence, if it is not a preparation for it. . . . For me there is no turning back. . . . I would rather be regarded as a fool but strong than as a knave and a coward." [2]

Nehru was soon moved back to his old cell in Dehra Dun jail, where he had the company of Ranjit Pandit, Nan's husband, whose fragile heart would allow him to survive only another three years. On "Complete Independence" Day, January 26, 1941, Nehru rose early, recalling "the past eleven years—the first Purna Swaraj Day in 1930 . . . with all its enthusiasm & excitement & jail-going—1931—I was discharged on this very day to see my dying father and the ten days that followed . . . so my mind wandered to the other anniversaries . . . and Kamala's illness and after. Four years ago. . . . Where was I then? I forget. . . . Two years ago in Almora with Indu, and Nanda Devi and the snowy peaks of the Himalayas looking down . . . from their eternal solitude as we repeated the . . . pledge. Last year? Where was I then? Probably in Allahabad. I do not remember. . . . And now again in Dehra Jail where I am likely to remain for a long time . . . no ending of this prison business this time till the cause of it goes."[3] In honor of the day and to keep his promise to Gandhi, Nehru spun one thousand yards of khadi thread, more than he had ever spun in a single day, though he was later able to double that. His sister Nan was now incarcerated in Naini Central Women's prison. She soon became something of a school nurse to seven toddlers aged one to three, whose mothers were jailed for murder, robbery, and prostitution. "There is a hierarchy in jail as elsewhere," Nan recalled, "and the murderess is at the top. She despises the woman who is in for theft and hardly acknowledges the prostitute."[4]

Indira left Switzerland in December, en route from Geneva to London, via Lisbon. From London, however, there was no easy passage back to India. On the last day of January Nehru wrote her again, "I looked long at the envelope which contained your letter. . . . Like a treasure which has come my way unexpectedly . . . your handwriting was not quite the same. . . . It was yours of course and yet it made me imagine that someone was trying to copy it . . . then, I wondered, does this indicate some change in you also? We all change as the days and years roll by and I am myself conscious of very definite changes in my own make-up. As I look back I seem to see a procession of different personalities merging into each other and yet each with its own distinctive features. Whether the changes are for good or not I do not know. Or perhaps it is not so much a change of personality that takes place as the emergence of different aspects of the same personality. We are, each one of us, a group of different individuals, all tied up together with no hope of release, and sometimes they quarrel amongst themselves and we feel the tension and the pain."[5]

Nehru was moved from Dehra Dun to the Lucknow district jail on the last day of February. He had received word by then of Indira's departure from England but learned neither the name of her ship nor the anticipated date of its arrival in India. German U-boats continued to take their toll of shipping in the Atlantic, so strict censorship was imposed on all shipping news. Nan was released from prison on April 2, and came to visit her husband and her brother a few days later. "Nan looks . . . thinner," Nehru noted. "Had a long talk—am inducing her to go to China for two or three weeks."[6] Madame Pandit and Madame Chiang Kai-shek soon became good friends and symbolized the beauty and courage of Asian women for much of the Western world throughout the war years. Later, as president of the United Nations General Assembly, the tiny white-haired Nan attained even greater glory, having personally experienced the widest range of hardship and fame of any woman in her lifetime, transported in little more than a decade from prison to presiding over the world's highest assembly.

On April 18, 1941, Nehru was awakened after midnight by the commanding officer of his jail to report that a telephone message just came from Bombay that "my daughter had arrived. Odd time to telephone to a prison," was his only comment on that news.[7] He had for the past few weeks been preoccupied with T. E. Lawrence's *Seven Pillars of Wisdom.* "The book has held me, not only because of its fine writing but also because of his problems and difficulties with himself," Nehru confessed to his diary. "Sometimes—not always—that problem was not unlike mine in some ways. And yet of course there is little in common between him and me," he quickly added, for whenever he lifted the veil that hid his true nature—his deepest secret self whose continued existence terrified him—he hastened to deny whatever he saw, closing his eyes to that most painful truth.

To Padmaja, who had been living at Anand Bhawan for some time, volunteering to arrange his library and other papers, and who probably knew him as well as any woman, with the possible exception of Indira, he wrote: "I spent quite a long time over Lawrence's *Seven Pillars of Wisdom.* It fascinated me. I had read his *Revolt in the Desert* [an abridged version of *Seven Pillars of Wisdom*] long ago. . . . Reading this bigger book now became almost a mental adventure. The fine writing of course attracted but even more so his digressions and self-analysis led me into all manner of trains of thought. He irritated me often enough because he appeared morbid or a poseur or an exhibitionist of the mind. Yet he made me think and his problems were not always unlike mine basically, though they might differ in outer dress. I suppose I am very different from Lawrence—still sometimes we touched."[8]

Indira came to see him in Dehra Dun Prison, to which he and Ranjit had been returned earlier in April 1941. "She had not been keeping well on board and since arrival," Nehru noted. "She was not well now and I was filled with anxiety about her future. Apart from health, other difficulties. I was very happy to see her, and yet my mind became engrossed with these difficulties. Some years back I would have worried more. Now I . . . am not easily upset. We do our best and take things as they are. All good luck and happiness to her in life's way. But it is not going to be a soft way. She has determination and self-reliance, which is good. But she is so . . . immature and perhaps tends to take things superficially. Yet she must have depths. She will reach them slowly . . . there may be shocks. Delicate and sensitive people can hardly avoid shocks and upsets. My mind was full of her and of life's queer ways after she left, and at night also I had curious dreams."[9]

Indira had finally told him about Feroze and their plan to be married. But he was so "filled with anxiety" over that news that he dared not even mention it to his diary, treating it merely as one of several "difficulties" concerning Indira that he would have to sort out. He was less worried about Feroze now, not as upset about her threat to marry him as he would have been earlier. For he believed that she would soon wake up and become more mature, understanding that Feroze Gandhi was hardly a suitable match for her. How could she not see that? She had "depths," after all! He knew how much she had read, how sensitively she resonated to all he wrote to her, told her, taught her, all these long years of his private tutelage! Naturally, it was a shock to Indira to see how negatively her father had reacted. But she would soon see the light, he felt confident, although for him the shocks and upsets stirred by his daughter proved much greater than any he could inflict on her. He understood much about himself and about "life's queer ways," but had no idea as yet that this delicate sickly girl would not only match the steel of his mind and resolve with her own but would prove to be much tougher, firmer, and stronger. Years ago in Paris she had promised herself to Feroze, and nothing Darling Papu would ever say or threaten to do, or undo, could make Indira change her mind.

Indira had written to the superintendent of her father's jail to say that her "medical adviser, Dr. Atal, would like to have an interview with me next Sunday," Nehru noted on May 2. "Also that Feroze Gandhi will accompany him."[10] The new superintendent of Dehra jail was an old Indian surgeon, Dr. Ram Sarup Srivastava, called back into service from retirement, of whom Nehru had written just a few weeks earlier in his diary: "He seems to be very anxious to be helpful in every way."[11] Since Feroze was perhaps the last person on earth Nehru wanted to see at this time, "the Superintendent told me this morning that he had allowed the interview with Madan [Atal] as he was coming as your medical advisor," Nehru wrote the next day to Darling Indu, "but he had not allowed it to Feroze as there was no special reason for this." To Feroze and Indira the reason was very special, of course, as Nehru understood too well. "The weather is getting hotter and hotter," he now added, for to him the mere threat that Feroze might attempt to see him sent his temperature rocketing, "—this is the worst time of the year and it will last for nearly two months. It will be a great trial for you to take long journeys during this weather and a risky business. I do not propose to write any letters for the present." It was his politest way of telling her that he preferred not to see her again either—at least for the next two months, the "cooling-off" period he mistakenly imagined might bring her back to her senses.[12]

Madan Atal came to see him alone on May 4, and they discussed Indira at length. Her "condition" was "frail," and "precarious," they agreed, Nehru suggesting that Madan keep his eye on her for the next few months at his hill station retreat near Simla, called Solan. "Madan Bhai will be there to advise you," Nehru wrote Indira that evening. "You can decide later what to do afterwards. . . . I do not know how to unlock the doors of my mind and heart, which burst with things to say. . . . I wonder sometimes if my love for you has been so wanting in something as to prevent those doors from opening. I seek to find the error, to purge the fault. Meanwhile we grow older . . . and I draw into my own shell. . . . In the quiet of Solon . . . [t]ry to think out things for yourself." Three days later he confessed that his mind remained "troubled and uneasy," mostly about Indira. His nights were restless, for he feared the worst.[13]

All the next week he heard not a word from her. Now he hoped she might return to visit him, but nothing happened. He felt so frantic that on May 15 in desperation, near panic, he started to write, "Indu . . . This letter is not meant to be sent to you. . . . Yet the writing of this has suddenly become an imperative need, an urge I cannot resist. . . . But no one likes to undress his mind and soul in public. So . . . all these long years, whenever I took pen in hand to write a letter, subconsciously I kept a check on myself, feeling that strangers would see that letter. . . . Perhaps even before these iron bars of censorship enveloped us and made me retire a little more into my shell, I had developed a measure of restraint . . . a way of self-protection against a fear I always had of being swept away by too much sentiment. . . . Yet, I fear, this hardness is only at the surface and underneath lies a sea of sentiment which has often frightened me. A life-time of disciplined living and deliberate training of the mind and body . . . has thrown a hard shell over this turbulent mass and . . . has given me a certain degree of self-confidence and usually a crisis or difficulty makes me clearer-headed and calmer. Yet on occasions the shell bursts . . . there was another reason which induced me to fortify my shell—that mighty Maginot Line which could after all be so easily turned. I realised that any slackening on my part produced far-reaching reactions on others . . . I was alarmed at the consequences. I could not live up to them and indeed I had no intentions of doing so. Thus I caused needless pain to others . . . so again I retired into my shell and peeped out of it." Padmaja and Bharati knew how painful any "slackening" on his part could prove to be.

"You will say . . . I really do not know myself. My mind cooped up for months past is just bursting. . . . How cribbed and confined and imprisoned we are by these iron bars of the spirit. . . . Can anyone be ever really frank about oneself, one's own emotions and mental struggles, one's urges and desires and those half-conscious imaginings which float, dream-like through the mind?

"My *Autobiography* is, I think, about as frank and truthful a document . . . as I could make it. . . . And yet . . . all the restraints and inhibitions were there, and I suppressed much that filled my mind and heart. . . . It was impossible for me to lay bare my heart before anybody . . . I have suffered greatly, experienced many hard knocks in my personal as well as my political life . . . hardened, matured, call it what you will. I am not just one and fifty years old . . . I feel as if I was hundreds of years old in mind and the weight of these centuries lies heavily upon me. If this is the beginning of wisdom . . . I would barter this wisdom . . . so dearly bought, for the lighthearted unwisdom of my younger days . . . one takes many things for granted in one's younger days . . . [I]n the twenties I . . . had a flame-like quality, a fire within me which burned and consumed me and drove me relentlessly forward; it made me almost oblivious of all other matters, even of intimate personal relations. I was in fact wholly unfit as a close companion of anyone. . . . Gradually I woke up . . . I realised then, and I realise now even more, what an impossible person I must have been to get on with. My very good qualities which made me an efficient instrument for political action, became defects in the domestic field. . . . As my awakening proceeded I yearned . . . for those closer human contacts of the spirit with those I loved. . . . Unfortunately long and trying periods of jail came. . . . It was in those days of the early thirties that I wrote those hundreds of letters to you . . . *Glimpses [of World History]*. . . . Dadu [Motilal Nehru] was dead. Dolamma [Swarup Rani Nehru] was . . . frail, ailing, enveloping me with the overwhelming love of a mother for her son. I was very fond of her, but she could take no part in the life I was living. . . . There were my sisters . . . so much younger than me that my relation to them was partly paternal. . . . My family life revolved and centred round two persons—Mummie and you . . . I dared not . . . expose all my inmost feelings and torments. . . . Not that there was any great secrecy about them . . . I am not a secretive person but there are limits even for me. . . . But even to you I could not say everything. It would hurt me and it might hurt you."[14]

No matter how hard Nehru tried to reveal himself to his daughter, it proved too painful. He dared not venture too close to the edge of that crevasse that he feared could swallow them both. "So a feeling of suppression grew upon me. In jail this is always so and I try to find a way of escape in hard work. . . . I looked forward to your coming back. You filled my mind and . . . kept me more or less calm. . . . You arrived. . . . That was an event for me . . . I was happy. But soon a cloud fell on my happiness . . . [I]mmediately I grappled with the new problem and gave you advice as to what you might do. You accepted it for the moment and, though I had a dull feeling inside me about your health, I felt that we were going to face the situation in the most effective manner."[15]

"Darling—please don't worry. Everything will be all right," Indira replied a week later from the new cottage she had rented in Mussoorie. "In some of your letters you have talked of joy and fulfilment. I have found mine. I have a serene happiness surging up from within, that no one and nothing can mar or take away from me. All troubles—illness and discomfort and disputes—just seem to sail away on the surface without really touching me. Most people spend their lives waiting for happiness but the cup always seems to be just a little beyond their reach and they have not the courage to stretch their arms to grasp it. I took it in my two hands and drank deep into it—and it entered into every nerve and tissue of my mind and body, and bathed me in its rich warm calmness. I have this now and forever. Happiness is indefinable—how can my feeble attempts at description suffice when the great writers of the world have not succeeded? . . . I love you a lot and I am worried about you just as much as you are about me. So keep your sunny side up & look a lot better when I see you next."[16]

Feroze had sailed home with her from England, and they had never felt more sure of their love or of each other. He was now faced with his own Parsi family's concerns, however, and Indira knew how angry and disappointed her father was.

"The whole thing was fantastic, absurd in its folly," Nehru wrote of his daughter's decision to rent a little cottage in Mussoorie rather than staying where Madan Atal could keep his eye on her. "So I raged for two days and life became a terrible burden. I could do no work, could not even spin. . . . I could sleep [only] with difficulty and would suddenly sit up in the middle of the night unable to control or direct the ideas that battered my mind. . . . I got up about 4 A.M. and decided to write to Madan. . . . But still my mind was ill at ease. I longed to meet you and talk to you . . . for my mind was with you. I must have behaved oddly." He was losing control, he feared, of the one person he loved, his sole heir to all the power he knew he would claim someday.[17]

"Madan came to see me," Nehru confided to his diary a few days later. "He brought a letter from Indu. An angry, agitated letter, angry at me, angry at events. All the long screeds I had written to her . . . had a very different effect on her from what I had desired. So much for my insight into human beings and my general competence! Not learning from this, I suffered myself to seize hold of pen and paper and write on and on to relieve my fevered brain. . . . It is terribly sad how this jail business prevents me from getting nearer to Indu just when I want to so much. And yet, would I succeed if I was out? . . . I seem to think that cold reason and logic must ultimately convince . . . which leads to false hopes and wrong actions. How many times I have failed in this endeavour during past years, both in private & personal life as well as in public life. As my experience of life grows I feel more and more perplexed. . . . I am losing confidence in my ability to cope with it in its personal aspect. I cannot even gain the confidence of my daughter! . . . What is to be done? What is to be done? Nothing. One cannot cheat the fates in personal matters or make them hurry." He was, of course, quite right in noting that "one cannot cheat the fates in personal matters," but mercifully, Nehru did not live to witness the tragic end of his daughter's life or that of both his beloved grandsons.[18]

"Darling, I am in the throes of remorse and regret," she wrote to him from Mussoorie on June 2, "—as usual, when it is too late to remedy what has or . . . has not been done. I am writing this because I believe one should always admit one's mistakes. . . . How simple, practical and obvious was your advice! It was only my blind prejudice which prevented me from following it. Truly it is feeble-minded to let oneself be so influenced by prejudice. . . . I am amazed at my selfishness. I should have lumped my feelings and thought a little of your feelings. I seem to have hurt you on purpose. . . . I can only hope and pray that this will be a lesson for me to be less stubborn . . . how utterly lost I am without you. I have been so arrogant and stupid. I tried to sail out on my own before I knew the rudiments of managing a boat. I deserve to sink . . . yet that would cause you so much pain. . . . In some of your letters you say that you do not wish to impose yourself or your advice. . . . Far from being an imposition, your advice is the strongest prop on which I can lean—for my own legs are pretty shaky . . . so the erring child asks for forgiveness, and asks too that you believe her when she says that she loves you. . . . One feels so inferior when you are about and I suppose that unconsciously one resents it."[19]

Nehru urged her to wait at least until he was released from prison; Indira and Feroze wanted to marry immediately. To Nehru, of course, the suggestion was his last delaying tactic, hoping as he did that almost four years (until his sentence expired) might be enough to cool their passion. At any rate it would give Indira's Aunts ample time to introduce her to many more eligible young men, not only Brahmans, but men of real talent and promise, with Oxford or Cambridge degrees, or at least the scions of great families. Hundreds of letters reached him each year asking for the mere chance to court her, for what Indian pandit politician, or prince would not want to marry Pandit Nehru's only daughter?

But now that Feroze was back home also, his own family went to work on him with equal "logic," for what on earth was the *rush?* Did he want to alienate Indira's father? Didn't he realize how much he was sacrificing for the nation? For Congress? How important, how great a man he was? To go against the expressed wishes and desires of such a man would be sheer madness! So Feroze himself went to Indira in Mussoorie and, in his sheepish way, hardly daring to look at her , he muttered something about how it might be best to wait, after all. So Indira wrote that contrite letter to her father.

"Bhai was upset by Indu's desire to marry Feroze," Betty recalled. "Though my brother was so modern in his ideas, in his distress he fell back on ancient tradition. He told Indira that since he was in prison and her mother was dead she must ask permission to marry from her aunts. . . . Indu went first to Nan, and, of course, got exactly the same answer her father had given her; she must meet more young men before she made up her mind."[20] Indira never forgave Nan for that response, and after becoming prime minister she never invited Madame Pandit to fill any post in her government, not the lowest ambassadorship, not any cabinet ministry, nor did she ever ask her for advice on any matter, public or private.

The German invasion of Russia came as a "thundering surprise" to Nehru.[21] A few days later, Indira came to visit him again, and the next day he wrote her passionately: "I want to make up for all your long absences from me. There is so much I want to ask you, to tell you, to discuss with you. Not argument and debate . . . I have enough of it in my life outside and I am weary of it. . . . Fate in many ways has been kind to us. Yet in some ways it has been most unkind and its greatest unkindness to me has been the enforced separation . . . between you and me. All these long years in prison . . . and then your years in Europe. . . . Some little time after you left . . . I searched the skies (it was rather misty) for the new moon and discovered it bashfully peeping through a veil of mist . . . it must have been three days old, but it was lovely enough and so arch! You know of course, your pet name Indu means the moon. I told you this when you were almost a baby. I have always been fascinated by the new moon . . . it might be worth while for you to start on a voyage of discovery of India. . . . It is a fascinating journey, not so much in just sight-seeing but in its mental aspect, when the past and present get strangely mixed together and the future flits about like an insubstantial shadow, or some image seen in a dream. The real journey is of course of the mind . . . full of pictures and ideas and aspects of India . . . old monuments and ruins . . . old song and ballad . . . the way people look and smile, and the queer and significant phrases and metaphors they use—whisper of the past and the present and of the unending thread that united them and leads us all on to the future. . . . So I try to understand and discover India."[22]

Nehru had started to write *The Discovery of India*. It was published after the war by the John Day Company in New York, which had just published an abridged edition of his autobiography, entitled *Toward Freedom,* in the United States. His *Discovery* in some ways was more ambitious, an Indian nationalist's attempt at integrating both the ancient and modern history of India, with his own life threaded through it, adding vivid personal embroidery to the texture and color of its grand, though doctrinaire, design. "The real voyage of discovery cannot be confined to books," he told Indira, yet books were "essential" for "they tell us of the past," so he listed many for her to read. Perhaps he hoped to distract her thoughts from Feroze. But more important, he wanted to train her, to bring her along with him, so that when he grew too old to steer properly she could take the wheel from his tired hands and guide their ship of state by the stars of political freedom and economic development and progress in which they both believed.

But less than two weeks later she was stubbornly fighting him again. "I have made it clear to you repeatedly that in no way will I obstruct you in following your own decisions about yourself," he wrote her on July 9, 1941. "Naturally . . . I have made innumerable plans about you; I have thought again and again how to help in making you a person who can face life and its problems serenely and with confidence. . . . About your marriage I never worried as that, above all, would depend upon your own choice. . . . I had hoped that after your formal education at a university was completed, you might supplement it by some travel . . . I wanted you to go to Russia to see things there for yourself. . . . Then . . . I expected you to return to India and discover the fascinating thing that is India. . . . I wanted to help you personally and I expected you to help me. . . . Hundreds and thousands of young men and girls have wanted to serve with me as secretaries. . . . I have never encouraged anyone and have shouldered my burdens alone, for I had always imagined you to occupy that niche. . . . No one else could take your place. . . . [W]ith this idea ever hovering in my mind I wrote piles and piles of historical and other letters to you. I wanted gently, slowly but yet surely to train your mind in that wider understanding of life and events that is essential for any big work. Of course . . . I knew you would marry. . . . I only wanted to give you some special training which would stand you in good stead in later life. . . . So that when the time came you could with assurance tackle any big national job. And the time for this will surely come for our people before long. . . . I mention it so that you may realise how I have pictured you to myself and woven tales about your picture in my mind. It takes a good deal of adjustment for me to throw all these . . . into the scrap heap. . . . Our sense of values seemed to differ vastly. That hurt. . . . What pained and surprised me was the casual way in which you were prepared, and even eager, to discard very precious traditions and heritage.

. . . It was rare good fortune for us to have this heritage and I, for one, was proud of it."[23] How ironic that might have sounded to Motilal, yet how pleased he would have felt to know that his sermons had at last taken root in the mind of his rebellious son.

"You have been seriously unwell," Nehru continued, desperately trying every argument he could dredge up to persuade her of the wisdom of what he wanted. "You require physical and bodily rest. Above all peace of mind. . . . Marriage is an important thing in life. It may make or mar one's life. And yet marriage is . . . smaller than life. Life is a much bigger thing . . . difficult enough to understand . . . still one has to try. . . . If you want to marry Feroze, well go ahead and do it. No one will stop you. So why worry about it? . . . As it happens even your present health indicates, I believe, an avoidance of marriage for some time . . . months at least. But that is for doctors to say. Apart from that . . . there is an element of absurd haste in your returning from Europe in frail health and suddenly marrying." If in our attempt to solve a problem we create half a dozen new problems, we have not acted very wisely. . . . To create irritations and ill will in others is never worth while. . . . Avoid also breaking as far as possible with old contacts and ways . . . you might well be landed high and dry. I am not referring to Feroze but life's other contacts, including Feroze's family. . . . I know nothing about his family. . . . There is too much of casualness in your approach."[24] He urged her to visit Aunt Betty in Bombay and Gandhi in Wardha.

"She told us that she truly loved Feroze and earnestly enumerated his good points, his fine character and kindness," Betty remembered. "My first answer was . . . 'Meet some more boys first.' With a flash of Nehru temper . . . Indu said, 'Why? It took you only ten days to make up your mind to marry Raja Bhai; and I have known Feroze for years. So why should I have to wait, and why should I have to meet other young men?' I had no answer to that, for I was happy with Raja; so smiling I said, 'Darling, if you think he is the right person go ahead and marry him.' "[25]

Nehru briefed Gandhi at length about Indira's desire to marry Feroze and explained that he had urged her to seek his guidance. Then he waited impatiently for her to write or come to visit again, but by the month's end had heard nothing. "I have had no news from you since we met last," he wrote. "I hope you are keeping well and putting on weight. . . . Shall I ever go wandering again in these mountains . . . and climb the snows and feel the thrill of the precipice and the deep gorge? And then lie in deep content on a thick carpet of mountain flowers and gaze on the fiery splendour of the peaks as they catch the rays of the setting sun? Shall I sit by the side of the youthful and turbulent Ganga . . . and watch her throw her head in a swirl of icy spray in pride and defiance, or creep round lovingly some favoured rock and take it into her embrace? And then rush down joyously over the boulders and hurl herself with a mighty shout over some great precipice? I have known her so long as a sedate lady, seemingly calm but, for all that, the fire is in her veins even then, the fiery vitality of youth and the spirit of adventure. . . . I love the rivers of India and I should like to explore them from end to end, to go back deep into the dawn of history and watch the processions of men and women . . . going down the broad streams of these rivers. . . . Heigh Ho! How many things I would like to do, how much there is to see. . . . What wonderful dreams we can fashion out of the past and out of the unknown future that is still to be! Men come and men go but . . . when failing hands can no longer hold the torch, others, more vigorous and straight, take hold of it, and they in their turn pass it on to yet others still. How I begin musing when I write to you."[26]

Thus, with the beauty of his words and his promise to pass the torch to her youthful grip, he tried to lure her from her passionate focus on Feroze and marriage. But she had grown weary and wary of his words. Hers was the symphony of youth; his, of advancing age. She would, of course, follow him; but first she had to create sons, who were to take the torch from *her* trembling hand.

Indira was with her father in Dehra jail when the news of Gurudev Tagore's death was wired to the world on Thursday, August 7, 1941. He and Motilal had been born on the same day, just a few months more than eighty years earlier. "I loved his love of life and all things beautiful," Nehru told his diary that night, "with him I was a pagan. How different is Bapu!"[27] Bharati came to him a week later and reported all the intrigues and conflicts among the Congress leaders in Bombay and Gujarat and indeed throughout the country. "The inevitable and preordained price to pay for Bhulabhai [Desai]," Nehru noted. "How he has throttled all vital & progressive elements and held on like a vice to his office of president B.P.C.C. The surprising part is that Bapu and Vallabhbhai should have pushed him on all these years, especially Bapu. How I remember my objection to the inclusion of Bhulabhai in the W.C. after the Lucknow Congress. I was on the point of resignation from the presidentship. . . . But I surrendered as I usually have done to Bapu. . . . I still think I acted wisely. . . . How comic & ridiculous it all is. . . . With all our faults, . . . the U.P. shines. . . . Curiously enough . . . I worry very little. I have accustomed myself to think in terms of long perspective and the petty conflicts . . . do not trouble me. . . . At any rate the old order in the W.C. is over. . . . If the right person appears and, throwing discretion to the winds, acts with faith and energy the clouds vanish and people roll up."[28]

Nehru felt certain that he was the "right person," India's man of destiny. He had begun to notice crowds gathering every evening at the gates of his jail "just to have a glimpse of me when I go out for my . . . walk in the jail compound." He would greet them silently with pressed palms raised in the *namaskar* salutation of Hinduism, adopted by Christianity for prayer. He watched them in the monsoon rain, drenched to the bone, "just for that distant glimpse. . . . Why do they all come day after day? . . . There is something very significant about it. . . . It gratified me . . . strengthens me also to see all this faith."[29] For to those silent Hindu crowds he was a living god and simply to see him, doing *darshan* in this way was worship enough. It enhanced his own view of himself, helped convince him of how right he had been to hold on with such tenacity to his hard line against the British. "Changes took place in my mind, and conflicts, but in action I carried on doggedly without giving in," he wrote.

Nehru's old Muslim comrade Mian Iftikharuddin came to visit him a few weeks later. Long an ardent communist, Iftikharuddin was so confused by the swift changes in Europe, especially Germany's invasion of the Soviet Union, that he had lost faith in his socialist doctrine and came to seek illumination from Nehru. "Iftikhar—disillusioned Iftikhar—seeing no light except in a compromise with the Muslim League—which of course enrages me and I shout at him till I am hoarse. . . . It seems to me that those who have been backing a Communist ticket . . . are the most helpless today. They are losing the very foundation on which they stood and so they clutch feverishly at every straw. I was really amazed to hear Iftikhar advance arguments in criticism of Congress policy during the past year or more. They were puerile, absurd, completely foolish. . . . [T]he Communists . . . have shouted so much at the top of their voices that they have lost all power of reasoning. . . . I spoke so . . . loudly to Iftikhar that I felt quite tired."[30]

But Nehru's days of enforced withdrawal, of obligatory retreat and literary contemplation, reflections on his own strange past and India's antiquities, the sort of scholarly isolation he found so congenial, were now numbered. In many ways prison proved more soothing to his turbulent temperament than relentless engagement in political action and argument without end, rallying his army of peasants and princes in their struggle to capture Simla and New Delhi. He loved the fight itself, of course, but resisted and resented most of the petty conflicts that preceded the battle and all the vulgar wrangling, backbiting, and bureaucratic turf protecting that were part of the politician's life. If only they would do as he said, and simply get on with it! Far more efficient, and so much less enervating! He had recently been rereading *The Republic,* recommending it most highly to Darling Indu.

"My main occupation has been writing," he told his diary. "I have so far written about 110 foolscap pages. . . . I call it *The Discovery of India*. . . . At this rate, I imagine, I shall finish . . . writing work by the end of November, provided there are no big gaps or stops. . . . I have just been given a phone message that Indu is coming to see me tomorrow."[31] But the visit was not helpful, and after it he was "feeling rather odd." His mental equipoise was disturbed by the sight of how delicate and feeble she looked and how depressed she obviously felt, wishing she were married. The more he urged and argued the wisdom of waiting, the more depressed she became, withdrawing into her shell, which was made of the same tough steel as his own. A week later she returned with Kamala's mother, frail and aged "Amma," who gave her blessing to Indira's desire to marry, knowing what a "good boy" Feroze was, how dear he had been to her daughter, how much Kamala had loved him. "Indu came and Amma. . . . Certain remarks of Indu rather upset me," he noted. "What a tangle we have got into! Amma is staying with her at Mussoorie."[32]

Gandhi was worried, having heard by now of how unhappy Indira was, how depressed and unwell. That news, conveyed to Nehru by Mahadeva Desai, Gandhi's secretary, made him decide to "have a long frank talk" with her.[33] He hoped that his explanation of how badly he felt as a result of her failure to take his advice and obey his commands might suffice to clear the air and solve their problem. He still believed she would wake up and see the light—as he saw it.

"Darling, Mahadeva Bhai saw me yesterday . . . he will visit you. Bapu is rather worried to learn . . . that you were not keeping well or sleeping well . . . and I would like to speak to you about this. . . . I think Amma had better not come with you next time. This is tiring for her."[34] Indira knew for whom it was more tiring! She wanted and needed her grandmother's support and insisted on bringing her, so her father agreed. This time she wanted him to see Feroze as well, and quite soon. Feroze had rightly argued that he should be allowed to press his suit personally. So shortly after that second visit with Amma, Nehru wrote Indira, "I should like to see Feroze. . . . [T]he best thing will be for him to get in touch with Mohanlal Saxena. I have promised him my next interview in November and Dr. Mahmud . . . will probably come too. Feroze can be the third."[35] He thus started to treat his future son-in-law as he generally dealt with him after becoming prime minister: at a distance—politely of course, as long as Feroze remained polite—but never warmly. He never forgave Feroze for having been so loving and devoted to Kamala and then luring Indira away from him.

He now wrote in strict confidence, to his old friend Dr. Jivraj Mehta, about his desire to turn Anand Bhawan into a Nehru family museum. The Kamala Nehru Memorial Hospital had been officially opened earlier in the year by Gandhi, and the old original Nehru house remained the Congress headquarters, named Swaraj Bhawan. But Indira never wanted to live in Anand Bhawan—that she had made clear to him—and it was really too large for him to keep just for the occasional night he spent there alone. Currently Nan and her children were using it, but even she was anxious to leave, so he asked Mehta to draw up the papers that would turn the old "haunted" house into a national trust. But when his sister Betty, who lived close to Mehta in Bombay, got wind of what he was planning, she was reduced to tears at the thought of losing her birthplace, her childhood home.

"Darling Betty. . . . About Anand Bhawan nothing is going to be done in a hurry though, for my part, I came to these conclusions long ago. . . . Do not think of it in terms of our losing Anand Bhawan—we shall not lose it. Our lives have become part of the larger life of the nation and we go up and down with it. If that is so, why not share the house also?"[36] He felt quite confident now that his political stock was destined to rise to the summit of his nation, whose importance in the world could only grow swiftly now that Great Britain and the rest of Europe were tearing one another to shreds. Even China had not escaped the slaughter and devastation. India—alone of all the great centers of civilization other than America—had eluded the horrors of this war. Nehru believed that India, China, and the United States would gain the greatest power after the war. He had established strong ties with Generalissimo Chiang Kai-shek and his wife, and his autobiography had been published to rave reviews in New York. His name was now known by as many people in the world as Gandhi's, thanks to his prolific pen and Krishna Menon's indefatigable energy in ensuring that Nehru's work was published and assiduously promoted in every possible country. Although Krishna's health always remained marginal, his iron will and neurasthenic energy drove him day and night to work as Nehru's unofficial agent, soon to be Nehru's most trusted ambassador (Nan being the other). His luxury-loving little sister Betty never quite understood her brother, however, any more than she had understood Gandhi when he had first come to Anand Bhawan.

"My dear, how can I advise you or Raja about family affairs?" Nehru wrote Betty from prison, for she had started feeling guilty at how comfortable they were in Bombay while he was incarcerated. "It depends entirely on one's outlook on life as well as one's reaction to what is happening in India and the world. . . . Going to jail is a trivial matter in the world today. . . . As a mere routine it has no doubt some value and I think does one good. But that value is not very great unless there is an inner urge to do it. If an inner urge is present then little else matters. . . . Raja and you should try to be clear in your minds about the kind of future you are arriving at and try to realize that future. . . . The large and joint families of the past no doubt served a useful purpose and fitted in with the social structure we had evolved through long ages. But that structure is cracking up now and it cannot survive. . . . It does not fit in with the thought and elemental forces that move the world today. So it must fade away. . . . But . . . with such deep roots . . . major changes take time. . . . I do believe that the family . . . is important, especially the smaller family. . . . It will survive. But the economic bonds that tie up large numbers of persons in a joint family tend to become . . . bonds. . . . There is a feeling, and you have mentioned it, that the burden falls on some and not on others . . . that some sow while others reap. . . . One cannot argue against this for it all depends on our sense of values. Is money-making our test or some other also?"[37] He knew as well as Betty and Raja did that he had never made much money—virtually none, in fact, until he started earning royalties from his books. Yet he alone had inherited Motilal's entire fortune, thanks to the mere accident of his gender at birth and now he was turning it all over to the government in trusts! Wasn't that risky? Or crazy? Only to someone not as yet aware of how firmly resolved he was to take charge of that government, to lead his nation as he led his family. For only he and Gandhi were strong enough, wise enough, firmly resolved enough to do it, Nehru now believed. And Gandhi was now too old.

"A time comes in the life of every individual when there should be some certitude of his way of life," he told his often uncertain little sister. "I do not know what life holds for me but I am not afraid of it and I do not think anything is likely to happen to India or the wide world which will bowl me over. At least so I think in my conceit. I am slowly developing a measure of serenity, of poise, of strength of purpose which is impersonal. . . . The smaller and more personal problems gradually lessen their hold on me and I feel more detached. I want to be unburdened . . . [but] I do not know that I shall succeed. . . . I am attached enough to life and its diversity and richness. I am perfectly willing to face it as pleasantly as I can and to take such joy from it as is possible, subject to my own mental limitations. But I want to be equally . . . prepared for the full stop when that comes. . . . And if, as at present, I cannot indulge in activity I prepare myself for it, physically and mentally, and store up energy."[38]

Rumors of his imminent release soon proved most irritating to his mental state, for they disrupted his writing and made him restless instead of tranquil. He knew how hectic his life would become once outside again. He was counting on almost three more years in enforced isolation, his only real retreat in India, the only place where he could be as unsociable as he liked without seeming either misanthropic or snobbish. His I.C.S. cousin Rajan Nehru sent him a warm message of congratulations on his fifty-second birthday, for which he responded thankfully, adding: "Madame Chiang Kai-shek sent me a month or two ago a pot of very delicious marmalade which she had made herself. She said . . . that this marmalade was symbolic of life with both its sweetness and bitterness—and without the bitterness would not life be a dull and sloppy affair? . . . The winter months here . . . are delightful. With love to you and Ratan [Rajan's wife]."[39]

"The birthday has come and gone," he told his diary. "Also a visit from Indu. . . . I had suggested . . . when she came to see me last . . . some celebration of this, in a small way, might take place on her return to Allahabad—a tea party etc. This would give her a chance to meet some people. . . . She seemed to agree. . . . I wrote the same day to Nan. . . . Yesterday I had a note from Indu. . . . [S]he did not want to go to any party . . . particularly, she did not want a party for her. . . . She hardly wants to stay in Anand Bhawan . . . just to pick up clothes . . . go to Wardha and then to Bombay. Anywhere but Anand Bhawan!"[40] Auntie Nan had invited every eligible bachelor in Allahabad and Lucknow to the "tea party" that she and Nehru had hoped would help bring Indira out of her shell, into the proper circle of society, from which she could choose anyone she liked. Indira understood them both and told them as much, hurrying off to Gandhi and thence to Bombay, to Auntie Betty, who was at least on her side, and to her dearest Feroze, who awaited her there.

That week the London *Times* editorial noted that "the destiny of India is closely linked with that of the English-speaking democracies," arguing that there was thus "ground for supposing that release [from prison] might soon . . . be extended to Mr. Jawaharlal Nehru."[41] "This obviously means that there is something behind it," Nehru wrote to himself, still frustrated by "this release business." He feared that "in spite of my efforts to the contrary, I might . . . be discharged. An unsavoury prospect with all that is happening outside, especially among this crowd of gallant and irrepressible speechifiers . . . led by the redoubtable Bhulabhai. . . . [I]f it comes, it will have to be faced with composure. . . . And so to bed."[42]

On December 1, Nan's husband Ranjit was released, and next day Nehru wrote that his cell felt "rather odd . . . without Ranjit. . . . In the afternoon I felt tired and slept rather heavily for an hour. . . . I am beginning to think of going back to the writing job. . . . Indu has not written to me since she left nine days ago. How very casual she is."[43] Two days later his own release order arrived, and on December 5, 1941, he met the press and issued "greetings" to "all my comrades, to Congressmen, to the people of the United Provinces. . . . It is good to see the wide fields and the crowded streets. . . . But it is not good . . . to come out of the narrow confines of a jail into the larger prison that is India today. The time will come surely when we break through and demolish all the prison walls that encompass our bodies and minds, and function freely as a free nation. . . . In this world of infinite suffering . . . there is no rest or avoidance of travail. In this India, where foreign and authoritarian rule oppresses and strangles us, there is no peace for us . . . no rest for us but to carry the burden of the day and hold fast to our anchor."[44]

Now that Nehru was out, everyone expected to see him. Padmaja Naidu, obliged to leave Anand Bhawan after Nan and her family moved in, could hardly wait for him to come to her in Bombay. "Bebee dear—Life is a difficult business and the dull routine of jail sometimes seems simpler and preferable . . . already I have a feeling of suffocation and distress. . . . Your mother wrote to me. Perhaps I shall see her soon. . . . Indu is accompanying me. We shall be three days in Bombay and then Bardoli. Possibly I may come back to Bombay."[45]

Darling Indu prevailed, of course. The eloquent silence of her mournful eyes alone sufficed to melt her father's heart, though he never changed his mind about Feroze. Mahatma Gandhi liked Feroze, however, and blessed the alliance, even though it would raise a hue and cry from orthodox Brahmans. Nehru announced his daughter's engagement to the Parsi Feroze in February 1942. "A marriage is a personal and domestic matter, affecting chiefly the two parties concerned," he stated. "I have long held the view that though parents may and should advise in the matter, the choice and ultimate decision must lie with the two parties concerned. That decision, if arrived at after mature deliberation, must be given effect. . . . Feroze Gandhi is a young Parsi who has been a friend and colleague of ours for many years and I expect him to serve our country and our cause efficiently and well. But on whomsoever my daughter's choice would have fallen, I would have accepted it. . . . The marriage will take place in about a month's time at Allahabad."[46]

Four days later Gandhi wrote in response to "several angry and abusive letters" attacking Indira's engagement: "I have been, and I am still, . . . an opponent of either party changing religion for the sake of marriage. . . . He nursed Kamala Nehru in her sickness. He was like a son to her. During Indira's illness in Europe he was of great help to her. A natural intimacy grew up between them. The friendship . . . ripened into mutual attraction. But neither party would think of marrying without the consent and blessing of Jawaharlal Nehru. This was given . . . I had also talks with both. . . . It would have been cruelty to refuse consent. . . . As time advances such unions are bound to multiply with benefit to society."[47]

"It was an intercommunity and an interreligion marriage," Indira later recalled. "And it did 'raise a storm.' Yet it was not the first mixed marriage . . . many people, including my own family, were very upset . . . though the person whom I would have expected to be angry—my maternal grandmother—was the one who accepted my marriage the most readily."[48]

Nehru journeyed to Wardha a few days before the wedding was scheduled in late March to consult Gandhi about the appropriate Vedic ceremony to be performed by Brahman Lachhmi Dhar Shastri, a Sanskrit scholar and priest. "Mahatmaji is of opinion that . . . as far as possible, normal procedure should be followed," Nehru wrote his family Brahman, explaining how many people "are objecting to this marriage" and that therefore it would be "desirable to adhere to the normal procedure. . . . I am entirely with you that we should try to encourage the use of verses which proclaim national ideals, but perhaps it is desirable not to mix up two things at this stage. The marriage itself is in the nature of a reform and it is disapproved of by many conservatives. That is a big step and for the moment we should be content with that. . . . As you wrote to me, I expect the ceremony to be performed in Sanskrit and Hindi."[49] Thus Nehru, the radical pagan, felt himself obliged to urge a Sanskrit teacher to officiate over a traditional Hindu wedding, chanting ancient vedic mantras, while the agnostic bride and Parsi groom, tied together, took their seven barefoot steps around the sacred fire, sealing their union in the most ancient of Hindu rituals.

Indira and Feroze were married on March 26, 1942, at Anand Bhawan. Indira wore a pink wedding sari made "of fine khadi yarn spun by her father," Auntie Nan recalled.[50] The day was "quite perfect, with the clear and brilliant light of a March morning," Betty remembered. "When the 'auspicious' hour approached everyone gathered in front of Anand Bhawan . . . Father [Motilal] had built a marble platform on the lawn, designed for such ceremonies, with removable marble slabs where a fire could be built, to please my orthodox mother [Swarup Rani]. The traditional canopy of brocade supported by tall poles had been set upon it, and the sacred fire burned, tended by priests. Bhai escorted his daughter to the platform where she sat beside Feroze, while he took a place on the other side of the fire. . . . So Indira and Feroze were married in the same ancient Vedic rite as her father and mother and all the generations before them . . . and took the seven steps while the fire blazed up as the priests threw ghee on it and chanted . . . blessing."[51]

Indira and Feroze had rented a small house in Allahabad not far from Anand Bhawan, but they had little time to themselves, as Feroze was put in charge of erecting the giant tent for the next session of the A.I.C.C. to be held in Allahabad. "Immediately after my wedding . . . [m]y husband was in charge of the construction of the *pandal* where the meeting was to be held," Indira recalled. "And I was put to work with the volunteers . . . we didn't see each other at all. . . . We had . . . a very small house, but since it was close to the Congress session, they requisitioned half of it for use as an office. We had only two small rooms for ourselves. By this time it was getting hotter and hotter. My father was going to the Kulu Valley, but Feroze couldn't get away because he had to complete the accounts. . . . He stayed behind and I went with my father to Kulu. Then . . . Feroze joined me for a trip to Kashmir . . . and we did some trekking."[52]

So Indira won the husband of her choice, defeating her powerful father in their first real test of wills. Yet by accepting that defeat gracefully, Nehru retained his daughter's undying devotion and love, keeping her mostly at his side—at least at his beck and call—grooming her to take up the torch when his hand should falter, and soon neither of them noticed the objections of her groom.

22

From Cripps's Mission to "Quit India"

1942

JAPAN'S ATTACK on Pearl Harbor in December 1941 and on Malaya and Singapore in February 1942 jolted the British cabinet into reassessing its India policy. Churchill initially favored no change, but President Roosevelt pressed him to rally Indian support for the war in the spirit of their inspiring Atlantic Charter declaration of the previous August. Clement Attlee was promoted to deputy prime minister in the war cabinet, and Sir Stafford Cripps was appointed leader of the House of Commons. Attlee and Cripps had long favored a conciliatory policy toward Congress and India. Attlee now chaired a cabinet committee on India, which included Cripps as well as Secretary of State Leo Amery and Lord Simon and Sir James Grigg of the India Office, as well as Sir John Anderson. This India Committee agreed to a new declaration of policy, offering what they hoped would be "a just and final solution" to India's political problems.

But Churchill warned them, "When you lose India, don't blame me!" After mid-February, however, when Singapore surrendered its garrison of sixty thousand Indian troops, anxiety in London that General Tōjō's army might well reach Calcutta before Linlithgow made another overture to Congress pressured Prime Minister Churchill to announce on March 11, 1942, that the war cabinet was flying Sir Stafford Cripps to New Delhi with Britain's "just and final" offer. Some cynics later said it was Churchill's clever way of impeding the brilliant Cripps's remarkably rapid rise in popularity. Even more trenchant savants said it was Attlee's shrewdest move, for India had proved itself the graveyard of more than one promising British politician's career. But Cripps volunteered for his mission. He understood India's complex political geography almost as well as anyone in the cabinet, and he was better than most at negotiating agreements among those whose differences seemed irreconcilable. Cripps was a missionary at heart, a man of near genius and integrity, and he thought he understood Nehru and even Gandhi. He really believed he could pull it off, even though he knew how high the odds were against him.

"The question before me is how my country will be free," Nehru told his Congress Working Committee colleagues at Wardha in mid-January. "In our country there is indignation in the heart of every Indian against the British. In U.P. the eyes of every cultivator become red with anger when he thinks of the events which took place in 1857. In 1857 the rivers flowed with blood. I thought that the eyes of Englishmen would open. . . . It is impossible that the problem of India would be solved by Englishmen. . . . Some people are talking about returning to parliamentary activity. The whole idea is ridiculous."[1] The only member of Congress's Working Committee courageous enough now strongly to urge support of the Allied war cause was sixty-two-year-old Chakravarti Rajagopalachari (C.R.) of Madras, who became the only Indian governor-general, briefly succeeding Mountbatten. "My dear Rajaji, I have been reading the reports of your speeches," Nehru wrote C.R. "Some parts of these have distressed me. . . . I am not challenging your right to say what you have said. But a continuous approach to the British Government . . . makes people think that behind all our resolutions there is . . . a compromise."[2] This was precisely what C.R. had in mind, but Nehru's anger soon drove him out of the Congress, though he never lost his unique popularity in Madras.

Nehru returned to Gorakhpur to repeat what he had said a year earlier—for which he had been sentenced to four years in jail—arguing that "as far as the British Government is concerned it is the duty of every individual in this country to oppose it and revolt against it." He added that the real question for India was "how to organize our country on the eve of the coming revolution. . . . If the British win the war it will be because of China and Russia and not because of their own strength. After the war they will not have the strength to stand on their own legs or be able to maintain any position in the world."[3]

On February 9, Chiang Kai-shek and his wife flew into India for a brief visit that included several private meetings with Nehru and Madame Pandit. Madame Chiang brought a gift of pearls for Indira, and the Generalissimo heartened Nehru by reminding him that "India and China, with a common land frontier of 3,000 kilometres, had lived at peace with each other for a thousand years."[4] This became the eastern cornerstone of

Nehru's foreign policy, which was based on his faith in enduring Sino-Indian friendship and brotherhood, until the last two years of his life, when China's invasion of India rudely wakened him out of that dream. The Chiangs did their best to convince Nehru, as well as every British official they met, to reconcile their differences and join in helping win the war against Japan. Linlithgow wrote Amery that their visit was a success, though "Nehru and his sister . . . monopolised the greater part of the afternoons that they were in Delhi, and made them difficult of access. . . . Madame [Chiang] is a very clever and competent little lady. . . . When they are on a big job she starts with the family trousers firmly fixed on her limbs, but by the final stage . . . Generalissimo is invariably discovered to have transferred the pants to his own person."[5]

Cripps flew into New Delhi on March 23, 1942, the second anniversary of Pakistan Day, which was celebrated by a mile-long procession of Muslims, who were addressed by Quaid-i-Azam Jinnah that afternoon in Urdu Park. "We are asking for justice and fairplay," Jinnah told his Muslim League followers, some fifty thousand strong. "We want to live in this land as a free and independent nation. We are not a minority but a nation."[6]

"I have come to India to discuss with the leaders of Indian opinion conclusions which the War Cabinet have unitedly reached in regard to India," Cripps told the press that evening. "I am here to ascertain whether these conclusions will, as we hope, be generally acceptable to Indian opinion."[7] Cripps met with the viceroy and his Executive Council and with the British commander in chief, General Sir Archibald P. Wavell, soon to replace Linlithgow as viceroy. Cripps then moved with his staff from the viceroy's house to a less imposing building nearby, where he could meet Indians in a more relaxed environment. Cripps better appreciated by then how difficult a task loomed ahead. Governor Sir B. Glancy of Punjab had warned that "if there were a hint of secession," Punjab's martial Sikhs "would concentrate on getting ready to fight the Moslems."[8]

Cripps met first with Azad and next with Jinnah on March 25. Azad wanted greater control over the Defense Ministry than the war cabinet would given him. "I pointed out to him that strategically India had to be regarded as a part of a much greater theatre of war," Cripps noted. "My general impression was that Congress wanted the appearance and name of an Indian Defence Minister."[9] The cabinet's offer was to promise a new "Indian union" after the war's end, a dominion "equal in every respect to the United Kingdom" and other dominions of the Crown, free to remain in or separate itself from equal partnership in the British Commonwealth of Nations. Any province of British India "not prepared to accept the new constitution could "retain its present constitutional position," allowing for subsequent separate negotiations with His Majesty's Government for a new constitution. The British government would retain "full responsibility for India's defence" during the war but otherwise "desire[d] and invite[d] the immediate and effective participation of the leaders of . . . the Indian people in the counsels of their country."[10]

Jinnah was "substantially only concerned with the first part of the document, which I think rather surprised him in the distance it went to meet the Pakistan case. . . . [W]e had a long discussion as to its effect, especially upon Bengal and the Punjab . . . he thought that the plebiscite was the only absolutely fair idea. . . . He was extremely cordial and when we parted expressed the view to me that the one thing that mattered was to be able to mobilise the whole of India behind her own defence and that he was personally most anxious to achieve this. . . . I was hopefully impressed by his general attitude and his lack of pernickety criticism." Jinnah had scored an important advantage, for until this meeting Cripps had viewed him in a much less favorable light, closer to Nehru's and Krishna Menon's negative image of him.[11]

Two days later Gandhi arrived, "and he impressed upon me that he had not, of course, anything to do with Congress officially," Cripps reported. He then "expressed the very definite view that Congress would not accept the document." Gandhi read Cripps's offer as "an invitation to the Moslems to create a Pakistan." Cripps asked him "as a friend" what "the best method of proceeding [was]. He said he thought it would have been better if I had not come to India with a cut and dried scheme to impose upon the Indians." Cripps planned to publish his offer in a few more days, but Gandhi asked him not to do so. "I . . . asked him how, supposing Jinnah were to accept the scheme and Congress were not to, he would . . . advise me to proceed. . . . He said that in these circumstances the proper course would be for me to throw the responsibility upon Jinnah and tell him that he must now try to get Congress in either by negotiating direct with them or by meeting them . . . with myself. He thought that if it was pointed out to Jinnah what a very great position this would give him in India if he succeeded, that he might take on the job and that he might succeed." This was the first, but not the last, time Gandhi advised the British to offer Jinnah responsibility for running a free and united India. Gandhi saw it as the best chance, perhaps the only chance, of averting the disaster of partition inherent in the demand for Pakistan. He understood Jinnah well enough to know that if Cripps "pointed out to him" what "a very great position" success in such a negotiated settlement would give him, Jinnah might indeed "take on the job" and he "might succeed."[12]

But Cripps did not know either Gandhi or Jinnah well enough to appreciate what golden advice Gandhi had given him, so he never took it seriously. "I told him [Gandhi] . . . that I should have to make up my mind as regards acceptance or not within the next few days and that, if this scheme was not accepted, there would be no question of any other scheme . . . before the end of the war, and that those people who had taken the Congress point of view in the past, like myself, would not be in a position to exercise further influence in England as regards the solution of the Indian problem."[13] This was true. But Cripps mistakenly believed it would suffice to bring Congress around, for he still thought he knew Nehru and imagined that Nehru would fall into his arms, thanking him profusely. Gandhi had tried his best to warn him otherwise, without being overtly disloyal to his former disciple whom he had only just recently named as his successor to Congress leadership, reiterating again his own withdrawal from power and responsibility for leading Congress through the countless political mines on the hardest stretch of road ahead.

The day after meeting Gandhi, Cripps met with C.R., who advised him to use the term *free member state* instead of *dominion*, a word that had become such a red flag to Congress. C.R. also urged Cripps to do something about an "Indian Defence Minister if we hoped to get the consent of Congress." C.R. himself favored the entire scheme but told Cripps that "the crucial question was whether or not Nehru would accept. "He begged me to try to make some adjustment in the final paragraph in order to meet what he knew would be Nehru's reaction."[14]

The next morning, "Jawaharlal Nehru came to breakfast and I was glad to find that he met me in the same completely friendly atmosphere in which we had last parted," Cripps noted. Nehru told him that he had had "no conversations with his Congress colleagues yet because he had only just seen the document at Allahabad and had then to go into strict isolation in bed for two days to try to get over his fever and had seen no one."[15] Maulana Azad had, however, gone to Allahabad immediately after meeting with Cripps, to attend Indira's wedding. If he did not discuss his conversation with Cripps and the document Cripps showed him and read aloud to him, it would have been the first time he ever failed to report anything so important to Jawaharlal. Azad returned, in fact, to meet with Cripps a second time on March 28, "raising with me further explanatory points prior to the meeting of the Congress Working Committee tomorrow afternoon. He was depressed."[16] The Maulana had by now heard enough of Nehru's harsh reaction to know that there was no hope that Cripp's offer would be accepted by Congress's Working Committee.

Sir Stafford stressed to Nehru "the need of using this opportunity to arrive at a settlement," hoping by such frank advice to convince his friend of the wisdom of agreeing to what Cripps knew was the best, the only, deal that Churchill was willing to make at this time. He thought after breakfast ended that the only real difficulty would be the question concerning the minister of defense, which Azad had first raised. Cripps thought he saw victory on the horizon as he and Jawaharlal drove off to Birla House, where Gandhi was staying in New Delhi and where Azad was waiting for them with the rest of the committee. But Gandhi had nothing to add to his advice and so sent Cripps off for another few hours of discussion with Govind Pant and other Congress leaders, who tried to convince him that it was "undesirable to encourage . . . the Muslim League."[17] Cripps was much better at debate than any of them (except Nehru), and so he believed after an hour and a half of futile argument that they had begun "to see that . . . there was something to be said for the scheme." But they were only being polite. The real power of Congress remained in the other room, where Gandhi, Nehru, and Azad shook their heads sadly, deciding that this soft-spoken, pipe-smoking, well-meaning Englishman did not have enough power to give them what they wanted, which was nothing less than *purna swaraj*. He was Churchill's messenger, this Cripps, and had come out to try to shore up the crumbling, dying empire by offering them "promises" of half a loaf of *swaraj*. Later in the day, after the Working Committee met, Cripps met again with Nehru and Azad.

"I had about two hours with these two," Sir Stafford noted, "then explained to Nehru the general picture which I had given to Maulana Azad. . . . [F]our points that were raised were, first, . . . use of the word 'Dominion.' I pointed out that this was not a question of substance but of phraseology . . . to stop objections by the House of Commons. . . . I think they attached psychological importance to this but it was in no sense a major point." Cripps still thought he was virtually home with a deal in his attaché case. Then they went on to the question of the Indian states, "I repeated the arguments I had used to Mr. Gandhi and . . . said if they wanted these States to come in, as apparently they did, this was the only way of inducing them." He thought that they were all on the same wavelength of Oxford Union discourse, indeed, that they were speaking one language. He had as yet no idea that he would soon have to go home a dejected, defeated, worn, and weary man. "They then passed to the non-accession point. I explained the method of deciding this . . . [and] they seemed to accept . . . [it]. [W]e had a long argument . . . I pointed out that Nehru and other Congress leaders had said they were prepared to envisage the possibility of Pakistan and that was all the scheme was doing." He tried to sway them with logic, hoping to win them over by wisdom, for surely they were speaking the same language? "The general attitude of Nehru, who was tired and not well, was mild and conciliatory and he left me in complete doubt as to whether Congress was more or less decided not to accept it and that it was not worth arguing or pressing for any alteration or whether he was not inclined to press his particular objections in view of the general character of the scheme and its grant of free self-government in India."[18]

The next evening Nehru came to dinner. "I have never known him more serious and more worried about the Indian situation," Cripps noted after that long and painful meal. "He was very fully conscious of the acute dangers that would arise if the Indian leaders were not to participate at the present time in the rallying of India . . . but he stressed the very dangerous state of opinion arising from a multitude of causes all of which had exacerbated Indian opinion against the British. . . . The principal of these were . . . treatment of Indian refugees coming from the eastern seaboard . . . in comparison to the treatment of the European refugees . . . the growing unemployment in certain industries such as the weavers at Benares. . . . Fourthly, the growing disbelief in the capacity of Great Britain to make any defence effective in the light of . . . Malaya and Singapore . . . [and the] consequent lack of respect for police. . . . Fifthly, the tendency for a reversion to sympathy for Japan . . . on the ground of fellow Asiatic nations, though this was moderated by the pro-Chinese feeling in India. He was afraid that these various factors would make for a general breakdown. . . . He then told me of the difficulties in the Congress Working Committee, and conveyed to me the impression that they would not accept the proposals, largely, I think, though he did not say so precisely, due to . . . Gandhi. I gathered that he [Nehru] was doing his utmost to gain support for acceptance but felt that he was fighting a losing battle. I naturally stressed to him again the hopelessness of the situation if nothing was now done."[19]

Cripps continued to labor, meeting again and again with all the parties, earnestly, carefully, even cheerfully at times trying to explain the good offer he had brought from what had once, not so long ago, been deemed the world's most powerful body of leaders, with whom he also kept in touch by wires sent to Churchill, Attlee, and Amery in early morning or late at night. He never slept for more than four hours, and then fitfully. He thought that by convincing the viceroy and the commander in chief of the need to add an Indian member to the Viceroy's Executive Council for "defence supplies" or by the creation of a "ministry of defence coordination" he would win Congress support, for Azad made him think that mattered most to them. After his dinner with Nehru, Cripps even broadcast his views to all of British India, explaining that "we wanted to make it quite clear and beyond any possibility of doubt or question that the British Government and the British people desire the Indian peoples to have full self-government, with a constitution as free in every respect as our own. . . . We ask you therefore to come together . . . as soon as hostilities are over to frame your own Constitution. . . . I consider it a high honour that has fallen to my lot to be the messenger of the War Cabinet . . . accept my hand, our hand of friendship and trust. . . . Let the dead past bury its dead! and let us march together side by side through the night of high endeavour and courage to the already waking dawn of a new world of liberty for all the peoples."[20]

On April 2, 1942, Congress's Working Committee voted to reject the war cabinet's offer, and Nehru and Azad returned to Cripps to tell him that "the Committee recognise that future independence may be implicit in the proposals, but the accompanying provisions and restrictions are such that real freedom may well become an illusion," adding in more of Nehru's inimitable words: "The complete ignoring of the ninety millions of the people of the Indian states and their treatment as commodities at the disposal of their rulers is a negation of both democracy and self-determination."[21]

In Washington, D.C., that day, Sir G. S. Bajpai, the I.C.S. agent to the governor-general in the United States since 1941, wired Linlithgow, "I sought to get the President's reaction to . . . Cripps' proposals. Mr. Roosevelt seems to think plan . . . does not go far enough. His idea . . . seems to be that complete autonomy, including power to raise armies, should be given to provinces. I tried to explain dangers in time of war . . . but the President is not a good listener."[22]

"I am afraid it looks as if Gandhi had once again persuaded them that wrecking is the best policy," Amery told Linlithgow next day from the India Office in Whitehall after reading the Working Committee's resolution. "I am not sure that these people really want responsibility, and if we offered them the moon they would probably reject it because of the wrinkles on its surface. . . . [T]he more I look at the Resolution the more doubtful I am whether people of that type would ever run straight." Amery knew how sick and tired of Cripps Linlithgow felt by now. From its inception, the viceroy considered Cripps's mission an intrusion onto his turf, a diminution of his powers.[23]

On April 5, Nehru called on Colonel Louis Arthur Johnson, President Roosevelt's personal representative to India from March through June 1942, telling him that Congress wanted to hitch "India's wagon to America's Star and not Britain's. Colonel Johnson . . . told him that it was the President's determination . . . to support Great Britain to the end of the war and to preserve the integrity of the British Empire and that there must be no doubt in anybody's mind in India that America would see the war through. If America was convinced that Congress was solidly supporting the war effort, the sympathy she had . . . for Congress would continue: if, on the other hand, it appeared that Congress was . . . hedging or taking action to slow down the conclusion of the war . . . America would hate Congress. . . . Nehru then spoke of his belief . . . that Indians, particularly villagers, would make fine guerrillas. . . . Colonel Johnson . . . was much impressed by Nehru's charm of manner, grasp of history and logic and wide intellectual gifts." Sir Olaf Caroe reported what Johnson told him, in a "most secret" message to the viceroy on April 6.[24]

"As I told you yesterday," Nehru then wrote Colonel Johnson, "the new proposals made by Sir Stafford Cripps . . . were entirely unsatisfactory. Both the approach and the allocation of subjects were, in our opinion, wrong, and there was no real transfer of responsibility for Defence to representative Indians in the National Government. Such transfer is essential . . . for on it depends the full mobilization of the war-potential of the country." Cripps had worked day and night to get Wavell, Linlithgow, and Churchill and his entire cabinet to agree to add an Indian defence member to the viceroy's council. "The approach you have made," Nehru now told Johnson, "seems to us a more healthy one. With some alterations . . . it might be made the basis of further discussions. . . . [A] very great deal depends on the allocation of subjects between the Defence Department and the War Department. . . . There are many other important matters which have to be considered." Nehru was losing no time in hitching Congress to America's rising star. "We presume that the independent status of India will be recognized by the United Nations . . . [and that] it will greatly help our common cause and strengthen our bonds."[25]

Johnson now called on Cripps and they agreed to a new proposal for merging Indian and British defense. They went together to present their proposal to the viceroy. "Sir Stafford . . . said that he thought Congress would come in on this formula," Linlithgow reported. "I asked how Congress had come to know about this formula. Cripps replied that Johnson had shown it to them."[26] The viceroy was shocked. His own position "might well be rendered intolerable." He now viewed Cripps and Johnson as virtual co-conspirators of Nehru. First Nehru called on Johnson, then Johnson called on Cripps, and now suddenly everyone had agreed to a restructuring of his council and the entire chain of British Indian military command that the viceroy had not even seen. Cripps replied that "matters had reached a climax in which something had to be done." Johnson said that Congress was going to "settle on this formula . . . tonight." Linlithgow was livid. The Defense Department was to be handed over to a "representative Indian," and a new War Department would be created under Commander in Chief Wavell, whose functions were limited to carrying on the larger war. The Indian Defense member would be given charge of all other matters relating to Defense.[27]

Churchill convened the war cabinet, and the cable wires hummed between New Delhi and Whitehall all day and night, April 9–10, 1942. The Cabinet was "concerned about the Viceroy's position." Churchill feared that Cripps "in your natural desire to reach a settlement with Congress" was being drawn into "positions far different from any the Cabinet . . . approved before you set forth."[28] Cripps offered to resign. "Unless I am trusted I cannot carry on with the task."[29]

Then the Working Committee met and Maulana Azad wrote to Cripps on April 10, "The picture . . . placed before us is not essentially different from the old one. The whole object which we and I believe you have in view, that is to create a new psychological approach to the people to make them feel that their own National Government had come, that they were defending their newly won freedom, would be completely frustrated when they saw this old picture again with even the old label on. The continuation of the India Office which has been a symbol of evil to us would confirm this picture. It has almost been taken for granted . . . that the India Office would soon disappear, as it was an anachronism. . . . [T]he National Government must be a Cabinet Government with full power. . . . [T]he suggestions we have put forward are not ours only but may be considered to be the unanimous demand of the Indian people."[30]

After reading that letter, Cripps wired Churchill, "There is clearly no hope of agreement and I shall start home on Sunday."[31] He then wrote Azad: "I was extremely sorry. . . . Nothing further could have been done . . . without jeopardising the immediate defence of India. . . . You make . . . suggestions . . . for the first time . . . nearly three weeks after you had received the proposals. . . . [Y]ou suggest 'a truly National Government' be formed. . . . Without constitutional changes of a most complicated character . . . this would not be possible as you realise. . . . The proposals of His Majesty's Government went as far as possible."[32] Feeling much older, if no wiser, Cripps packed his bags.

Nehru, who had drafted Azad's letter to Cripps, now wrote another over his own signature to Franklin Roosevelt: "I know that you are deeply interested in the Indian situation today. . . . The failure of Sir Stafford Cripps' mission . . . must have distressed you, as it has distressed us. . . . [W]e have struggled for long years for the independence of India, but the peril of today made us desire above everything . . . to organize a real national and popular resistance to the aggressor and invader. . . . [T]he least that we considered essential was the formation of a truly National Government. . . . I do not wish to trouble you with details . . . I only wish to say how anxious and eager we were, and still are, to do our utmost for the defence of India . . . [and] of freedom and democracy. . . . We would have liked to stake everything . . . to fight with all the strength and vitality that we possess. . . . We are a disarmed people. But our war potential is very great. . . . A government divorced from the people cannot get a popular response. . . . To your great country . . . we send greetings and good wishes for success. . . . And to you, Mr. President

. . . assurances of our high regard and esteem." He sent it in Colonel Johnson's pouch.[33]

Cripps was exhausted but not embittered. On the eve of his departure from India he broadcast his farewell: "We have tried by the offer that I brought to help India along her road to victory and freedom. But for the moment past distrust has proved too strong to allow a present agreement. . . . Our effort has been genuine. No responsible Indian has questioned the sincerity of our main purpose—complete freedom for India. Such an effort . . . will prove to have been the first step along the path of freedom for India and of friendship between our two countries." Three years of war and frustration, sabotage and prison were to ensue because of that psychological barrier of "past distrust." Gandhi labeled Cripps's offer "a post-dated cheque on a bank that is failing." And in the aftermath of the failure of his mission, many Indians looked back with regret for having let that single dove of peace fly away, Britain's only offer of friendship and political cooperation to Congress for the remainder of the war. But it was too late for tears of regret; soon the days of iron and misery would begin.[34]

Roosevelt made a final attempt, through Harry Hopkins on April 12, to save the aborted mission, wiring Hopkins to inform Winston Churchill that "public opinion in the United States believes that negotiations have broken down . . . due to the British Government's unwillingness to concede the right of self-government to the Indians notwithstanding the willingness of the Indians to entrust to the competent British authorities technical military and naval defense control. . . . Should the current negotiations be allowed to collapse because of the issues as presented . . . and should India subsequently be invaded successfully by Japan with attendant serious defeats . . . it would be hard to over-estimate the prejudicial reaction on American public opinion. Would it not be possible, therefore, for you to have Cripps' departure postponed on the ground that you personally transmitted instructions to him to make a final effort to find a common ground of understanding?"[35]

Churchill wired Roosevelt from Chequers: "About 3 A.M. . . . Harry [Hopkins] and I were still talking. . . . [Y]our message to me about India came through from London. . . . Cripps had already left and all the explanations have been published. . . . You know the weight which I attach to everything you say to me, but I did not feel I could take responsibility for the defence of India if everything has again to be thrown into the melting-pot at this critical juncture. That I am sure would be the view of Cabinet and of Parliament. . . . Anything like a serious difference between you and me would break my heart and would surely deeply injure both our countries." Roosevelt let it go at that.[36]

"So blood and tears are going to be our lot whether we like them or not," Nehru told the press on April 15. "Let us not be afraid, let us not lose our anchor. . . . The turn of fate's wheel has brought this new ordeal to us, the last ordeal that can face a nation. . . . Our blood and tears will flow; it may be that the parched soil of India needs them so that the fine flower of freedom may grow again and its fragrance envelop the land. We shall pay the price." The final act in India's long struggle was about to begin.[37]

"Let us be realist," Nehru told a large meeting at Gauhati a week later. "Sir Stafford is mistaken in his impression that the negotiations have improved the situation in India. . . . [A]ll prospects of friendship have been lost! I know the English people . . . it is a natural habit of the English people to feel that God has given them the right to give paternal treatment to others. . . . In fact, Sir Stafford knew nothing about India. . . . If I would have been in charge of Defence, I would have armed the whole nation. . . . We must get prepared . . . it will be a real war effort."[38] Nehru told friends he had "liked Cripps as a man" but considered him "a somewhat muddleheaded politician" and was "surprised at his woodenness and insensitiveness, in spite of his public smiles."[39] It was never easy for Jawahar to respond positively to paternal treatment.

The A.I.C.C. met in Allahabad for the last few days in April and on May Day agreed on a resolution that Nehru had drafted. "The proposals of the British Government . . . have led to great bitterness and distrust of that Government. . . . They have demonstrated that even in this hour of danger . . . the British Government . . . refuses to . . . part with any real power. . . . If India were free she would have determined her own policy and might have kept out of the War. . . . A free India would know how to defend herself in the event of any aggressor attacking her. . . . Not only the interests of India but also Britain's safety . . . demand that Britain must abandon her hold on India. . . . In case an invasion takes place, it must be resisted. . . . The Committee would therefore expect the people of India to offer complete non-violent noncooperation to the invading forces." Congress thus fell back on what had long been Gandhi's program and position. Had Nehru accepted Cripps's offer, he might well have been asked to become India's first minister of defense, a job that he knew was anathema to the Mahatma. Now there was no need of military cooperation at any level, and soon he would be back in prison, able to finish his partially written *Discovery of India*.[40]

Gandhi was pleased. He had encouraged Vallabhbhai to abandon Congress if it did not pass his noncooperation resolution. C.R., however, felt obliged to resign from the Working Committee and left Allahabad for Madras, where he rallied provincial support for the British war effort and also conceded that "Pakistan in principle" made sense, thereby winning the support of the Muslim League in Madras, where he pressed for the restoration of a responsible provincial coalition. But the rest of the committee closed ranks behind Mahatma Gandhi, who called on the British more openly now to withdraw entirely. "I am convinced that the time has come," Gandhi reported in early May, "for the British and the Indians to be reconciled to complete separation from each other. . . . It will be the bravest and the cleanest act of the British people."[41]

From Madras C.R. reiterated that the Muslims' right of self-determination must be recognized by Congress if it wished its own demands for self-determination to be granted by the British. Nehru responded sharply. "Rajaji has acted most irresponsibly . . . dividing the Congress into two camps. My only satisfaction is that Rajaji has not succeeded. . . . The very idea of dividing India is revolting. . . . [T]hose who talk of Pakistan these days are befooling the people. . . . Being a socialist, I think that in the present age no one can afford to live separately; [rather] . . . countries must act unitedly. . . . I have been looking forward to the day when we shall have a federation of Iran, Afghanistan, India and China."[42]

The governor of Punjab was now worried about Master Tara Singh and his Khalsa Sikh followers raising the cry of "Khalistan," a Sikh "land of the pure," in emulation of the Muslim League's call for Pakistan. "As you are well aware," Governor Glancy wrote the viceroy, "the practical objections to 'Khalistan' are even greater than those which lie in the path of Pakistan. . . .[T]here is not one single district in which the Sikhs command a majority. . . . Sikhs are still clamouring for what they profess to regard as their due representation on the Governor-General's Executive Council. . . . [N]ames most commonly mentioned as suitable candidates . . . are . . . Sardar Baldev Singh . . . occasionally Sardar Buta Singh." Baldev became Nehru's defense minister, and Buta later became Indira's home minister.[43]

By the end of May, Nehru was in Lucknow, attending his United Provinces Congress meeting, which endorsed his and Gandhi's Allahabad resolution. "I am so sick of slavery," Nehru told his comrades that afternoon, "that I am even prepared to take the risk of anarchy."[44] He had just been to Wardha, where he and Gandhi discussed plans for the "Quit India" Satyagraha campaign that would soon be launched. At a public meeting in Lucknow later that same evening, Nehru asked, "Where do we stand? We in India . . . stand in the midst of a river surrounded by crocodiles with their mouths open. . . . We have become helpless."[45] That same day in Wardha, Gandhi's *Harijan* insisted that "withdrawal of the hated power is the only way to rid the land of the debasing hatred."[46] Asked by anxious British friends to whom they should "leave India" if they were to take his advice and simply withdraw, Gandhi told them, "Leave India to God.

If that is too much, then leave her to anarchy."[47] The viceroy was now informed that the "air is thick" with "rumours that Gandhi will start a new movement."[48]

Amery told Linlithgow, "If Nehru and Co. are really prepared, Gandhi consentient with them, to embark on a policy of real mischief, then I hope you will not hesitate or lose a moment in acting firmly and swiftly. Don't refer to me if you want to arrest Gandhi or any of them, but do it and I shall back you up." With such firm support from the India Office in Whitehall, the viceroy was ready to act swiftly on the eve of the new Satyagraha campaign.[49]

In June, Nehru alerted his closest comrade on the frontier, Mohammad Yunus, that "the next two weeks will . . . bring far-reaching developments in India, . . . I hope Badshah Khan [Abdul Ghaffar Khan, the "frontier Gandhi"] realises this. We have to keep ready."[50] The "red shirt" Pathans of the North-West Frontier, Badshah (Emperor) Abdul Ghaffar Khan's tribal followers, would, Nehru hoped, rise up whenever he gave the signal. Yunus was Nehru's messenger to Badshah Khan and remained a close friend of the Nehru family for the rest of his long active life. On July 1, 1942, Gandhi wrote to President Roosevelt, "I hate all war. . . . I venture to think that the Allied declaration that the Allies are fighting to make the world safe for freedom of the individual and for democracy sounds hollow, so long as India and, for that matter, Africa are exploited by Great Britain, and America has the Negro problem. . . . If India becomes free, the rest must follow . . . I have suggested that, if the Allies think it necessary, they may keep their troops, at their own expense in India, not for keeping internal order but for preventing Japanese aggression and defending China. So far as India is concerned, she must become free."[51] Gandhi sent his letter with Louis Fischer, who delivered it to the White House. Fischer had been interviewing Gandhi, Nehru, and others for his biography of *Mahatma Gandhi.*

In early July the Working Committee met again in Wardha. "On 5 July . . . Gandhiji spoke to me for the first time about the Quit India Movement," Azad recalled. "I felt that we were facing an extraordinary dilemma. . . . [T]he Japanese had occupied Burma and were advancing towards Assam. . . . It seemed to me that the only thing we could do was to wait upon the course of events . . . Gandhiji did not agree. He insisted that the time had come when . . . the British must leave India."[52] Gandhi's resolution calling on the British to withdraw from India was debated by the Working Committee for five days and passed on July 13, 1942. "In making the proposal for the withdrawal of British rule from India, the Congress has no desire whatsoever to embarrass Great Britain or the Allied powers in their prosecution of the War," Nehru wrote two days later. "Nor does the Congress intend to jeopardise the defensive capacity of the

Allied powers."[53] A meeting of the A.I.C.C. was called for Bombay on August 7, to which Nehru would go with Indira and Feroze, recently returned from their Kashmir honeymoon.

"The only course open to the country is to fight British imperialism," Nehru told a large crowd in Meerut on July 19.[54] He was tired of all the meetings and endless hours of wrangling, listening to C.R. and Bhulabhai and Sarojini express their anxieties and fears and doubts. It was time for action, he urged Gandhi. The old Mahatma's feet hurt, and he was weak, but he agreed. Krishna Menon kept cabling messages of "gravest" concern, for he had become an air-raid warden in London, and argued that since the Soviet Union's invasion it had been transformed from an "imperialist war" to a "people's war." He could not understand what was wrong with Congress and Nehru. "Earnestly desire to give every help but the very gravity of the situation demands complete reversal of policy in India," Nehru tried to explain to him. "Otherwise progressive deterioration and desperation . . . British policy statements have infuriated people."[55]

"We will do or die!" *(Karenge ya marenge!)* was the slogan Gandhi coined for the "Quit India" movement he launched in early August. "Quit India" was used by thousands, later millions, of Indians whenever they saw an Englishman, an American soldier, or any foreigner during the remaining years of the war. Small children would race after them, shouting "Quit India!" It became the most popular slogan ever coined by Congress, the culmination of more than half a century of Indian nationalist agitation and conflict.

"We have been demanding freedom for many years past, and it is our birthright," Nehru told a cheering crowd in Bombay, echoing the words of Lokamanya Tilak. "People talk of giving us a promissory note . . . of future freedom. But . . . [t]he forthcoming meeting of the A.I.C.C. is going to be the most momentous and important session. . . . The Working Committee has taken a decision that this country should no longer live in slavery. . . . [A] struggle for freedom should be launched. It is a question of life and death . . . for the entire country. . . . We have now very rightly decided that it is much better to fight with valour and go down rather than keep quiet."[56] Two days later at the A.I.C.C. meeting, Nehru argued that "the movement contemplated is not for merely achieving national ends but for achieving world freedom. The Congress is plunging into a stormy ocean and it would emerge either with a free India or go down. Unlike in the past, it is not going to be a movement for a few days, to be suspended and talked over. It is going to be a fight to the finish. The Congress has now burnt its boats and is about to embark on a desperate campaign."[57]

Gandhi spoke last that night of August 8. "The Quaid-e-Azam [Jinnah]) himself was at one time a Congressman," Mahatma Gandhi said, still hoping to secure Hindu–Muslim unity, knowing it to be the prerequisite to full freedom. "If, today, the Congress has incurred his wrath, it is because the canker of suspicion has entered his heart. May God bless him with long life, but when I am gone, he will realize and admit that I had no designs on Musalmans. . . . Where is the escape for me, if I injure their cause or betray their interests?" In that outpouring of his heart, Gandhi said of the Pakistan demand and possible partition: "To demand the vivisection of a living organism is to ask for its very life."[58]

The Congress meeting ended after midnight. Gandhi, Nehru, and the entire Working Committee all were arrested before dawn on Sunday, August 9, 1942. "Everyone is free to go the fullest length under ahimsa," Gandhi left as his final message for millions of followers, before he was driven off in police custody. "Complete deadlock by strikes and other non-violent means. Satyagrahis must go out to die and not to live. They must seek and face death. It is only when individuals go out to die that the nation will survive. *Karenge ya marenge.*" He himself soon embarked on a "fast unto death" in the Aga Khan's Poona Palace, where Gandhi, his frail wife Kasturbai, Sarojini Naidu, and Admiral Slade's daughter Madeleine "Mirabehn" (beloved sister), as Gandhi called her, were driven together. Nehru, Vallabhbhai, and the others were taken to Ahmadnagar Fort.[59]

"The Palace was constructed in one long line, with big verandahs all around," Mira recalled. "The ground floor . . . was for us, and we were asked to select our rooms. Bapu chose a small room next to the bathrooms, and refused to consider any of the larger ones." Churchill and Linlithgow favored deporting Gandhi and Nehru and several other leaders of the Congress, but the Viceroy's Council and leading Indian police advised against such a move, apprehensive that it would cause even more violence than was now anticipated, and which started very soon after those predawn arrests. The violence continued destroying vital infrastructure and individuals who supported the Raj throughout the war. Gandhi's faithful secretary, fifty-year-old Mahadeva Desai, had also been brought to the Aga Khan's palace-prison but died of a heart attack within days of arriving there. "Everyone was speechless," Mira remembered, as they all gathered around the dead body, "but the thought of the irreparable loss it would mean to Bapu rent our hearts . . . Bapu remained almost completely silent, as if passed into a world of prayer."[60]

Nehru had slept in his sister Betty's home in Bombay, with Indira and Feroze taking Betty and Raja's bedroom. At 5:15 A.M. on August 9, "Indu comes into my room and wakes me: 'The police have come.' So that's that. The invitation has come. . . . What was I to do? . . . Should I submit to this jail going quietly or refuse to go? Refusal would ultimately lead to being carried out—an undignified exit. . . . So I thought rapidly—

meanwhile, anyway, I might as well shave and have a bath." Nehru then wrote a few letters and had some tea. An hour later he went to meet the police officer, who had been waiting with his warrant in the sitting room. This time all the Congress leaders were arrested under wartime Defence of India Rules, which eliminated any need for a trial or even an appearance before a magistrate. "And so goodbye to Indu & Betty & Feroze. . . . When would we meet again I wondered. When would I see Indu again? . . . That morning I was to have unfurled the National Flag at the Congress *pandal* and later to have addressed the students there. Immediately after there was to have been a workers' meeting at Birla House. . . . Then the Working Committee. In the evening . . . a public meeting in Shivaji Park." He was driven in a waiting taxi to Bombay's central station, Victoria Terminus, where a special train was waiting, with "plenty of police . . . a big haul . . . Bapu was in the train." It left at 7 A.M. and sped toward Poona, "hardly stopping," Nehru recalled. "We had breakfast in a restaurant car attached!"[61]

Nehru was taken with nine other members of the Working Committee to the Old Mughal Empire's Fort, a cheerless rock pile in sweltering Ahmadnagar. Some barracks inside the fort had been "converted for our use." They were large rooms with thin wooden partitions, more of a dormitory than the isolated prison cells he had been used to and had expected again. To Nehru, having so many companions was annoying, "seems quite a crowd," he reflected. "This leads to many common services and time is spent in talks, games etc. . . . not as much time available for quiet reading or like work." For the first month of this incarceration there were no newspapers, no letters, and no visitors. The very venue of their incarceration was itself a secret, as the government feared that if news of their being in Ahmadnagar Fort was broadcast, a mass march to the prison might follow. "We were hermetically sealed. . . . We fretted of course and each one of us reacted according to his temperament. . . . Some wanted a limited fast. . . . I had three books with me when I came! Plato's *Republic,* Proust's *[A la recherche du temps perdu]* and Lin Yutang's *With Love and Irony.* In a slow & leisurely way I read through Plato again."[62]

Nehru received his first letter, from Betty, on September 4, learning then that Indira was living in Anand Bhawan. He did not feel up to writing to her, however, but was thinking of it on September 12, 1942, when a newspaper was brought to him by the jailer, reporting her arrest and that of Feroze as well. "So that's that. I was a little surprised to find that Feroze had survived for so long. Churchill's speech also in today's paper. Straight from the shoulder. Offensive to the Congress . . . his usual style. . . . On the whole I am pleased. Let there be no compromise. There can be none and it is about time that people realised it."[63]

23

Last and Longest Incarceration

1942–1945

NEHRU REMAINED locked away in the sweltering, remote Ahmadnagar Fort prison from August 9, 1942, until March 28, 1945. It was his longest, harshest period behind British bars, ending only on the eve of Germany's surrender. But his fortitude, stamina, and spirit remained uncrushed by all the imperial pressure that Churchill, Linlithgow, and Wavell brought to bear on him.

"How does one spend one's time in prison?" he wrote, in his first letter to Darling Indu. "There is reading and writing of course. But apart from this there are innumerable other activities . . . watching ants and wasps and various insects. And then there is the sky . . . Although we live in a kind of Plato's cave, yet we have this sky over our yard and a lovely sky it is with fleecy and colourful clouds . . . and, now, brilliant star-lit nights. What a fascinating world this is if only we kept our eyes and ears open."[1] A month later he wrote again "From the Unmentionable Place!" inquiring as to how she occupied herself in prison, advising "gardening. Even apart from the beauty that flowers give . . . the joy of tending them and seeing them grow . . . is fascinating business. I love to play about with the soft warm earth. . . . [T]here is a certain psychic satisfaction about the earth, and we, who have cut ourselves away from it, miss this very essential thing; if I read or write all day, there is something that I lack, and this contact with the earth goes some way to supply it . . . I suppose you can easily get flower seeds. . . . Sweet peas are, I think, the most suitable flowers for one in jail. They appeal equally to the sight and the sense of smell."[2]

"Tomorrow is my birthday and I feel a little overwhelmed and somewhat resentful of the fact that I have completed 53 years. Time sweeps on and the energy and vitality of youth slowly ebb away. . . . What of the big things and brave ventures which have filled my mind these many years? Shall I be capable of them when the time comes?"[3] This was the eighth birthday he had been forced to celebrate in jail. His comrades gave him bouquets, and Mahmud, his old Cambridge friend, also gave him a bag of dried fruit and a "long letter . . . full of affection, rather pathetic and sentimental."[4] Another prison companion managed to bribe the jailer to buy him a cigarette holder, since Nehru was hooked on smoking. Govind Vallabh Pant presented a huge garland he had made, and "the Maulana [Azad] garlanded me." They even decorated the bare table with a bowl of roses into which they stuck a paper national flag and sat down to his special birthday meal.

On January 26, 1943, they celebrated the thirteenth anniversary of Independence Day. They all stood up for a few minutes and silently repeated the pledge. "We did not have the wording . . . nor was it necessary. Vallabhbhai then spoke for a few minutes and I followed. . . . I thought of . . . what was happening all over India today. Surely there would be a celebration in many places in spite of all the terrorism of government . . . of lathi-charges, shootings, deaths, injuries and arrests. What a terrible burden our friends and colleagues outside have to carry!"[5]

In the streets of Bombay, throughout Gujarat, in every town and village of the United Provinces and Bihar, from Delhi to Calcutta, violence and counterviolence had become daily occurrences. The government's predawn preemptive arrests had paralyzed young Congress cadres and loyal followers for the first few days in August, but as whispered word of what had happened to their Mahatma and Panditji and Vallabhbhaiji and the Maulana reached the streets and student hostels and raced through crowded bazaars and village fields, unarmed Indians everywhere clenched their fists and vowed "To Do or Die" in this last, longest struggle to force those "British bastards" to "Quit India!" None of the viceroy's slickly written rationales fooled any one. Linlithgow summoned to Delhi all the elected members of his Central Legislative Assembly in early September to explain the government's policy and position.

"Your excellency has, no doubt, convened the present session to get an endorsement of the repressive policy of your Government and to have it proclaimed to the world that all was quiet on the 'Indian front,' " wrote Mohan Lal Saksena, one of the few Congress members of the Assembly not yet jailed, being so mild and soft spoken a gentleman. "This is indeed the truest reproduction of . . . Hitlerian tactics. The promulgation of lawless laws and ordinances, the gagging and throttling of all self-respecting and independent newspapers, the suppression of all news and views, except those doled out by the provincial . . . Goebbels, the internment without trial of thousands of patriots, the banning of Congress and . . . meetings, lathi charges, firing, public flogging, shooting resulting in loss of life and limb to thousands and now summoning an attenuated legislature to ditto Your Excellency's Government are but Maxwellian rendering of Nazi methods designed to terrorise and cow down a whole people fired with the spirit of freedom and democracy and which the allies profess to be fighting for."[6]

By December 1942, Governor Lumley of Bombay reported a food crisis. Bombay's wheat supply was down to just a few days, and all the bakeries in Poona were likely to close down in a day or two. The wartime overloading of all communications was blamed, but the governor also noted growing civil disobedience and disturbances. He reported in confidence to the viceroy that even if Bombay received the promised but as yet undelivered fifty thousand tons of wheat, there was a very real danger that by February, "we shall not be able to feed the population of Bombay City."[7] Bombay barely survived. And in 1943 Bengal suffered its worst famine in recent history, caused more by the total breakdown of transport and communications in that vast province during the war than by monsoon rains. An estimated three million Bengalis had died of starvation by December 1943. Agatha Harrison appealed, though in vain, for Gandhi's release, but Linlithgow did not feel justified in making a "special exception at this stage," he told Amery in January 1943, "and I propose therefore to mark time."[8]

Nehru and Vallabhbhai kept healthy, but the rest of the Ahmadnagar Fort dozen had started to break down. "Mahmud is a bundle of diseases but his chief trouble now relates to his eyes. Pantji & Kripalani also have their troubles," Nehru reported, also noting that Maulana Azad had lost twenty-two pounds. In the other prison, Gandhi now decided it was time for him to risk losing more than that.[9]

On New Year's Eve 1943 Gandhi had written to Linlithgow explaining that he had waited patiently for six months and adding that he was now losing his patience and that "the law of *Satyagraha* . . . prescribes a remedy in such moments of trial. . . . [I]t is 'crucify the flesh by fasting.' That same law forbids its use except as a last resort. I do not want to use it if I can avoid it. . . . [C]onvince me of my error or errors and I shall make ample amends."[10] Linlithgow drafted a reply on January 6, which he first sent to Amery for approval. "I was glad to have your letter," the viceroy wrote. "I have been profoundly depressed during recent months first by the policy . . . adopted by the Congress in August, secondly, because while that policy gave rise . . . to violence and crime . . . no word of condemnation for that violence and crime should have come from you, or from the Working Committee. . . . When I think of these murders, the burning alive of police officials, the wrecking of trains, the destruction of property, the misleading of those young students, which has done so much harm . . . to the Congress . . . the story is a bad one . . . . But if I am right in reading your letter to mean that in the light of what has happened you wish now to retrace your steps and dissociate yourself from the policy of last summer, you have only to let me know."[11]

On January 19, 1943, Gandhi replied: "Your letter gladdens me to find that I have not lost caste with you. . . . The inference you draw from my letter is, I am afraid, not correct. . . . If I could be convinced of my error or worse . . . I should need to consult nobody . . . to make a full and open confession and . . . amends. But I have not any conviction of error. . . . Of course I deplore the happenings . . . since 9th August last. But have I not laid the whole blame for them at the door of the Government of India? . . . I am certain that nothing but good would have resulted if you had stayed your hand and granted me the interview, which I had announced, on the night of the 8th August, I was to seek. But that was not to be. . . . If you want me to make any proposal on behalf of the Congress you should put me among the Congress Working Committee members. I do plead with you to make up your mind to end the *impasse*."[12] But Linlithgow's mind was made up. "I am still, I fear, rather in the dark," he wrote Gandhi a week later. "I am very glad to read your unequivocal condemnation of violence. . . . But the events . . . even the events that are happening today show that it has not met with the full support of . . . your followers. . . . [This is] no answer to the relations of those who have lost their lives . . . lost their property or suffered severe injury. . . . We are dealing with facts in this matter, and they have to be faced."[13]

Gandhi now decided to "fast to capacity" for three weeks, starting on February 9, 1943. He would take no solid food, only water and fruit juice. Churchill considered Gandhi a "charlatan" and by now hated him almost as much as Hitler or Tōjō. On the very eve of Gandhi's fast, Churchill sent a "most secret" encripted message to the viceroy urging him against any show of leniency toward Gandhi, "which I fear would bring our whole government both in India and here at home into ridicule. . . . I ask this as a friend and also because I am convinced that such an episode would be a definite injury to our war policy all over the world which is now moving forward victoriously after so many perils have been surmounted by British resolution."[14] Linlithgow agreed with his prime minister and felt no compunction about letting Gandhi "fast to death," but his council was unanimously opposed to keeping the Mahatma in detention once he started fasting. On Sunday, February 8, the war cabinet met and

"Winston . . . launched out on the Gandhi subject at once," Amery informed Linlithgow. "At first he [Churchill] continued, as is often his habit, muttering away his dissatisfaction, but giving me the impression that he was going to agree with a shrug of the shoulder. Presently, however, he warmed up and worked himself into one of his states of indignation over India. . . . [O]ur hour of triumph everywhere in the world was not the time to crawl before a miserable little old man who had always been our enemy."[15]

Seventy-three-year-old Gandhi began his fast on February 10, 1943, remaining in the Aga Khan's palace-prison. A few days later Nehru, Azad, and others learned of the fast, and Nehru drafted a letter to Linlithgow for the Maulana to sign as the Congress president. "There is a vital difference of opinion between you and us in regard to many matters. . . . You . . . say that you 'have ample information that the campaign of sabotage has been conducted under secret instructions circulated in the name of the All-India Congress Committee.' What your information is we do not know, but we do know . . . that the A.-I.C.C. at no time contemplated such a campaign, and never issued such instructions. . . . You mention that an underground Congress organization exists now and that the wife of a member of the Congress Working Committee is a member of it . . . 'actively engaged in planning the bomb outrages and other acts of terrorism.' Congressmen, no doubt, consider it their duty to carry on civil resistance."[16]

Asaf Ali's wife, Aruna, was the underground "terrorist" that the British sought in vain until the war's end. Asaf Ali had been worrying himself sick about her, Nehru confided to his diary: "Aruna has gone up greatly in my estimation. What she has done or is doing, I do not know. But deliberately she has chosen the harder, riskier, more dangerous path when it was easy enough for her to go to prison. She has shown grit & determination & rare courage. May it be well with her! If there are many others like her, it will be well with all of us."[17] Whenever negotiations broke down, Nehru had long favored violence as the best strategy for getting rid of the British. After drafting and sending that letter to Linlithgow, Nehru wrote in his diary, "For my part I got a little tired of the discussion and felt that it did not matter very much after all what was written. . . . We have gone far beyond the letter stage. Other forces will decide the issue. . . . I can only guess but it seems to me that India is going the way of Ireland in the later Sinn Fein days."[18]

The government now published Gandhi's prison letters to the viceroy and Linlithgow's responses. To Nehru, Gandhi's tempering letters were "disappointing. . . . What a difference in tone from his pre-arrest utterances! There was fire then and sparks of electricity flew about from his words and phrases. Now there is . . . a tone of justification, of defence, of legal reasoning. It is not the ringing language of defiance, come what may. As usual, perhaps he reflects the mood of the country." Nehru, as usual, marched to his own drum, militant, eloquent, passionate, closer in some respects to Winston Churchill's than Mahatma Gandhi's. "What will happen if Bapu dies? An end of an era in India and a bitterness that will eat into the very soul of India," Nehru mused. "It will be war in every way then, war continuously whatever the ups and downs might be. . . . If he survives . . . what then? Anti-climax!"[19]

That same day, February 17, 1943, Sir Homi Mody, N. R. Sarkar, and Dr. M. S. Aney all resigned from the Viceroy's Executive Council, to protest Gandhi's continued detention while his health was so precarious. Dr. B. C. Roy came from Calcutta and was permitted to visit the Mahatma, to check his vital signs. On February 19, William Phillips, President Roosevelt's new personal representative to India, handed the viceroy a telegram from the U.S. secretary of state, Cordell Hull, which stated: "President Roosevelt and I suggest that you . . . convey to him [the viceroy] . . . our deep concern over the political crisis in India. Please express . . . our hope that a means may be discovered to avoid the deterioration of the situation . . . if Gandhi dies."[20] Gandhi's health had deteriorated overnight, and on that same day Linlithgow wired to Amery that "the position is now definitely serious. In the event of his death I am asking Governor of Bombay to telegraph to you code word 'EXTRA.' "[21]

Gandhi had lost fourteen pounds since he started his fast; his pulse was faint; and he could not leave his cot. Madame Chiang Kai-shek, "over whom all America has gone crazy," was "staying at the White House" at this time, Amery reported to Linlithgow, and he feared that she was strongly influencing Roosevelt in favor of India and Gandhi, as well as Nehru. "*Cherchez la femme.*"[22] In a subsequent "private and secret" wire, Amery told Linlithgow, "I am irresistibly drawn to the conclusion that Madame Chiang Kai-shek and Mrs. Roosevelt between them have got at the President. . . . [I]f . . . . it once leaks out that the President has intervened and that we have turned him down—which we must do whatever happens—the result will be most unfortunate . . . like Gandhi's death. . . . I have of course immediately sent on your telegrams to Winston urging him to take up the matter at once with the President. The trouble is that the poor man is in bed with a really bad chill: his temperature was 102 last night. . . . On the other hand, Anthony [Eden] is off for America . . . and I think he can be trusted to handle the matter firmly as well as tactfully. . . . So we must hope for the best."[23]

"Three days ago the end seemed very near," Nehru wrote on February 25. "I came to the conclusion that there was no hope of survival. Indeed, one evening, a vague rumour spread that he had actually died. It was soon contradicted. . . . So we spent those long & weary days almost prepared for the worst and yet dreading the future. It was no good finding fault with what Bapu had done in the past. . . . Then suddenly. . . [h]e seemed to have turned the corner."[24]

On February 28, Gandhi looked "more cheerful," his bedside disciples reported. Sapru and Jayakar convened a meeting in Delhi, attended by most of India's leading politicians and industrialists still out of jail, appealing to the viceroy to release Mahatma Gandhi "immediately and unconditionally."[25] Sapru wired the message to Churchill a day later.[26] Neither Churchill nor Linlithgow budged. Churchill cynically insisted that Gandhi's fast was a hoax, its "climax" staged from the start of the "eleventh day onwards." "With all those Congress Hindu doctors round him it is quite easy to slip glucose or other nourishment into his food."[27] To Field Marshal Jan Smuts in South Africa, Churchill wired, "I do not think Gandhi has the slightest intention of dying, and I imagine he has been eating better meals than I have for the last week."[28]

On the morning of March 3, 1943, Gandhi broke his long fast, sipping six ounces of orange juice from a glass handed to him by his frail wife while his disciples gathered around, reciting stanzas from Tagore's *Gitanjali* and singing *Lead Kindly Light*. "The imprisonment of Gandhi is the stupidest blunder the Government has let itself be landed in by its right wing of incurable diehards," George Bernard Shaw commented. "The King should release Gandhi unconditionally as an act of grace . . . and apologize to him for the mental defectiveness of his cabinet."[29] In November 1942, Churchill had replaced Cripps as leader of the Commons with Eden, removing Cripps from the war cabinet entirely and appointing him minister of aircraft production.

All of India's moderate leaders now urged the government to reconsider its policy and to set Gandhi free unconditionally. When Nehru learned of their request he wondered: "How far has Bapu approved of the steps that are being taken? There appears to be more than a hint in Rajaji's [C.R.'s] utterances that he [Gandhi] has come to some understanding with him. I do not like all this. There is trouble ahead—Plenty of trouble, I reckon."[30] Any one else's negotiating with the British always troubled Nehru. He feared an imminent retreat to nonviolence or no resistance. "It hardly seems possible that we shall ever go back now to our ways & methods of the past 23 years, even if we want to," he noted. "The war itself breeds the mentality of violence. How can we ask people to defend the country against a foreign aggressor . . . and, at the same time, ask them to be nonviolent in their struggle for national freedom? . . . What of Bapu? . . . I have a feeling that I shall not see him again. I do not think he will survive this imprisonment. . . . Fate—destiny—Am I growing into a fatalist?"[31]

Two weeks later Nehru lost his famous temper with Mahmud, who was teaching him Urdu and happened to answer "one or two unthinking questions by me . . . about the political situation. What Mahmud said hurt and angered me suddenly and I flared up, using rather hard language. I proclaimed further that as we differed so radically in regard to basic political questions. . . . I shall in future not discuss politics with him. . . . I was sorry for my outburst. Nevertheless it is amazing how weak and flabby Mahmud is. (As for Asaf he has completely gone to pieces). A curious irony of fate to put us all together." Maulana Azad continued to shrink away, losing over thirty pounds and mourning in silence now for his wife, who died in April in Calcutta. Madame Chiang had been reported to have urged Nehru's release, calling him a man of "world vision," and that cheered him a bit, but then the *Times of India* had the temerity to suggest that the government might try to be "soft & kind to me—I am a good boy and all that . . . hinting that some Congressmen should form a new party. . . . A faint suggestion that I might do this job! . . . I would sooner consign myself and everybody else to hell before I did such a thing."[32]

In late April Nehru heard a rumour that Gandhi planned another fast, this time "unto death," starting on August 9, 1943, unless the situation changed. He read a report of Jinnah's speech to the Muslim League, finding it "blatant, vulgar, offensive, egoistical. What a man! And what a misfortune for India and for the Muslims that he should have so much influence! I feel depressed about it. . . . There is no way out, so far as I can see, except a real bust-up in India with all its horrid consequences. We cannot build anew on the old foundations." So the die of partition was cast by April 1943, four years and four months before it actually took place. Nehru was no more willing to try to work in harness, administratively, with Jinnah, than he was ready to work with Linlithgow.[33]

Nehru's peculiar tolerance of physical pain was also tested during April, as he wrote to inform Indira. "You ask me about my arm . . . I tried various massages, ointments . . . but the pain continued. . . . There is electric current here and . . . [i]t was suggested that I might . . . indulge in this for my arm. . . . [A]lways agreeable to having new sensations & experiences I readily agreed. . . . So a mild current was passed through my arm for a few minutes. No obvious results . . . a few days later, a stiffer and a larger dose. . . . I was asked: Can you bear it? An odd question . . . I can bear a good bit in the way of pain. . . . So I bore it without a whisper. When it was all over and my arm was unwrapped, it was discovered that my skin and some tissues had been burnt up. . . . It was entirely my fault for quietly submitting to this ordeal. . . . Anyway it took about three weeks for this burn to heal and I have got a biggish mark on my forearm which I am likely to carry to the end of my days . . . a permanent souvenir of this place."[34]

He read many books in prison, including Sri Aurobindo Ghose's *Essays on the Gita.* "Am I going metaphysical?" he mused. "I do not think so. Yet there is a desire to go behind the veil . . . a growing feeling that . . . all this war and conflict, is just puppet show. How little we influence events! We are carried by forces we do not understand and hurled hither and thither. Is there nothing behind all this—no rhyme or reason? . . . I feel that this war, and all its secondary consequences, are all part of a mighty revolution. . . . And yet, and yet—what an odd way of doing things, what terrible waste & misery, what hatred. . . . Here in India the iron has gone deep down into the soul of the people. . . . [It is] fortunate that I am physically well and even mentally fairly healthy. . . . Aurobindo talks of the present, that razor's edge of time & existence which divides the past from the future, and is . . . as 'the pure and virgin moment.' I liked the phrase. The virgin moment emerging from the fog of the future in all its purity, coming in contact with us, and immediately becoming the soiled & stale past. Is it we that soil it and violate it? Or is the moment not so virgin after all for it is bound up with all the harlotry of the past?—Determinism! . . . Karma—"[35]

Lord Halifax, then Britain's ambassador to Washington, wired Amery to tell him about having lunched with President Roosevelt, who "took the opportunity" to speak of Linlithgow's refusal to let Phillips see Gandhi. Halifax defended the viceroy's refusal, reporting that "the President did not dissent, and indeed appeared to agree. I then told him that I had had a report . . . that Gandhi was going to have another fast, this time unto death. President said he had had this too. . . . President did not seem at all excited about the possibility. I also told him that the Viceroy had complained very strongly of alleged remarks of Madame Chiang Kai-shek. . . . President also told me that he had advised Madame . . . if she went back to China by India, not to stop there and ask to see Gandhi."[36]

Amery had just proposed to Churchill that Anthony Eden be sent out as the next viceroy, when Linlithgow's term expired next year. Or "if you feel that . . . an impossible solution . . . the next best choice is Sam Hoare."[37] Churchill was not happy with either, however, and "dashed off again," as Amery soon reported to Linlithgow, "taking Wavell . . . with him, and I imagine the whole question of the Burma and Beyond campaign will be considered over again in consultation with the Americans."[38] On that dash to Washington, India's fate was sealed for the rest of the war, with Nehru destined to remain behind bars until its end in Europe, for Churchill decided then that Field Marshal Archibald Percival Wavell, his iron-jawed, one-eyed commander in chief of India, was precisely the sort of no-nonsense viceroy that India needed, not the honey-voiced Sir Anthony or the ambassador to Spain Sir Samuel Hoare, whom Churchill "regards as an arrant appeaser."[39]

In early May Gandhi wrote to Jinnah, taking up an invitation issued by the president of the Muslim League at its annual meeting in late April: "What is there to prevent Mr. Gandhi from writing direct to me? . . . . Let us sit as two equals and come to a settlement."[40] "I welcome your invitation," Gandhi wrote Jinnah on May 8. "I suggest our meeting face to face rather than talking through correspondence. But I am in your hands. I hope that this letter will be sent to you."[41] Linlithgow favored sending it, but Churchill vetoed the idea of allowing an "interned person" to talk with anyone for "the purpose of uniting" to drive the British out of India.[42]

"Were all our dreams then shadowy nothings?" Nehru now inquired of his diary. "Our flaming enthusiasm smoke and hot air? Our sufferings and sacrifices mere futility? Is this world the happy hunting ground only for the violent, the bullies, the vulgar and the opportunists? . . . Life seems to have lost its savour and the day's routine is without meaning and value. I have lost the desire to go out of India. . . . [It is] best to remain here . . . how? Not as one submitting to alien and arrogant authority . . . as a rebel? Perhaps."[43]

The sixty-eight-year-old Vallabhbhai was dreadfully ill, his clogged digestive system giving him such pain and keeping him awake virtually all night and in agony most of each day, that he tried fasting, not for political, but for physical reasons. Yet even that barely helped, so Maulana Azad asked the superintendent to send for Patel's own doctor or, better still, to send him to the Bombay hospital. Mahmud also was worse, bringing up "daily quantities of blood." They all thought it was his lungs, but the superintendent insisted it was only his "throat." "We are supposed to be living in an unknown place although everybody knows exactly where we are kept," Nehru noted, explaining their superintendent's reluctance to send out for proper medical assistance. "If anyone is sent out to hospital . . . this very transparent veil of mystery is torn up."[44]

At the end of May 1943, Nehru learned of the government's refusal to forward Gandhi's letter to Jinnah, reading Jinnah's press-quoted comment that "this letter of Mr. Gandhi can only be construed as a move on his part to embroil the Muslim League . . . into a clash with the British Government solely for the purpose of helping his release."[45] Nehru commented, "It exposed Jinnah more than anything. . . . [I]t gives an intimate glimpse into his mind . . . opportunism raised to the nth degree, pomposity and filthy language, abuse, absence of a single constructive idea, a capacity for what is considered 'clever' politics, vulgarity, incapacity for any action apart from shouting. . . . One would have thought that only a bloody fool could be taken in by all this. Yet he carries it off! . . . Was Hitler's analysis correct—that the masses are just fools who can be made to do anything if your lie is big enough? Certainly Jinnah has been an apt pupil of Hitler's." The violence of Nehru's reaction to much of what Jinnah said reflected his frustration at Jinnah's coldly conservative legal mind, in many ways reminiscent of the pre-Gandhian Motilal.[46]

On July 1, 1943, Nehru woke his cell mates after midnight with such loud groaning and howling that several of them thought either Vallabhbhai or Mahmud was dying. "This nightly moaning is . . . a nuisance," he admitted. "Evidently it is an old habit . . . Kamala used to tell me about it. . . . I remember, nearly ten years ago in Dehra jail, my shouting in my sleep and two jail wardens rushing to find out what had happened. Last year, during the Cripps' talks, when I was . . . in Delhi, Rajan [R. K. Nehru's wife] dashed in one night because she heard me moaning loudly. . . . Last night it was much worse . . . Pattabhi says it sounded as if someone was being murdered! Horrible thought! What have I got in my sub-conscious self, or wherever it is, which comes out in sleep when the restraints of wakeful consciousness are absent? . . . The nightmare is not vivid and I forget it soon after waking. . . . Usually it has to do with struggle and conflict and a certain inability . . . to reach the person or thing which is troubling me. I then shout out either for help or as a warning. . . . This peep into some inner depths within me troubles me. What kind of life do I lead under the conscious covering of self?"[47]

Indira and Nan were released from Naini Prison in May. Betty had been released earlier. Indira went from Anand Bhawan to Bombay in June and sent her father *langra* mangoes, which he loved the most of all varieties. Betty had earlier sent him some *alphonsos* (grown in Goa, named for Alphonso the Great of Portugual), which were larger.

In early July Nehru was alarmed by published reports that Gandhi had written to Linlithgow, offering unconditionally to withdraw the "Quit India" resolution of the previous August. "If this is so, it is a cruel and heart-breaking surrender. All the brave words he said last year . . . empty verbiage. . . . With all his very great qualities he has proved a poor and weak leader, uncertain and changing his mind frequently. . . . It is very very sad, this deterioration of a very great man. . . . Have I any big role to play in the future? Vanity says yes. And reason says no: You are too squeamish, idealistic, proud, unbending and aloof, . . . totally unfitted for the political game."[48]

Those reports about Gandhi were completely false, however, and Nehru's vanity proved a much better prognosticator than his reason about his own future political role. By mid-1943 with 200,000 fresh American and British troops pouring into India, shortages of every variety mounted swiftly. Inflation was rampant, and silver rupees were taken out of circulation as fast as they appeared, leaving millions of peasants and landlords to start hoarding their grain rather than shipping it off to urban or overseas markets for "worthless paper." The viceroy and his governors feared that food shortages and rising prices posed the gravest current wartime danger, not a Japanese invasion or the internal political situation. To add military fuel to inflationary fears and impending famine in Bengal, Netaji Subhas Chandra Bose suddenly appeared in Tokyo, after leaving Hamburg in a German submarine. Tōjō placed Singapore's surrendered Indian garrison under Bose's command, and Bose vowed to liberate India with his Indian National Army. Military intelligence warned the viceroy and Britain's war cabinet that "Bose's great drive and political acumen, his prestige in Indian revolutionary circles, his understanding of both Indian and English character, will be of real value to the Japanese. . . . [U]nder Bose's direction subversive activities and espionage in India will be greatly intensified."[49] Bose had been broadcasting daily from high-power transmitters in Germany and would soon be heard from Burma, in several Indian languages, urging Indian soldiers, "Lay down your arms! Stop serving British Imperialists! Liberate India now, Comrades! Join our Indian National Army! Comrades, give me your Blood and I will give you Freedom!" Were it not for the massive injection of American personnel and matérial into India and the monsoon rains that came none too soon, Netaji Bose and his army might have moved beyond Manipur's Imphal, where his army bogged down, to Bengal, where he was all but worshiped by tens of millions.

But the vanishing food stocks worried Linlithgow more than Subhas Bose did. "We are sparing no effort . . . to get things straight," the viceroy wired Amery in mid-July 1943. "But the problem is a very serious one. . . . the cultivator has every incentive to hold for higher prices. . . . [O]ur military effort over the next two years will be in serious danger unless . . . [a] solution is found to the food problem."[50] The war cabinet refused, however, to believe Linlithgow's warnings of how dangerous the looming famine in Bengal really was. Instead it criticized India's failure to deal with its hoarders and control its raging inflation, committing no more than a meager portion of barley from Iraq and some wheat to Colombo, for use in both Ceylon and India as demanded. By now the roads of rural Bengal were cluttered with carcasses of peasants forced to abandon barren fields and homes without food for their last barefoot trek to oblivion.

In Europe, however, the tides of war were turning in favor of the Allies, with the threat to Moscow and Leningrad finally broken and Hitler's army going the way of Napoleon's in the deep snows of Russia's continental ice trap. Mussolini's downfall and hanging in late July started bells of liberation ringing all over Italy, which surrendered unconditionally in early September 1943. Alhough Hitler was too mad to believe he was finished, his brightest generals, including Rommel, understood it, as the Allies prepared to mount an invasion of France.

Feroze also was released from prison in July and joined Indira in the hills, but she felt more depressed outside prison than inside, for she was becoming increasingly restless and frustrated at being able to do so little to help those dying in Bengal and Orissa. Her father tried to cheer her. "If you feel depressed . . . the next day you are sure to get over this—it is just a passing mood. Depression usually comes from uncertainty and doubt—what to do, what not to do? That is a difficult question . . . however, it is better to engage oneself in some activity of mind and body . . . for thus we maintain a certain poise. . . . So cheer up, my dear, and let the old and decrepit, with no vitality and no sense of the future, brood and brood."[51]

But it was less easy for him to practice what he preached, as he confessed two weeks later. "Two dull and depressing weeks. Such moods . . . I have been feeling less energetic. . . . Partly, I think . . . the stories of starvation and death in Bengal. A few thousand deaths by starvation are bad enough. . . . But to think of millions . . . slowly starving away, emaciated, stunted children."[52] Reading helped restore his spirits, however, as did his writing. "I am reading Faust again," he reported to Darling Indu in early September. "Recently I read Nietzsche. I remember reading him rather carelessly when I was in Cambridge. . . . [T]here is much that is attractive in what he says. . . . [O]ur main trouble is a lack of organic connection with nature. We have gone off at a tangent. . . . Why does one do anything? . . . Our moods depend even less on reason and the smallest things affect them. . . . I have felt sometimes extraordinarily exhilarated by the sight of a sunset sky, or the deep blue patch between the monsoon clouds, or even a flower which I had missed . . . I have felt at one with nature. Why does one act? Impossible to answer . . . I have been reading Virginia Woolf *(To the Lighthouse)*. The more I read her the more I like her. There is a magic about her writing, something ethereal, limpid like running water, and deep like a clear mountain lake. What is her book about? So very little that you can tell . . . yet so much that it fills your mind, covers it with a gossamer web, out of which you peer at the past, at yourself."[53]

Roosevelt and Churchill agreed on the need for an overall Supreme Allied Commander of the Southeast Asian area, which was to include eastern India and Ceylon, in order to reverse Japan's victories in the region as well as in southern China. What was less easy to agree on was the choice of the new commander, for he would have to win the respect and cooperation of not only Viceroy Wavell and his powerful new commander in chief, General Sir Claude Auchinleck, but also of so strong willed an American as General "Vinegar Joe" Stilwell, and Generalissimo and Madame Chiang Kai-shek. Churchill finally decided to appoint King George's second cousin, Queen Victoria's forty-three-year-old great-grandson, Rear Admiral Lord Louis Francis Albert Victor Nicholas Mountbatten (known among family and friends as Dickie) to that most prestigious job. Mountbatten's only experience in India had been the brief trip he had made with the Prince of Wales, his cousin David, in 1922, when he became engaged to the lovely heiress Edwina Cynthia Annette Ashley. Dickie and Edwina would play key roles in helping Nehru achieve his premier political power in India. Krishna Menon had known Edwina before the war, first "introducing" her to Nehru by giving her a copy of his autobiography. He also gave her Nehru's other books that he had edited and helped publish and urged Dickie to meet with Nehru as soon as possible after his appointment. In 1944 Mountbatten tried to see Nehru while he was still in Ahmadnagar Fort, but the governor of Bombay refused to permit it, even for the king's cousin. No Englishman in the Indian Service imagined at the time that Dickie Mountbatten would be the next viceroy of India.[54]

On October 8, Nehru noted "Jaya Prakash's arrest in Lahore . . . nearly eleven months after his escape from Hazaribagh. This news made us sad . . . Govt. will take their revenge on him . . . in prison—Or is it going to be death for him?" His young disciple, J.P., had "directed the Congress underground . . . was captured in Nepal, escaped again, and was finally arrested in Punjab in September 1943." J.P. was not killed in jail, but left it prematurely gray and in several ways physically impaired, though his spirit could never be crushed, not even by Indira, who later dared imprison him again, during her so-called emergency Raj, from mid-1975 through 1977.[55]

Indira and Feroze had moved into Anand Bhawan by now, and Nan decided shortly afterward to rent a smaller bungalow in Allahabad for her family. She and Indira rubbed "each other the wrong way," Nehru noted to his diary, after Indira reported her aunt's decision to move.[56] Such friction between Madame Pandit and Indira went back to Indira's earliest years, when she always took her mother's side against her father's brilliant and beautiful sister. But Nehru rarely noticed such things in those years when Indira, Nan, Betty, and Kamala all vied daily for his mere attention, not to speak of affectionate regard. "I am curiously blind to many things that happen right in front of my nose," he now realized. "Bebee [Padmaja Naidu] told me so once, so also Indu—both very pointedly and bluntly. Several others have hinted at it. I think they are all right and there is that blind spot. . . . May be I am self-centered. . . . Some family problems are apt to be almost insoluble."[57] The jealousy and distrust that divided Indira from Auntie Nan only grew worse over time. "I don't understand why she hated me so," Madame Pandit confided shortly after her niece's assassination. "I always loved her and treated her as my own daughter."[58]

Reflecting on such family squabbles may have made Nehru a bit more sensitive at this time to the larger conflicts that kept India so tragically divided and led daily to arguments—conflicts of personality—among his fellow Congress inmates. "I asked Maulana . . . what advice he would give to Gandhiji about the present situation. . . . Maulana's answer was quick . . . rather too eager I thought. He said he . . . would advise that owing to the new situation . . . in the N.E. of India, the appointment of Mountbatten to the South East [Asian] Command . . . also because of the famine . . . the time had come . . . to withdraw formally the A.I.C.C. resolution of August 8th. . . . C.D. [civil disobedience] was anyway dead and it would be right to admit this and formally end it."[59] Nehru was shocked at Azad's swift resolve to reverse the Congress's policy and argued fiercely against any change. The next day the Maulana argued with greater self-assurance that it was time to change, speaking with Vallabhbhai Patel as well as Nehru. "Maulana wanted us to take the initiative in informing Govt. that we had decided to . . . withdraw the C.D. part of the August Resolution in order to help in famine relief and not to hinder in the forthcoming operations against the Japanese. . . . Maulana added that he was sure that Gandhiji would approve of this action on our part. . . . My own reaction was instinctively against his proposal . . . Vallabhbhai said that he too sensed dangers . . . and we should not act in a hurry."[60] The men who in just a few years would be India's first prime minister and deputy prime minister both agreed it might be best to wait and see what Gandhi would do next, for they felt intuitively that he was "bound to take some step." Azad replied that the only "step" he could take would be to "fast" and that would not "do much good." Meanwhile Hindu–Muslim conflicts were getting more violent, with Jinnah and his league gaining in prestige and power.

So next they all met to discuss the question of what to do or not do. Nehru forcefully argued that "if we surrendered and went out of prison in humiliation & sackcloth & ashes, we would have little power to achieve anything. . . . [O]ur very release would shift responsibility on us to solve all these problems. . . . No sackcloth & ashes . . . . I feel that we are serving the cause of India even by remaining in prison."[61]

But the Maulana was still the Congress president and now he wanted to write to the government, to offer the Congress's support for the famine relief work and against the Japanese. He was, after all, from Bengal. "I spoke for an absurdly long time," Nehru noted, "in opposition to Maulana's suggestion. On the whole I carried on quietly . . . but occasionally broke out into heroics! Kings discrowned become outlaws not ordinary citizens etc . . . I think some phrases of mine rather irritated Maulana. . . . Vallabhbhai followed me briefly. . . . He has got a lucid mind, though it may not be deep. He . . . has a strong practical sense—He opposed Maulana's proposal." Then the others spoke, none daring to talk as long as Nehru had, for an hour and a half! Afterward Maulana Azad took an hour and forty minutes to reply. "He was obviously angry . . . said some hard things—repeated himself—was rather bullying (the worst possible attitude if one is trying to convince). . . . And then threw out a very remarkable . . . hint about the communal problem . . . that we should tone down a little towards Govt . . . and then with its help we shall be able to solve the problem easily. This took my breath away."[62]

As Congress's leading Muslim, Maulana Azad tried his best to solve the communal problem before it was too late, for he knew far better than Nehru ever dreamed how much damage would be done by partition. But Nehru's mind, which vacillated on so many issues, was made up on this one, at any rate about Jinnah and his Muslim League. He resolved never to budge an inch in trying to win him over. He condemned and hated Jinnah and had by now grown bitter against the British. The only English leader he still admired was Churchill, the one who hated Congress most. "I have got contempt for official leadership of the British Labour Party. Winston Churchill I consider an honourable enemy. He is implacable but he obviously has fine qualities. . . . One knows where he is. But what is one to do with the humbugs of the British Labour Party?—weak, ineffective, pedestrian and singularly ignorant . . . Cripps? A total failure. . . . He means well. But what is the good of meaning well if this leads to . . . wrong doing? . . . [H]ow he has fallen in my estimation." He had no desire now to visit England again, or so he thought in this long night of imprisoned frustration.[63]

Before year's end he found a moment's respite from his depression in Wendell Wilkie's *One World,* which reached his cell from America. He read that short book with "delight" and considered Wilkie more of a statesman than Roosevelt, "the champion of freedom," who had fallen in Nehru's eyes to little more than "a canny politician." Not so low as Jinnah, however, who "produces an impression on me of utter ignorance and lack of understanding . . . not a single thought. . . . I think that it is better to have Pakistan or almost anything if only to keep Jinnah far away and not allow his muddled and arrogant head from interfering continually in India's progress."[64]

The new year brought no peace and little respite. Nan's husband, Ranjit, died on January 14, 1944, after an agonizing few months of pain from choking asthma compounded by pleurisy and a failing heart. "How could death come to him so early?" Nehru reflected. "How we delude ourselves. I feel desolate and very lonely."[65] Indira also was "ill" again, but this time her "indisposition" was pregnancy. Nehru had as yet not been told the reason for his daughter's discomfort; her successor to India's highest office, Rajiv, would not be born until August. Aunt Betty invited Indira to come to Bombay, where the baby would be delivered. But her father grew more anxious when he received no letters from Indira, reflecting: "Poor little girl! She is not physically strong & well and mentally she cannot adjust herself to the dead-ends and frustrations which life seems to offer at present to most people in India. Who can adjust himself fully to this? Very very few I imagine."[66]

A few days later Kasturbai Gandhi, who had been the Mahatma's wife and companion for more than sixty years, died in the Aga Khan's palace-prison and was cremated just outside its grounds. "Dear Friend," Gandhi wrote Viceroy Wavell, "I send you and Lady Wavell my thanks for your kind condolences. . . . Though for her sake I have welcomed her death as bringing freedom from living agony, I feel the loss more than I had thought I should. . . . [I]n 1906 . . . we definitely adopted self-restraint as a rule of life. To my great joy this knit us together . . . she became truly my *better* half. She was a woman always of very strong will which, in our early days, I used to mistake for obstinacy. But that strong will enabled her to become . . . my teacher in the art and practice of non-violent non-cooperation. . . . You are flying all over India. . . . May I suggest . . . a descent upon Ahmadnagar and the Aga Khan's Palace in order to probe the hearts of your captives? We are all friends of the British, however much we may criticise the British Government and system in India. If you can but trust, you will find us to be the greatest helpers in the fight against Nazism, Fascism, Japanism and the like." But Wavell was no more ready to trust Gandhi than Nehru was to trust Maulana Azad, Jinnah, or Cripps. So the deadlock and stalemate continued.[67] Distrust, fear, hatred, and doubt immobilized them all.

Indira wrote to tell her father that she was pregnant and felt much better. "I am glad she is going to have a baby, though this must involve a great strain on her," he noted. "But she can never be happy unless she is a mother. She loves children and I think she has rather fretted at the possibility of not having any. . . . Last year she wrote to me that she almost thought of adopting a baby she met in Naini Prison."[68] He wrote to assure her that "this act of creation, full of mystery as it is . . . is the fulfillment in some ways of a woman's being." Then he asked her to send him three sets of really "decent" knives, forks, dessert spoons and teaspoons, having "suddenly begun to feel that there is no particular reason why I should not have better spoons & forks. . . . China plates are breakable . . . I do not want you to send them."[69] He listed a number of other required dishes, however, and of books he wanted to read.

By April 1, the silverware had not arrived, so he wrote again: "Darling Indu . . . I am annoyed at having asked you to send me spoons and forks and knives. I had no idea you were so hard up for them. . . . Do not trouble about the teaspoons. . . . I get these brainwaves and ask you to send various articles for no particular rhyme or reason, except . . . change. I hate falling into a rut. . . . From time to time I make petty changes in the arrangement of my bed and table & chair or bookshelves. Or I rearrange some flowerpots in our dining room, or shift about something. . . . My attempts at these minor changes are not always appreciated, sometimes, to my secret annoyance, they are hardly noticed!"[70] Two days later the new silverware arrived, and Nehru wrote in his diary, "It was pleasant to use them."[71]

On April 14, 1944, an explosion in Bombay's harbour killed five hundred and injured another two thousand people, sinking or seriously damaging no fewer than sixteen ships and putting Victoria Dock "out of action for a long time."[72] Indira, who was living at Aunt Betty's house, "lying in bed reading" at the time, just "before tea," about 4 P.M., thought it was "an earthquake." She soon learned better, but "I don't think the censors will allow me to tell you more about things," she informed her father.[73] "The explosion in Bombay was indeed a terrible affair and a vivid reminder of the kind of thing that happens suddenly. . . . That gives some faint idea of happenings elsewhere," he replied, wondering perhaps which of his comrades had ignited that deadly blast.[74] The dreadful tragedy was compounded because in addition to so many lives lost, between fifty thousand and sixty thousand tons of precious grain that had just been brought at great risk and expense from Australia to feed "not only Bombay City . . . but deficit districts such as Ratnagari" were destroyed in the conflagration.[75]

Wavell wanted to try a constructive new political approach and so decided it might be best to release Gandhi early in May, after his health had deteriorated because of malaria. The viceroy had initially wanted to transfer him to Ahmadnagar Fort, but Amery was not so sure about the wisdom of doing that. Churchill preferred to keep him locked up where he was. Amery knew how irrational Winston was about Gandhi, however, and Dr. B. C. Roy urged Gandhi's removal from the Aga Khan's Palace, since it was "a malarial area." Gandhi therefore was released unconditionally on May 6, and went first to the house of his old friend Lady Thackersey, in Poona, then on to Juhu Beach in Bombay, where he recovered quite swiftly. In a few days Gandhi's health seemed miraculously better. He had many visitors, all of whom came to talk and pay their respects. When Amery learned of this, he reminded Wavell of what Lord Byron had said of his mother-in-law, that she had been "dangerously ill; she is now dangerously well." He also reported that Agatha Harrison advised him to ask the viceroy to "visit Gandhi and effect a death-bed change of heart—I am not quite sure whether it was yours or his that was to be effected!"[76]

Nehru was rather alarmed at the news of Gandhi's release and of all the visitors who rushed to see him. "I wish people would restrain their desire for *darshan,*" he wrote Indira, who was one of the first to see Gandhi in Poona. "We are a terribly inconsiderate people where illness is concerned. . . . We overwhelm the person who is ill with too much consideration. I think one of Bapu's chief troubles has been this and he is seldom allowed privacy or rest. Crowds gather and gape . . . imagining that they are not interfering. . . . The worst offenders are those whom he encourages himself by his affection. . . . He has built up such a vast family and given a bit of himself to so many that each one of them considers it a right and duty to hover around him. . . . But it is his age that troubles me . . . nearly 75 . . . an age which I certainly do not expect to reach."[77]

"He [Gandhi] asked so many questions about you and sent his love," Indira had written. "When I arrived he was sitting spinning and gave me a big grin and the usual whack—only much much milder. He was looking very . . . weak and tired."[78] "Very foolishly I try to dole out good advice," Nehru wrote his daughter again that same month of May. "Probably it just irritates. And yet I cannot quite get out of my school-masterish habit. Each person has to find his or her own way out of the problems that beset one. . . . Usually the troublesome things are the little things of life. . . . My own way out for present worry over some small matter is to switch off my mind from it and think of some big thing. That restores balance and perspective. That is why . . . a major disaster, like a war, does one good service. . . . We have been here now for 21 months and 3 weeks . . . my longest term."[79]

The old nightmares persisted. "I shouted and moaned more than usual, disturbing my neighbours," he noted in mid-June. "I am told that I make ghastly noises. . . . My waking hours have not been bad, except for occasional depression, a malaise, a nostalgic feeling for all that is unattainable. And . . . at night I go off the deep end into some unknown, fantastic and rather horrible land. What is this that troubles me? What possesses my unconscious self? . . . [I]s it some childhood memory or shock which I cannot recall? I do not remember anything that affected me so strongly in childhood. . . . There is an old childhood dream which I remember. It came to me repeatedly. . . . Some horrid creature was pursuing me, something that I could not clearly see or visualise, and I ran my fastest away from it. I ran and ran but always it was close behind me. Ultimately I entered a building and mounted a high tower. There near the top I locked myself from the inside in a room. But there was a high window or a skylight and to my exceeding horror I saw a huge hand come through this window and stretch itself out towards me. It was only a hand with an arm attached, nothing more. There was no escape possible for me from that room. The hand approached, came nearer and nearer, and then I awoke in a cold sweat, shivering with fright and horror. . . . I want to move and my limbs do not obey me. I want to shout and warn someone and my throat is parched and dry and no voice comes out of it, or just a hoarse grunt. . . . What is all this due to? What old suppressions? Is it sex—Partly perhaps. . . . Is it the feeling of helplessness and oppression due to political causes? Perhaps. . . . What is going to happen to India—to the world—to me? Where do we go in this devil dance of ours?"[80]

Nehru shared his prison cell with his old Cambridge friend Mahmud, who was almost blind. Jawahar read the daily newspapers aloud to him. Although "I do not take kindly to newspapers," but "I cannot . . . do without them. I want naturally to know what is happening, and yet. . . . Reading to Mahmud, I have to read not only what interests me but what . . . might interest him. So . . . if . . . Churchill has been delivering one of his long rhetorical speeches, I am quite exhausted by the time I have done with it. . . . I think of Virginia Woolf's book: *A Room of One's Own* and of the advantages of having a room to oneself."[81]

Outside, the "devil dance" continued. C.R. proposed a new formula for Congress and the Muslim League to share responsibility for a provisional central government, informing Jinnah that Gandhi might agree. Then in mid-July Gandhi wrote Jinnah and offered to "meet whenever you choose. . . . Do not disappoint me."[82] Jinnah was then in Kashmir but replied that he would be "glad to receive" Gandhi in Bombay in mid-August. Nehru felt "very much put out, angered and out of temper" at that news of negotiations carried on without him. "The very frequent utterances of Rajagopalachari have overwhelmed me . . . and I feel stifled and unable to breathe normally. For the first time in these two years I have a sensation of blankness and sinking of heart."[83] He was so enraged that night that he "created quite a disturbance by my shouting in my sleep," waking all of them up with his tantrum. He reread *War and Peace* and unburdened his heart to Indira about this world in which "all of us seem to drift into a strange condition which is neither war nor peace—a kind of unreal existence in an unreal world."[84] But the dance resolved nothing, healing none of the wounds, curing none of the fatal diseases that were to consume each of them, and all but destroying India herself before that darkest decade of war and peace was over.

On the morning of August 20, 1944, Indira Gandhi gave birth to her first son. "So you have made me a grandfather!" Nehru wrote her the next day. "I felt very happy that this tension was over . . . I hope it is well with you and the little one."[85] But she kept running a low temperature for several weeks and had a persistent "cough and cold. . . . The worst of it is that I keep on giving them to baby."[86] The question was what to name him. Aunt Betty sent a long list to Nehru, who was to choose the name of his grandson, ultimate heir to his fortune and power. Indira had "a special liking for the name Rahul. . . . I like Rajeeva and Rajat and Karna. . . . You had better choose . . . rather nice is Sanjaya. Also Nakul."[87] Nehru consulted his colleagues and with the help of Narendra Deva and Maulana Azad he came up with Priyadarshi Birjees, the latter being Persian for Jupiter. Gandhi's suggestion of Motilal was instantly vetoed by Jawahar. Indira liked Rajiv Ratna best, and Feroze agreed. So Rajiv "Jewel" it would remain. His horoscope predicted great things, and the astrologers scanned the heavens in high excitement, for there were five planets in the house at the moment of his birth, 8:11 A.M., presaging royal power. Then Indira recalled that several months before Rajiv's birth, some Parsis had reported "that it was written in their ancient book that a Hindu girl of high family would marry a Parsi and their son would do great things . . . reincarnation of the Shah Behram of Persia!"[88] None of those seers warned Indira, however, of how high a price Rajiv would pay for donning that ancient cloak of imperial power.

Nehru finished writing his most influential book, *The Discovery of India,* in September 1944. He had pondered tackling this ambitious project ever since he finished his autobiography, deciding to integrate the entire history of India into the frame of his own experience, to retell it so that the Western world as well as his own compatriots might better appreciate the brilliant antiquity and modern genius of India, the nation he was destined soon to lead. For years he had gathered books and quotes, doing his research mostly in his book-filled jail cells, where he pursued his literary scholarship in the isolation he required and cherished. The actual writing took him only five months, from April to September, prodigiously covering 998 foolscap pages in his neat and small-lettered hand. "What is my inheritance? To what am I an heir?" Nehru asked in this most popular of his works. "To all that humanity has achieved during tens of thousands of years . . . to its cries of triumph and its bitter agony of defeat, to that astonishing adventure of man which began so long ago and yet continues and beckons to us. . . . But there is a special heritage for those of us of India . . . something that is in our flesh and blood and bones, that has gone to make us what we are and what we are likely to be . . . and it is about this that I should like to write."[89]

"Ever since I finished writing my interest in it largely evaporated," Nehru noted in his prison diary on September 20, 1944. "Thinking of it now it does not appear too good . . . I am weary of it. . . . Perhaps it is my mood that is at fault. I feel rather empty—not exactly depressed but just empty. Not hoping nor fearing—just not feeling anything acutely—Not caring perhaps. My colleagues here are greatly worked up about the Gandhi–Jinnah talks. They discuss them without end. . . . Whatever the outcome I doubt I shall be excited about it. . . . Fed up—that I suppose describes my state of mind. Fed up with myself, with our people, with India, with the world, with the horror of war, with the worse horror of men's littleness and selfishness, with the Congress, with almost everything."[90] Gandhi and Jinnah talked and talked for almost three weeks, posing for photographers before and after each meeting in Jinnah's elegant house atop Malabar Hill in Bombay. "The Gandhi–Jinnah talks are dragging on and the latest rumour is that they have broken down," Wavell wrote Amery on September 27.[91]

"What a damp squid after all the shouting and praying and hou-ha!" Nehru noted when he learned of the failed summit. "What a quagmire it all is! . . . I was not greatly excited . . . less worried than others here. . . . I think this must be some unconscious way of protecting myself from continuous irritation. I keep my mind occupied with other matters and the newspapers produce only a temporary ruffling of the surface. Not perhaps a healthy sign. Certainly it would be unhealthy if I behaved in this passive, escapist way outside. But here in confinement perhaps it is not so bad. It saves energy and worry." That same passive, escapist habit of protecting himself from the constant daily irritations that faced him after he became prime minister of India, however, allowed Nehru to survive in that most demanding, frustrating job for more than a decade and a half with little more than any temporary "ruffling" of the surface of his mind. But even though his old "escapist" habit certainly saved him much "energy and worry," it was hardly very healthy for India's troubled polity and poor economy.[92]

Near year's end, Nehru knew that the war would not last much longer and that perhaps quite soon they all would be released. It was always easy for him to reflect on how "wrong" others—outside and free to act—were in trying to solve the political problems that faced India and the world. "Gandhiji issues statements. . . inspiring . . . yet somehow they lack some basic reality," he noted. "Rajaji [C.R.] goes on talking, talking imagining that he is the only possessor of a clear head. . . . Now the Sapru Committee of jurists and wise men has been formed—What a collection of notables of a bygone age to tackle the problems of this revolutionary epoch! But what would I do myself if I was out? I do not know . . . I would not and could not remain passive and silent. Nor could I just repeat old slogans and phrases. What then?"[93]

Nan flew off to New York. She had hoped to visit her brother before leaving, but he wanted no visitors, neither her nor Betty, not even Indira and Rajiv, feeling as he wrote Nan, "rather uncomfortable with tiny babes . . . a little afraid of hurting them. . . . It may be I have hardened; certainly I have changed . . . with the passing of years. . . . I do not fancy being treated like a wild beast in a cage . . . I dislike being the plaything of others."[94] Madame Pandit went to America with a "small delegation" of Indians chosen by Sir Tej Bahadur Sapru, then president of the Indian Council for World Affairs, to attend the Pacific Relations Conference in Hot Springs, Virginia. "My first view of New York was when Pearl Buck took me shopping," Nan recalled. "I was . . . overwhelmed by what I saw. Great tall buildings, glittering shop windows, cars, well-dressed men and women. . . . We went . . . to Saks Fifth Avenue, and I was dazzled. . . . The stores were all decked for Christmas and I had never before . . . been subject to such temptation . . . an unreal world. . . . The past, with its jail-going, was more real, and so were the filth and stench of the famine-stricken area from where I had so recently come." Nan was unable to meet President Roosevelt, who was then at Yalta and died soon after that summit ended, but she did have lunch with Mrs. Roosevelt. "I was surprised by the near austerity of the meal itself," she wrote. "That afternoon was the beginning . . . of a friendship that I have valued and from which I gained a great deal."[95]

Nehru felt weaker and more impatient now with each passing month, noting on New Year's Eve, "I have developed a pain in my left shoulder & back . . . some kind of neuritis. . . . [I]t is a great nuisance. . . . What worries me even more . . . is the frequent occurrence of petty troubles and pain . . . trouble with my calf, my forearm, my hand—also my eyes and teeth. . . . Age, age, advancing steadily and weakening the whole system, in spite of all my exercises and other efforts!"[96] He was eager to reread *Anna Karenina,* first asking Betty to send him her French copy and then urging Indira to send the English translation. He also was interested in Aldous Huxley's novel *Time Must Have a Stop*. And finally, the long desired cigarette holder reached him. "It is just right," he told Indira in early January 1945 (but in February it "fell on a stone floor and cracked up"). "Do not trouble to send the honey. . . . Honey has gone up . . . I suppose, ever since sugar was rationed. . . . We have been getting here an ounce a day per person. . . . For me it is ample as I use it with tea & coffee only—and a pinch in my dal. Is this habit of putting a little sugar in dal a Kashmiri habit or just a Nehru vagary?"[97]

That winter Betty planned a spring trip to Kashmir and invited Indira to join her. Nehru urged his daughter to go. "You know that for me there is no place like Kashmir," he wrote,[98] and when she took his advice he said he was glad, adding: "Kashmir must always remain first choice for a holiday. Your going there means that a bit of myself also goes and I feel a sense of exhilaration at the prospect."[99] Nehru never lost his romantic attachment to Kashmir. Although India would fight several wars with Pakistan, Kashmir remained divided, shattered by communal conflict and devastation, its once beautiful vale no longer a magnet for tourists but, rather, the front line of partition's most prolonged legacy of death and destruction.

"Last night I had again a nightmare accompanied by shouting. . . . I shouted or groaned as usual under some compulsion or fright or both. . . . The bedding itself seemed to be pressing me at the sides, crushing me almost, and a vague terror continued. So I tried to call out. . . . My voice must have sounded strained. . . . I have an idea that I was saying or shouting in English—get away! go away! . . . I woke up slowly. I felt very exhausted as after some struggle. . . . What is all this business? What frightens me during these nightmares and compels me to shout 'go away'? I have no clear recollections. . . . [S]ome pictures remain in the mind but they fade away very soon. Only a feeling of pressure & compulsion remains—something that strangles & throttles and crushes the body. . . . Certainly it is not a healthy sign and I am beginning to get a little worried."[100]

In mid-March the imprisoned leaders met again to discuss developments in their country and possible future strategies for the Congress. Maulana Azad first "praised Bapu for his wise leadership" but then argued that he had erred in sponsoring the "Quit India" resolution after the failure of Cripps's mission. Nehru basically agreed with the Maulana's analysis, finding his mind "keen," although his language was "over-rich."[101] Vallabhbhai, however, was "full of suppressed anger, pain and bitterness" when he spoke, following Azad and Nehru. "He said that he had long suspected that Maulana and others [Nehru] had felt the way they had spoken. . . . [H]e [Vallabhbhai] did not agree with Maulana's analysis and . . . was firmly convinced that the . . . steps taken by Bapu had been correct and inevitable. Any other course would have meant . . . annihilation of the Congress . . . so Vallabhbhai went on referring to the 'two parties' in the Congress. . . . There had always been trouble between them. . . . [H]e added that because of references to guerrilla warfare in a speech in Assam in April 1942, he had sent his resignation from the W.C. to the Maulana."[102]

It was the closest that Vallabhbhai Patel had come to breaking with Nehru before 1947, for the speaker in Assam here referred to had been Nehru. Vallabhbhai, like Gandhi, believed that nonviolence was even more important than freedom, but he well understood that Nehru felt no hesitation about using any means to attain swiftly India's complete independence from the British. "I was amazed at this outburst," Nehru noted, after Vallabhbhai spoke. "I was a little annoyed. . . . It all seemed due to some misunderstanding. . . . Ultimately I said that it was direct insult to me. . . . Thus the meeting broke up, and I am presently hardly on speaking terms with Vallabhbhai. . . . I think I was less wrong than usual . . . yet why can I not keep my temper?" That night he "shouted" again in his sleep, "my usual agonized cry."[103]

Nan's visit to America proved so successful that she agreed to a lecture tour of the United States, which kept her out of India another year. Nehru was rather surprised to learn of her change in plans, for he had asked Nan to bring him a "Parker '51' pen & a Schick dry electric shaver," he told Indira.[104]

He kept trying to teach his daughter how to behave in a more socially acceptable manner, for she was by nature even more introverted and reclusive than he had been. "Brothers and sisters may be very different from one another and yet they often have a common fund . . . a private world of their own which they alone share. . . . Your personal experiences have been far more individual than communal. . . . You have had no experience . . . of a joint Hindu household. . . . Indeed you have had little experience of any normal family life . . . [so it is] not surprising that you should feel . . . isolated. . . . This development of the individuality has its good points, but it also definitely has its drawbacks. . . . After all we have to live in this world and mix with human beings, such as they are. . . . Otherwise we dry up. The so-called social life of Allahabad is, I agree with you, rather deadly. Even so we can always find oases of friendly feeling. . . . And then we are not limited to Allahabad, our home is the whole of India. It is desirable therefore to develop the extrovert sides of our nature from the point of view of . . . the work we do."[105]

On the night of Wednesday, March 28, 1945, Nehru was driven out of the old stone fort, where he had been locked away for more than two years and seven months, to the railway station at Vilad, where a very comfortable tourist saloon coach took him in "luxury" back to Naini Prison in Allahabad and then on to the United Provinces's Bareilly Central Prison. Govind Ballabh Pant came with him. The Maulana and Vallabhbhai and most of the others remained behind but were taken elsewhere a few days later. Word of his transfer had moved on ahead of his coach, and by the time he reached the Naini prison gate, a cheering crowd had gathered outside to greet him, including Indira and Feroze. "It was exciting," he wrote her a week later, "you looking just as you looked three years ago, a dainty and lovely slip of a girl, apparently unchanged by . . . motherhood."[106] Wisely she had left Rajiv asleep. Nehru was not ready to face his grandfatherhood yet. He had been allowed to be with her for a few seconds before he had to enter the waiting police car. She thought him "thin and shrunken." The thin part he considered "not always a disadvantage," for his weight was now back down to what it had been in his late twenties. The "shrunk-up appearance," however, distressed him, a clear sign of age. She urged him to drink more milk, and he agreed to try.

By the end of April, when Berlin was conquered and Hitler committed suicide, Nehru wrote in his prison diary, "So the great European War is already . . . history. . . . What stark horror Europe has passed through . . . what a war! . . . This is the end of all the pomp & glory and proud boasting. . . . What now?"[107] Indira and Rajiv were off to Kashmir with Betty and the Nehru clan that gathered there around Srinagar and in the surrounding hills. "So you are in Kashmir now," he wrote Indira. "What flowers are blooming, what fruits hang from the over-burdened branches? I suppose the peach blossoms are over now. . . . The lotus must be still in bud. . . . It blossoms on the Dal Lake in July. . . . Cherries will soon be out and apricots and apples. . . . And the Nishat and Shalimar *baghs* [Gardens]—what do they look like now? . . . I think of all these scenes treasured in memory's chambers, but even more I visualize the . . . snows and glaciers, with their ice-cold brooks gurgling and rushing down. . . . Kashmir attracts . . . crowds of visitors. . . . Many of our friends may be there now." [108]

As the war ended in Europe, the viceroy tried to turn the war cabinet's attention to India, finally getting Amery's permission to fly home in late March to meet with the India Committee, chaired by Attlee and again including Cripps. Wavell hoped to get Gandhi and Jinnah to agree to sharing power over his Executive Council, and the war cabinet authorized him to try to bring them or members of their parties together. Churchill knew that his own days in power were numbered, for the collapse of Germany meant general elections in England, and every political straw in London's wind blew toward Labour. In late April the first meeting of the United Nations was held in San Francisco, and Madame Pandit was there, not yet as India's representative, but as the spokesperson for "the real India. . . . The press . . . gave us good write-ups," she recalled of that eventful time of hope, lobbying for peace. "Very often . . . one of the delegates did attend our meetings. . . . General Romulo of the Philippines, M. Schubert of the French . . . and most important to us, Mr. Molotov of the Soviet Union." [109] Nan's speeches on "the challenge of Asia" were so well attended and warmly received that President Harry Truman soon invited her to the White House to meet with him. She later returned to the Oval Office as India's ambassador.

"The summer advances and brings with it the hot winds and sultry weather," Nehru wrote Darling Indu in late May. "But far at the back of those winds those sea-changes are taking place which precede the monsoon—water being sucked up into the clouds . . . across the sea to far lands which crack with the heat and pant and thirst. . . . So the seasons follow one another and after the bad comes the good and the better . . . can spring be far behind? . . . I am often with you, not only in mind but almost it seems to me as if it were a physical transference, and I experience the climate of the Kashmir valley and look out through your window in the attic at the snows and the green hills and lakes and the Vitastha winding slowly on its leisurely . . . journey before it enters the gorges. . . . So I live a double life, or why double only? . . . a succession of many different and overlapping lives in many places far away from each other." [110]

On June 10, 1945, Nehru was moved from Bareilly to the Almora district jail, where it was cooler, back to his old familiar cell. "As I was looking round this old-new environment, faint memories began to stir within me. . . . There was a dream quality . . . or perhaps it was rather like that middle state which lies between waking and sleeping when vague ideas float through the mind and pass away leaving no trail behind. . . . The gradual reduction in my weight . . . has brought my weight back to what it was . . . in my middle twenties—a slim lad who once coxed a boat. This going back to the past has its advantages . . . and I am not sure that I would not like to go back. But there is one very serious drawback—there would be no Indu then," he wrote his daughter, "—she would just be a hope and an aspiration."[111]

On June 14, 1945, Nehru noted as his final entry in that prison diary, "Perhaps we might be discharged tomorrow . . . after 2 years, 10 months and 6 days. . . . And we are expected to rush about immediately . . . perhaps to Delhi—I am greatly put out at this prospect. I did so much want to have at least three weeks to myself—to rest—visit Indu in Kashmir & generally to meet old friends quietly." Much to his chagrin, however, at 8 A.M. on Friday, June 15, Nehru was "unconditionally released" from his longest term behind bars.[112]

24

From Prison to Power

1945–1946

INDIRA WAS ALONE in Uncle Birju's big house in Kashmir when she heard the radio report that her father was about to be released. "I rushed out to book a seat . . . to Allahabad," she recalled. "At that moment, I forgot the baby. When the family returned from the [maharaja of Kashmir's] party they wondered how I could help [Papu] by going. I said: 'I don't know but I just have to go.' . . . So I left the baby. . . . I returned to Allahabad."[1]

Their reunion was all Nehru had imagined it would be. His daughter rushed to his side, forgetting not only Rajiv but Feroze as well. They would have to learn to cope without her. To Indira, her life's mission was clear by now, as it had long been willed by Nehru. Her destiny was to be at his side, or just a step behind, discreetly hovering in his shadow, always at his beck and call, ready to do whatever he asked, to travel with him whenever he ventured abroad, to serve as his official hostess in the prime minister's house, and to pick up the torch as his faltering fingers lost their grip. Her mind and soul belonged to him, perfectly attuned to his mind, each serving as the other's alter ego.

"My first thoughts on coming out of prison are with those who are still rotting in prison," Nehru told his welcoming friends outside. "It is a matter for shame and sorrow that so many of our comrades are still behind prison walls."[2] His still fearless defense of human rights would later prove an embarrassment to his daughter, after she took "emergency" power, and to himself, during his own long tenure at the top. But now he knew that the viceroy was eager to reach a political agreement with Congress and that Wavell wanted him to help run India as efficiently and quietly as possible. Wavell hoped that he and Azad and others, including Jinnah, might actually agree to join his Executive Council and had announced in his broadcast the day before Nehru was released that His Majesty's Government had authorized him to propose a new plan to India's political leaders, designed to "advance India towards her goal of full self-government."[3] It was the first time a viceroy had used that term, and so many Congress leaders, including Maulana Azad, felt optimistic. But Nehru initially declined to comment, insisting that he had not read Wavell's offer, and "even if I have . . . [s]ending merely fifteen men to the Executive Council will not solve the problem. . . . India needs a surgical operation. We have to get rid of our preoccupation with petty problems and concentrate on the fundamental problem of slavery and poverty. That is the touchstone on which every new offer has to be tested."[4] For another year he continued using such rhetoric when confronted with offers of settlement from the British Raj.

Congress President Azad, however, called his Working Committee to Bombay on June 21, 1945, to consider Wavell's proposal. Germany's unconditional surrender on May 8 had cleared the skies over Europe, and so Wavell felt hopeful that in inviting the leaders of India's major parties to meet at his viceregal lodge in Simla he might be able to break the deadlock, which had paralyzed India's political progress for more than six years. Cripps wired "best wishes." Wavell even asked Amery to fly out and join him in Simla before the end of June, but Leo Amery was faced with his own political "mud bath" that would topple him, together with Winston Churchill, when England's general elections in July resulted in a Labour landslide. Krishna Menon played an important electioneering role, helping get out the mass vote that brought Attlee to Downing Street with a two-to-one majority of Labourites in the House of Commons. Krishna Menon's communist comrade R. Palme Dutt destroyed Amery in Birmingham, and Harold Laski, who was now deputy chair of the Labour Party, addressed a packed St. Pancras Town Hall crowd at Krishna's behest, insisting to thunderous cheers that "Indian freedom is inevitable and inescapable." Krishna also convinced Laski that Nehru alone was the man to lead India to freedom. Attlee appointed old vegetarian Pethick-Lawrence as secretary of state for India. Pethick was Agatha's good friend and had long been an admirer of Mahatma Gandhi.[5]

Maulana Azad invited Nehru to meet with him in Bombay that June, before the entire Working Committee met. A week of freedom with Indira helped Nehru view Wavell's offer in a more favorable light, especially since Azad insisted that Jawaharlal be put in charge of External Affairs on the Viceroy's Council. Nehru always enjoyed focusing on foreign policy and preferred external affairs to domestic difficulties. "He [Azad] appeared to accept the main principles underlying the proposals," Wavell reported in delight, after meeting the Maulana on the eve of his Simla summit.[6] "I met the Viceroy at ten in the morning," Azad noted. "He received me courteously. . . . He appealed to me to trust the Government. It was his sincere desire that the problem of India must be solved. . . . I told him clearly that an agreement with the [Muslim] League seemed very doubtful. Those who were in control of the League seemed to be under the impression that they had the support of the Government and they would not therefore accept any . . . terms. The Viceroy emphatically said that . . . [the] Government was and would remain neutral."[7]

Jinnah came to see the viceroy the day after Azad had been there. "He began by saying that the Muslims would always be in a minority in the new Council because . . . Sikhs and Scheduled Castes would always vote with Hindus," Wavell reported to Amery. "I said I doubted his assumption. . . . He then claimed that Muslim League had the right to nominate all Muslim Members to the new Council. I said that I could not accept this . . . I thought him depressed."[8] Jinnah was depressed. His smoke-saturated lungs were riddled with tuberculosis compounded by the cancer that claimed his life in a little more than three years. Jinnah refused to meet with Azad, whom he contemned as a "show boy" Muslim, chosen by Gandhi as Congress president only to refute Jinnah's claim that his Muslim League represented all Muslims. Wavell now considered Jinnah "the main stumbling-block" to any progress.[9] Jinnah argued that there should be only fourteen members on the new council, two English, five Hindus, five Muslims, one Sikh, and one scheduled (untouchable) caste. He insisted that the Muslim League nominate all Muslims.

Nehru left Indira sick in bed on June 30, taking the train from Bombay to Simla and meeting the viceroy there on the morning of July 2, 1945. Azad had briefed him on everything that had happened in Simla to date, and Nehru met the viceroy with his most gentlemanly Harrow–Trinity manners. Wavell was pleased to find Nehru "very pleasant throughout" their hour and a half discussion, sensing "little bitterness" in this most cultivated of all congressmen. "He described India as suffering from industrial 'arrested growth.' . . . He then spoke of the Russian system . . . and said that the Russian methods were not really applicable to India. . . . [I]t should be done peacefully and without revolution. . . . [T]hough he described himself as a Socialist, he did not believe in pure Socialism." Wavell liked Nehru enough after that meeting to be willing to appoint him to his council. He was charmed and impressed, preferring Nehru to Gandhi or Jinnah, thinking more highly of his intellect than he did of that of Azad or Patel.[10]

Wavell asked Azad and Jinnah to submit lists of those they would like as members of the new Executive Council. Azad met with Nehru and then submitted a list of fifteen, including Jinnah and two other members of the Muslim League and the president of the Hindu Mahasabha, Dr. Shyama Prasad Mukerji. Topping the list, however, were himself, Nehru, Patel, and Prasad. Rajkumari Amrit Kaur was the only woman, and Tara Singh the only Sikh. Jinnah refused to submit a list. "Azad is deeply hurt at Jinnah's refusal to treat with him and I have seen an Intelligence report of attempts by Azad to consolidate the minor parties with the Congress against the Muslim League," Wavell told Amery. "He is said to have offered Tara Singh full Congress support for the Sikhs . . . if an agreed Sikh name were sent in through the Congress. . . . Azad's line with the Sikhs was that if the League stood out, the other parties must prevail upon the Viceroy to go ahead. . . . He believed that on this basis Jinnah and the League could be broken. . . . Sikhs have not accepted the Congress offer . . . Tara Singh's list was sent in separately."[11]

Wavell's first choices for the new council were Nehru, Patel, Rajendra Prasad, Dr. M. S. Aney, and Sir B. N. Rau. Sir Benegal Rau, Nehru's old Cambridge contemporary, who had entered the ICS after graduating with a first, had just resigned from being prime minister of Kashmir state. Aney had been on the Viceroy's Council, but was now the Indian government's representative in Ceylon. Wavell wanted five Muslims as well but did not even list Jinnah, topping the league's candidates with Liaquat Ali Khan and Khaliquzzaman of the United Provinces. He kept Tara Singh on his first list and would have added Ambedkar as well as another untouchable had he had the opportunity to appoint a new council at this time. His slot for Nehru, however, was Planning and Development, not External Affairs, which he would have offered to Aney, giving Defense to Tara Singh, Commerce to Patel, Labor to Ambedkar, Law to Rau, and War Transport to Liaquat. None of those offers would ever be made, however, nor was the viceroy's list made public. He met Jinnah again, finding him "distinctly depressed and rattled . . . afraid that he may be made the scapegoat for the failure of the Conference . . . yet unwilling to give up his claim that no one but himself represents the Muslims."[12]

Lame duck Leo Amery reported his own electoral defeat on July 11, advising Wavell in his parting letters to "carry on with your existing Executive and try to hold fresh elections in India during the coming cold season." Such elections, of course, would test Jinnah's claim to represent all Muslims. Amery also felt that "the attitude of Congress can hardly be as bitter against us now as it was before our proposals."[13] On July 14 the viceroy addressed the final meeting of the Simla Conference of 1945, taking full blame for its failure. Azad congratulated Wavell and praised his sincerity but insisted that the blame for the breakdown lay with the Muslim League, not the viceroy. Jinnah thanked the viceroy and said that it "suited the Congress to come into the scheme for . . . once they came in they would strangle Pakistan."[14] Therefore, the Muslim League chose to stay out and would remain out until it could, by "right," nominate all the Muslim members. Tara Singh stated that if Muslims wanted Pakistan, Sikhs must also have a separate state of their own, called Sikhistan or Khalistan.

After that final meeting, Wavell met Nehru again. "He was quite friendly," the viceroy noted. "His main theme was that Congress represented a modern Nationalist tendency—the League a medieval and separatist one. He showed no special bitterness against Jinnah and the League, admitted that there was a psychological fear of Hindu domination, but claimed that it was unreal and unwarranted though he admitted that there was a section of Hindus out for complete Hindu domination. He did not put forward any special solution of the problem. He is more of a theorist than a practical politician but earnest and I am sure honest."[15]

Wavell was not nearly as stupid or insensitive as many Congress leaders believed. But he lacked subtlety and had no understanding of the complexity of a mind like Nehru's, to say nothing of Gandhi's. Rather, his mind was more nearly attuned to Jinnah's. "Jinnah is narrow and arrogant," he wrote Amery, "and is actuated mainly by fear and distrust of the Congress. Like Gandhi he is constitutionally incapable of friendly cooperation with the other party. Azad is an old-fashioned scholar with pleasant manners, but I doubt if he contributes very much to Congress policy. His main object is to get even with Jinnah and the League Muslims who despise him as a paid servant of the Congress. Nehru is an idealist and I should say straight and honest."[16]

Wavell quite rightly blamed "Jinnah's intransigence about Muslim representation and Muslim safeguards" as the immediate cause of his first Simla summit's failure. But then he added, "The deeper cause was the real distrust of the Muslims, other than Nationalist Muslims, for the Congress and the Hindus. Their fear that the Congress, by parading its national character and using Muslim dummies will permeate the entire administration of any united India is real, and cannot be dismissed as an obsession of Jinnah."[17]

From Simla, Nehru stopped in Punjab before heading up to Kashmir, where Indira was waiting with his grandson. Azad went along with him, eager to spend the month of August in Gulmarg, to regain some of the weight he had lost in prison and to enjoy his last months as Congress president in Kashmir. Nehru now felt ready to take back formal control of his party, as he soon would of the government. Yunis also was in Gulmarg at this time, awaiting their arrival at that cooler altitude, where Indira stayed with Rajiv. Nehru found a "glorious" azure sky there and two generations of himself reincarnated waiting to welcome him.

On August 6 and 8, 1945, two atomic bombs destroyed Hiroshima and Nagasaki, convincing Japan to surrender. For India, August brought happier news of elections set for November and December. Nehru sprang into action before month's end, galvanized by the Labour Party's victory in London and the defeat of Japan. His only potential rival for Congress leadership, Netaji Subhas Chandra Bose, was killed that August after escaping from Saigon, when his overloaded plane crashed in Taiwan.

Nehru left Kashmir in midmonth and headed for the frontier, where Abdul Ghaffar Khan and his brother, Dr. Khan Saheb, whom Nehru first met when they were students in Cambridge, organized his frontier and Punjab campaign tour. Accompanied by those two Pathan "kings," Nehru was warmly welcomed on the frontier, yet many Muslim League supporters were there to heckle him as well. Though the "Frontier visit was pleasant," he reported to Indira, it was "rather depressing . . . the realisation once again how immature human beings are and how they fall out on the most trivial of matters. Temperament is usually more powerful than logic. Yet my visit I think did good all around . . . and so to Pindi, where a vast meeting of about 100,000 was waiting for us." But the largest crowd, close to a quarter million, awaited him in Lahore, the capital of the Punjab, where Jinnah and his league had launched their Pakistan movement half a decade earlier.[18]

"What a meeting!" Papu wrote his Indira. "I do not think I have ever seen anything like it. . . . The loudspeakers of course broke down and even the platform partly collapsed. Fear invaded the people round about the platform . . . an infectious affair. I stood on the collapsing platform and surveyed the seething mass of humanity. Many people urged me to go but I refused to budge. . . . Men & women & children fainted and I was told that two children had died. . . . But I was in a savage mood and said I did not care what happened; I would stick on. Piteous appeals were made to me to go but I was harsh in my replies and . . . [y]et I was perfectly cool and laughing at the fear and discomfiture of others. How hard and cruel I was! I was myself surprised at this savage aspect which revealed itself unawares. Ultimately I jumped into the crowd and was pushed and tossed about by it. . . . How easily such a crowd can become a savage & aggressive mob! But, being expert at the job, I managed to add to my popularity with it. . . . [W]e finished after midnight and we returned with thousands of frenzied people escorting us."[19]

By now Wavell had received enough reports of Nehru's "injudicious speeches" to revise his earlier assessment of his character. "The Congress leaders are difficult people to deal with. They are outwardly very reasonable when one meets them, but in dealing with their followers they have no balance. . . . I think Nehru is trustworthy in the sense that he would not deliberately deceive, or break his word, but he is unbalanced and unreliable. The other Congress leaders I have met are, I should judge, neither trustworthy nor scrupulous, as we understand such terms," Wavell wrote Pethick-Lawrence. "I am not surprised that Jinnah is apprehensive of them."[20]

The Working Committee of Congress met in Poona in mid-September and agreed to reaffirm its policy of calling for a "democratically elected constituent assembly to prepare a constitution" for India. Nehru drafted Congress's rejection of "Pakistan" or any attempt by the new Labour government to resurrect Cripps's "opt out" formula of 1942. On September 21, 1945, the A.I.C.C. met in Bombay, its 283 members approving the Working Committee's resolutions while some 25,000 spectators looked on and cheered. Next to Gandhi, Nehru remained the favorite of the crowd, with Patel and Azad hardly noticed by more than a few dozen of their faithful followers.[21]

All the important resolutions were drafted by Nehru. The first of them conveyed "greetings and congratulations to the nation for the courage and endurance with which it withstood the fierce and violent onslaught of . . . British power. . . . Three years of frightfulness have left their long trail behind them, of death and agony and suffering and avoidable man-made famine which took its toll of millions. . . . Yet these years have also demonstrated the courage of the Indian people . . . and have steeled and hardened them in their resolve to gain freedom and deliverance from foreign rule."[22]

Wavell flew back to London to confer with Pethick-Lawrence and the cabinet, whose India and Burma Committee was chaired by Attlee and included Cripps, now president of the Board of Trade. Attlee and Cripps knew that India expected much from the Labour government, but Wavell warned them that Jinnah probably spoke for "99 per cent of the Muslim population," who feared "Hindu domination."[23] He also warned of "bloodshed on a wide scale" if Pakistan appeared to be an "imminent reality" to the Sikhs of Punjab, who would not submit peacefully to a "Mohammedan Raj."[24] Cripps and Attlee thought their old soldier exaggerated the dangers inherent in Pakistan, which remained as yet untested, little more than a nebulous dream in Jinnah's mind.

With Japan's surrender, the remnants of Bose's Indian National Army (I.N.A.) were taken back to India, several of their officers to be tried for treason in Delhi's Red Fort. Nehru became one of their defenders, the most famous barrister in a trial that excited national passion and attracted nationwide attention. Congress resolved in mid-September that "it would be a tragedy if these officers . . . were punished for the offence of having laboured, however mistakenly, for the freedom of India."[25] It was a bitter pill for British India's loyal soldiers to swallow. To most nonmilitary Indians, however, Netaji Bose remained their war hero, the one ex-Congress president who actually fought against the British imperial Raj in battle, not from prison or in the press and on platforms alone. Millions throughout Bengal believed he was still alive, that he could not, indeed, die but had escaped to Soviet Siberia, through China, and was now raising an army to liberate India. The myth of Bose's immortality would not be put to rest even after Prime Minister Nehru's government sent a special parliamentary delegation to Japan in 1957 to "study Netaji's ashes," returning to Delhi to report that they were the true ashes of Subhas Chandra Bose.

Wavell returned to New Delhi in late September to report that as soon as possible after the winter elections it was "the intention" of His Majesty's Government to convene a constitution-making body and that until its work was done, "the Government of India must be carried on." Therefore, he would "take steps to bring into being an Executive Council which will have the support of the main Indian parties." Wavell hoped—as Attlee, Cripps, and Pethick-Lawrence all believed—that this was "the best way . . . to give India the opportunity of deciding her destiny. . . . I for my part will do my best, in the service of the people of India, to help them arrive at their goal."[26]

To Nehru, however, such promises sounded paternalistic, and although Wavell meant well—indeed, would try his best to help all Indians reach their goal—nothing he did or said would ever win Nehru's trust or support. Winter elections would be held, but none of Wavell's other hopeful plans came to fruition. There was no single goal on which the leaders of India's most powerful parties could ever agree, so they rushed toward the abyss without political fences long or strong enough to keep them from falling over into it. Faster than any of the others, Nehru rushed ahead, rallying masses, jumping into crowds, "cooled" by the "savage mob" he so loved.

"Darling Papu, You come and you go, almost like a flash of lightning. It is wonderful to have a glimpse of you, but even before one has got used to the fact of your being there, it is time for you to leave. I can't tell you how empty everything feels when you are not there. . . . [I]f it weren't for Rajiv, I could hardly bear it," Indira wrote from Anand Bhawan that November. She and Feroze lived with Nehru although he rarely remained at home more than one night, for it was election time.[27]

Wavell warned Pethick-Lawrence that "Nehru is very bitter against the British regime in India and against the Muslim League. . . . Some people believe that Nehru's plan is to make use of the I.N.A. . . . both to train Congress volunteers and as a Congress striking force; and also possibly to tamper with the Indian Army. He is said to have had conversations about the use of the I.N.A. for subversive purposes during his visit to Lahore. It is always difficult to say what Indian politicians really mean . . . but Nehru's uncompromising attitude implies that he is not opposed to a violent mass movement. . . . It is said that Gandhi is unhappy." Now that Bose was dead, Nehru was ready not merely to defend his followers in the Red Fort trial but also to take up his mantle as the "netaji" of the I.N.A. Nor did he waste time consulting Gandhi, for Nehru now considered him "out of touch" with postwar "reality" and a "spent force" whose preoccupation with nonviolence had only led them, sheeplike, into years of incarceration.[28]

As the preelection rhetoric heated up, so did the communal conflict, with Hindu–Muslim riots rocking Bombay's slums before the voting started in December. Pethick-Lawrence was "very sorry to hear" that "Nehru is turning his mind towards violent courses."[29] Krishna Menon had been lobbying Pethick-Lawrence as well as Labour's backbenchers, many of whom had joined his India League on behalf of Nehru and Congress. "I was approached the other day by [William] Dobbie [M.P.], who is chairman of the India League," Pethick-Lawrence wrote Wavell, "with the suggestion that a party of Labour Members, who are members of the India League, and Krishna Menon should be given facilities to go to India. . . . I have known Menon for some time . . . I do not feel that he is at all likely . . . to be a helpful factor. . . . He is, I believe, primarily Nehru's man though he purports to purvey the views of Congress as a whole."[30]

Wavell invited Nehru to the viceroy's house to talk for an hour in early November, first assuring him that His Majesty's Government was "genuinely anxious for a settlement, but by constitutional methods and compromise, not violence."[31] The viceroy argued that the future of India must depend on some sort of compromise between Hindus and Muslims. "He [Nehru] . . . said that Congress could make no terms whatever with the Muslim League under its present leadership. . . . [I]t was a reactionary body . . . Hitlerian in its leadership and policy, and tried to bully everyone. . . . [T]he Congress would never approach the Muslim League again, because of Jinnah's rudeness to their leaders." Wavell told Nehru he had had reports of his recent speeches, which called for reprisals against officials in several provinces. "He particularly admitted that he was preaching violence, and that while he deplored violence, he did not see how violence could be avoided if legitimate aims could not be attained otherwise. I warned him that . . . violence must . . . lead to violent counter-measures."[32] Nehru had never liked Wavell and certainly felt no fear of British violence at this point, for as he told a cheering crowd in Lahore a week later:

"India is on the brink of a mighty revolution. . . . Therefore, to vote for the Congress is to vote for the freedom that is coming soon. By your vote you have to declare whether you stand for freedom or slavery. The elections will decide the fate of the Red Fort and the Viceregal Lodge.

. . . Those who try to go against the great gushing torrent of Indian nationalism will be swept ashore, lifeless as a log of wood."[33] He warned Punjabi officials against trying to influence voters in favor of non-Congress candidates. Nehru felt as invincible as he was fearless, never dreaming on the eve of these elections that in less than two years this great and once beautiful city of Lahore would be lost to Pakistan.

Gandhi remained mostly silent at this time, his health reported to be very poor, and Wavell feared that he might no longer be able to restrain Nehru and Patel, both of whom continued to make violent speeches. The viceroy expected a serious attempt by the Congress, probably in the spring of 1946, to subvert by force the British administration of India. He warned the cabinet that half measures would be of no use in dealing with such a mass movement. The only options then would be "capitulating to Congress" or "using all our resources" to suppress such a movement. He now believed that nothing less than "the grant of immediate independence to India under a Government selected by the Congress . . . would satisfy Nehru and Patel."[34]

Pethick-Lawrence dared not take the viceroy's warning lightly, reporting it to cabinet colleagues in mid-November: "Nehru's utterances . . . have made specific reference to 'revolt' and adopted a threatening tone. . . . I do not take all Nehru's speeches at their face value, and recognise that much may be set down to the ebullience of electioneering. . . . But . . . I am told for the first time that there are signs of a demoralising effect not only among the civil services but also in the Indian Army."[35]

In Bombay that October Nehru had called on a cheering audience to "prepare . . . for a mass battle for freedom, which may come sooner than people expect."[36] Pethick-Lawrence agreed with Wavell that "Nehru is . . . going to be the most difficult element in our problem. . . . [H]e has pretty well made up his mind to force the issue without any . . . compromise with Jinnah." The secretary of state took some comfort, however, from the fact that elections were soon to be held, reminding Wavell that "Nehru has a fairly sound sense of international publicity and I should think he would be careful to play his cards in such a way as to win the maximum support from outside for such a campaign."[37]

Rajkumari Amrit Kaur, one of Gandhi's intimates, at whose Simla palace he had stayed during the viceroy's summit, flew to London to inform Cripps that Gandhi was "ready and willing" to try again to "influence Indian opinion towards moderation." She advised Cripps to urge the viceroy to see Gandhi as soon as possible, and Sir Stafford chaired a meeting of the cabinet's India Committee to pass on that suggestion to Pethick-Lawrence and the earl of Listowel, then postmaster general but soon to become Britain's last secretary of state for India. "Gandhi still exercised an immense influence on Indian public opinion," Cripps assured them. "He might not have as much control over Nehru, but Nehru would be influenced by the results on Indian public opinion of action taken by Gandhi."[38]

Wavell was too nervous to invite Gandhi to meet with him, however, fearing that any good it might do would "immediately do corresponding harm with the Muslims and tend to redouble their suspicions." He also informed Pethick-Lawrence that "Nehru is so angry with Jinnah" that he might not even talk to him if, as Cripps hoped, they were both invited to meet with the cabinet in London, to try to break India's political deadlock. Wavell's fears about a possible mutiny in the Indian army or a revolt and civil war to be launched by Congress or between Congress and the Muslim League kept his mind focused on questions of how best to secure India or put down any uprising during his watch. The Delhi trial of less than a dozen I.N.A. officers had already stirred threats to kill "twenty English dogs" for every "Indian patriot" of the I.N.A. executed.[39] Although twenty thousand captured I.N.A. soldiers had been set free and only the worst war criminals were put on trial, Nehru's spirited defense of them as Indians—whose only crime was to have "loved their country too well"—roused such a wave of sympathy for "Netaji's boys" from Delhi to Calcutta, Bombay, and Madras that both Wavell and Auchinleck felt quite apprehensive about the "inner feelings of the Indian soldier," which British officers had been notoriously bad at accurately gauging, at least since 1857.[40]

Governor Richard Casey of Bengal met with Gandhi in early December, although he was never quite sure of why Gandhi came and then came back to him again and again, but Casey was politically sensitive and astute. Gandhi felt much more at ease talking with Casey than with Wavell. After many meetings with Gandhi, Azad, Patel, and Nehru also came to see Casey in Calcutta, which had recently experienced preelection communal riots as bad as those that rocked Bombay. "Clearly Patel and Nehru are firebrands of the outfit and are suffering from suppressed frustration, indignation and a rather hysterical impetuosity particularly on Nehru's part," Casey wired Wavell after that meeting. "Patel is coldly vindictive, Azad is more detached and theoretical."[41]

The day before meeting with Casey, Nehru's newspaper, the *National Herald,* attacked the United Provinces' governor Maurice Hallett's attempts to "pulverize" Congress. "If Sir Maurice and his political tribe will not quit India . . . there will have to be a revolution in this country. . . . As a matter of fact, we are already in the midst of one."[42]

Governor Arthur Hope of Madras, where riots had just begun, warned Wavell of the incendiary impact of the I.N.A. trial. "Thanks to Nehru & Co., whose example is being followed down here, a tremendous attempt has been made to make national heroes of the I.N.A. and the attempt has had considerable success among a large and emotionally unstable section of the public."[43] Wavell met with Gandhi the next day in Calcutta but found that he had "nothing special to say. . . . He admitted danger of violence and indicated that he was trying to reduce temperature."[44]

The secretary of state and the viceroy continued to write and speak of one India to which the British government hoped to transfer its powers. Cripps—alone, however, among those cabinet ministers on the India Committee—kept in touch with both Nehru and Krishna Menon, and before the end of December 1945, Sir Stafford informed Attlee and Pethick-Lawrence that "we might have to contemplate a division of India into Hindustan and Pakistan as the only solution. It would in that case be necessary to contemplate two Pakistans, one in the west and the other in the east." That first British cabinet warning as to South Asia's fragmented future reflected in part Nehru's own feeling, as expressed to Cripps, that it would be impossible to work in harness with Jinnah, and he would rather be rid of him than spend the rest of his life trying to change his "antiquated" mind or "awaken" him to "modern" reality.[45]

Neither Pethick-Lawrence nor Attlee was quite ready to accept Cripps's pessimistic prognosis, however, so the caravan lumbered on, and soon a parliamentary delegation was be flown to India to test its political climate. Labour's avuncular professor Robert Richards led the ten-member delegation, which included Agatha's friend Muriel Nichol and Krishna Menon's India League comrade Reginald Sorensen, as well as the brilliant young Major Woodrow Wyatt, who returned in 1946 as Cripps's private secretary. "The immediate reaction of the ever-suspicious Nehru and other Congress leaders was that they would not see us," Wyatt reported. "All Britain had to do was quit India, and Nehru saw the Parliamentary Delegation as a device for further delay. But Gandhi was curious to see the strange ragbag of British MP's." Wyatt later met Nehru, finding him "handsome, sharp, full of life, argument and strength . . . also irascible. I made a bad start with him. He was explaining that the Muslim League was a chimera. . . . [S]upport for the Muslim League would melt away. I disagreed. I explained that while he had been in gaol the previous year I had travelled in the Punjab and Bengal . . . Muslim League support was solid . . . I expected him to be irritated. I did not expect him to jump away from the chimney-piece he had been leaning on and shout at me, furious in the face. I was frightened but I held my ground. Gradually he became more rational and we talked sensibly for four hours. By the end he conceded that the British might have to agree to Pakistan."[46]

That December Congress won only fifty-five general seats in the Central Assembly, four less than it had before, whereas Jinnah's league swept all thirty Muslim seats. "The day is not far off . . . when Pakistan shall be at your feet," the triumphant Quaid-i-Azam promised his jubilant followers. As to Nehru's "thunder," Jinnah now mocked that "impetuous Pandit who never unlearns or learns anything and never grows old . . . nothing but Peter Pan."[47]

Reports by the parliamentary delegation, as well as other news of continuing shouting and violence whenever Congress and Muslim League speakers came within sight or sound of each other, yelling "Jai Hind!" (Victory to India!), the I.N.A.'s battle cry, and "Pakistan Zindabad!" (Victory to Pakistan!), the anthem of the Muslim League, convinced Pethick-Lawrence by mid-January 1946 that the time had come to launch another cabinet mission. The secretary of state would lead the three-man team himself, joined by Sir Stafford Cripps and the first lord of the admiralty, A. V. Alexander. These "three wise men," as they were soon called, would have full authority to decide on the spot all points at issue, and they planned to fly to New Delhi toward the end of March.[48]

Cripps had written to ask Nehru "what action I would lay down . . . after the election, if I happened to be the Viceroy." Nehru responded: "The first thing is for the British Government to declare in the clearest terms possible that they accept the independence of India and the constitution of free India will be determined by India's elected representatives without any interference from the British Government. . . . Further that the British Government . . . considers any division of India harmful . . . it would weaken the defence of India at a time when defence is a paramount necessity. . . . If a definite area expresses its will clearly in favour of separatism . . . no compulsion will be exercised to force it to remain in the Federation or Union. But it cannot take other areas away with it against their will, and there must be a clear decision by plebiscite of all the adult voters of that area. . . . Jinnah refuses both the plebiscite and the demarcation of the area according to the wishes of the inhabitants. It seems clear that he is not after Pakistan but something entirely different, or perhaps he is after nothing at all except to stop all change and progress. . . . Pakistan is just a fantasy. . . . Jinnah appears to be wholly intransigent and threatens bloodshed and rioting if anything is done without his consent. I do not think there is much in Jinnah's threat. The Muslim League leadership is far too reactionary . . . and opposed to social change to dare to indulge in any form of direct action. They are incapable of it, having spent their lives in soft jobs." It was Nehru's most serious error. He thought Jinnah was bluffing, that once confronted with the threat of force or the deflating reality of a much smaller, truncated Pakistan than what he wanted, Jinnah would simply back down.[49]

Nehru knew that most Englishmen in London were as sick of India as India was of the British Raj. He had learned from old Reggie Sorensen, Krishna Menon's friend who had stayed with him in Anand Bhawan during a week of the parliamentary delegation's tour of India, that the "smart set" back home were betting on Nehru to save both India and Britain from a further waste of manpower and money in hanging on to an empire no one but Churchill and Wavell wanted any longer.[50] Krishna Menon's labors in boosting Nehru's stock, not only with Cripps and Labour's backbenchers, but also with Edwina and Dickie Mountbatten and their set, were paying off. Mountbatten had tried unsuccessfully to visit Nehru in Ahmednagar Fort prison in 1944 but now was more anxious to meet him in Burma or Singapore, to discuss how most swiftly and efficiently to solve India's problems. Mountbatten had set his own sights on the viceroyalty of India even before the end of the war and asked Wavell to permit Nehru to leave India in order to help him placate the still restless and disturbed overseas Indian communities in Burma, Malaya, and Singapore.[51] Wavell had no objection, though the governor of Burma, Sir Reginald Dorman-Smith, refused to allow "rabble-rousing" Nehru to set foot in Burma as long as he remained in charge. So Singapore became the fallback venue for that first exciting meeting in March 1946 of the brilliant and beautiful trio, who would so swiftly change the nature of India's political destiny.

One month before Nehru left India for his rendezvous with the Mountbattens, some of his activist disciples almost inflated a Royal Indian Navy minimutiny in Bombay into a major revolt. On Monday evening, February 18, 1946, a "strike" was started by young ratings on the HMIS *Talwar* (Sword). They went ashore next morning, "seized a number of military trucks," and drove around Bombay, "shouting slogans," Governor John Colville reported to Wavell. Several officers and many petty officers aboard the twenty-two naval ships in Bombay harbor were soon "out with the men." Then the army and the air force were called in. The Royal Navy's cruiser *Glasgow* was on its way but "not expected to arrive till Saturday."[52] By Wednesday many of the ships were pointing their four-inch quick-firing guns at the city, having hauled down their Union Jacks and raised the Indian tricolor. By then the rioting had spread to the streets of Bombay. "There are crowds surging around. They are breaking windows. We have been asked to close our offices," sister Betty's husband Raja told her on the phone. "For three days and three nights the shooting and the rioting went on as the city rose in sympathy with the sailors," Betty recalled. "My brother, who was due in Bombay, sent messages to the mutineers. . . . The ringleaders must have known he was coming to stay with us, for one night the doorbell rang and we found four of them outside our flat. They looked terribly young, very earnest. . . . Raja was sympathetic to them as were most of the people of Bombay, in spite of their shooting cannons at us. He took them to see Sardar Vallabhbhai Patel. . . . It was Patel and Raja who finally persuaded these young lads to give up the ships and surrender, by telling them that it was a disgrace for Indians to behave like that. They must have discipline."[53]

S. K. Patil, then president of the Bombay Provincial Congress, told Governor Colville that "Nehru had come to Bombay on the invitation of the more fiery members of Congress, and against the advice of Vallabhbhai Patel."[54] Vallabhbhai and Nehru both addressed a large rally on February 26, which remained peaceful, thanks to Patel's stressing "the folly of disorder and violence." "My brother's first reaction was . . . sorrowful," Betty recalled. "At the same time my brother sympathized with the sailors who had genuine grievances."[55] Nehru told the Bombay crowd that "the first duty of an Indian soldier is to associate himself with the forces fighting for India's freedom. . . . [I]t is not a breach of discipline for an Indian soldier to refuse to quell riots." Another minimutiny in Karachi followed Bombay's. Four-inch shells were fired from the HMIS *Hindustan* at Sind's capital port while rioters in the streets attacked Grindlay's Bank and two post offices. The police opened fire, leaving eight dead and eighteen injured. Sind's governor, Sir Francis Mudie, placed the ultimate responsibility for the Karachi mutiny and riot on Nehru, who "has succeeded in communicating his own pathological hatred of the West and his indifference to the consequences of his action to the masses."[56]

Nehru returned to Allahabad in early March to find a letter from Cripps. "I suppose we shall meet before the end of the month," Jawaharlal responded. "I am going to Malaya on the 16th on a brief visit but I am anxious to be back by the time you come. . . . I am troubled at some indications of the British Government's policy. . . . I find a strange reluctance to use the word 'independence' with all that it conveys . . . that rubs people up in the wrong way. This is not a minor matter."[57]

That day Wavell wrote Pethick-Lawrence that "Azad has taken quite a responsible line recently. . . . Nehru, on the other hand, though advocating restraint for the moment is using the same sort of language that led to the Calcutta and the Bombay disturbances and that may have even more serious results in the future. It is difficult to know what to do about him, and I have no doubt that he embarrasses the Working Committee almost as much as he does the Government. . . . [H]e is practically inciting the services to mutiny. . . . Nehru's idea seems to be to build the largest possible bonfire into which he can throw a lighted torch if negotiations with the Cabinet Delegation do not go well."[58] If the cabinet mission failed to solve India's pressing problems, Nehru warned a large gathering in Jhansi on his way back to Anand Bhawan, "a political earthquake of devastating intensity will sweep the entire country. . . . We are sitting on the edge of a volcano which may erupt at any moment."[59]

Before leaving India on his flight east, Nehru addressed cheering crowds in Bengal and wrote to Indira from Calcutta's jute mill suburb of Howrah. "I have lost my Parker 51 to which I was getting rather attached. . . . This disappeared from my pocket in the crowd. . . . Before I go to

Malaya I should like to have another pen. Will you please send with Hari [his old valet] . . . the new Parker 51 sent by Kesh Naoroji for me. You will find this in the top left-hand drawer of my writing desk. . . . I leave by train tomorrow for Bombay." His last two days of electioneering in Calcutta had been "very tiring." He felt it more now "perhaps because I am growing older." He was thinking of founding a new society—the SPCJN—"Society for the Prevention of Cruelty to Jawaharlal Nehru" and wryly asked Indira next morning, "Will you be one of the office-bearers?"[60] Dr. Bidhan Roy, the Nehru family physician and soon to be elected the chief minister of Congress's government in Bengal, "has given me his pen and I am using it now," Nehru reported a day later from the train that took him back across all of India to Bombay. He was suddenly bored with campaigning, feeling restless and tired of waiting to take charge of India.[61]

Nehru flew out of Delhi, headed for Singapore on the eve of the cabinet mission's departure from London, headed for Delhi. His mission to Mountbatten, though much briefer and far less formal than the mission of Pethick-Lawrence, Cripps, and Alexander to India, had more momentous and terrible consequences for India's future than that of the three wise men. For the mighty cabinet mission trio were on their way out, whereas Jawahar, Dickie, and Edwina were on the way in, though few others knew it besides Krishna Menon, Indira, and a handful of Dickie's closest friends.[62]

The Mountbattens landed in Singapore one day before Nehru arrived. Dickie sent his own open-topped limousine to the airport to pick up Pandit Nehru, whose Royal Air Force plane touched down shortly after noon on March 18, 1946. Captain Ronnie Brockman was waiting to drive Nehru to Government House, where the handsome, charming Dickie welcomed him warmly. They hit it off beautifully at first handshake and had a most civilized chat, Nehru magically transported back to his halcyon years at Cambridge and in London as they drove through lines of cheering, flag-waving crowds of adoring Indians, Malays, and rather startled English troops, who lined the road six deep as they rolled passed the high open gates into Singapore's YMCA compound, where Edwina awaited them upstairs.

Forty-four-year-old Edwina Cynthia Annette Ashley Mountbatten was one of the richest, thinnest, fastest, most fearless women in Great Britain. She had devoted herself since the war began to volunteer services and was currently superintendent in chief of the St. John Ambulance Brigade. She flew everywhere, inspecting hospitals and visiting the sick and wounded the world over, and was checking up on the St. John Ambulance Indian Welfare Center recently opened inside Singapore's YMCA compound when Dickie and Nehru arrived. She and Nehru barely had time to greet each other before the surging crowd of excited, hero-worshiping Indians broke into the room, shouting "Nehru ki Jai!" (Victory to Nehru), knocking Edwina off her feet as they rushed to touch their hero.

"Lady Mountbatten was flat on the floor when my father and she met in Singapore," Indira recalled. "So the first thing they had to do, Lord Mountbatten and my father, was to rescue her and put her back on her feet."[63] Dickie and Nehru locked fists and linked arms, doing the "Rugger scrum" charge that Nehru had learned at Harrow to muscle the mob out of Edwina's way. Nehru lifted her up in much the way Feroze had once lifted up Kamala and fell just as swiftly in love with her. She was the perfect English lady, after all, the countess he had always dreamed of having, holding, keeping locked away in a secret chamber of his heart, all to himself. He admired Dickie as well, indeed, loved him in a different way, for Lord Mountbatten was to Nehru an ideal reincarnation of his most charming tutors and of his Cambridge and Oxford friends.

Dickie was not only the supreme commander of Southeast Asia but second cousin to the king and Krishna Menon's first choice to be the last British viceroy of India. They dined together that evening, those three who were soon to rule India, "and we made lifelong friends," Dickie recalled.[64] So many promises, spoken and unspoken, were communicated among them that night in the eloquent language of trust and love from Nehru's hypnotic eyes to the beautiful faces of Dickie and Edwina, each so sensitively attuned and receptive to the other that words were almost redundant. The next morning Dickie and Edwina flew off to Australia; Jawahar back to India. Their rendezvous in Singapore, though brief, proved most potent.

"Mac" (M. O.) Mathai was waiting for Nehru back home in Allahabad. Mac had volunteered his services as a secretary when he first met Nehru in Assam late in 1945. The tiny, young, militant Kerala-born confirmed bachelor—who immediately told Nehru that he had "no intention of marrying" and expected no salary, indeed, would take none—soon became Nehru's closest secretary and all-around aide-de-camp. Mac's rivals tried to poison Nehru's feelings for him by whispering that he was a CIA agent sent by the Americans to spy on Nehru and India. Nehru never believed such gossip, however, and Dickie came to like Mac almost as much as Jawahar and Krishna did. Mac was fearless, efficient, brilliant, and totally committed to his job and his chief, for whom he was ready at any time, day or night, to lay down his life, if need required it or Nehru requested it. He was the perfect gentleman's gentleman, in most respects more English than Indian, yet also an Indian patriot. He was always reading and had a gift for languages and a remarkable memory. Though much younger and smaller than Nehru, he was an interesting companion and conversationalist and together with Nehru's equally hard-working public secretary, Dharma Vir, soon filed most of Nehru's personal papers, answered many of his letters, and arranged his daily calendar for him. He did those things so well that Indira, who would have done them if it had not been for Rajiv and her second son, Sanjay, who was born before year's end, became intensely jealous of him. "The first impression Indira made on me," Mac recalled thirty-one years later, "was that of conceit. . . . Indira had a constant complaint against her father—that he always kept quiet at mealtimes, when they were alone, and never gave satisfactory answers to her questions. I advised her . . . to tell him amusing stories and jokes and make him laugh. This she could never do . . . Indira's taste in art bordered on the grotesque."[65] Indira resented her father's intimacy with anyone but viewed Mac as a singular rival for his affection as well as attention, much the same way she had always been jealous of Nan and soon became of Edwina.

The cabinet mission flew into Karachi on March 23, 1946. "We have come with but one purpose in view," Pethick-Lawrence announced to the awaiting press. "It is, in conjunction with Lord Wavell, to discuss with the leaders of India and her elected representatives how best to speed the fulfilment of your aspirations."[66] Nehru could hardly keep from yawning as he read Pethick-Lawrence's careful statement, promising "to complete the transfer of responsibility with pride and honour," though "the precise road towards the final structure of India's independence is not yet clear." Not clear to whom?

Before the end of March the cabinet mission met with the viceroy and his Executive Council in Delhi. They next met with the governors, each of whom flew into Delhi to see the wise men. Cripps met with Jinnah on March 30 and found him "calm and reasonable but completely firm on Pakistan."[67] Nehru reached Delhi on April 2 and stayed with his I.C.S. cousin Ratan (R. K. Nehru). Maulana Azad came to see him that evening and went next morning to meet the cabinet delegation and viceroy. Azad insisted, as Nehru instructed him, that Congress proceed only on "the solid basis of India's independence."[68] That afternoon, the delegation met Gandhi. Gandhi advised them immediately to release all political prisoners (J. P. Narayan and many others were still in prison) and to abolish the salt tax. Such actions would "produce a hearty friendship," the Mahatma argued. He then explained that he had met with Jinnah for eighteen days, but Jinnah had given him no concrete definition of Pakistan, which Gandhi called "a sin." "Jinnah is sincere but his logic is utterly at fault," Gandhi argued, "a kind of mania possesses him. . . . Let Mr. Jinnah form the first Government and choose its personnel from elected representatives in the country. . . . If he does not do so then the offer to form a Government should be made to Congress. After all, it is no light responsibility."[69] Gandhi thus proved his selflessness and the sincerity of his desire to save

India from more violence and conflict, the traumas of partition. He clearly believed that Jinnah deserved to be offered a chance to govern India, hoping such administrative responsibility and duty would mollify his "mania" and possibly save India from "vivisection." Nehru's name was never mentioned that afternoon by Gandhi or by any of the wise men or Wavell.

Jinnah came to meet with the delegation and Wavell the next morning on April 4, 1946, and argued that there "had never been any single Government of India" before to the British conquest and unification of the subcontinent. He insisted that "Muslims have a different conception of life from the Hindus" and therefore that Pakistan was imperative for them. The delegation questioned Jinnah about India's defense against external aggression as well as internal peace. "Mr. Jinnah said that they must assume that they would be handing over power to responsible people. The Muslims had not decided that they would have nothing to do with the British Commonwealth. . . . Hindustan and Pakistan must have common defence arrangements." Jinnah agreed to a "defensive alliance" and "mutual consultation" on foreign policy matters with "Hindustan." He also said that "inter-running communications" between the two States could be arranged. He insisted on a five-province Pakistan, which would not be "carved up" or "mutilated," rather, a "live State economically." He could not contemplate losing Calcutta because it had a Hindu majority, insisting that much of its Hindu population had been "brought there from outside."[70]

The cabinet delegation then met with several maharajas, representing the Chamber of Princes, and discussed the constitutional future of the some 570 princely states and the complexity of their relations with Britain and how these states would relate to a constitution-making body. It was all arranged to give the wise men a clearer picture of the problems confronting them on this most important of all British missions to India. Nehru sat alone in his cousin Ratan's bungalow nearby, angry at the "vulgarity" of the cabinet delegation's impolitic procedure. Only Cripps had called him privately to tea, and Sir Stafford's paternalistic advice, as usual, was to "trust us" and "be patient." "Jawaharlal is by nature warm-hearted and generous," Maulana Azad wrote of his colleague, with whom he had spent so many years in close prison association, adding, "Jawaharlal is however very vain and cannot stand that anybody else should receive greater support or admiration than he."[71] To receive recognition from Pethick-Lawrence, Cripps, and Alexander, as well as Wavell, Nehru now realized it would be necessary for him to take back the Congress crown, which he had discarded long ago. Much as he recoiled from the thought of presiding over future Working Committee meetings, Nehru knew that as prime minister he would have to get used to many longer, more boring, more frustrating meetings.

The Congress Working Committee met in Delhi during mid-April to choose its president for the next crucial year. There were only two candidates, Nehru and Patel. By now it was clear to all that the next president of Congress would be the first prime minister of India. Azad, who had never felt close to Vallabhbhai Patel, nominated Nehru, who was thus elected without much controversy. "I acted according to my best judgment but the way things have shaped since then has made me realise that this was perhaps the greatest blunder of my political life," Maulana Azad recalled, "not to stand myself . . . [or] not support Sardar Patel. We differed on many issues but I am convinced that if he [Patel] had succeeded me as Congress President he would have seen that the Cabinet Mission Plan was successfully implemented. He would have never committed the mistake of Jawaharlal which gave . . . Jinnah the opportunity of sabotaging the plan."[72]

The cabinet delegation worked virtually round the clock and came remarkably close to solving India's complex constitutional problems, constructing a three-tiered federal plan that was something of a work of confederational art, almost good enough to achieve the impossibility of reconciling Congress and Jinnah's league, granting India's Muslims the essence of Pakistan without putting all of South Asia through the horror of partition. It almost worked, for Pethick-Lawrence, Cripps, and Alexander were wise men, indeed, and had Attlee's full trust and the confidence of their entire cabinet in Westminster.

In early May they all went up to Simla, to the viceregal lodge. Nehru felt very depressed on the eve of that second Simla summit which Wavell hosted.[73] He considered the past month "entirely wasted," as he told the viceroy, finding no clear-cut solution emerging from all three wise men's talk, everything seeming to him to become "vaguer and more nebulous." He wrote that evening to Clare Boothe Luce, whom he had met in Delhi in 1942 and liked very much. "I have often thought of you. . . . I was unhappy in Delhi in 1942 and I suppose I am not too happy even now. . . . Yet these four years have made a tremendous difference to all of us and . . . I believe I have grown somewhat more detached than I used to be. . . . I wonder when we are likely to meet . . . I should like very much to do so, but most of us seem to be prisoners of circumstance and can seldom do exactly what we want. . . . In this world of turmoil and conflict and hatred, it is not easy to look forward with confidence and optimism. I am afraid that we are going to have a hard time in India as elsewhere. But I do believe that India will emerge not only as an independent nation but as something different and at the same time fundamentally as she has been. I think she has a part to play in the world. . . . At present the outlook is not good. One thing is certain: that India cannot remain where she is or go back. Changes and big changes must come.

Whether those changes will be for the good or not I am still unable to say."[74]

Nehru stayed with Maulana Azad at "The Retreat" in Simla. It was a big house with many bedrooms, and soon his adoring Padmaja and her garrulous mother, Sarojini Naidu, moved in as well. Indu, who was pregnant again, went with Feroze and Rajiv to Naini Tal, where Nehru had hoped to join them after the Simla summit ended, but that proved impossible. Jinnah remained intransigent, refusing even to talk to Azad. After five futile plenary sessions, they all decided that perhaps the best hope of breaking the deadlock was to ask Nehru and Jinnah to meet alone, to try to resolve Congress–League differences. But nothing changed. Jinnah refused even to accept the principle of turning the question over to an "international umpire." Nehru became more bored, tired, impatient, enraged at the stupidity of the entire process. Jinnah grew weaker physically and withdrew behind his wall of defensive silence. Each of them by now hated the other with cold and burning fury that reflected the violent passions of communal hatred soon to be unleashed all across north India from the frontier through Punjab to Bengal, claiming more than a million lives in less than a year.

After a seventh plenary session in Simla, they all agreed it was hopeless. Pethick-Lawrence announced that the delegation would put out a statement before week's end and asked both parties to refrain from publishing any documents or making any public press announcements until then. Pethick, Cripps, and Alexander then agreed to make one desperate final effort, presenting both sides with their own formula, which would have to be accepted in its entirety or rejected. That final scheme would embody all points of agreement earlier submitted in writing by Jinnah and Azad, which the wise men tried to integrate into a single federal formula.

Commander in chief Auchinleck now informed the cabinet delegation that there was no hope of keeping the Indian army intact if one part of India separated. He thought that if the political leaders refused to accept a cabinet mission scheme, they would call on the army to "join them in a war of independence. Jai Prakash Narain was already making speeches which went beyond all bounds."[75] J.P. had learned well the techniques of revolution from his mentor J.N..

On May 16 the cabinet mission released its scheme to the press, and Lord Pethick-Lawrence broadcast its meaning to India. "There is a passionate desire in the hearts of Indians . . . for independence," he announced. "His Majesty's Government and the British people as a whole are fully ready to accord this independence." He then explained the three-tiered plan that would allow Muslims to "secure the advantages of a Pakistan" without suffering its risks and the disadvantages and pain of partition to all of India. As for princely states, which covered one-third of

South Asia and included almost one-quarter of its some 400 million people at this time, Pethick-Lawrence admitted that they would "wish to take part in the constitution-making process" but that his delegation had not yet decided anything about just what action could be taken to bring princely states into the new union. Hearing Pethick-Lawrence say that, Nehru decided it was time to take preemptive action of his own. What better time, after all, than before another British mission arrived, full of well-meaning Labour leaders who would bore him with all their long-winded talk about how important it was to mollify the princes? "The constitution for India has to be framed by Indians and worked by Indians," Pethick-Lawrence concluded. "The responsibility and the opportunity is theirs and in their fulfillment of it we wish them godspeed." Soon Nehru would act with godspeed, dashing off to Kashmir with Mac to court arrest, as he had earlier done in Nabha.[76]

Gandhi's reaction to the cabinet delegation's scheme was that "whatever the wrong done to India by British rule, if the statement of the Mission was genuine, as he believed it was, it was in discharge of an obligation . . . to get off India's back. It contained the seed to convert this land of sorrow into one without sorrow and suffering."[77] Jinnah could not come to any decision about the statement without first consulting his Working Committee. He pleaded for time, arguing that several initial reactions from Muslims to the statement were "very strong." He begged the delegation not to be "too hasty."[78] Jinnah sent Liaquat to Delhi to ask for more time, since he felt too sick to leave Simla himself.

Cripps also was sick by now, and both Pethick-Lawrence and Alexander were exhausted. Wavell was daily growing more anxious about Congress's preparations for a mass civil disobedience movement, which made him focus on a fallback plan for a British withdrawal to anticipate what he feared might become a South Asian Dunkirk. On May 22, 1946, he briefed the delegation on "dangerous possibilities," noting especially that "outside . . . agitators" were being sent into Faridkot state in Punjab and that J.P. might "at any time start trouble"[79] in Kashmir. Faridkot was close to Nabha, and although the viceroy was right about its dangerous potential, he did not realize that it was not only J.P. orchestrating the State Peoples Conference agitation in both northern states. The popular forty-year-old Kashmiri Sheikh Mohammad Abdullah, Nehru's other most active political disciple, led a vigorous national conference protest against the maharaja of Kashmir and his prime minister, Pandit Ram Chandra Kak, on May 20–21, 1946. The charismatic sheikh, hereafter hailed as the "Lion of Kashmir," rallied such crowds in Srinagar to his passionate speeches in emulation of his mentor, Nehru, that the military was called out, opened fire, and killed one person. A curfew was then imposed on the capital of Kashmir, and all outsiders were banned from entering the state.

Three days later, on the eve of his departure from Delhi to Faridkot, Nehru wrote Wavell to "let you have a glimpse of what I have in mind. . . . I feel profoundly depressed and disappointed at the turn events have taken. We came to Delhi two months ago with some hope that at long last a way might be found for a peaceful settlement of the Indian problem. . . . We began these long and interminable discussions. Repeatedly we were held up, as we are today, because someone wanted more time. . . . Our people have hardened . . . their expectations have risen. . . . I wrote to you about Faridkot some days ago . . . nearly a month since this affair started and. . . . chaos there. It shames me. . . . I have decided to go to Faridkot. I shall leave Delhi tomorrow."[80] Wavell called Nehru in to see him before leaving Delhi next day, and on May 27, when Nehru reached Faridkot state, the raja met with him and agreed to lift his ban on entry by members of the All-India States Peoples' Conference, also releasing those who had been jailed.

Nehru appealed to the viceroy regarding Kashmir, protesting the arrest of "my friends and colleagues," primarily Sheikh Abdullah, who was vice president of the States Peoples' Conference, over which Nehru himself presided. He called the viceroy's attention to incitement to communal antagonism by troops who dared occupy the "inner shrines of the mosques" and demolished a wall of the Jama Masjid in Srinagar to "make a passage way for military lorries." Wavell promised to look into the matter of Kashmir as well as Faridkot. By mid-June, however, Abdullah was still in jail, and Nehru was anxious to go to Kashmir, as he wrote Indira. He reserved seats for Mac and himself on a flight from Delhi to Rawalpindi on the morning of June 19, 1946, and planned to drive with Yunus from Pindi to Srinagar.[81]

"After weeks of exhausting talks and debates and a gradual oozing out of spirit and vitality, I have had a little tonic today," he wrote Indira from Domel in Kashmir. "I had hoped that this very brief Kashmir visit would provide a diversion. My hopes have been amply justified. . . . So we must offer thanks to the Kashmir Govt for the astonishing folly with which it conducts its affairs. It tried to stop me at Kohala. It did not succeed. I spent five hours there getting more & more bored. Then we sallied out and faced the Kashmir police and pushed forward. I felt in my element as I always do when there is a question of forcing a barricade. . . . Ultimately the police retired and we drove on. . . . What an odd mixture is my life. There is talk of a Provisional Govt and at the same time I function as a law-breaker!"[82]

From Domel they drove up to Uri, where he and Yunus and Mac spent the night in a Dak bungalow, cooled by winds from the world's highest peaks. "Today Uri, tomorrow where?" Nehru wrote his daughter that night. "We have been served with orders of detention under the Defence of Kashmir Rules. We are detenus. . . . I imagine I have managed to upset many an applecart. Well, what am I to do about it? If the Kashmir Govt wants to behave with crass stupidity and discourtesy, things will happen. I offered them an opportunity to adopt a correct course gracefully and without loss of dignity. They were too conceited. . . . And so the Kashmir forces are on the move and the whole political situation in India is affected. Frankly, I am not at all sorry. Let people realise the forces that are astir in India and not imagine that everything can be settled by using a few soft words."[83]

Nehru's daring dash to Kashmir did not secure Sheikh Abdullah's release, however. The brave Lion of Kashmir was sentenced to three years for calling on the playboy maharaja Hari Singh to "Quit Kashmir" in May 1946. A year later, Nehru, whose power was second only to the viceroy's, assured his friend's wife that "in spite of heavy preoccupations with vital problems, my mind has frequently turned to Kashmir. . . . I have thought often of Sheikh Saheb suffering imprisonment and I have felt distressed . . . Kashmir is dear to me. . . . Being a Kashmiri, I can never forget it and I am passionately attached to its mountains and . . . scenery. . . . Kashmir and its people . . . will ever remain a first priority with me."[84]

Mountbatten helped Nehru win the early release of his Kashmiri disciple, but a few years later, when the Lion roared too loud, showing Muslim claws, Nehru put Sheikh Abdullah back behind bars, keeping him there for more than a decade of India's freedom.

25

Provisional Prime Minister

1946–1947

"YOUR MAJESTY," Viceroy Wavell wrote King George VI from Simla in early July 1946, "I do not think any men could have worked more wholeheartedly and with greater patience and good temper than did the Mission. . . . They came near success. . . . But at the last moment Gandhi . . . threw a spanner in the works at the Congress end; and Jinnah chose that moment to give the Press an intemperate letter he had written to me about the attitude of Congress. . . . And Nehru at the same critical juncture went off on a quite unnecessary and provocative expedition to Kashmir, mainly for reasons of personal prestige and vanity. . . . I have seen much of Nehru and cannot help liking him. He is sincere, intelligent, and personally courageous. But he is unbalanced."[1]

While Wavell was in Simla writing that report, the A.I.C.C. met in Bombay to vote for its new president, Pandit Nehru, to replace Maulana Azad. "I told Jawaharlal Nehru that he must wear the crown of thorns for the sake of the nation and he has agreed," Gandhi announced to his Congress friends. "The constituent assembly is going to be no bed of roses for you, but only a bed of thorns. You may not shirk it. . . . It is not a prize to be sought as a reward for sacrifices, but a duty to be faced, even like mounting the gallows."[2]

On July 10, 1946, two days after he was elected Congress president, Nehru spoke to the press. He dropped many veils that fateful midsummer's day, insisting that Congress had made "no commitment" to the cabinet mission or the viceroy concerning the Constituent Assembly. "We agreed to go into this Constituent Assembly," Nehru informed the battery of reporters, "and we have agreed to nothing else. . . . What we do there, we are entirely and absolutely free to determine."[3] At the heart of the cabinet mission's three-tier plan were "groups of provinces," the northwest group, essentially Pakistan (Punjab, Sind, the Frontier, and Baluchistan); the northeast group, dominated by "Bangladesh" (Bengal and Assam), would also have a Muslim majority. The rest of India would be a third group, overwhelmingly Hindu. After months of careful formulation, the mission had finally convinced Jinnah to accept their confederal plan, thanks to its vital middle group in the formula. Each group would have virtual autonomy over its internal affairs, the top center a relatively weak body charged with external affairs and overall defense. Nehru, however, now predicted "the big probability" that there would be "no grouping," since a majority of the elected members in the Constituent Assembly, to be controlled by Congress, opposed the idea.

The next morning when Jinnah read in every Bombay newspaper what Nehru had said, he resolved that he could not trust this mercurial "Peter Pan" Pandit, whom he had never liked but who had seemed of late somewhat more responsible thanks perhaps to the combination of age and hard prison years. Now Jinnah sensed that there was no change in Nehru's negative attitude toward himself or his league. He had believed ten years ago that separate nationhood might be the best solution for India's Muslims and had led his league so to resolve in Lahore three years later. Then the cabinet mission's clever plan seemed a potentially plausible alternative, assuring him the essence of Pakistan without harsh surgery. But now he was jolted rudely from that dream of ever working in harness with Nehru and his Congress. Jinnah knew he was dying but resolved before he did so to bring Pakistan to life. "All efforts of the Muslim League at fairplay, justice, even supplication and prayers have had no response of any kind from the Congress," Quaid-i-Azam Jinnah told his Muslim League Council on July 28 in Bombay. "The Cabinet Mission have played into the hands of the Congress. . . . Pandit Jawaharlal Nehru as the elected President . . . made the policy and attitude of the Congress . . . clear. . . . Congress was committed to nothing."[4] The following day, Jinnah's council unanimously voted to empower him to "resort to Direct Action to achieve Pakistan."

"We have taken a most historic decision," Jinnah announced, after the final league vote. "To-day we have said good-bye to constitutions and constitutional methods."[5] Friday, August 16, 1946, was chosen by the league as "Direct Action" Day, launching a year of violent civil war in which countless thousands of innocents would be slaughtered in anticipation of the birth of two dominions, India and Pakistan. Many years later, shortly before he died, Nehru reflected that it was a mistake for him to have said what he did, with such arrogant self-assurance, at that press conference in early July 1946, when all of India was still so calm and remarkably peaceful. What a precious moment that was, what an opportunity for friendly, generous overtures of cooperation by the new president of Congress to old Mr. Jinnah and his cadre of anxious Muslim League lieutenants, all nervous at the prospect of facing a hostile "Hindu Raj" the day after the British troops went home. But there was no turning back.

"Why is Mr. Jinnah perturbed and angry?" Nehru replied in an editorial of his *National Herald* in mid-July. "Had he any doubt at any time about the Congress attitude in regard to this or any other matter? Surely in the forest of letters and talks during the past three or four months this Congress attitude has emerged clearly enough. The Congress stands for a sovereign constituent assembly with no external limitations whatever. . . . The minority problem is with us certainly and no one can just wish it away. That is the real limitation in the way of our progress, and it might give trouble. . . . Lawyers and constitutionalists may ponder over these problems, but there is something beyond the lawyer's textbook and precedent. . . . [V]ital forces are at play and sometimes in conflict. . . . For the old order changeth yielding place to new and Caesar too passes into the story of things that have happened and ceased to be."[6]

To his comrades in Delhi a few days later, Nehru announced that "the transfer of power may be peaceful or may be the result of another struggle. We should be ready for every contingency. . . . The British Government has been forced by the circumstances to start negotiations for a settlement. . . . At the same time, it is making preparations for crushing any movement that might be launched by us. . . . We must also be prepared for a fight, if a fight comes."[7] Wavell met with him again a day late and found him "very quiet and sensible" and "very friendly throughout" that forty-minute meeting.[8] Wavell was now ready to invite Nehru to join his council as the leader of a provisional "national government," which would be inaugurated in less than a month. First, however, Nehru went to Kashmir on a brief visit and called a meeting of his Working Committee at Wardha for August 8–10, which best suited Gandhi's convenience.

Nehru went to Kashmir with two of Subhas Bose's leading I.N.A. officers, Major General Shah Nawaz and Colonel Habibur Rahman, the latter of whom soon opted for Pakistan. When their car reached Srinagar, Nehru came out and was "almost mobbed by friendly people," riding on the running board and accepting their cheers with obvious pleasure. "A few dozen persons followed the car shouting slogans. Repeatedly the police drove them away with lathis. Twice I got down from the footboard and tried to intervene. . . . [A] few persons surrounded me and a policeman and one member of the crowd slapped the policeman. I remonstrated with him." No other such incident occurred, however, and Nehru was permitted to meet four times with Sheikh Abdullah alone in his prison cell. They talked at length, without interruption or visible monitoring. Nehru still liked and trusted Abdullah but later came to fear him and suspect him of trying to lure Kashmir away from India, either to keep it independent or to turn it over to Pakistan. Nehru also wanted to meet the maharaja on this trip, but the elusive ruler claimed to be "unwell."[9]

Back in Allahabad in August, Nehru addressed a youth rally of volunteers near his home, stating: "I want every boy and girl of India to become a soldier in the cause of the independence of the country. By soldier I mean a disciplined and honest worker."[10] He knew that Congress would need all the help it could muster in the arduous months ahead as he prepared to take charge of the provisional government in Delhi. On the eve of the Working Committee meeting he wrote to all members, advising them of the long and heavy agenda he had drafted for them. He planned to revitalize Congress and expand its base of support and economic as well as political activities nationwide. He proposed recruiting a volunteer army for Congress, which he put under the military command of Major General Shah Nawaz, Bose's lieutenant, whose life sentence in the Red Fort trial had been commuted. Nehru also suggested attaching a women's section to that corps. More money would be required to achieve all that he had in mind, of course, but his colleagues in Congress were happy to support him in everything he requested, knowing that he soon would be running the country.

Indira, five months pregnant with Sanjay, who was born that December, was back at Anand Bhawan to take care of her father during his brief stay there before leaving for Wardha. She worried about his tiresome pace and hectic life on the brink of assuming power, and as soon as he reached Wardha she wrote to remind him of how important he was to the entire nation as well as herself and Rajiv. "I wish you would also think of . . . yourself," She wrote. "Try to drink orange juice every day, especially when you are tired. Much love from Rajiv."[11]

The Great Calcutta Killing started in Maniktolla on Friday, August 16, 1946. The Muslim League's chief minister, H. S. Suhrawardy, proclaimed the day a public "holiday"; hence there were no police or other officials on the streets of India's most populous city when the bloodletting began. Mobs of murdering Muslims ran wild in every Hindu neighborhood, killing or maiming any person they could find. Soon the retaliatory mobs of frenzied Hindus lashed out in opposite directions, and with the brutal murders came panic, with thousands fleeing Kipling's "City of Dreadful Night," whose stench of blood would never leave its Bengali palaces or darkest slums. "It was unbridled savagery and homicidal maniacs let loose to kill and kill and to maim and burn," General Tuker reported. "The underworld of Calcutta was taking charge." In the next few days some ten thousand people were murdered in Bengal alone.[12]

On September 1, Wavell accepted Nehru's list of members to administer the provisional national government of India, headed by himself as minister of External Affairs and Commonwealth Affairs, with Vallabhbhai Patel in charge of Home Affairs, and Rajendra Prasad handling Agriculture and Food. The viceroy appointed Nehru as vice president of his new council, expecting him "to act when I was absent," and Wavell also suggested calling it an "interim government." Nehru accepted his suggestions. Wavell then noted that J.P. had just become a member of the Congress Working Committee and asked Nehru if that meant J.P. was becoming "constitutional" or the committee was becoming "revolutionary"? "Nehru laughed and said 'Both, I hope.' " Wavell invited Nehru and his sister Nan to dinner and found him "quiet and friendly" all evening. As he was about to leave Government House, Wavell reported, "I congratulated him on his courage and statesmanship in coming into the Government."[13]

"The door to Purna Swaraj has at last been opened," Mahatma Gandhi told his New Delhi prayer meeting at Birla House the next day. He then called on the new government to "remove all vestige of the salt tax."[14] Wavell would not be alone among those in his interim government who considered Gandhi "a tiresome old man." The swearing-in ceremony of the new government went off well, Wavell told Pethick-Lawrence. "Nehru seemed a little depressed and overwhelmed but he produced no fireworks except a very quiet 'Jai Hind' at the end of the last oath." That "Victory to India" slogan had been the battle cry of Bose's I.N.A. Agatha Harrison suggested to Pethick-Lawrence that Nehru might now be invited to broadcast a message to Britain, but Wavell was strongly opposed to any such idea, arguing that it was best to get Nehru to "say as little and as seldom as possible," since "crowds and press conferences always go to his head, and the microphone might do the same."[15]

Attlee, Pethick-Lawrence and Cripps still hoped it might be possible to bring Jinnah into the interim government, thus averting further communal disasters. But none of them quite understood how incompatible Nehru and Jinnah were. So they planned to invite Nehru and Jinnah to London before year's end, to try another summit session at 10 Downing Street and in Whitehall. Nehru insisted that he and his colleagues in the new government be called a cabinet, rather than a council, and Wavell agreed, though he was not yet willing to concede to Nehru the title of prime minister. Administration weighed heavily on Nehru's mind. He wrestled almost daily with temptations to resign. He found Wavell almost as difficult to deal with as Wavell found him. Before the end of his first month in power, Nehru sounded like a British character out of a book by Kipling, complaining to Pethick-Lawrence that "for the present the burden is heavy."[16]

The following week, Nehru wrote Krishna Menon that "Jinnah is certainly anxious to come in or, at any rate, to send his men in. I think he has got thoroughly alarmed at the prospect of continuing riots."[17] Nehru sent this letter for hand delivery to Krishna Menon by his cousin Ratan K. Nehru, whom he had appointed to lead the Indian delegation to the preparatory session of the International Trade and Employment Conference that was to be held in London. Nehru was sending Krishna Menon to the United Nations and wanted him to go on from there to Russia, to speak with the representatives of all Asian countries and the Soviet Asian Republics at the General Assembly, urging them all to come to Delhi to attend the Inter-Asian Relations Conference, which Nehru was eager to host. Krishna served as his unofficial personal envoy all over the world and was soon appointed high commissioner in London. He went first to Paris, where he met with Molotov, asking the Soviet foreign minister for food grain for India. Molotov was ready to develop friendly relations with India but regreted that no food was available at this time.

"Nehru looks very tired," Wavell reported on October 1. "I understand he is working extremely hard. . . . Like most Indian politicians when they first come into office, he is besieged by callers who make all sorts of unreasonable demands on him. . . . He has a mercurial temperament, and unless he is able to organise his life so as to reduce the strain on him, I foresee a breakdown." He took the night train to Allahabad most weekends, enjoying hours alone with Indira and Rajiv at Anand Bhawan more than any other moments in his newfound power. He also got a good night's sleep on the train back to Delhi from Allahabad.[18]

Wavell had pressed Jinnah to join the interim government, and by mid-October Jinnah agreed to submit a list of five names as Muslim League candidates. When the viceroy told Nehru of Jinnah's decision, Nehru found it "frankly amazing."[19] He was chagrined as well as surprised at this turn of events and now had to decide which of his own cabinet choices to remove in order to make room for Jinnah's team. Fresh reports of the killings of many Hindus in Noakhali District in East Bengal left Nehru "greatly perturbed," so he told Wavell that "we must face this issue somehow or else we retire from the public scene."[20] He felt ready to give it up, finding the prospect of trying to handle a Congress–League coalition cabinet daunting.

Nehru insisted on taking a tour of the frontier, despite Wavell's strong advice against it. The tour was a disaster. Governor Sir Olaf Caroe called it "the unfolding of a new act in a Greek tragedy on the old theme of hubris followed by nemesis. I had never met Nehru before our meeting last week in Delhi, but . . . one feels . . . his intellectual arrogance, and I could not help noticing how like he is to his friend Madame Chiang Kai Shek. In a sense during his visit here he showed courage, but it was courage better described as bravado, with something feminine in its composition." Nehru called the tribesmen "pitiful pensioners," making the governor wonder whether if Congress decided to end those payoffs the tribes would rise. Returning from Waziristan to the Khyber, Nehru found no Afridis waiting to welcome him. He reviewed the Khyber Rifles at Jamrud, however, and went to the Afghan frontier, but on his way back, near Landi Kotal, the stone throwing started. One of the stones broke a window in Nehru's car. He was unharmed, but soon after that his Khyber Rifles escort opened fire, and the tribal mob disbursed. Farther along, however, in the Malakand, a second batch of league-inspired stone throwers did injure Nehru (bruising one ear and his chin), as well as Dr. Khan Saheb and his brother Abdul Ghaffar Khan, forcing their guards to open fire to avoid disaster, and obliging Nehru and his party to leave by another road. At Abdul Ghaffar Khan's home base in Sardaryab, however, Nehru was warmly welcomed by thousands of cheering "red shirts." Caroe cautioned Wavell that keeping Nehru "or any other Hindu" in charge of tribal affairs would prolong the disaster and "probably lead to tribal risings." India's North-West Frontier Province soon became part of Pakistan.[21]

Before the end of October, the league's Liaquat Ali Khan was sworn in as finance minister, and four other of Jinnah's nominees took less important portfolios in the interim government. A few days later, Nehru could not keep an appointment with Wavell and wrote vaguely that "something happened to me which does not usually happen. Various factors combined together with accumulated fatigue of mind and body knocked me over last night."[22] He said "mental strain" had caused his breakdown but then explained it was really compounded by the difficulty he felt in daily dealings with "our new colleagues." He also informed Wavell that "the attitude of HMG has been rather frigid to us" and intimated that he might decide to resign. A few days later Nehru resolved to stay on. "Having come to this decision my mind feels clear and light," he wrote Indira from Bihar, where communal rioting, ignited by news from Noakhali's killings, claimed hundreds, if not thousands, of Muslim lives. "Hindu peasantry . . . have been terribly cruel—the lust for murder is a horrible thing and to it is added arson and looting. . . . Mass misery has a curious numbing effect."[23]

"The riots in Bengal, Bihar and elsewhere make dreadful reading," Pethick-Lawrence wrote that first week in November. "They have not been without effect on Nehru. He is an incalculable person and I confess I was astonished to see that he had referred to the possibility of bombing [the] area in Bihar where disturbances continue. On this Gandhi seems to be in conflict with him."[24] Years later, after he became prime minister, Nehru was again angry enough to order the bombing of Bihar. Then the Congress president, his former disciple J. B. Kripalani, confronted him, shocked at the idea, shouting, "Are you crazy, Panditji, wanting to bomb our own people? How can you *think* of such a thing?"[25] Pethick-Lawrence also believed by now, from the personal experience of listening to him, that "the influence of Krishna Menon . . . is likely to be an unfortunate one."[26]

On November 11, Wavell called in Nehru and Liaquat to urge them to try harder to support the provincial ministries and, if possible, to form coalition ministries in communally troubled provinces. "At mention of coalitions in Provinces Nehru suddenly blew up in characteristic fashion and denied the existence of a Coalition at the Centre since the Muslim League members declined to recognise him as de facto Premier or to attend his daily 'Cabinet meetings.' I reminded him of the Constitutional position," Wavell reported, "whereupon he proffered his resignation. I took no notice of this or of its subsequent repetition. . . . Liaquat remained calm and said that Muslim League members had every intention of cooperating. . . . Nehru eventually calmed down." Meanwhile, the communal disturbances continued in many parts of India, the daily death toll mounting. "Nehru's government" was now being criticized by many of those Hindus in Calcutta and East Bengal, who had thought that with Congress leaders in high office they would be protected.[27]

Nehru urged Wavell to convene the Constituent Assembly in early December for its first meeting, to establish procedures and appoint committees for drafting the constitution. Jinnah refused, however, to consider entering the Constituent Assembly at this time, insisting that Congress had provoked the massacre of Muslims in Bihar, where he claimed to have reliable estimates of about thirty thousand killed, though later official estimates were between five thousand and ten thousand. The Muslim League now considered itself faced with civil war.[28] At his next meeting with Nehru, Wavell "questioned the wisdom of sending Krishna Menon on his tour of European countries without first discussing the matter with the Muslim League, since foreign relations were a Central subject, and it seemed inadvisable to send someone who was not only a noted Congress propagandist, but had hardly been in India at all for many years."[29] Little did Wavell realize at this time that his own days as viceroy were numbered.

"The Viceroy has failed to carry on the Government in the spirit in which he had started," Nehru was reported to have told the Subjects Committee of Congress, adding: "After coming into the Government, the League had been endeavouring to establish itself as the King's Party. . . . I warn the Viceroy that our patience is fast reaching the limit." Nehru now disclosed that he had thought of resigning no less than "fifty times" but had mentioned it only "twice . . . to the Viceroy." Nehru's decision to stay in government, however, made him resign as Congress president, passing the crown at this Working Committee meeting to Kripalani.[30]

Cripps volunteered to go out himself to replace Wavell, but Attlee refused to agree, nor would Churchill or the opposition support such a move. Attlee invited Nehru and Jinnah to London in early December for one last try at bringing them together on the eve of the scheduled first meeting of the Constituent Assembly on December 9, 1946. Nehru decided to take his Sikh war minister, Baldev Singh, with him to London. During that trip, Nehru resolved to keep Baldev as minister of war in his first cabinet. Jinnah went to London with Liaquat Ali Khan, although until their departure it was not clear whether or not they were willing to go so far for what seemed to the ailing Jinnah a hopeless attempt at reconciliation with a party he now considered the worst enemy of the Muslim nation.[31]

Mountbatten offered to be Nehru's host[32] in London, but Nehru decided it was best for him to stay at the Dorchester, where Edwina[33] kept a suite for herself, overlooking Hyde Park. It was a more restful hideaway after those long, boring sessions with Attlee, Pethick-Lawrence, and Cripps. Wavell flew to London for this summit, but it would be his last. He still had no intimation, however, that he was on his way out of high office. Pethick-Lawrence met with Nehru in London on December 3. "Pandit Nehru said that though there was no personal unfriendliness between him and Mr. Jinnah he had never succeeded in getting any response from him. . . . He saw no hope of reconciling Mr. Jinnah and thought it would be wrong to try to appease him as a result of violence."[34]

Pethick-Lawrence found Nehru "dis-spirited; no doubt the fatigue of the journey played some part in this." His journey back to London was, however, a much longer one for Nehru than the old secretary of state could imagine. He also met with Jinnah that day, finding him "very bitter and determined . . . like a man who knew that he was going to be killed and therefore insisted on committing suicide."[35] Cripps met with Nehru, and Alexander with Jinnah and Liaquat at Claridge's, where they stayed. Attlee met with them all every day that first cold week in December, when none of the ice dividing Congress from the Muslim League was broken, no magic formula was found for bridging the ocean of mistrust that separated Nehru from Jinnah.

Nehru flew home with Baldev Singh to attend the first meeting of the Constituent Assembly. Jinnah and Liaquat stayed on in London. Attlee now warned his cabinet of the possibility of "civil War in India, with all the bloodshed which that would entail."[36] Vallabhbhai Patel, still angry at being left home by Nehru during the London summit, announced in Bombay that "whoever else might resign from the Government, he had no intention of doing so."[37] The "Sardar," as Congress's "strongman" was called, was determined to stay and solve whatever problems remained, rather than running away from them. He had long viewed Nehru as a "weak sister" and often wondered why Gandhi thought so highly of him.

The first meeting of India's Constituent Assembly was convened in New Delhi on December 9, 1946, with the Muslim block of seats conspicuously empty. Dr. Rajendra Prasad was elected first president of the assembly but turned that job over to Dr. B. R. Ambedkar after independence, himself soon to become the first president of India. "Untouchable" Dr. Ambedkar saw to it that untouchability was "abolished" by India's federal constitution, which became the highest law in India after January 26, 1950, the thirtieth anniversary of Purna Swaraj Day, since then celebrated throughout India as Republic Day.

Wavell informed Attlee and the cabinet on December 11, 1946, that if "Congress leaders" were now to "stimulate mob violence, the destruction of railways and Government buildings," the best course of action would be to withdraw British troops from all Hindu provinces, leaving them to protect the northwest Muslim provinces, which would become Pakistan. Attlee and Cripps rejected the idea, knowing how negatively it would be viewed in Washington and Moscow. Wavell warned, however, that with fewer than five hundred British ICS and another five hundred British police left in India, there was "not the machinery of government sufficient" to permit the Indian government to function for more than eighteen months. The spring of 1948 thus became the British cabinet's deadline for the final transfer of British power in India. But to whom?

On December 13, 1946, Nehru addressed the Constituent Assembly for the first time, rising to introduce the objectives resolution, which stated the basic principles underlying the constitution yet to be drafted. "We should . . . always keep in mind the passions that lie in the hearts of the masses of the Indian people and try to fulfil them," Nehru told his Congress audience, "sorry" to see "so many absentees" in that room devoid of Muslim League or princely state delegates. "We shall have to be careful that we do nothing which may cause uneasiness in others," he cautioned. "It has ever been and shall always be our ardent desire to see the people of India united together." But even as he was speaking, the civil war raged on, reaping its early harvest in remote villages of Bihar and Bengal, and in crowded slums of Calcutta and Bombay.[38] No resolutions, no rhetoric would suffice anymore to calm the winds of communal hatred.

Less than a week after Nehru spoke to the assembly, Attlee called Dickie Mountbatten to 10 Downing Street and asked him to take up another impossible burden, Wavell's job. The Labour government hoped to leave India in proper "royal style," rather than "scuttling" the Raj as Wavell proposed, running off with troops protecting the last civilians to depart. What better image than Mountbatten's as the last viceroy? Though

Nehru later told his good friend, the U.S. ambassador to India, John Kenneth Galbraith, "I'm the last Englishman to rule India!"[39] He meant it, of course, and was proud of it.

Cripps was in Attlee's office on December 18, 1946, when Dickie was offered the viceroyalty. Sir Stafford eagerly offered to go out with him as his chief of staff, but Dickie shuddered at the prospect, politely insisting it that would be "too great an honour."[40] Dickie wrote his cousin Bertie (King George VI), "I don't want to be hamstrung by bringing out a third version of the Cripps offer!!!"[41] Instead, Mountbatten's choice as his chief of staff was "Pug" (General Sir H. L.) Ismay, Churchill's right hand at Defense throughout the war. He also insisted on taking his close friend Alan Campbell-Johnson with him as press attaché, knowing what a good job Alan would do on his world press briefing "mission with Mountbatten."[42] He also took along Sir Eric Mieville, the king's assistant private secretary, whose experience in India went back more than a quarter century, urging him "to chuck everything and come with me to start the last Chukka [Polo game] in India—12 goals down!"[43]

Before leaving, Dickie told Attlee that "my wife and I would wish to visit Indian Leaders . . . in their own homes and unaccompanied by staff; and to make ourselves easier of access than the existing protocol appears to have made possible."[44] He and Edwina wanted to be sure that none of the viceregal police followed them into Nehru's bungalow whenever they felt like dashing off there for a few hours. Attlee granted all of Dickie's requests, hoping only that whatever he and Edwina did in India would help him and his cabinet to achieve their "escape from empire."[45] Even the king was glad that Cousin Dickie had agreed and would be taking Pug and Eric out with him to do the job right. "Such a good pair of friends should never let you down," Bertie wrote Dickie, adding, "Edwina of course will do wonders with all her knowledge and experience gained in the war years . . . and I shall follow your dealings with N. [Nehru] and J. [Jinnah] with the greatest of interest."[46]

"I saw Mountbatten before he left," Woodrow Wyatt recalled. "He was chiefly concerned with what he should wear on arrival. 'They're all a bit left wing, aren't they? Hadn't I better land in ordinary day clothes?' He was delighted when I said, 'No, you are the last Viceroy. You are royal. You must wear your grandest uniform and all your decorations and be met in full panoply and with all the works. Otherwise they will feel slighted.' And that is what he did, to everyone's pleasure."[47]

In January 1947, before Mountbatten's appointment was made public, Nehru called Wavell's attention to the sad "state of affairs in Kashmir where for the last eight months a bitter struggle involving severe repression of the people has gone on . . . [yet] the full machinery of the State has not succeeded in repressing the popular movement there. This failure itself is evidence of the futility of the methods employed."[48] Indeed, none of India's problems seemed easy to solve, even though Nehru himself was now so close to the pinnacle of power. Six weeks after he had introduced his objectives resolution to the Constituent Assembly, he returned to prod them, bogged down as they were in debating amendments, shouting, "How long are we to wait now? Many of us . . . are nearing the end of [our] lives. We have waited enough and now we cannot wait any longer." His anger grew as he watched them wince and slump lower in their seats, too many of them overfed and self-satisfied, until he started to yell at them, reminding them of the "starving people" outside and the "naked masses" waiting in the cold to be clothed. Blood rushed to his face, his voice strained as he shouted: "Look at India today. We are sitting here and there is despair in many places, and unrest in many cities. . . . [T]he greatest and most important question in India is how to solve the problem of the poor and the starving. Wherever we turn, we are confronted with this problem. If we cannot solve this problem soon, all our paper constitutions will become useless."[49]

Nehru had to preside over the thirty-fourth annual session of the Indian Science Congress and also of the A.I.C.C., which also met in Delhi, simply to suit his convenience. He was expected each day to see and be seen by everyone, even though Jivan Kripalani now presided over Congress. Burma's Aung San, president of that neighboring state's Anti-Fascist People's Freedom League, visited Delhi, so he could not escape a long meeting. "I have never been quite so desperately busy as during the past fortnight," he wrote Krishna Menon, who kept sending long letters, followed by many telegrams. "You have written at some length about your visit to the Russian Ambassador in Washington and you were evidently distressed at various developments. . . . I wish you had not pressed the matter of an invitation being sent to me from the Soviet Union. . . . We cannot make ourselves too cheap or too eager."[50]

Nehru's second grandson, Sanjay, had been born in mid-December 1946, yet Nehru was so busy that he now barely found a weekend to spend with Indira and the boys. "I was really hoping for a daughter," Indira recalled. "In fact, we had kept only girls' names ready."[51] Betty was staying at Nehru's house in Delhi on York Road that December evening. They dined at 9:30, Indira "in good spirits," and Nehru, "despite his cares, was like his young self, joking and teasing all of us." At about three in the morning Indira awoke in labor, and Feroze took her and Betty to the hospital. After Sanjay's birth, Betty called Nehru to tell him the news. He rushed over, but exhausted Indira had fallen asleep by the time he arrived. "I won't ever forget his face when he saw his daughter lying semiconscious and whiter than her pillow with all her blood drained away," Betty recalled.[52] Not even Nehru's acute premonitory powers could have made him imagine, however, what a dreadful end Indira would meet thirty-eight years later, when she was gunned down by two trusted Sikh guards in her Delhi garden or that baby Sanjay would have died in an airplane crash four years earlier.

On February 1, 1946, Attlee told Wavell it would soon be time for him to go home. He was promised an earldom to sweeten that bitter pill, but Wavell never fully recovered from the shock of what he considered Labour treachery. When Nehru rushed to him in early March to ask the outgoing viceroy's advice about what to do to stop all the horrible killing in Punjab, Wavell coldly asked Nehru whether he had any suggestions.[53] For now there was civil war in Amritsar as well as murder and arson all round Lahore. Tara Singh had just called for a "mass Sikh rising."[54] None of it surprised Wavell, nor did he offer any advice to his foreign minister, who decided to fly to Lahore for the weekend, "to see things for myself."[55]

In London that weekend, Krishna Menon was busy briefing Dickie Mountbatten. "In the new situation, the role of the Viceroy is by no means a negative one," he advised Dickie on the eve of his departure for India. "His influence would be very potent. . . . His role would be historic." Krishna Menon proposed as a realistic approach to the Muslim League that Mountbatten accept the demand for "Pakistans" as Muslim "homelands" but to be limited strictly to those areas where the Muslim population was "predominant" and where the League had won "an appreciable majority of elected seats." The "Western Pakistan" he told Mountbatten to accept would include only the Muslim-majority districts of the Punjab, together with Sind, allowing an "approach to the sea at Karachi." Krishna Menon's "Eastern Pakistan" would contain the Muslim districts of Eastern Bengal, without Calcutta. He insisted on keeping Calcutta for India's West Bengal but recognized Eastern Pakistan's need of a port, offering "Chittagong, that is, provide the money for it however many millions it may cost."[56]

Mountbatten liked Krishna Menon enough to take his advice on many sensitive points concerning India. Moreover, Dickie knew just how close he was to Nehru. The new viceroy thus left London well briefed and took with him in Krishna Menon's own writing, the formula he would spring on all parties in India as his own solution to the hitherto impossible puzzle that had baffled many more brilliant minds for decades. That solution, of course, would meet with Nehru's full approval and could hardly be rejected by Jinnah, since it was what he had always wanted, was it not? Chittagong, destined to become East Pakistan's only major port (and remain so for Bangladesh) was but a kerosene-lit village at this time. Krishna Menon's sharp mind, like Nehru's, always focused on future planning, and he knew that no matter how many millions it cost to build a decent port there, thousands of millions more would be made by India in Calcutta, destined to remain the brightest Eastern jewel in independent India's crown. In that nocturnal briefing of Dickie by Krishna, moreover, India's top secret decision to remain in the Commonwealth, thus avoiding a termination of relationships with Britain, was also promised. Without that agreement from Krishna Menon, Mountbatten knew he would never be able to win Churchill's opposition support for the final Transfer of Power Act that would have to muster a majority to pass through Parliament. But as Nehru had already impatiently informed his own Constituent Assembly, times were "fast changing," and "friendly" relationships with the British people and Commonwealth were certainly "preferable" to a "trail of hostility." His father Motilal had often told him that, long long ago.

In Punjab, Nehru tried to convince Governor Sir Evan Jenkins, with whom he met after a busy day in Lahore, that "the solution to our immediate problem in seriously disturbed areas . . . was to hand over to the Military Commanders." The British governor explained, however, "the difficulties about Martial Law and said that as I understood the position it could no longer be imposed by proclamation and Ordinance." It must have sounded odd to Sir Evan, being asked to call out British troops to put down Indians fighting one another, by the leader of Congress, who had spent many years behind British bars. "Nehru said that he was not well up in the legal technicalities, but the 'short point' was that the people should feel that really firm measures were being taken to suppress the agitation." But Jenkins insisted on adhering to the principle of minimum force, whereas Nehru wanted more action and urged "the carrying of weapons, etc. to the rural areas." Nehru also told Jenkins that "some sort of partition was inevitable." During their conversation Nehru phoned Dr. Khan Saheb in Peshawar, wanting to visit the frontier the next day. While he was still with Jenkins, however, Wavell wired him, strongly "asking him not to go to Peshawar and he agreed."[57]

The Royal Automobile Club, of which Dickie was president, gave the Mountbattens a proper farewell party that included every celebrity in London, as well as the duke of Gloucester, the duchess of Kent, and Clement Attlee, "standing alone and talking to no one," Campbell-Johnson recalled. "I exchanged a few words with Noel Coward. He deplored Mountbatten being landed with such a tremendous task. 'The position having become impossible, they call on Dickie.' "[58] Dickie and Edwina flew to India on March 20 with daughter Pamela, Mountbatten's private secretary Ronnie Brockman, and Peter Howes, Mountbatten's senior aide-de-camp. They took off from London at dawn and landed in Delhi in the afternoon on Saturday, March 22, welcomed by Nehru with a bouquet of roses for Edwina as they stepped out of their plane. They drove together to the viceroy's house (now called Rashtrapati Bhavan): Dickie and Jawahar, Edwina and Pamela, seated in an open landau, with cheering and rose-throwing Indian children waving a warm welcome. The youthful Mountbattens walked up the red-carpeted stairs to be greeted by the solemn, sad, somber Wavell at the top. Wavell briefed Dickie, Pug, and Eric that afternoon and evening. Wavell's private secretary, George Abell, stayed on to serve Mountbatten. Attlee had assured Dickie that he could retain any members of Wavell's personal staff he liked, and this facilitated an efficient transfer of British power from one viceroy to the other. The next morning the Wavells said good-bye, and Campbell-Johnson felt some "panic" over the swearing-in ceremony planned for the following day, the first such ceremony ever to be filmed.

"I went round to Mountbatten's suite and had a discussion with the Viceroy designate, clad in his underpants and vest, on the implications of letting in all the local news-reel and camera-men. . . . He proposed that a large platform should be built for them. . . . He showed me this morning's masterpiece on the front page of *Dawn*. It is a photograph of Ronnie Brockman and Elizabeth Ward, Lady Mountbatten's private secretary, . . . described as 'Lord and Lady Louis arriving'!"[59]

The viceregal trumpeters blared out their call to start the big show inside Durbar Hall, launching Dickie's year in Delhi next day. Resplendent in his admiral's dress whites, gleaming medals, and the blue sash of the Order of the Garter, Montbatten looked more king than viceroy.

"I am under no illusion about the difficulty of my task," he announced in his statement at the ceremony, winning the warm support of Indians when he modestly added, "I shall need the greatest good will of the greatest possible number, and I am asking India today for that good will." Nehru had been captivated since their first meeting in Singapore, but after three more hours that afternoon closeted alone with Dickie in the viceroy's house, he was totally charmed and enthralled. "I want you to regard me not as the last Viceroy winding up the British Raj," Mountbatten told Nehru as they shook hands on parting, "but as the first to lead the way to the new India." To which Nehru replied, "Now I know what they mean when they speak of your charm being so dangerous."[60]

For the first time, Nehru felt perfectly at home with a British viceroy. His two worlds had magically merged, now that Dickie and Edwina were with him in India.

26

Interim Raj

1947

MOUNTBATTEN MET with Maulana Azad the day after he became viceroy. Azad told him that if he had remained president of Congress, his party would have "accepted the Cabinet Mission's plan."[1] The Maulana blamed himself as much as he blamed Nehru for the tragedy of Partition and its aftermath. But Dickie was less interested in the roots of problems than their solutions. Moreover, he would never have been able to work as closely with any Maulana, Mahatma, or Sardar as he did with Jawaharlal, for they spoke the same language and shared that ineffable experience of proper public schooling. After the Mountbattens' first garden party on March 28, 1947, to which all members of India's Constituent Assembly were invited, Nehru took Edwina and her daughter back to his own house at 17 York Road for a nightcap in that more modest and intimate setting. It eased the whole transition process, having India's prime minister–in-waiting as virtually a member of the viceregal family.

Mountbatten met Gandhi for the first time on March 31. The Mahatma stayed for two hours. "He talked of his life in England, of his life in South Africa, his recent tour of Bihar, his discussions with former Viceroys," Dickie reported. "I felt there was no hurry and deemed it advisable to let him talk along any lines."[2] They met again on April 1. "Finally, he gave me the first brief summary of the solution which he wishes me to adopt: Mr. Jinnah should forthwith be invited to form the Central Interim Government with members of the Muslim League. This Government to operate under the Viceroy . . . this solution coming at this time staggered me. I asked 'What would Mr. Jinnah say to such a proposal'? The reply was 'If you tell him I am the author he will reply 'Wily Gandhi.' I then remarked 'And I presume Mr. Jinnah will be right'? To which he replied with great fervour 'No, I am entirely sincere in my suggestion.' " And he was. But Dickie had been briefed by Krishna Menon, and Edwina had just told him how "wonderful" and "perfect" a prime minister Nehru was going to make, and he agreed, so he asked Gandhi's "permission to discuss the matter with Pandit Nehru . . . in strict confidence."[3]

"Nehru was not surprised to hear of the solution which had been suggested, since this was the same solution that Mr. Gandhi had put up to the Cabinet Mission. It was turned down then as being quite impracticable; and the policy of Direct Action by the Muslim League, and the bloodshed and bitterness in which it had resulted, made the solution even less realistic now than a year ago," Dickie recorded after his chat with Nehru that afternoon, sealing a destiny of partition for British India. Nehru told Mountbatten that afternoon how "immensely keen" Gandhi was to keep India "unified," "at any immediate cost." They both appreciated the noble work that Gandhi was doing in Bihar, recognizing its high purpose, but "as Pandit Nehru so aptly pointed out, Mr. Gandhi was going round with ointment trying to heal one sore spot after another on the body of India, instead of diagnosing the cause of this eruption of sores and participating in the treatment of the body as a whole. I entirely agreed, and said that it appeared that I would have to be the principal doctor in producing the treatment for the body as a whole."[4] Dickie and Nehru agreed that a swift surgical cure would be best for the eruptions of communal strife that were consuming India's body politic. The partition "cure," of course, would only escalate the bloodletting.

Gandhi returned to meet with Mountbatten the next afternoon. "Mr. Gandhi came down firmly for his great plan. . . . He wants me to invite Mr. Jinnah to form a new Central Government for India, which will be the Government to which I am to turn over power. He suggests I should leave it to Mr. Jinnah to select the Ministers, if necessary entirely from the Muslim League, but if he feels so inclined he can of course then make it a coalition Government by including Nehru and other Congress Ministers as well as representatives of Minorities." Mountbatten listened patiently, but then he "twitted" the Mahatma, insisting again that this "offer to Jinnah was merely a manoeuvre." But Gandhi tried his best to "assure me with burning sincerity that this was so far from being the case that he then and there volunteered to place his whole services at my disposal in trying to get the Jinnah Government through first by exercising his influence with Congress to accept it, and secondly by touring the length and breadth of the country getting all the peoples of India to accept the decision." Finally,

Mountbatten did believe him, but instead of doing what the wise old Mahatma advised, he convinced Gandhi of the "supreme importance of complete secrecy" of this plan, especially regarding the press.[5]

Gandhi had, indeed, correctly assessed India's most explosive problem, deciding that Jinnah alone could rule a unified independent India, after June 1948, the date that the British cabinet set as its absolutely final month of rule in India. Only Quaid-i-Azam Jinnah could control most Muslims, and although he was seventy-one and in poor health, Gandhi was seven years older and felt almost as frail. Like himself, Jinnah grew stronger when he was needed, wanted, and appreciated for all his remarkable talents. Gandhi understood his Muslim "adversary" much better than he ever understood Nehru, for Jinnah was a fellow Gujarati. They spoke the same language, and both were men of integrity, whose word was their bond. Gandhi loved India and understood what would happen if it were divided by partition. So he was ready to do anything to avert that disaster. He was even ready to admit to Jinnah that he, Gandhi, had been wrong in wresting control of the Congress from him more than a quarter century ago but was now ready to offer him the premiership he deserved. For Jinnah alone could rule India, a united India, as the best British viceroys had done, honestly and impartially, without fear or favor, unflappable, impeccable, never stooping to nepotism, never seduced by passions, never tempted by vainglorious promises. He was above or beyond all such wretchedly common human weaknesses and here, too, he and Gandhi were much alike.

Nehru was a younger man and liked to look and act even younger than he was. Gandhi still believed that he could manage Nehru, as he long had. Nehru was Motilal's son, after all, and despite his strange, mercurial, weak, and erratic behavior, Nehru could always be brought to see the light, after it was clearly shown to him, or so Gandhi thought. He did know that he could always count on Sardar Vallabhbhai to keep Congress in line and hold down the more fanatical Hindus. Jivan Kripalani was also a good man, a Congress president whose judgment Gandhi trusted. He was more the idealist than Vallabhbhai, for the Sardar was a pure realist; that was why he could keep the Congress machinery running. Then there was Maulana Azad; he too could be trusted to keep all the Congress Muslims in line. These men would be enough to do the job, Gandhi believed. For he was personally ready to tour the land himself, telling those millions of worshippers, who came to take dust from his feet and garland his naked neck, that from now on they must listen to and obey "Jinnah Saheb," who was a good man and would now be the great leader for all of India.

Two hours after Gandhi left him, Maulana Azad arrived for another audience with the viceroy, and "he staggered me by saying that in his opinion it was perfectly feasible of being carried out, since Gandhi could unquestionably influence the whole of Congress to accept it and work it loyally. He further thought that there was a chance I might get Jinnah to accept it, and he thought that such a plan would be the quickest way to stop bloodshed, and the simplest way of turning over power."[6] Azad had lived in enforced intimacy with Nehru for years and knew him perhaps even a bit better than Gandhi did. Although Azad still was angry that Jinnah had refused so much as to shake his hand in Simla or acknowledge him as anything more than a "showcase Muslim" of Congress, he was wise enough to know that Gandhi was right. His plan alone could save India, if Jinnah accepted, and Azad believed that Jinnah might accept it. Why, then, did Mountbatten never suggest Gandhi's plan to Jinnah? If it could have saved all the horrors of partition, why was it kept a secret from the one man who should have been told of it immediately?

Even to Gandhi's old friend, Pethick-Lawrence, the lame-duck secretary of state now at Whitehall, Dickie wrote on April 2, 1947, only that "I should like to be able to paint an encouraging picture . . . but fear it would be misleading. . . . The scene here is one of unrelieved gloom. . . . [T]he whole country is in a most unsettled state. There are communal riots and troubles in the Punjab, N.W.F.P. [North-West Frontier Province], Bihar, Calcutta, Bombay, U.P. [United Provinces], and even here in Delhi. . . . [U]nless I act quickly I may well find the real beginnings of a civil war on my hands. . . . I have had three meetings with Mr. Gandhi . . . it would be too early for me to comment on the solution which he has in mind."[7] Gandhi returned on April 3. "We continued our talks on Mr. Gandhi's great scheme for the All-India Jinnah Government. He informed me that those of the leaders of the Congress he had spoken to had all agreed that it was feasible and would support him, but that he had not yet had time to talk to Pandit Nehru, which he intended to do that evening."[8]

Nehru's totally negative reaction buried Gandhi's plan. The last thing Nehru wanted to do was turn over his hard-won power to the man he hated most in all of India. Without Nehru's approval, Mountbatten would not consider the idea further, for even if Gandhi's prognosis concerning it were right, Dickie knew, and his staff agreed, that it could only mean more hard work and greater trouble for all of them. The viceroy would have to preside over countless hours of meetings, first with the Muslim League cabinet and then with the Congress leaders of the assembly, each side arguing against the other. Krishna Menon, who had just flown to Delhi to stay with Nehru, also did not think the plan could possibly work, for his mind was tuned to Nehru's. Krishna Menon, "whom I saw twice in London before coming out, came to see me at 12:15 and stayed to lunch," Mountbatten reported on April 5. "I asked him categorically whether Mr. Gandhi's scheme of turning over the Central Government to Mr. Jinnah could be made to work. . . . Menon replied emphatically

. . . that not even Mr. Gandhi could put this particular scheme through; even if Mr. Jinnah could be made to accept it."[9]

Mountbatten also never liked Jinnah, who finally came to dine with them on the night of April 5. Dickie found him "most frigid, haughty and disdainful."[10] Edwina found his sister Fatima almost as cold as her chauvinist brother. After meeting him, Mountbatten asked Ismay to remind Gandhi not to mention his plan to Jinnah. Mountbatten was worried about what would happen to Gandhi's plan if it were launched and then Gandhi died.[11]

Gandhi knew that Nehru no longer wanted him in Delhi, that he considered him a nuisance and no longer listened to or sought his advice. And although the young viceroy was polite and more charming, Gandhi knew that he was no more ready than Nehru was to accept the one plan that might keep India united. So on April 7 Gandhi wrote to Lord Mountbatten, "Dear Friend, I have pressing letters from friends in the Punjab asking me to go there . . . Pandit Nehru agrees. Nevertheless I would like you to guide me too."[12] Did he still believe it possible that Dickie would listen to him? Mountbatten did not, of course, try to detain the Mahatma in Delhi. So Gandhi left and the last chance to keep India united left with him. Nehru remained to rule Delhi, with the help of his younger brothers, Krishna and Dickie. The closest that Mountbatten ever came to informing Jinnah of Gandhi's brilliant plan was near the end of their meeting on April 9.

"I told him [Jinnah] that I regarded it as a very great tragedy that he should be trying to force me to give up the idea of a united India. I painted a picture of the greatness that India could achieve—four hundred million people of different races and creeds, all bound together by the central Union. . . . I finally said that I found that the present Interim Coalition Government was every day working better and in a more cooperative spirit; and that it was a day-dream of mine to be able to put the Central Government under the Prime Ministership of Mr. Jinnah himself. He said that nothing would have given him greater pleasure than to have seen such unity . . . it was indeed tragic that the behaviour of the Hindus had made it impossible for the Muslims to share in this. . . . Thirty-five minutes later, Mr. Jinnah, who had not referred previously to my personal remark about him, suddenly made a reference out of the blue to the fact that I had wanted him to be the Prime Minister. There is no doubt that it had greatly tickled his vanity."[13]

But Mountbatten never posed Gandhi's plan as a serious offer to Jinnah, for Nehru's vanity would not permit it. By April 11, moreover, Mountbatten himself had concluded that "Mr. Jinnah was a psychopathic case," for he had brought "all possible arguments to bear" against Jinnah's demand for Pakistan but got the impression that Jinnah "was not listening. He was impossible to argue with."[14] As Jinnah thus seemed to grow more "insufferable" to Dickie, Nehru became more charming, convivial, and politically accommodating. He agreed to virtually everything Dickie asked of him and devoted himself, night and day, to whatever either Mountbatten wanted of him.

Communal violence continued to escalate in Punjab and spread all along the frontier and throughout Bengal, Assam, and Bihar. Mountbatten managed to persuade Jinnah and Gandhi to sign a joint appeal in mid-April, "deeply deploring" the "acts of lawlessness and violence," denouncing the "use of force to achieve political ends," and calling on "all communities" to "refrain from all acts of violence."[15] It was a noble statement but did little to avert the murder and madness that spread its contagion near and far. Private armies were tramping through rural as well as urban centers, half a million Muslim League "green shirts" and Congress "red shirts," the Hindu mahasabha's "army of Shiva," and Sikh "soldiers of Khalistan" each marching under a banner of "God" to murder their "godless enemies," who were mostly helpless innocents. "I should say that among the dead there are 6 non-Muslims for every Muslim," Governor Jenkins of Punjab reported that blood-soaked April, "One of my troubles has been the extreme complacency of the League leaders in the Punjab, who say in effect that 'boys will be boys.' "[16]

In Bengal the situation was almost as bad. Chief Minister Suhrawardy complained, however, that "he had no army of his own," fearing that "one day Mr. Nehru, on a big black horse, might lead a Hindu army against Muslim Bengal." When the Englishman who heard this complaint and fear replied that Nehru hardly seemed the type "cast for this role," Suhrawardy agreed but suggested that Nehru would instead "try to induce" a Gurkha army from Nepal to do the job for him. Suhrawardy hoped to rule a separate nation of Bengal, Hindu and Muslim combined, 65 million intelligent, enterprising people with their great capital of Calcutta, but Mountbatten and Attlee never seriously considered creating a Bangladesh (land of Bengal) at this time, any more than they did a Sikhistan or Khalistan (Sikh land of the pure). Suhrawardy thought of emigrating if Bengal were partitioned, but where would he go? What would he do? "Was England a good place,—or Ireland? . . . [H]e believed hall porters at hotels in New York did very well." He remained in Pakistan, however, and, clever politician that he was, briefly climbed to the top rung of prime minister, only to be toppled by West Pakistan's Pathan–Punjabi–army coup and to die six years later of heart failure on a street in Beirut.[17]

"I hate violence, more especially of the brutal and vulgar type that we have seen lately in India and I would go very far indeed to stop it," Nehru wrote Mountbatten at this tragic juncture. "The question is how best this can be done." The best solution he could think of was to partition Punjab and Bengal, but he was wise enough to note that "in attempting to solve one problem we might well have to face a number of graver problems and even that one problem may not be solved."[18] Nehru left Mountbatten to ponder this thought, flying off the next day to Gwalior to address the annual meeting of his States People's Conference. "Our aim at the moment is to liberate whatever part of India we can and we shall then deal with the question of getting independence for the rest," Nehru told his cheering audience. "As you know, Sheikh Abdullah was elected to preside over the session but he could not do so because he is behind prison bars. When I think of it I hang my head in shame. All I can say now is that Kashmir is like a flame in my heart."[19]

"Krishna Menon reminded me that he was staying out here specially in the hope of being of use to me personally as a friend," Dickie noted, meeting with Krishna at length on the eve of Nehru's flight to Gwalior. "He offered to stay as long as he was of use." Krishna Menon not only kept Mountbatten informed of every nuance of Nehru's thinking but was especially useful in convincing Nehru and other Congress radicals of the need for India to retain Commonwealth ties to Great Britain after full independence was achieved, for reasons of defense. With a potentially hostile Pakistan emerging on both sides of India, the value to India's military power of being assured that British officers would remain in command of its three services could hardly be exaggerated. Mountbatten explained that "by India deciding not to quit the Empire and not to sever the link with the Crown," it would retain the services of "men like Field Marshal Auchinleck."[20] Before the end of 1947, Auchinleck's continued command of both new dominion armies saved India from losing military control over Kashmir, thus keeping alive the "flame" in Nehru's "heart." Krishna Menon, who had long agreed with Nehru that India must leave the British Commonwealth, now coined terms like "Free Nation of the Commonwealth" and "Union of India" to allow "Free India" to retain its vital military and economic "link with the Crown."

To the tragic end of his life, Mountbatten considered India's unbroken "link to the British Crown" and its retention within the Commonwealth "my most important achievement."[21] He was a missionary and something of a visionary about the Commonwealth, viewing it as the best global structure for security as well as for trade and political support, taking great personal pride in its continued strength in South Asia in the aftermath of independence, believing that the renamed and reborn British Empire and the old British Crown were the best—indeed, the wisest—institutional structures for leading, if not ruling, the world.

"Finally he said to me," Dickie reported of his "top secret" and "specially restricted" conversation with Krishna Menon, " 'Unless you take the first step and approach us, nothing will be done.' I replied: 'Then nothing will be done, because it is entirely your loss, and I am not going to allow any sentimental reasons make me pull your chestnuts out of the fire. If you do not take the first step, you will have a rotten army; you will lose all the benefits of the Commonwealth; and you will save the other nations of the Commonwealth the expense, anxiety and responsibility of your defence.'

"He then said that he did not mean there could not be private off-the-record discussions; he meant that the first step could not be taken publicly by the leaders [Nehru] without reversing everything that they had preached to their people. I then asked him what he proposed. He said: 'If the British were voluntarily to give us now Dominion Status, well ahead of June 1948, we should be so grateful that not a voice would be heard in June 1948 suggesting any change, except possibly to the word dominion if that had been actually used up to that date.' "[22]

Krishna Menon knew that as soon as India's independence—now to be called dominion status—was announced he would be able to move into the high commissioner's office and residence in London, since Nehru had promised him that top diplomatic post. Nehru's sister Nan would have to be content with the second-choice plum, becoming India's first ambassador to Moscow, since Krishna Menon would be a far more discreet asset in helping him visit Edwina back at Broadlands in top-secret privacy. "I said that I was in favour of taking steps if it was . . . feasible," Dickie told Krishna, for he was anxious to accelerate this frustrating and enervating process of seeing so many Indians, all of whom wanted different "solutions" from him. "If the Muslims were to stay in a Union of India, I would certainly recommend Dominion Status next month; but . . . I could not possibly recommend the present Interim Government being given dominion powers; the Muslim League would violently object. . . . He asked me whether I could not propose equal dominion status to Hindustan and Pakistan. . . . I said: 'Certainly provided I could retain full powers over defence, since I would have to coordinate the use of the single army for both. . . .' He said that Dominion Status without control over the Army would be laughable and would never be accepted by the Indians. I told him to go away and think of any solution by which dual dominion status could be granted."[23]

Five days later, Mountbatten had another "long and friendly talk" with Krishna Menon, Nehru's living channel of communication to the viceroy. "We properly let down our hair together," Dickie recorded of that important interview. "I found that he had very shrewd views on the future trend of governments in the U.K. and America, and on world-wide politics. He expressed his fear to me of American absorption from every point of view. . . . [The] Americans' object in India was . . . to capture all the markets, to step in and take the place of the British . . . that their aim might even be to get bases in India for ultimate use against Russia. . . . In fact, backed by British and American arms and technique, Pakistan would in no while have armed forces immensely superior to those of Hindustan. . . . [P]laces like Karachi would become big naval and air bases. . . . [H]e absolutely shuddered, and said 'How can we prevent it'? I said 'By the simple expedient of being in the Commonwealth yourselves.' "[24]

"Dear Bapu," Nehru wrote Gandhi the next day. "You know that the Congress Working Committee is meeting here on the 1st May. I should very much like you to be here . . . to see the formulation of Mountbatten's scheme for the future."[25] Nehru tried to distance himself from "Mountbatten's scheme," as he hoped Gandhi would view partition. Nehru had just met with Mountbatten to discuss the princely states and warned that "Kashmir might produce a difficult problem." Dickie expressed "great disappointment" at Nehru's "demagogue"-like attacks on the princes, but Nehru denied being "a hot-headed leader," insisting that he was, in fact, only trying to keep "extreme elements in order."[26] There was, indeed, a more radical fringe of Congress, always calling for swifter, more violent action against all princes, and a growing fundamentalist party of Hindu extremists, demanding death to all Muslims. Nehru tried to keep such distant and usually irreconcilable leaders from bolting Congress and India's union entirely. He found the realities of power more difficult and demanding than he had imagined they would be, when behind British bars.

Gandhi returned to Delhi on May 1 to attend the Working Committee meeting. He amended Nehru's draft of the Congress response to Mountbatten's proposal, the full extent of which Nehru denied knowing. Lord Ismay was flying that proposal to Britain for the cabinet's approval the next morning. "Our whole policy, even in regard to conflict, has been peaceful," Congress insisted, but the policy of the Muslim League "has deliberately encouraged violence and disorder and has resulted in murder, arson and loot."[27]

Julian Huxley had invited Nehru to contribute something to a volume he was preparing on human rights for the United Nations at this time, but Nehru declined. "Just to write some pious sentiments will serve little purpose," he rightly noted. "Apart from this, we have to face at present very difficult and intricate problems in India and I have the misfortune to be tied up with these problems."[28] He no longer found any time for "quiet consideration" of broader issues and reflective "writing." Mac Mathai, who lived with him at 17 York Road, recalled innumerable nights without sleep, Nehru's telephone ringing throughout the night and Nehru himself often working upstairs until three or four in the morning.[29] Krishna also was living with Nehru then and urged Dickie to take his friend to Kashmir on holiday to keep him from "breaking down from over work."[30]

In early May Dickie decided to take both Jawaharlal and Krishna Menon up to Simla's mile-high retreat. His staff all came along, of course, as did Edwina. Campbell-Johnson and his wife had flown up a bit earlier to open the viceroy's old weekend lodge in Mashobra, a lovely isolated garden spot surrounded by lush orchards, "quite glorious and completely remote," as Edwina remembered it.[31] There she and Nehru strolled and climbed together, alone at last, their rarest luxury—privacy together. She was very much like him, a lonely heart, introspectively passionate, a soul long seeking its other half. Each of them had mystic insights and saw in the other the ideal mate that each had always longed to find yet never had before. Cheerful, handsome young Dickie had proved almost as much of a disappointment to Edwina as Kamala had to Jawahar, though not in the same ways, of course. Each of them long looked elsewhere for that friend their much publicized marriages had failed to provide. Yet here in the viceroy's retreat beyond Simla, strolling arm in arm around those lush and glorious floral paths, they finally found each other, while India lay smoldering far below.

For Nehru that week's retreat in May 1947 was much more than a holiday or a break in the murderous pace of his work. It was his homecoming. Home to Harrow. Home to Trinity and the Cam. Home to London parties and every stroll through Hyde Park. Edwina was all those early joys and pleasures he had loved, now, at last, miraculously at his side. For Dickie and Krishna Menon, however, Simla was all work, trying to revise the plan to meet a number of Congress criticisms and to work out a formula to allow India to accept its dominion status within the Commonwealth, without totally abandoning Congress's commitment to *purna swaraj*. The reforms commissioner, V. P. Menon, also joined the Simla party that week, helping draft and redraft vital paragraphs and planning how best to bring all the princes into the new dominions. V. P. became Sardar Patel's right hand in that awesome task of ego-soothing and proper pensioning that convinced all but three of the some 570 princes, that they must opt to join either the Dominion of India or the Dominion of Pakistan. Kashmir and Hyderabad, and briefly Junagadh, hoped to elude the imperative of rapid decision, but the Sardar and India's Army persuaded the latter two that Indian paramountcy was at least as powerful as the departing British. Kashmir proved a more persistent problem.

Nehru assured Mountbatten in Simla that "the Congress-majority part of India would be able to take over power almost immediately."[32] He suggested, in fact, June 1947 rather than 1948. Mountbatten's plan, which Ismay took to London, had called for a meeting with Jinnah and Liaquat as well as Nehru and Vallabhbhai and Baldev Singh, on May 17, 1947, to confront one another and "admit openly" that there was "no possibility of securing agreement to a unified India." Once that was done, the plan could then be announced both in London and in India, and the final transfer of power machinery, especially the job of drawing the new boundary lines through Punjab and Bengal, as well as dividing the accumulated assets of an empire that had lasted for ninety years, would begin. Thus June seemed a bit early, even to Dickie.

The left wing of Congress was already causing much anxiety in Simla, for J.P. had just been arrested in Hyderabad, sentenced to two years in prison by the nizam's government, for preaching revolt against tyranny. Nehru assured Mountbatten and his staff, however, that even though "Jai Prakash Narain had been behaving in a very irresponsible manner recently . . . he was an intelligent and honest man."[33] Nehru had no doubt that J.P. would "play an important part in India in the future" and that he could bring him around and keep him in line. And he always did, unlike the less patient Indira, who found an older J.P.'s agitation impossible to handle as adroitly as her father had, and so imprisoned him for more than a year during her dreadful "emergency" Raj three decades later.

How strange it must have sounded to Nehru, trying to explain to that handsome viceregal team of elegant Englishmen in the dim, velvet-draped sitting room of a Victorian palace in Simla that "over the last many years, there had been a tremendous sentiment in India in favour of complete independence. The words 'Dominion Status' were likely to irritate because of past associations—although in theory it could be shown that Dominion status was equivalent to complete independence. Such fine points were not, however, . . . understood by the majority of the people. . . . He wanted to prepare the ground. The world was changing."[34]

The closer they all came, however, to the awful realities of what partition would mean, the more nervous, anxious, and uncertain they grew. On the afternoon of May 11, Mountbatten and Nehru met with Punjab's Governor Jenkins in Simla's lodge. Mountbatten noted that "the greatest snag as he saw it was in connection with the Sikhs. . . . [The] Sikhs were prepared for war and . . . would fight." Nehru now admitted that when Congress agreed to the partition of Punjab, it "had not gone into any great detail. . . . [T]he rough line of demarcation must be between Muslim and non-Muslim majority areas . . . but there were Sikh shrines in some of the predominantly Muslim areas. This point should also be borne in mind." Jenkins said that that "would be very difficult." The viceroy agreed but suggested that a boundary commission "could be instructed to take Sikh religious interests into consideration."[35]

Nehru suddenly cautioned Dickie that the "present proposed timetable was too much of a rush."[36] He suggested waiting until June to call the leaders to a round table, wanting "some days" before that session to meet with his Working Committee of Congress. Mountbatten told Nehru that "he did not consider that . . . desirable." Dickie suspected by now that Nehru was a veritable Hamlet in Indian garb, almost incapable of decision. "I have very carefully considered, in the short time at my disposal, the papers shown to me," Nehru wrote Mountbatten in Simla. "The picture presented by the proposals was an ominous one and the whole approach . . . appeared to me . . . dangerous. Not only do they menace India but also they endanger the future relation between Britain and India. . . . [They] would encourage disruptive tendencies everywhere and chaos."[37] Mountbatten was staggered by Nehru's "bombshell" and wired Attlee that he would have to postpone meeting the top leaders to ask for their final opinions, from May 17 to June 2, after Parliament returned from Whitsuntide recess.

Now Bengal's governor, Sir Frederick Burrows, former president of the Railwaymen's Union, warned that "Calcutta . . . will become a battleground after our departure" if Bengal was, in fact, partitioned. "I have sufficient troops, if I am allowed to utilize maximum fire-power . . . to deal with any situation likely to arise in Calcutta. I have not sufficient troops . . . to deal with widespread rising in East Bengal."[38] This, perversely enough, only served to convince Mountbatten as he wired Ismay that night that "speed is the essence of the contract. Without speed, we will miss the opportunity."[39] Dickie and his staff were "a bit rattled by Nehru's *volte face*," as Eric Mieville wired Pug. "I cannot help thinking that his party must have got at him."[40] More than his "party," Nehru's heart and mind had awakened—albeit too late—to the dark realities of that death-dealing double partition.

Mountbatten returned to Delhi on May 14, deciding that he would have to fly to London himself. Nehru's bombshell had so puzzled Attlee and Cripps that the latter and the new young secretary of state for India, Lord Listowel, would otherwise have flown to New Delhi themselves to find out exactly what was happening. Nehru sent Krishna Menon flying back to London on the same day that Dickie left Delhi. He wanted Krishna to meet with Listowel and wire back to inform him of any changes of heart on the part of the Attlee cabinet about any part of the plan.

Eric Mieville wired Mountbatten on May 20, after Jinnah informed him, "I beg you to tell Lord Mountbatten once again that he will be making a grave mistake if he agrees to the partition of Bengal and the Punjab."[41] That day Mountbatten met with Attlee and the cabinet in London, reporting on Jinnah's request that a referendum be held in Bengal and Punjab to determine whether or not they should be partitioned. Mountbatten argued that any such attempt to ascertain the popular will of those two provinces could "achieve no useful purpose and would merely result in delay."[42] Having flown to London in record time, Dickie was determined to avoid any delay.

Nehru drove to his sister's "Cosy Nook" in Mussoorie the day after Dickie and Edwina flew to London, to spend a relaxed week in the hills with Indira and the boys. Vallabhbhai joined him for two days, and he tried luring Gandhi up as well, but the Mahatma felt much too depressed to indulge in "a few days of rest," as Jawaharlal suggested, since "nothing much is likely to happen in Delhi until the Viceroy returns."[43]

At his prayer meeting in Calcutta, Gandhi had just been asked, "When everything at the top goes wrong, can the goodness of the people at the bottom assert itself against its mischievous influence?" Bapu's answer was that "Pandit Nehru was at the top. But, in reality, he was sustained by them [the people]. If he went wrong, those at the bottom could remove him without trouble." Nehru was, therefore, anxious to speak with Gandhi at some length, to explain to him what was happening at the top. As Gandhi moved across north India from Calcutta to Delhi, he daily addressed enormous crowds at prayer meetings in towns and villages where death and arson had generated flood tides of fear, hatred, and boundless suffering. Time and again he was told that "Congress was fast becoming an organization of selfish power-seekers and job-hunters. Instead of remaining the servants of the public, the Congressmen had now become its lords and masters."[44]

Gandhi reached sweltering Delhi on May 25, 1947. Nehru and Vallabhbhai came down from the mountain to meet with the Mahatma, who always preferred India's bottom to more exalted altitudes of any variety. Mountbatten had just gone to meet the ailing Churchill in bed, and Churchill told him that "he hoped to get Indian matters dealt with on a bi-party basis. . . . He then asked me if I foresaw any difficulties . . . with Mr. Gandhi. I told him that Mr. Gandhi was unpredictable, but that I doubted whether he would create any difficulties which could not be dealt with by Patel and Nehru. . . . I had received a letter from Nehru accepting Dominion status if power was transferred this year. . . . [Churchill] authorised me to give Mr. Jinnah the following message: 'This is a matter of life and death for Pakistan . . . accept this offer with both hands.' "[45] Churchill's strong support assured Attlee that his Transfer of Power Bills would pass through Parliament that summer. Mountbatten had thus taken all the "hurdles" in record time. Krishna Menon met with Mountbatten in London advising him, at Nehru's insistence, that "if Mr. Jinnah wants a total separation . . . and if we agree to it . . . we want to be rid of him, so far as the affairs of what is left to us of our country are concerned."[46] To get rid of that Jinnah "headache," Nehru was ready, with surgeon Mountbatten's assistance, to cut off India's head.

27

Freedom and Partition

1947

"FRIENDS AND comrades," Nehru broadcast on the night of June 3, 1947, "soon after my assumption of office, I spoke to you from this place. . . . Nine months have passed, months of trial and difficulty, of anxiety and sometimes even of heartbreak. . . . [T]he burden on the common man still continues to be terribly heavy and millions lack food and clothes. . . . [M]ost of our dreams about the brave things we are going to accomplish have still to be realised. . . . [Y]et . . . at no time have we lost faith in the great destiny of India." The richly cultivated timbre of his voice was as comforting to most Indians as that of Roosevelt had been to Americans. Mountbatten had spoken just before him, announcing his plan to quit India in little more than two months, leaving the Muslim-majority areas to opt for a separate Pakistan, an offer that Nehru and Congress now accepted, though it had been rejected and blasted as "insulting" five years earlier, when Cripps made it. "We are little men serving great causes," Nehru confessed to his nation now. "There has been violence—shameful, degrading and revolting violence. . . . This must end . . . *Jai Hind.*"[1]

The next day, Krishna Menon sent a "very urgent, strictly personal" note to Dickie from Nehru's house. Gandhi was very disturbed, and Nehru wanted Mountbatten to know that "it is important that he should be assured that the perils on which he is distressed" were exaggerated. They all had to do whatever they could to allay the Mahatma's anxieties, Krishna cautioned Dickie, adding that "Jawaharlal also had talks with me about the 'hereafter' and wants me to talk them over with you."[2] As Nehru's high commissioner–designate in London, Krishna Menon would have to ask Mountbatten's advice about India's purchases of vital British military equipment and shipping, for Dickie not only knew everyone but usually managed to cut most expeditiously through the red tape in the business world as well as in politics. Moreover, Krishna Menon rubbed most Englishmen and virtually all Americans the wrong way, so Mountbatten was quite helpful at introducing him to the right people.

Mountbatten met the press that morning and found that many Indian reporters were shocked to hear that Nehru had accepted the idea of dominion status, since, as one of them noted, until very recently Nehru had "made it emphatically clear" that Congress would not allow any "foreign power to have bases in any part of India." Dickie smoothly allayed any such anxieties, explaining that "dominion status" meant "absolute independence in every possible way." He had always been a great communicator, at his very best in fielding questions from reporters. "I am really sincere in my desire to help the Sikhs," he also told them that busy day.[3] The *Daily Herald* was "stunned" by Mountbatten's brilliant performance, and the *Times* dubbed it a "*tour de force.*"[4]

Then Nehru faced the grim reality of dividing the assets of an empire between the Congress-led majority and the Muslim League quarter of British India, soon to be called Pakistan. Jinnah and Nehru barely spoke to each other, agreeing on virtually nothing. Indeed, Liaquat Ali Khan and Nehru almost came to blows in the interim government's cabinet, when Nehru named his sister Nan as India's first ambassador to Moscow. Liaquat was livid at such autocratic blatant nepotism, but his protests fell on deaf ears. Nehru yelled louder and threatened to resign immediately if Dickie supported Liaquat in this matter.

Nan had hoped for Washington, where her dearest old friend and former lover, Syed Hossain, then presiding over the Committee for India's Freedom, had reentered her lonely life two years ago. Hossain returned to Delhi soon after Nan did, moving into the then still grand Imperial Hotel, where they met "as much . . . as possible," Mac Mathai recalled. After sending Nan to Moscow, Nehru sent Hossain to Cairo as India's ambassador there, where soon afterward "death put an end to an unhappy and tortured life."[5] "I had no desire to be sent abroad," Nan recalled, after learning of her brother's decision. "For some still unknown reason I was to leave for Moscow before the day on which the transfer of power was to take place. . . . If Bhai . . . realized how little importance the Soviet Union attached to this day we might have been permitted to celebrate it at home."[6] Nehru's "unknown reason" for rushing his sister out of Delhi had more to do with the Imperial Hotel, however, than with Moscow.

By mid-June Delhi was blistering, Punjab burning, and the frontier ablaze with rhetoric on the eve of a referendum that would transfer it from a lame duck Congress ministry to the Muslim League, from India to Pakistan. Famine also loomed now in Madras and Bengal, and Secretary of State Listowel tried in vain to get 670,000 tons of food grains shipped to India, as both a political and a humanitarian gesture. Mountbatten was eager to prove to Nehru and Patel how valuable in real terms India's continued membership in the Commonwealth would be. The trouble, however, was that Attlee's minister of food had just shipped 42,000 tons of wheat to Germany, turned down a French request for 20,000 tons urgently needed, and saw no hope of getting the 350,000 tons promised from the United States because of a shipping strike that loomed in America. The best that Britain's cabinet could do for India, therefore, was to allow 10,000 tons of Australian wheat destined for the United Kingdom in July to be diverted to Calcutta.[7]

"I am . . . writing this to you rather late at night because I am distressed," Nehru wrote Mountbatten on June 22. "Yesterday I went with Gandhiji to Hardwar and visited the numerous refugee camps there. There were . . . about 32,000 refugees there from the Frontier Province and the Punjab. . . . Daily some 200 or so fresh arrivals came. . . . But this letter is mainly about the city of Lahore where fires are raging and consuming hundreds of houses. . . . The human aspect of this is appalling. . . . Amritsar is already a city of ruins. . . . Lahore is, of course, a much larger city than Amritsar. If you will forgive a personal touch . . . my mother came from Lahore and part of my childhood was spent there. . . . [H]orror succeeds horror and we cannot put a stop to it." It was barely the beginning.[8]

The next day Mountbatten met with Jinnah, telling him what he had heard about Lahore. Jinnah's response reportedly was "I don't care whether you shoot Moslems or not, it has got to be stopped." The viceroy also met with Nehru later that day and then wired Governor Jenkins in Lahore, to suggest that he declare martial law.[9] Jenkins wired back, however, that he was against any declaration of martial law and that Major General J. G. Bruce agreed, "although he will take on anything he is told."[10] Their stated reasons were that "cloak and dagger activity" in Lahore and Amritsar was so widespread that any decisive action by troops was considered out of the question. They feared that they could not restore order quickly and that soldiers would be exposed to the same communal attack as the police. Then, of course, all Englishmen in the Punjab would be in as much danger as the Hindus, Muslims, and Sikhs were. What Jenkins suggested instead of declaring martial law was that leaders of the Muslim League tell their followers to stop all "burning and stabbing" and that leaders of the Hindu extremist group R.S.S.S. tell its followers to stop all "bombing." Jenkins and Bruce and most Englishmen still in India were eager to be headed home by now and preferred to leave the natives to fight over their spoils alone.

So when Mountbatten met with Nehru that day, "I gave him my painting of a proposed flag for the Dominion of India which I had designed. This consisted of a Congress flag with a small Union Jack in the upper canton," Dickie noted. Designing flags had long been one of his favorite hobbies. Nehru took the picture and promised to let Dickie know what Congress decided. They also talked about the frontier referendum, which Congress had decided to boycott, since it seemed clear that the vote would go strongly in favor of joining Pakistan rather than Hindustan, the only two options to the single question posed. Nehru said nothing more about the situation in Lahore, but they talked at length about Kashmir. Nehru was eager to go there to visit Sheikh Abdullah, still kept in prison by Maharaja Hari Singh. "He thought he would soon have to go to Kashmir to take up the cudgels on behalf of his friend and for the freedom of the people," Mountbatten noted, pleading with Nehru not to go, fearing that he might start a riot in Srinagar or get arrested. But Nehru remained adamant, so Mountbatten reminded him of "his duty to the Indian people as a whole. He would soon be the Prime Minister . . . ruling at least two hundred and fifty millions; and I would consider it highly reprehensible of him to desert his most important duties at the Centre to interest himself on behalf of four millions who might very well be going to join Pakistan." Nehru reluctantly agreed but, after becoming prime minister, focused all of India's might and his own time and attention on Kashmir.[11]

Then Jenkins reported at length in a "secret" letter to Mountbatten that "Lahore and Amritsar . . . have never really settled down since the first week in March. . . . [When] people seem to have discovered . . . how easy it is to burn the average building in an Indian city. The expected stabbing campaign began . . . accompanied by an entirely new campaign of incendiarism. . . . Most of it was done by Muslims. . . . The flight of Hindus from Lahore . . . made our task more difficult."[12]

"Viceroy has just had a very difficult time in Cabinet over Lahore," Secretary George Abell wrote that afternoon. "He explained why martial law was not likely to be effective, and asked for suggestions. Nehru blew up and said that the situation must be controlled, and that officials concerned from top to bottom should be replaced. Viceroy replied very strongly that this was a totally irresponsible suggestion and that he could not consider anything of the sort. . . . Liaquat Ali Khan was against martial law. . . . It was unanimously agreed that . . . Viceroy should request Governor [Jenkins] to call leaders again . . . and ask if they could agree on a set of local officials in whom they would all undertake to repose complete confidence."[13]

By June 30, 1947, the British government's "Indian Independence Bill" was ready for unveiling before the leaders of Congress and the Muslim League. Nehru and Patel insisted that Gandhi come with them to Mountbatten's house on the morning of July 1, and Krishna Menon showed up on the evening of June 30 "to warn me that Nehru intended to come with an ultimatum tomorrow that unless he was allowed to take away copy of Bill he would resign," Dickie wired Attlee that night. "Nehru is in very difficult state and maintains that it is gross insult to his people not to be allowed a copy of the Bill at this historic moment." Mountbatten, who was by now beginning to understand Nehru's mercurial nature, pleaded with Attlee for "your authority to use my discretion to avert a crisis. . . . [I]t would be tragic if Congress at this stage were to refuse to cooperate."[14] Attlee agreed, of course, but urged that Jinnah be given "similar facilities."[15]

Winston Churchill was "much concerned to hear" on July 1 that the proposed legislation was called "the Indian Independence Bill," writing Attlee to complain that "the essence of the Mountbatten proposals and the only reason why I gave support to them is . . . Dominion status . . . not the same as Independence. . . . It is not true that a community is independent when its Ministers have in fact taken the Oath of Allegiance to The King."[16] Churchill was not as easily fooled as most Indians were by Mountbatten. He never forgave Dickie for being so tricky and angrily snubbed him after Mountbatten returned to London. When Anthony Eden threw a big party to welcome Dickie home, Churchill glowered and pointed a threatening finger at him as he strode toward Winston with outstretched arms: "Dickie, stand there! What you did to us in India was like whipping your riding crop across my face!"[17] Then Churchill turned on his heel and walked away, not speaking to Mountbatten again for seven years. Fortunately for Attlee's government and Mountbatten's plan, Churchill was recuperating at this time from a very painful hernia operation. Attlee wrote to Churchill that it would not be possible to hold up the bill's progress through Parliament until Churchill was strong enough to come and speak against it. Attlee closed his letter with "I hope that you continue to make good progress."[18]

As time raced on, however, Mountbatten felt increasingly frustrated with Jinnah, who refused to invite Dickie to be the governor-general of Pakistan. He also did not trust Gandhi. "My private opinion is that Gandhi is adopting his usual Trotsky attitude and might quite well like to see the present plan wrecked," he confided to Attlee on July 2, 1947. He kept using Krishna Menon to keep Nehru pacified, and Reforms Commissioner V. P. Menon (no relative of Krishna Menon) to mollify Sardar Patel. V.P. worked most efficiently with Vallabhbhai in bringing most of the princes into India's union in record time. Mountbatten tried in vain to convince Jinnah of the monetary and military advantages to Pakistan of accepting

Mountbatten as joint governor-general of Pakistan, in the way that Nehru had agreed to his remaining in that job for India. "He is suffering from megalomania in its worst form," Dickie noted of Jinnah, "for when I pointed out to him that . . . as Prime Minister he really could run Pakistan, he made no bones about the fact that his Prime Minister would do what he said."[19]

Sir Cyril Radcliffe, a distinguished London barrister, agreed to chair both boundary commissions, which now started their hopeless jobs of drawing new international borders through Punjab and Bengal. Congress named two of the judges who served on each commission, and the Muslim League named the other two, so Radcliffe, who knew nothing at first hand of the geographic or ethnographic realities of either huge province, would have to decide every critical point of disagreement. It is no wonder that he never returned to India after that bloodiest of all jobs was completed in little more than one month. Radcliffe started work in New Delhi on July 8, 1947, which was "celebrated" as Sikh Protest Day throughout India. Most of India's 5.7 million Sikhs, many of whom went to pray on that day in Gurdwaras, wore black armbands to protest against the threat to split their community in Punjab, Reuters reported. "Sardar Baldev Singh, Defence Member of the Interim Government said Sikhs should be prepared to make all sacrifices if the verdict of the Boundary Commission went against them." Prepared or not, more than a million Sikhs, who woke up after August 15 to find themselves living in Pakistan, would be forced to sacrifice all their land, most of their property, and, for many, their lives.[20]

Gandhi met with Mountbatten on July 9, reiterating his fear that "the British would leave a legacy of war." Mountbatten assured the Mahatma that he was wrong. Gandhi also asked Mountbatten "to do everything in my power to ensure that the British did not leave a legacy of Balkanisation."[21] Again, the viceroy dismissed those fears of the much wiser and older man, whose worst fears were soon to be exceeded by the horrors of reality. Punjab's governor wrote Mountbatten next day that "my personal view is that the Boundary problem cannot be solved in any rational way."[22]

Krishna Menon left for London before mid-July, to take over as high commissioner, urging Mountbatten before he flew out to "keep in touch with Panditji [Nehru], not least on States. He is your P.M. now and obliged to give you advice on all matters."[23] Dickie conveyed "my warmest personal thanks" to Krishna Menon, concluding, "I am glad to think that I shall have a personal friend as the first High Commissioner in London."[24] Nehru sent a personal letter to Attlee, informing him that Krishna Menon "might prove helpful in explaining the situation here."[25] Nehru hoped that his high commissioner in London would win more English friends in high places for India. He was, however, destined to do quite the opposite, though "Krishna Menon phobia," which soon thrived in America, never reached the same epidemic virulence in Britain.

The Indian Independence Bill sailed through both houses of Parliament on July 17, 1947, enacted without a division and no opposition in the Commons and only minor complaints in the Lords. The king signed the act without delay, and Attlee wrote Dickie to congratulate him on "managing to jump a lot of awkward hurdles," adding that he knew that "Edwina has played a great part in creating the new atmosphere. . . . We are all very grateful to you. . . . I put you in to bat on a very sticky wicket to pull the game out of the fire. Few people would have taken it on and few, if any, could have pulled the game round as you have." The next day the Mountbattens hosted a grand dinner party in the viceroy's house to celebrate their silver wedding anniversary. Nehru and Indira came, of course, as did the entire cabinet.[26]

Nehru remained most anxious about Kashmir and was eager to go there. But Mountbatten insisted that he was needed more in Delhi, so Nehru agreed to send Gandhi to visit Kashmir on his behalf. The shrewd maharaja of Kashmir was almost as frightened of a visit by the Mahatma as he was of one by Nehru, so he decided to send his own prime minister, Pandit Kak, to Delhi instead, hoping that by meeting Gandhi and Nehru there he might persuade them both to remain away from the state that continued to refuse to make up its mind as to which dominion it wished to join. "I feel it would be both courteous and wise if you and Pandit Nehru could have a talk with Pandit Kak before deciding on the precise date and details of your visit," Mountbatten advised Gandhi.[27] Gandhi was persuaded after that to postpone his visit, so Nehru decided to go there in early August himself. But when Pandit Kak and his maharaja learned that instead of worrying about Gandhi they would soon be faced with a visit by Nehru, they both urged Mountbatten to try to convince Gandhi to come instead. The next day, therefore, Mountbatten met with Gandhi, Nehru, and Patel, the three of them trying to convince Nehru not to go to Kashmir. "Sardar Patel gave it as his view that neither of them should go, but that in view of Pandit Nehru's great mental distress. . . . 'I consider that Gandhiji's visit would be the lesser evil.' "[28]

But Nehru still wanted to go and "held forth at some length about his mental distress and defended his visit on the grounds that . . . he was over-worked; that he would like to go away for three or four days' rest somewhere in any case, and that Kashmir would be a delightful place in which to have a brief holiday."[29] "Ever since I arrived out Nehru has been hankering after a visit to Kashmir," Mountbatten reported to Attlee and the king. "He is obviously still suffering from an emotional upset. . . . Kak and the Maharajah hate Nehru with a bitter hatred and I had visions of the Maharajah declaring adherence to Pakistan just before Nehru arrived and Kak provoking an incident which would end up by Nehru being arrested just about the time he should be taking over power from me in Delhi!"[30]

Mountbatten wired Listowel in early August to invite the dominion prime ministers to the royal wedding of Princess Elizabeth and his nephew Phillip, which was scheduled for November. Dickie explained to his amenable secretary of state that "it would be a great thing if we could get Nehru to come to London at such a time of national rejoicing. . . . He is an inveterate sentimentalist, and I feel it would greatly help to strengthen Anglo-Indian bonds if he went." Although Nehru would soon look forward to those annual Commonwealth reunions with Dickie and Edwina as the most pleasurable event on his burdensome official calendar, war in Kashmir was to keep him in Delhi this November.[31]

Dickie found it much easier to mollify Nehru than Gandhi, of whose visit to Bengal he wrote: "Gandhi's absence from the celebrations in Delhi on the 15th August is, of course, intentional. He had never given the . . . plan his unqualified blessing and his position might be difficult. He also realises that it would not be possible to fit him into the programme in the way to which he would feel himself entitled." Gandhi had "announced his decision to spend the rest of his life in Pakistan," Dickie added, concluding that he now believed the Mahatma's "influence is largely negative" and "even destructive."[32]

Another "rather tiresome" problem for Mountbatten was the negative assessment of New Delhi's astrologers, who suddenly declared that August 13 and 15 were both inauspicious days. He thought, however, that he and Nehru had cleared the highest hurdles by convincing the Constituent Assembly to meet before midnight "on the auspicious 14th and take over power as midnight strikes . . . apparently still an auspicious moment." The real problem, however, was that superstitious members of the cabinet wanted their swearing-in ceremony finished before midnight, which would have "dreadfully eclipsed" the grand ceremony and photo opportunities Dickie had planned.[33]

Saturday, August 9, 1947, was a fateful day for South Asia. Sir Cyril Radcliffe and his Punjab boundary commissioners were ready to announce their award. Sir Cyril had reason to feel pleased with his speedy progress, for Mountbatten had urged him to try to be ready before the actual date of partition, in order to give everyone involved as much advance warning as possible. Governor Jenkins suggested that a week's early warning might help secure the new line and could certainly give the troops time to be redeployed. It might even allow for resettling entire villages of Sikhs or Muslims who found themselves on the wrong side of that line. Sir Cyril and the maps all had been moved into the comptroller's house inside the walled, steel-gated viceregal estate, so that there would be absolutely no

leaks to any person outside, for everyone knew how terrified and violent people caught on the wrong side of that new border might become.

But when Mountbatten learned of Sir Cyril's remarkably efficient work and saw the final line drawn by Radcliffe's own hand, he knew that the worst of Sikh fears might be broadcast the next day, as cries of "Betrayal!" echoed across the blazing dome of Punjab's sky. So Dickie wondered "whether it would be desirable to publish it straight away," reasoning now that "the earlier it was published, the more the British would have to bear the responsibility for the disturbances which would undoubtedly result." He therefore emphasized the "necessity for maintaining secrecy," and in order "to ensure the loyalty of the Sikhs in the Governor-General's Bodyguard," his staff were ordered to make it clear to every Sikh guard that the viceroy personally "had nothing to do with the award of the Boundary Commission."[34] Several members of his staff futilely argued that the governor of Punjab had "pointed out that there were administrative advantages from early publication."[35] Mountbatten, however, insisted he would much prefer to postpone the announcement of the Punjab award until after the Independence Day celebrations, because he wanted no grief to be "allowed to mar Independence Day itself."[36]

So instead of issuing the early warning to Punjab that might have allowed at least some of its trapped millions to move east or west before the new international doors slammed shut on them, the award, ready on August 9, was not released for another full week, leaving those doomed to death clinging futilely to the hope that springs eternal in the hearts of innocents. Radcliffe's award was thus kept under wraps as both Mountbattens flew to Karachi on Wednesday, August 13, to convey His Majesty's greetings to the new dominion of Pakistan, which was born the next day. Governor-Generals Mountbatten and Jinnah dined together that evening, and the next day Dickie and Edwina flew back to Delhi to a much more festive round of national birthday celebrations that awaited them there, the last hurrahs and wild cheering before the furies of fear and terror were unleashed.

By August 12 Radcliffe felt quite upset by the fact that his award—which Mountbatten had pressed him to finish as early as possible—remained under viceregal "embargo." Dickie was, however, adamant about not releasing a word officially until after August 15.

Nehru sent Mountbatten a copy of a message he wrote to mark the "departure of the first contingent of British troops from India," on August 13. "As an Indian I have long demanded the withdrawal of British Forces for they were a symbol to us of much that we disliked," Nehru told those departing troops, adding that "I had no grievance against them as individuals and I liked and admired many whom I came across . . . I wish them godspeed. . . . It is rare in history that such a parting takes place not

only peacefully but also with goodwill. We are fortunate that this should have happened in India."[37]

That very day, Governor Jenkins wrote from Lahore to report that to stop the escalating killing throughout Punjab "we should need at least two Divisions of full strength and on a War footing . . . about 20,000 effective fighting men." The police were totally unreliable by now; no roads or railways were safe. "Raiders" roamed the land from village to village, murdering, raping, and looting. Hindus all over Punjab were "thoroughly terrified."[38]

"Long years ago we made a tryst with destiny, and now the time comes when we shall redeem our pledge, not wholly or in full measure, but very substantially," Prime Minister Nehru told them that inauspicious mid-August midnight. "At the stroke of midnight hour, when the world sleeps, India will awake to life and freedom. A moment comes, which comes but rarely in history, when we step out from the old to the new, when an age ends, and when the soul of a nation, long suppressed, finds utterance."[39]

Mountbatten waited for Nehru at his desk in the viceroy's house, renamed Rashtrapati Bhavan (president's house) after India became a republic in 1950. Dr. Rajendra Prasad, who lived there as India's first president, now came with Nehru from the assembly, over which he presided as speaker, to invite Mountbatten to serve the newborn dominion of India as its governor-general. "Mountbatten and Prasad stood facing each other, with Nehru half sitting on Mountbatten's desk," Alan Campbell-Johnson recalled. "Prasad [speaker of the Constituent Assembly] began murmuring a formal invitation. However, he forgot his lines, and Nehru played the role of benign prompter." To no one's surprise, with cameras rolling and clicking, Dickie accepted their invitation.[40]

"I know well that the rejoicing which the advent of freedom brings is tempered in your hearts by the sadness that it could not come to a united India," Governor-General Mountbatten told them all the next morning as he was sworn in before many more cameras. "At this historic moment, let us not forget all that India owes to Mahatma Gandhi—the architect of her freedom through non-violence. We miss his presence here today. . . . In your first Prime Minister, Pandit Jawaharlal Nehru, you have a world-renowned leader of courage and vision. His trust and friendship have helped me beyond measure."[41]

Four guards of honor, each with one hundred of British India's strongest soldiers, barely managed to keep back the surging crowds long enough to permit the Mountbattens to get into their state coach. "Nehru went up to the roof and waved to the crowd to go back; the door was then opened and surrounded by our staff we fought our way through to the coach," Dickie reported. There was a grand luncheon, and then at 6 P.M. the great event of the day was scheduled to be the salute to the new dominion flag,

accompanied by a parade of all three services. But the crowds around Government House in New Delhi had by then grown to more than 600,000, and "we should never have been able to get on to . . . the Grand Stand. . . . Nehru fought his way to the coach and climbed in to tell us that our daughter Pamela was safe. George Abell . . . described how Nehru came to their rescue when they were overwhelmed by the crowd, fighting like a maniac, striking people right and left and eventually taking the topee [stiff turban] off a man who had annoyed him particularly and smashing it over his head."[42]

That evening Nehru broadcast "for the first time officially as the first servant of the Indian people, pledged to their service and their betterment. I am here because you will it so and I remain here so long as you choose to honour me with your confidence," he told those of his compatriots gathered around radios in Delhi, Bombay, Calcutta, and Madras. "The burden of foreign domination is done away with, but freedom brings its own . . . burdens. . . . Today there is no time for quarrelling or overmuch play, unless we prove false to our country and our people." Nehru ended his first brief broadcast to his nation with "Jai Hind!" his dead comrade Subhas Bose's old battle cry.[43]

28

Days of Darkness

1947–1948

"THE FREE FLAG of India is the symbol of freedom and democracy not only for India but for the whole world," Nehru told his cheering compatriots as he raised India's tricolor over Delhi's Mughal Red Fort on the morning of August 16, 1947. "We had taken a pledge that we shall lay down our lives for the honour and dignity of this flag . . . under the brilliant leadership and guidance of Mahatma Gandhi. . . . If credit is due to any man today it is to Gandhiji."[1]

But Gandhi was far to the east, in Bengal's Calcutta, trying to quell continued communal riots there. "Is there something wrong with me?" the Mahatma inquired in a letter he wrote that day to his disciple Amrit Kaur, now Nehru's minister of health. "Or are things really going wrong?"[2] In Punjab by now, everything was going wrong. Nehru and Liaquat Ali, Pakistan's prime minister, visited Punjab on August 17, meeting with the governors of East and West Punjab and their senior officers. "Every possible step must be taken immediately to put an end to this orgy of violence," Nehru insisted. "No government worthy of the name can tolerate such lawlessness and crime."[3] A day later he broadcast again from New Delhi that "nearly the whole of India celebrated the coming of independence, but not so the unhappy land of the five rivers. . . . [T]here was disaster and sorrow. There was murder and arson and looting . . . and streams of refugees poured out. . . . Each one of us who cares for this country must help in this business of restoring peace and security."[4]

A few days later Nehru wrote Gandhi, reporting that "all this killing business has reached a stage of complete madness, and vast populations are deserting their habitations and trekking to the west or to the east." Dr. Zakir Husain, destined to become India's first Muslim president during Indira Gandhi's premiership, had been forced to step out of his first-class carriage at Jullundur Station, and as Nehru told Gandhi, "He narrowly escaped being killed by an armed band . . . Indian soldiers standing by merely looked on. Ultimately a Sikh captain in the army rescued him. . . . What a terrible legacy! The Punjab will be a ruined province."[5]

Indira took her boys up to Mussoorie after the Independence Day celebrations ended in Delhi and wrote her father from that hill station retreat: "Rajiva was thrilled by your call yesterday. He got so excited . . . I think he enjoyed his birthday. He looked very grown up. . . . Bebee [Padmaja Naidu] sent a cake for Rajiv—just a little larger than yours."[6] The Nehru dynasty's third prime minister was only three and would have to wait thirty-eight more years before taking up the bullet-shredded mantle of his murdered mother.

Feroze was stuck in Lucknow, trying to run the family's *National Herald* there, and had just found a house for his wife and sons, "the only flaw of which is that it has some mutiny graves in the compound . . . rather gruesome, isn't it? Still I suppose one has to take what one gets—ghosts and all?"[7] Indira was haunted all her life by ghosts, though she never spent much time in that old Lucknow house with Feroze. "It's such a nuisance not having a phone," she complained to Nehru soon after moving in. "What a peculiar deadness there is in our provincial towns. And what makes the atmosphere sickening is the corruption and the slackness, the smugness of some and the malice of others. Life here has nothing to offer . . . middle-class young men. It is not surprising that the superficial trappings of fascism attract them in their tens of thousands. The R.S.S. [Rashtriya Svayamsevak Sangh] are gaining strength rapidly. They have been holding very impressive rallies in Allahabad, Cawnpore, Lucknow—except for very minor details following the German model. . . . Are we inviting the same fate to our country? The Congress organisation has already been engulfed—most Congressmen approve of these tendencies. So do Government servants."[8] In less than two months, one of those young Hindu fascists, named Godse, assassinated Mahatma Gandhi in New Delhi.

Nehru was too preoccupied with Punjab to worry much about that growing army of "brown shirt" Hindus mustering in every major city of north India, terrorizing Muslims and other minorities left in the nation they worshiped as "Mata (Mother) Bharat," the ancient Aryan name for India. "Our High Commissioner for Pakistan [Sri Prakasa]," Nehru wrote Mountbatten on August 27, "sent a message through the telephone to say that . . . a very large number of persons were being done to death daily.

. . . I do not mention the figure he gave because it is incredible. . . . I suppose I am not directly responsible for what is taking place in Punjab. I do not quite know who is responsible. But . . . I begin to doubt whether I have any business to be where I am. And even if I don't doubt it . . . other people certainly will . . . I am not an escapist or quitter. . . . But I wished to unburden my mind a little to you and hence this totally unnecessary letter."[9] By now half a million refugees from West Punjab had moved east, many of them settling in camps north of Delhi, around Meerut. Nehru flew back to Amritsar but was overwhelmed by the tragedy he witnessed there. "We are up to our eyes in work and worry," he wrote Vallabhbhai, urging him to come to Punjab as well. "Your presence will energise the people here and you will bring a fresh mind to bear on these problems. I must confess that I feel a little overwhelmed by these difficulties and the urgency which accompanies them."[10]

"I am sick with horror," Nehru wrote Mountbatten that day, after visiting Sheikhupura. "What we saw was bad, what we heard was worse. . . . There was still an odour of death, a smell of blood and of burning human flesh. . . . [W]e came across destitute wanderers trying to reach some haven. . . . I saw some old colleagues who were bereft of everything and were trudging along . . . in the long caravan. This Punjab business becomes bigger and bigger . . . killing and arson still . . . hundreds of thousands . . . on the move . . . in the nature of an uncontrollable natural phenomenon. . . . I imagine that quite a million people have been uprooted and have already changed from one place to another. Another million are in refugee camps either in West or East Punjab, or are wandering about without home."[11] Now he wired Gandhi as well, asking him to leave Bengal and come back to try to do something about the "Punjab problem overwhelming in extent and intensity."[12] For now that it was too late, Nehru understood why Gandhi was ready to have done anything to avoid this hideous disaster of partition.

Gandhi himself was almost killed in Calcutta. Hindu Mahasabha extremists vented their frustrations at this saintly old man because he continued to urge nonviolent love rather than preaching hatred of Muslims. Gandhi had become the symbol of "anti-Brahmanism" to many fanatical Hindus. "Someone received knife wounds," Gandhi wrote Vallabhbhai from Calcutta on September 1, 1947. "People brought him here . . . so their anger was turned on me. There was an uproar in the front yard. . . . Everyone suspects the Hindu Mahasabha . . . mischief-makers. . . . Please tell Jawaharlal about this . . . I feel totally lost. I pin my hopes on you two. . . . Blessings from Bapu."[13]

In response to that outrage, Gandhi started to fast that night in Calcutta, hoping "it may touch the hearts of all the warring elements even in the Punjab."[14] All of Bengal's leaders rushed to his bedside, urging him to break his fast, to save his life. On September 4 he agreed but warned his followers "not to be lulled into complacence" by the speed of his decision to stop, for if violence broke out again he would resume fasting, the next time "unto death."[15]

On September 7, 1947, Gandhi left for Delhi on his way to the Punjab where he hoped to achieve another miracle of reconciliation by his peace-inspiring presence. After reaching Delhi, he listened "the whole day long to the tale of woe that is Delhi today. I saw several Muslim friends who recited their pathetic story. I heard enough to warn me that I must not leave Delhi for the Punjab until it had regained its former self. . . . I must apply the old formula 'Do or Die' to the capital of India."[16] India's Conscience had come to the capital to remind his disciples of their promises and sacred vows to serve every Indian alike.

"Has the city of Delhi which always appeared gay turned into a city of the dead?" Bapu cried out next evening at his prayer meeting in Birla's garden. "Pandit Nehru and Sardar Patel . . . had declared that in the Indian Union . . . minority communities would receive the same treatment as the majority. . . . The leaders of the Indian Union . . . cannot fling their hands in despair and say that it is all the doing of the goondas. . . . Now they are not forced to do anything against their will under the crushing burden of Imperialism."[17] Such truthful barbs from Gandhi went deeper and proved more painful to Nehru and Patel than harsher words aimed at them by anyone else.

"During my rounds of the city I heard complaints that the refugees do not get their ration. And whatever is being supplied to them is not fit for human consumption."[18] Nehru broadcast again the day after Gandhi reached Delhi, "I went to see him [Gandhi], and I sat by him for a while wondering how low we have fallen from the great ideals that he had placed before us!"[19]

Nehru then flew to Lahore to meet with Liaquat Ali Khan and, on his flight back, reported seeing "two convoys of refugees on foot," stretching some forty miles.[20] One of those convoys was, in fact, fifty-two miles long. Subsequent estimates confirmed that about fifteen million people tried to move from one side of the partitioned Punjab and Bengal to the other in the year following this tragic birth of two dominions, although fewer than fourteen million ever reached their destinations. Nehru returned to inform his cabinet that many women who had been abducted in the Punjab could not return home, even if they tried, since most Indian families considered abducted women "damaged goods" and were usually unwilling to speak to them again.

Two days later Nehru wrote to Rajendra Prasad, "I must confess to you that recent happenings in the Punjab and in Delhi have shaken me greatly . . . shaking my faith in my own people. I could not conceive of the gross brutality and sadistic cruelty that people have indulged in. . . . There is a limit to killing and brutality and that limit has been passed during these days in North India. . . . Little children are butchered in the streets. The houses in many parts of Delhi are still full of corpses. . . . I am fairly thick-skinned, but I find this kind of thing more than I can bear . . . 50,000 or 100,000 people have been murdered."[21] That was in Delhi alone!

Gandhi visited the refugee camps around Delhi every day. "I have not seen anywhere the like of the squalor I found in that camp," he reported, after visiting the Muslim refugees. He felt ashamed, wondering how things could go so wrong in Delhi "in spite of our Government, in spite of Jawaharlal who is like a lion and with a Home Minister like Sardar Patel? Why should his authority not be accepted? If he sends out an order that a child has to be protected here, that child must be protected . . . whose Government is it after all? It is your own Government. You have made it."[22] Gandhi's barbs were becoming sharper, more painful. At his next prayer meeting in Delhi, when recitations from the Koran were begun, an angry Hindu shouted, "We will not let you recite these verses . . . Gandhi *murdabad* [death to Gandhi]!"[23] The meeting was soon called off, and Gandhi managed to return unhurt to Birla House.

"As long as I am at the helm of affairs India will not become a Hindu State," Nehru now assured millworkers in Delhi. "The very idea of a theocratic state is not only medieval but also stupid. . . . Remember the great lesson of nonviolence taught by the greatest man of the world and the Father of the Indian nation. One can either follow the lead given by Mahatma Gandhi or oppose it. There is no third way open . . . India is getting a bad name in countries abroad."[24] Nehru well knew that armed Sikh bands and Hindu bands who owed allegiance to the R.S.S. had "a great deal to do with the present disturbances, as he told his home minister, having heard that from many sources. Vallabhbhai also knew it, of course, since as home minister he received as many daily police reports as did the prime minister. "As far as I can make out, we have had to face a very . . . well-organised attempt of certain Sikh and Hindu fascist elements to overturn the Government," Nehru wrote Patel on September 30, 1947. "Many of these people have been brutal and callous in the extreme. They have functioned as pure terrorists."[25] Until they were jolted by the assassination of Mahatma Gandhi, however, both Nehru and Patel seemed unable to conceive of any way to stop such brutal slaughter in Delhi.

In the aftermath of those terrorist acts and mass migrations, relations between India and Pakistan deteriorated further, each dominion blaming the other. Nehru sent Nan to New York to lead India's delegation to the United Nations and now wired her to be "strictly formal" in her dealings with the Pakistan delegation. He also advised her to "make no comment on Mahatma Gandhi's statement" that "I hate the very idea of war." Gandhi, Nehru reminded Nan, "is wedded to nonviolence. . . . In a state run by him there would be no police and no military. But he was not running the Government of India."[26]

Nehru's mind now focused on Kashmir. He feared that Kashmir's predominantly Muslim population and direct, easy access to Pakistan would sway Maharaja Hari Singh in favor of opting to join Pakistan. Although Hari Singh was Hindu by birth, Nehru knew how conservative he was and how much he hated Congress socialists like himself and Kashmiri radicals like Sheikh Abdullah, whom he still kept in Srinagar's prison. The maharaja hoped most of all, Nehru knew, to see Kashmir state emerge independent of both its dominion neighbors, but neither Attlee or Mountbatten ever considered that a possible option. What Nehru feared most, as the winter snow and ice approached, was the imminent, permanent loss of India's largest, most strategically important state, which his ancestors all called home.

"It is obvious to me from the many reports I have received that the situation there [in Kashmir] is a dangerous and deteriorating one," Nehru wrote Vallabhbhai in late September. "The approach of winter is going to cut off Kashmir from the rest of India. . . . Therefore it is important that something should be done before these winter conditions set in . . . by the end of October or, at the latest, the beginning of November." Nehru therefore urged his home minister and minister for states to "take some action in this matter to force the pace and to turn events in the right direction. We have definitely a great asset. . . . It would be a pity to lose this. . . . [T]ime is of the essence . . . and things must be done in a way so as to bring about the accession of Kashmir to the Indian Union as rapidly as possible."[27] After receiving this powerful letter from his prime minister, Sardar Patel immediately convinced the vacillating Maharaja Hari Singh that it was time to win his people to his side by releasing the popular National Conference leader Sheikh Abdullah.

Two days later, "My dear Sheikh Saheb," as Nehru addressed him, was out of prison. Nehru wrote Abdullah to say how delighted he was. "But what a world it is which we have to face now! . . . People have gone mad and I do not know when sanity will return. . . . Undoubtedly the position in Kashmir is difficult. . . . One has to be careful now not to lose sight of the wood for the trees. Our old slogans do not always work as I know to my cost in Delhi. . . . [F]reedom itself is in danger not from outside, but from inside. But when the inside weakens, outside elements come into play."[28]

Nehru hoped Abdullah would be able to administer Kashmir as an integral part of India, for like himself the sheikh was a brilliant man. He was, of course, born Muslim, but his mind was basically secular, his outlook socialist, progressive, dynamic. He was the "Nehru of Kashmir," which was why Jawaharlal liked him so much.

"We have had to face here in India a reactionary and fascist upsurge," Jawaharlal told him. "The Punjab is a ruined province. . . . We agreed to the partition of India. . . . Yet even we never realised the extent of these dangers which are so evident today. We agreed because we felt that India's political and social life was being undermined and poisoned by continuous inner conflict and we wanted to put an end to this so that people may consider the questions facing us dispassionately. Our calculations were evidently wrong as events have proved." Nonetheless, Nehru assured his troubled friend that "Pakistan . . . is in a worse position and I doubt very much if it can survive at all. Financially it will be completely bankrupt."[29] For India kept the lion's share of sterling assets, having not yet released any of the 18.5 percent of all Reserve Bank money that was to have gone to Karachi in the immediate aftermath of partition. Mahatma Gandhi would have to undertake his final fast unto death a few months later, before Nehru and Patel relented and reluctantly agreed to unfreeze those Pakistani funds.

"All of us sit on the edge of a precipice and dangers surround us," Nehru went on to explain to Sheikh Abdullah, who would remain free only as long as he agreed to work with Jawaharlal in keeping Kashmir for India. "I think we in India have a firm hold and can keep ourselves going. . . . Pakistan has no such hold and no strength or resources. . . . We have to suffer for our own errors . . . not much good crying over spilt milk. . . . But we must understand clearly what has happened. . . . Kashmir is, of course, of vital significance to this picture of India. What happens in Kashmir will affect the rest of India. . . . For me Kashmir's future is of the most intimate personal significance. On no account do I want Kashmir to become a kind of colony of foreign interests. I fear Pakistan is likely to become that if it survives at all. It may well be that the Pakistan people look upon Kashmir as a country which can yield them profit."[30]

"For me it is both a personal and public matter. It would be a tragedy, so far as I am concerned, if Kashmir went to Pakistan," Nehru wrote to M. C. Mahajan, newly appointed prime minister. "They [the Pakistanis] are already a tottering state. They look to Kashmir for a means of recovering. They intend to raise capital in America on the strength of Kashmir by giving special privileges, leases etc. for development there to Americans. All their present policy is to get help from America. . . . [I]t will certainly mean the overrunning of Kashmir by adventurers and others. No Kashmiri can welcome this prospect." Nehru heard that Jinnah had promised Hari Singh virtual independence, if Kashmir opted to join Pakistan and that he was even "prepared to give the right of secession."[31]

On October 25, 1947, Nehru cabled Attlee that "a grave situation has developed in the State of Kashmir. Large numbers of Afridis and other tribesmen from the Frontiers have invaded State territory . . . and massacred large numbers of non-Muslims. . . . We have received urgent appeal for assistance from the Kashmir Government. We would be disposed to give favourable consideration to such request. . . . [The] security of Kashmir, which must depend upon its internal tranquillity . . . is vital [the] to security of India. . . . I should like to make it clear that question of aiding Kashmir in this emergency is not designed in any way to influence the State to accede to India. . . . [T]he question of accession in any disputed territory . . . must be decided in accordance with wishes of the people." Attlee wired back two days later, urging Nehru to "refrain from armed intervention in Kashmir as it would only aggravate the problem" and might lead to "an open military conflict" with Pakistan. By the time that cable was sent, however, India's First Sikh Battalion had already landed in Srinagar, and the first Indian–Pakistani war in Kashmir was well under way.[32]

Several thousand armed Afridi and Waziri tribesmen had been driven in British trucks from Pakistan's Pindi over the border of Kashmir state, taking Muzaffarabad and looting, shooting, and raping as they headed east along the Baramula Road toward Srinagar. Sheikh Abdullah flew to Delhi with Mahajan, both reporting immediately to Nehru on that critical situation. Maharaja Hari Singh still refused to join either dominion. Mahajan informed Nehru, however, that the maharaja was ready to opt for India if Nehru could assure them enough airlifted Indian military support to defend Srinagar from the advancing tribals. Nehru rushed over to alert Mountbatten to the situation, insisting that unless India acted swiftly it would be too late to save Kashmir. Mountbatten understood Nehru's concern but explained that without the maharaja's signed accession to India, airlifting Indian troops to Srinagar would be viewed as an invasion of Kashmir by neighboring Pakistan. The rest of the Commonwealth might well agree, moreover, as Attlee's cautionary cable clearly indicated.

But Nehru resolved not to let any legalities stop him from saving his ancestral motherland, and he told Dickie as much. Nehru then called a meeting of his cabinet's Defense Committee for October 25, ordering the Sikh minister of defense Baldev Singh to alert the First Sikh Battalion to be ready for combat. Every available airplane in the region was flown to Delhi's military airport. Within twenty-four hours, more than a hundred transport planes, civil as well as military, were fueled, serviced, and kept at the ready for predawn liftoffs to Srinagar. It was the sort of swiftly orchestrated military exercise that Admiral Mountbatten did best, though as governor-general he understood the perils of acting without legal cover and explained to his passionate prime minister that unless Hari Singh signed a document stating, first, that he wished to accede to the Indian Union and, second, requesting "help against raiders" who had entered Kashmir without any invitation from his government, India would be starting a war by flying its troops from Delhi to Srinagar.

Nehru reluctantly agreed to Mountbatten's suggestion that Vallabhbhai's assistant in the States Ministry, V. P. Menon, who had also been Mountbatten's reforms commissioner and public secretary, be flown to Srinagar immediately with the instrument of accession in his attaché case, meet the maharaja in his palace, get him to sign the document, and fly directly back to Delhi, thereby turning what whould otherwise be an aggressive invasion of Kashmir into the perfectly legal protection and defense of part of the Indian union. But V.P.'s mission to Srinagar proved futile, for when he flew back to Delhi that Sunday morning, October 26, he had no signature on his document and could report only that Hari Singh "had gone to pieces completely" and could "come to no decision." This time Nehru's temper flared to white heat, and both Mahajan and Abdullah urged him to "give us the military force we need. Take the accession."[33] To Nehru's mind that was legality enough, since the "crazy tribals" were headed toward Srinagar and almost all of India's air transports were fueled, ready to fly the first seven hundred Sikhs over the mountains into Kashmir to save Srinagar from death and destruction or, what to Nehru's mind was the same thing, accession to Pakistan.

"I think it is important for Sheikh Mohammad Abdullah to go to Srinagar immediately as the situation there requires urgent handling and his presence will be helpful," Nehru wrote Mahajan that fateful October 26, 1947. "He will come to Jammu whenever the Maharaja so desires."[34] V.P. had not been able to get the distraught Hari Singh to sign any document but did manage to convince him that it was too dangerous to spend another night in Srinagar, packing up his most precious jewels and other possessions and taking his entourage of family and servants in a motorcade of Rolls Royces to his southern palace and capital in Jammu. "In the early hours of the morning of the 27th," Mahajan later recalled, "I could hear the noise of the planes flying over Sardar Baldev Singh's house and carrying the military personnel to Srinagar."[35]

At 9 A.M. on the morning of October 27, Prime Minister Mahajan received a phone call from Srinagar reporting that India's troops had landed and secured the capital. He then flew off to Jammu with V.P. and the same instrument of accession, and this time both men managed to convince the weary and worried Maharaja Hari Singh of the wisdom of agreeing to join India now that Indian troops had control of his northern capital and more troops were being flown in every hour. As soon as the instrument was signed, V.P. flew it back to Nehru, who wrote to thank Mahajan and say, "I am glad indeed that the Maharaja Saheb signed the

Instrument of Accession. . . . I am sure this is a wise decision which will do good to Kashmir and India."[36] Nehru also wrote the maharaja to "congratulate you on the wise decisions that you have taken. . . . The way the people of Kashmir, Muslim, Hindu and Sikh, are facing the situation and preparing to defend their country is most heartening . . . in this defence we shall give a demonstration to all India and to the world how we can function unitedly and in a non-communal way in Kashmir. In this way this terrible crisis in Kashmir may well lead to a healing of the deep wound which India has suffered in recent months."[37] By retaining control of predominantly Muslim Kashmir, Nehru hoped to prove to the world that India was a truly secular union of peoples of many faiths and not just a Hindu state.

To Sheikh Abdullah in Srinagar, Nehru wrote that evening, "Maharaja is going to invite you to form an interim government after the Mysore pattern. In the emergency before us we cannot meticulously be careful about details or strict legal position. . . . You need not trouble to sit down and study. . . . the Mysore arrangement. Generally speaking it means that you form the Government including the present Prime Minister, Mahajan, who retains his title. . . . [Y]ou will be called the Chief Minister. . . . The Maharaja will in law retain all his powers, but in practice he should abide by the advice of his Ministry." More important than such legal details, Nehru reminded Abdullah, were the military realities. Indian troops had arrived "in the nick of time" and Nehru praised the whole operation as "a remarkably fine achievement." More and more Indian troops, weapons, and vehicles were being flown into Kashmir on the hour and half hour. He urged Abdullah to arm as many young men, Muslim and Hindu, as he could trust to form "a kind of home guard." "No looseness or weakness should be tolerated," Nehru cautioned. "We have taken on a tough job. But I am dead sure that we shall pull through. Ever since the decision was taken yesterday, and I heard today that our troops had reached Srinagar, I have felt much lighter in heart. We have taken the plunge and we shall swim across to the other shore."[38]

Mountbatten, on the other hand, "considered that it would be the height of folly to send troops into a neutral State, where we had no right to send them, since Pakistan could do exactly the same thing," Campbell-Johnson recalled. He had just flown back to Delhi from London, and Mountbatten woke him at 3 A.M. on October 28 to brief him on the latest in Kashmir. As Dickie explained it, "The essential prerequisite was accession, and unless it was made clear that this accession was not just an act of acquisition, this in itself might touch off a war."[39]

Nehru told Mountbatten that he would agree to a plebiscite in Kashmir as soon as law and order were restored. No such plebiscite was ever held, of course. Nehru sent Major General Hiralal Atal to Srinagar to supervise secret operations throughout Kashmir. "We hope to send some planes for strafing from the air in the Valley. . . . I am afraid we cannot do much beyond Baramula and up the Jhelum Valley Road as the valley there is narrow. . . . What about the two bridges at Kohala and over the Kishanganga near Muzaffarabad? [Air Marshal] Mukerjee tells me that it is next to impossible to bomb them from the air with the aircraft at our disposal. . . . At the same time we must make an attempt to destroy these bridges. I am sure you can arrange with Sheikh Abdullah a small party with necessary equipment. . . . It is a very risky job, but worth doing." [40]

"I am aware of what is happening in Kashmir," Gandhi told his prayer meeting at Birla House the night before. "It is reported that Pakistan is trying to coerce Kashmir to join Pakistan. . . . It is not possible to take anything from anyone by force. . . . If the people of Kashmir are in favour of opting for Pakistan, no power on earth can stop them from doing so. But they should be left free to decide for themselves. . . . If the people of Kashmir, in spite of its Muslim majority, wish to accede to India no one can stop them. . . . If the people of the Indian Union are going there to force the Kashmiris, they should be stopped, too, and they should stop by themselves. About this I have no doubt." [41] Nehru was too busy to attend Gandhi's prayer meetings. "Granted that we have all become barbarians—whether here or in Pakistan, no act of madness is left undone—should the people in Kashmir also turn barbarians and indulge in indiscriminate killing of women and children? Should Kashmir be reduced to such a terrible state? Pandit Jawaharlal Nehru and his Cabinet came to the conclusion that something should be done and those soldiers were sent. . . . God alone knows what the outcome will be." [42]

"This trouble in Kashmir has been thrust upon us and yet it may well be the saving of us in many ways," Nehru wrote to General Atal. "It may go a long way in settling our problem with Pakistan as well as in the Indian States. It may and I hope it will change the entire communal atmosphere in India. The fact that Hindus, Muslims and Sikhs are cooperating for the defence of Kashmir will tone up our whole system." [43]

Nehru cabled Attlee on October 28, telling him that "some 2,000 or more fully-armed and well-equipped men came in motor transport, crossed over to Kashmir . . . near Muzaffarabad, sacked that town killing many people and proceeded . . . towards Srinagar. . . . The Maharaja appealed urgently to us for help. He further suggested accession to Indian Union. . . . We decided at first not to send any troops. . . . But later developments made it clear that, unless we send troops immediately, complete disaster would overtake Kashmir. . . . We therefore elected to send troops to Kashmir. . . . [E]arly this morning one battalion . . . was flown to Srinagar, and has landed there. . . . Our attitude and policy have been, as I have stated to you, that in case of any disputed State territory, the problem of accession should be decided amicably and in accordance with the wishes of the people. . . . Our military intervention is purely defensive in aim and scope, in no way affecting any future decision about accession that might be taken by the people of Kashmir."[44]

Nehru also cabled Liaquat Ali Khan that day, similarly to assure him that the "action Government of India has taken has been forced upon them by circumstances and imminent and grave danger to Srinagar. They have no desire to intervene in the affairs of Kashmir State after raiders have been driven away. . . . In regard to accession . . . this is subject to reference to people of State and their decision. Government of India have no desire to impose any decision and will abide by people's wishes."[45]

That had been a busy October day for Field Marshal Auchinleck, who flew to Lahore that morning to explain to Governor-General Jinnah and General Douglas Gracey, Pakistan's commander in chief at that time, that Jinnah's order to "move two brigades of the Pak army into Kashmir" was unacceptable.[46] As the supreme commander of British forces in South Asia, Auchinleck threatened to issue an immediate stand-down order to every British officer serving in Pakistan's armed forces if Jinnah insisted on an invasion of another dominion's territory by his regular troops. Jinnah was forced to back off, finally seeing how high a price Mountbatten could still make him pay for not having agreed to Dickie's fondest desire to serve as governor-general of both dominions. Before leaving Lahore, Auchinleck promised to convey Jinnah's invitation to Mountbatten and Nehru to fly there on October 29 to discuss the Kashmir crisis. Mountbatten accepted that invitation, knowing how bitter a pill he and Auchinleck had concocted for the mortally sick, tired old Jinnah to swallow. Nonetheless, Jinnah had the grace to invite them all to meet in his governor's house of Punjab's once beautiful capital, still Pakistan's most glorious city. So Dickie dined with his staff that evening, and all of them toasted the imminent prospects of peace in Kashmir, when "Vernon [Erskine Crum] . . . arriving late for dinner after some harassing hours on the telephone, announced, 'It is the end.' The whole plan, he said, had broken down, as Nehru could not go to Lahore because of illness." Mountbatten felt "sure that his illness is genuine," Campbell-Johnson reported.[47]

Next morning Mountbatten phoned Jinnah to request a postponement, explaining how serious Nehru's illness was and inviting Jinnah and Liaquat Ali to Delhi instead. Jinnah, however, understood exactly what afflicted Nehru and hence declined Dickie's polite invitation. Mountbatten returned to Nehru's bedside to suggest that perhaps it might be best if both of them flew to Lahore a few days later, after Nehru felt better. For not only was Jinnah dying, but Liaquat also was sick, confined to bed with a bleeding ulcer. Nehru agreed, and Mountbatten called Jinnah again that Wednesday afternoon (October 29) to promise that Nehru would accompany him to Lahore on November 1. By then, however, Nehru's temper had flared up, and he refused angrily to discuss anything with Jinnah, so Mountbatten took Ismay with him to Lahore. They went to visit Liaquat, who was "sitting up with a rug round his knees, looking very ill." [48]

Mountbatten had lunch with Jinnah that November 1, their last meeting as governors-general. Jinnah charged that India had seized Kashmir state by "fraud and violence" and informed Mountbatten that Pakistan would never accept that fraud. Dickie tried his best to argue that it was all perfectly legal and that the violence had been initiated by Pakistan's frontier raiders. Jinnah repeated his charge, however, firmly believing it until his death less than a year later. As they were leaving, Jinnah told Mountbatten and Ismay that he had "lost interest in what the world thought of him since the British Commonwealth had let him down when he asked them to come to the rescue of Pakistan." [49]

"My dear Vallabhbhai," Nehru wrote his home minister, informing him that "information has reached me that R.S.S. volunteers have been organised in East Punjab to be sent to Jammu for a campaign against the Muslims. It is stated that 500 were sent . . . some days ago in special trucks. . . . I do not know what we can do about all this, but we should at least try to stop in so far as we can the activities of the R.S.S. and the Hindu and Sikh refugees in this direction. . . . The whole Kashmir position will crack up if in Jammu . . . an anti-Muslim drive takes place. . . . Also the Maharaja is in a state of nervous collapse and somebody should be there to keep him on track." [50]

The price to India as a whole of saving Kashmir remained very high, as Nehru explained to Mahajan: "I suppose you know that a whole brigade of the Pakistan Army was kept on the Jammu border and another brigade was kept at Kohala . . . expected to march into State territory as soon as Srinagar fell. This manoeuvre was foiled by Kashmir's accession to the Indian Union and our sending troops by air. . . . You will appreciate, we have made a tremendous effort to pour in troops and equipment into the valley of Kashmir. A brigade has also gone to Jammu. . . . I hope that soon our troops will take the offensive. This has been done at tremendous cost to us and holding up most of our other activities in India. All our air services have stopped and every available plane is going to Kashmir. . . . I see no reason why any of you should go to Lahore to confer with Mr. Jinnah or anybody else. Our position is perfectly clear and there is very little to discuss. The only point that might arise would relate to . . . the plebiscite. It is obvious that a plebiscite cannot take place till complete law and order have been established. I see no chance of this happening for some months." Nor, indeed, for half a century, although Nehru continued to insist that he was willing to hold an "impartial" plebiscite someday in Kashmir, possibly under U.N. auspices.[51]

"Now my mind turns to Qaid-e-Azam Jinnah," Gandhi told his prayer meeting on November 2. "I know him well. I used to go to his house . . . I used to have cordial talks with him. . . . That is the reason why I would ask him, Liaquat Ali and his Cabinet what had prompted them to accuse a man like Jawaharlal of fraud. Where was the need for him and his Government to act fraudulently in this? . . . Jawaharlal is not a man who will deceive anyone. . . . Can India or any country be saved by deceiving? Why then do they say such a thing? . . . With all humility I would request the entire Cabinet and people that they should all become good if they wish that India should not be destroyed."[52] A few days later Mahatma Gandhi was asked if he had abandoned nonviolence and approved of sending India's army into Kashmir. He replied softly and sadly: "I am a nobody and no one listens to me. People say that the Sardar is my man and Panditji [Nehru] also is mine. . . . I have never abandoned my nonviolence . . . it was acceptable till we attained independence. Now they wonder how they can rule with non-violence. And . . . they have taken the help of the army. . . . If I could have my way of nonviolence and everybody listened to me, we would not send our army as we are doing now. And if we did send, it would be a non-violent army. . . . It would be a non-violent war. . . . But to whom can I say this? Today poison has spread on all sides and people kill each other in barbarous manner."[53]

Nehru was not listening to the gentle voice of the Mahatma, however, but to louder voices and drums of battle raging in Kashmir. Sardar Patel and Baldev Singh flew to Srinagar and returned to report to their cabinet colleagues on the Defense Committee what they had learned first hand in Kashmir. "I feel that there has been a certain easy optimism and complacency on our side," Nehru informed Abdullah, telling him of the cabinet's decision to appoint a new "overall" commander, Major General Kalwant Singh. "He is able and strong . . . just the man for taking the offensive. . . . Our orders to him are to take Baramula at any cost and very soon. . . . Having cleared up the Valley completely we shall proceed along the Jhelum Valley Road to Kohala and clear that up. It is possible that a number of raiders might take refuge in the mountains. They can easily be dealt with . . . Every army has to face this kind of thing."[54]

"All dealings with Pakistan should be through the Indian Union," Nehru warned Abdullah. "Any direct contacts should be avoided."[55] Unnerved by the enormity of the Delhi airlift, Indian planes disgorging battle-ready Sikhs and Dogra Rajputs every hour into Srinagar, Sheikh Abdullah had tried contacting Liaquat Ali and had even thought of speaking personally to Jinnah. The growing conflict was hardly helping Kashmir and could engulf all of South Asia in disaster. Yet now Nehru cut all the lines of direct communication he had opened with his Muslim brothers in Pakistan. Nor could he ever speak with the maharaja, who was either "sick" or "hiding" in his palace in Jammu. Nehru sent at least four of his own "helpers" (most of them relatives) to Srinagar as well as Jammu with personal messages and detailed instructions to Abdullah, several of whom directly countermanded the orders of others. The army listened only to the orders of its own generals. So the chaos was compounded; Kashmir kept unraveling.

"We are all agreed here that the Maharaja's attitude has been bad and is bound to lead to trouble," Nehru wrote his personal emissary, Dwarkanath Kachru, on November 7. "The attempt to evacuate Muslims from Jammu was an amazingly stupid thing to do. . . . As the Maharaja is so much in the wrong, I should like Sheikh Saheb to be absolutely correct in his approach. Let him not take any hasty step or forgo a reference to the Maharaja where it is necessary. . . . We must not talk about or suggest a separation of Kashmir State into Kashmir and Jammu. This is dangerous . . . you must remain with Sheikh Saheb."[56] The difficulties only proliferated, however, so Nehru ignored Sheikh Abdullah's objections to his own visit to Kashmir, flying to Srinagar on November 11, 1947, two days after he was to have flown off to London with Dickie and Edwina for the royal wedding. Since Kashmir kept him from enjoying the wedding of the decade, he felt that the least he could do was to visit Srinagar.

"During the past two weeks, the eyes of India have been fixed on Kashmir, where a battle is in progress between freedom and slavery," Nehru told all the soldiers and officials lined up to greet him. "The people of Kashmir have written a page in . . . history which posterity will read with pride. . . . I am proud of you. I congratulate you all, officers and men belonging to all branches. . . . You have not only saved Kashmir, you have also restored the prestige of India, your mother country."[57]

Nehru met with Maharaja Hari Singh in Srinagar and told him that "the only person who can deliver the goods in Kashmir is Sheikh Abdullah." Nehru urged the maharaja to keep in "close personal touch" with Abdullah, arguing that the "path of wisdom" was to "adapt oneself" to change. In power, Nehru himself had grown more adaptable. His high regard for Abdullah, however, soon changed to fear that the "Lion of Kashmir" wanted to bite the hand of India that had fed and sustained him for so many years.[58]

Nehru was soon "horrified to learn in some detail" of what had happened in Jammu on November 5 and 6, when approximately five thousand Muslims had been forced out of their homes and sent off in convoys under the protection of Kashmir State Dogra troops.[59] The Muslim evacuees were gunned down outside the city, out of earshot of its residents, by Sikh and R.S.S. terrorists from East Punjab. Nehru now wrote of the R.S.S. as "an injurious and dangerous organization and fascist in the strictly technical sense of the word. We have known about it for many years. . . . They are very well organised but extraordinarily narrow in their outlook. . . . If by any chance the U.N.O. intervenes, it is Sheikh Abdullah . . . that will impress them and show them that the people of Kashmir are fighting for their freedom against the raiders. The R.S.S. will merely support Liaquat Ali Khan's case against us." [60]

Nehru's oft-repeated pledge of a plebiscite for Kashmir started to worry Sheikh Abdullah's National Conference supporters, some of whom feared they might lose a free and impartial referendum. "Dwarkanath [Kachru] writes to me that there is strong feeling in the . . . National Conference against a referendum," Nehru wrote Abdullah. "I share the feeling myself. But you will appreciate that it is not easy for us to back out of the stand we have taken before the world. That would create a very bad impression abroad and more specially in U.N. circles. I feel, however, that this question of referendum is rather an academic one at present. . . . There is no difference between you and us on this issue. . . . I would personally suggest to you not to say anything rejecting the idea of a referendum but to lay stress on the fact that the people of Kashmir, by their heroic resistance, are deciding the issue themselves. . . . [I]t is a little absurd for people to carry on a little war in Kashmir and, when defeated, to want a referendum." [61]

Attlee just wired Nehru to suggest that perhaps the "speediest and most satisfactory way" of arranging to test the will of the Kashmiri people as to which dominion most of them wanted to join was to appeal to the International Court of Justice. "I am grateful to you for your message regarding Kashmir," Nehru wired back. "We do not, however, consider the International Court of Justice to be appropriate organ for providing requisite machinery." [62]

Two days later, Prime Minister Nehru told India's Legislative Assembly: "Kashmir has gone through fire and I am sure that the House would like me to communicate their sympathy to the people of Kashmir for the tribulations they have been going through. . . . This fair land which nature has made so lovely has been desecrated by the people who have indulged in murder, arson, loot and foul attacks on women and children. . . . Whatever the future may hold, this chapter in the history of Kashmir will be worth reading." Yet even as he was speaking, his old friend and the current high commissioner of India to Pakistan, Sri Prakasa, was telling Mountbatten that "for the sake of peace all round," the best thing India could do would be to hand over Kashmir to Pakistan.[63]

"I was amazed," Nehru wrote Prakasa, "that you hinted at Kashmir being handed over to Pakistan. . . . If we did anything of the kind our Government would not last many days and there would be no peace at all. . . . [I]t would lead to war with Pakistan because of public opinion here and of war-like elements coming in control of our policy. We cannot and we will not leave Kashmir to its fate." The fate of Kashmir, Nehru believed, was tied to the fate of the Nehru family, their intertwined destiny. "The fact is that Kashmir is of the most vital significance to India. . . . There lies the rub. . . . We have to see this through to the end. . . . Kashmir is going to be a drain on our resources, but it is going to be a greater drain on Pakistan."[64]

Mountbatten tried to persuade Nehru to be more conciliatory regarding Kashmir, knowing how betrayed Jinnah felt and how angry Liaquat and his cabinet were at the speed with which India's army had responded to the maharaja's accession, which no Pakistani leader considered legal. Liaquat flew to Delhi on November 26 to meet with Mountbatten and Nehru and their Joint Defence Council at the viceroy's house. Nehru and Liaquat exchanged harsh words, but Mountbatten finally got Nehru to agree to consider a "plebiscite under U.N. auspices," leaving a "minimum number of Indian troops at vital points," "fair elections," and the "return of refugees to their homes."[65] After subsequent consideration, however, Nehru decided against all those sound proposals. Instead of reducing Indian forces in Kashmir, he "gingered things up," as Nehru informed Abdullah a few days later. "Several battalions are on the move and will be arriving daily in Jammu. Rifles are being sent. Other supplies are being arranged. We are collecting our aircrafts also . . . a very good day's work. . . . I have no doubt in my mind that we shall triumph and that Kashmir will function as a prosperous and free land. The tragic part . . . is that so many more might die in the near future, and vast numbers have been rendered homeless."[66]

"I do not know that I believe in the ultimate sanity of the human race," Nehru wrote Darling Indu that week, after having visited Jammu for a day with his sister Nan, back from New York briefly before returning to Moscow. "I have an idea that this depends greatly on one's physical & mental health. As I am fortunately by and large a healthy person, or so I imagine, my view tends to be more optimistic than facts warrant." Obsessed as he was with "saving" Kashmir, Nehru felt more vigorous than he had in many years.[67]

"Mountbatten has struggled with what I can only describe as heroic zeal to . . . prevent the whole sub-continent falling apart from a monomaniac obsession over the political future of a single Indian State," Campbell-Johnson reported at this time. Mountbatten often reminded Nehru that Kashmir's population was only four million, whereas India's was four hundred million.[68] Mountbatten insisted on December 8 that Nehru fly with him to Lahore to meet Liaquat Ali, emphasizing that "the whole future welfare of India depended on an agreement over Kashmir being reached between the two Dominions." Dickie had just met with Attlee and Churchill at the royal wedding, and knew too well what they both thought of Nehru's actions in Kashmir. He told Nehru in as kind a way as he could that "world opinion" now required a solution to this problem, short of war. Still Nehru insisted that "whatever might happen in the future," about which he worried less nowadays, "Kashmir was at the present time part of the territory of India." He then talked about how Pakistan had assisted the raiders and how evil they had been. He admitted that there had been atrocities on both sides, but argued that the raiders' "large-scale looting, destruction, massacre and abduction of women" was worse than anything done by Indian soldiers. Liaquat Ali disagreed, insisting that "many thousands of Muslims had been killed in Jammu."[69]

Mountbatten tried to get them both to agree on a plebiscite, hoping to hold one before he gave up his job, but Nehru and Liaquat Ali would agree to nothing. Then Governor-General Mountbatten suggested that India issue a unilateral statement, but "Pandit Nehru replied that he would not make such a unilateral statement. The question of a plebiscite in Kashmir did not arise until the raiders were thrown out. . . . If necessary, he would throw up his Prime Ministership and take the sword himself, and lead the men of India against the invasion. Nothing else in India mattered—until Kashmir was cleared up though it might take five years or ten." Mountbatten next suggested inviting the U.N. to "send out observers or advisers . . . to help the two Dominions solve the impasse." Liaquat agreed, but "Pandit Nehru said that he would entirely reject this idea."[70]

Back in New Delhi, Nehru wrote to Sheikh Abdullah to report on the Lahore deadlock: "As you well know the situation in Kashmir is extremely complicated. . . . Any false step may add to our difficulties. If it was a simple matter of using military force only, that would at least be a straight issue even though it might be difficult. But everything is so interlinked. . . . We are acting now in Kashmir on a world stage and the greatest interest is being taken by other countries, . . . especially the Great Powers."[71]

Gandhi addressed his Constructive Workers Conference that night in Delhi, admitting that "it is difficult to answer the question why constructive work is making so little headway, though the Congress has sworn adherence to it for years and men like Jawaharlal, Rajendra Babu and Vallabhbhai are at the helm of affairs. . . . Why is it then that our work is not progressing? It may be that we have no heart. Because if we were endowed with a heart we would have been sensitive to the pain of others. . . . [T]he freedom that came was not true freedom. . . . [M]y eyes have now been opened." At another meeting that night he said: "Today, everybody in the Congress is running after power. That presages grave danger."[72]

Nehru was busy writing to Maharaja Hari Singh that night, explaining that "Pakistan Ministers are bent on pushing out Sheikh Abdullah. Of course they want to push you out also. . . . Kashmir State offers us an extraordinarily difficult problem. . . . Nothing has exercised our minds so much during recent weeks as the problem of Kashmir. I have an intimate and personal interest in it and the mere thought that Kashmir should join Pakistan and become a kind of foreign territory for us is hateful to me."[73] He wired Attlee that evening as well, noting that his discussions with Liaquat Ali had not as yet resolved the Kashmir dispute but that "my colleagues fully share my desire that this major cause of potential conflict between Pakistan and ourselves should be settled by peaceful and friendly negotiations."[74]

Victory in Kashmir was taking too long. After their lightening initial action, India's troops had, in Nehru's opinion, settled down to much too slow a pace. "There has been a progressive deterioration and the initiative appears to have been with the enemy," Nehru, always impatient, wrote in a mid-December "Note on Kashmir" for his cabinet and for India's officers in Jammu and Kashmir. "It is perfectly clear . . . that what is happening in Kashmir State is not merely a frontier raid but a regular war." And this war had the full backing of Pakistan. "It seems to me that our outlook has been defensive and apologetic, as if we were ashamed of what we were doing. . . . I see nothing to apologise for. . . . We have to think afresh and adopt different tactics. We cannot go on bartering . . . and allow the situation gradually to deteriorate."[75]

Mountbatten asked his prime minister whether he intended to "repudiate the payment" of sterling balances due to Pakistan or to "delay them until a settlement was reached on Kashmir." Nehru insisted that there was no question of repudiation. Gandhi also was deeply concerned about this matter and would soon undertake his final fast unto death to hasten the release of those vital, frozen Pakistani assets that India retained. Liaquat, of course, was most eager for some kind of settlement of the Kashmir conflict as well as of the division of the sterling assets. Nehru confessed that he felt "it would be foolish for us to make those payments until this Kashmir business had been settled."[76]

Nehru was eager to attack Pakistan's support bases that fed Kashmir raiders with arms and other supplies, but Mountbatten finally convinced him that it would be improper to launch such an attack, which would be viewed the world over as initiating open war without prior reference to the United Nations, charging Pakistan with aggression. Sheikh Abdullah was with Nehru in Delhi when he wrote to alert the maharaja to their new strategy, "to keep you fully informed" as he put it to the remorseful Hari Singh.[77] Nehru remained obsessed with Kashmir. His frustrations and anger grew daily. He felt thwarted, surrounded by enemies or naysayers. His passion for Kashmir appeared to blind Nehru to every other problem that plagued the newborn nation over which he presided. Vallabhbhai Patel was the only cabinet minister strong enough to warn Nehru against wasting so many precious resources on Kashmir, and several times that December Vallabhbhai nearly resigned, sick of how unbalanced his prime minister had become. But Gandhi always talked the Sardar into staying, for the sake of India.

On January 3, 1948, India informed the Security Council of the United Nations that Pakistan had committed aggression against it, in violation of the U.N. Charter. Nehru's appeal called on the Security Council to "take immediate and effective action" to stop Pakistan from permitting "raiders" to "use its territory for operations against Jammu and Kashmir State" and to provide no military or other supplies to such raiders, or any other "kinds of aid."[78] Nehru also warned that unless Pakistan immediately stopped its "assistance to the raiders," India might feel compelled in self-defense to take "such military action as they may consider the situation requires."[79] Thus began what became one of the longest, most acerbic, inconclusive debates in the history of the Security Council, leading to the appointment of no fewer than three high-powered U.N. Commissions on India and Pakistan (UNCIP), seeking in vain for almost half a century to resolve this worst of all conflicts between India and Pakistan.

Nehru and Patel met with Gandhi on January 12, 1948, trying to explain to him why it was impossible for them to do any more to stop attacks against Muslims in Delhi and why it would only hurt India to pay Pakistan the 550 million rupees that India had agreed months ago to pay as Pakistan's fair share of the cash reserve balances that remained in the Reserve Bank of British India before partition. The three of them sat alone together for an hour that afternoon, Gandhi saying little, sadly listening to weighty explanations from India's most powerful leaders.

"When on September 9 [1947] I returned to Delhi from Calcutta," Gandhi told his prayer meeting on January 12, soon after Nehru and Patel left him, "Gay Delhi looked a city of the dead . . . I observed gloom on every face. . . . At once I saw that I had to be in Delhi and do or die. . . . I yearn for heart friendship between Hindus, Sikhs and Muslims. . . . Today it is non-existent. . . . My impotence has been gnawing at me of late. It will go immediately the fast is undertaken." Gandhi had decided moments after his discussion with Nehru and Patel that the only resource left to him was to fast unto death. "Death for me would be a glorious deliverance rather than that I should be a helpless witness of the destruction of India. . . . Just contemplate the rot that has set in in beloved India and you will rejoice to think that there is an humble son of hers who is strong enough and possibly pure enough to take the happy step."[80]

Nehru and Patel agreed two days later to release the 550 million rupees to Pakistan. They did so against their own better judgment, to "save"

Bapu's life. But in notifying his U.N. delegation of this action in mid-January, Nehru insisted "that this does not mean any weakening on our part on the Kashmir issue. Indeed we want it to have the reverse effect." He called Krishna Menon back to Delhi now as well, to brief him on Kashmir, knowing that he alone could articulate India's hard line at the United Nations once that endless debate over Kashmir started in earnest.[81]

Rajendra Prasad convened the Communal Harmony Committee of more than one hundred members of every faith, each of whom signed a pledge, delivered to Gandhi on January 18. The pledge promised to end discrimination against Muslims and to restore their many mosques in Delhi. Nehru was among those who crowded into Birla House that morning to see Gandhi and reassure him that they truly meant their pledge that "Hindus, Muslims and Sikhs . . . should once again live in Delhi like brothers and in perfect amity."[82] "Delhi is the heart," Gandhi told them, before breaking his fast. "Men had become beasts. . . . What greater folly can there be than to claim that Hindustan is only for Hindus and Pakistan is for Muslims alone?"[83]

Two days after breaking his fast, Gandhi told his prayer meeting that "Panditji [Nehru] is a man who will do everything for the refugees. . . . He says there is no room left in his house and still people keep coming. He is our Prime Minister. He has visitors, some of them Englishmen. Is he to turn them out? And still he says that he will spare for the refugees one or two rooms or whatever he can. . . . I congratulate Jawahar and I congratulate you on possessing such a jewel." Just then a loud explosion went off some twenty-five yards from Gandhi's platform in Birla's garden. It was meant to create enough panic to allow a designated Hindu assassin to rush up and toss a hand grenade at the Mahatma, but he lost his nerve, seeing how calm Gandhi had remained. There were half a dozen conspirators, all angry young R.S.S. and Hindu Mahasabha terrorists, who dashed from the scene of their intended murder, speeding off safely to Old Delhi in a waiting taxi.[84]

"I thought it was part of army practice somewhere. I only came to know later that it was a bomb," Gandhi told his prayer meeting the next day. "You should not have any kind of hate against the person who was responsible for this. He had taken it for granted that I am an enemy of Hinduism. . . . I am told that is what he said when questioned by the police. He was well dressed too. But I am sure God is not out of His mind to continue sending such men."[85]

Many urged Gandhi to leave Delhi, to rest or to go on a pilgrimage to the Himalayas, some saying that he had "done enough harm already." But he stubbornly refused to run away or to silence himself. "Someone came to see me today," he announced on January 29. "He mentioned peasants.

I said if I had my way our Governor-General would be a peasant; our Prime Minister would be a peasant. . . . If the man who produced food-grain out of the earth becomes our Chief, our Prime Minister, the face of India will change."[86]

On Friday, January 30, 1948, Vallabhbhai came to see Gandhi at 4 P.M., and they talked alone for more than an hour. There had been many disagreements between Sardar Patel and Prime Minister Nehru, and no one but Gandhi was fearless enough to try to resolve such conflicts, which threatened to paralyze India's government. Nehru was to come after the prayer meeting, at 7 P.M., to tell Gandhi his side of the growing gulf between himself and his deputy prime minister. Gandhi's loving disciple, Abha, one of his two "walking sticks" on whose shoulders the old Mahatma always leaned nowadays, brought his evening meal while he and the Sardar were talking, softly reminding him that it was time for prayers, holding up his watch to let him see that the hour had passed. At 5:10, Gandhi finally rose to leave Birla's palatial house for his last walk down that garden path, where assassin Naturam Godse was waiting to greet him with three deadly bullets aimed point blank at frail Bapu's torso. "*Hey Ram* [Lord Rama]!" cried the Mahatma as he crumpled, blood pouring from his body. Hindus believe that a person's final words determine his reincarnated destiny, and so after his last breath at 5.17 P.M., Bapu went directly to Lord Rama's heaven, attaining his eternal "release"—*moksha.*

"Immediately after the assassination, the telephone rang at 17 York Road [Nehru's house]," Mac Mathai recalled. "The call was from Birla House announcing Gandhiji's assassination. The caller [he does not name him] thought that Nehru would be at home . . . but he was still in his office. . . . I immediately rang him up and he rushed to Birla House."[87] Mountbatten had just flown back to Delhi from Madras and also headed for Birla House as soon as he heard the dreadful news. "By the time we . . . reached Birla House the crowd had gathered and was peering into the windows of our car," Campbell-Johnson recalled. "Everyone was in tears. . . . In the far corner was the body of Gandhiji . . . his head was being held up by one of about a dozen women who were seated around him chanting prayers and sobbing."[88] A few days later when people were still sobbing, Sarojini Naidu contemptuously asked them, "Did you want him to die of indigestion?"[89]

"Friends and comrades, the light has gone out of our lives and there is darkness everywhere," Nehru told his nation in mourning that night. "Our beloved leader, Bapu as we called him . . . is no more . . . we will not see him again as we have seen him for these many years. We will not run to him for advice and seek solace from him, and that is a terrible blow, not to me only, but to millions and millions. . . . The light has gone out,

I said, and yet I was wrong. For the light that shone in this country was no ordinary light. The light that has illumined this country for these many many years . . . represented . . . the living, the eternal truths, reminding us of the right path, drawing us from error, taking this ancient country to freedom. . . . A madman has put an end to his life, for I can only call him mad who did it, and yet there has been enough of poison spread in this country during the past years and months, and this poison has had an effect on people's minds. We must face this poison, we must root out this poison. . . . The first thing to remember now is that none of us dare misbehave . . . nothing would displease his soul so much as . . . any violence. So we must not do that . . . we should, in strength and in unity, face all the troubles that are in front of us. We must hold together and all our petty troubles and difficulties and conflicts must be ended in the face of this great disaster."[90]

It was Nehru's finest speech, his strongest moment in the face of the darkness that descended on Delhi and threatened now to enshroud all of India under its mantle of madness. The Hindu fundamentalist conspiracy, led by the R.S.S. and Mahasabha, which had silenced the Mahatma for preaching love of Muslims and tolerance of all faiths, had captured the allegiance of countless frustrated, disappointed, angry young Brahmans, like Godse and his cohort from Pune, turning nonviolent people into the killers of India's saintly father. His own father's death had left Nehru speechless, but the murder of his adopted Bapu gave him strength to speak out before it was too late for him ever to utter a sane or secular statement again. By now, reports had reached him that he was the second target on their hit list.

Gandhi's body, covered with flowers and draped in the Congress flag, was drawn the next day from New Delhi to the banks of the Jumna River in Old Delhi, to the Raj Ghat (Royal Steps) cremation ground, almost six miles from Birla House, followed by the mourning millions who came for a farewell glimpse, to pay their final tribute to the saint who had visited India for more than three-quarters of a century. "Sudden and overwhelming tragedy has befallen India and the world," Nehru wrote the next day. "The greatest and noblest of India's sons has passed and the world mourns him and pays homage. . . . His life was one long struggle for justice and tolerance. . . . Mahatma Gandhi's death is a grim and urgent reminder of the forces of hate and violence that are at work in our country and which imperil the freedom of the nation and darken her fair name. These forces must be swiftly controlled and rooted out. . . . There is no place today in India for any organisation preaching violence or communal hatred. No such organisation will, therefore, be tolerated. No private armies will be permitted."[91]

On February 2, 1948, Prime Minister Nehru rose to speak in New Delhi's Constituent Assembly: "A glory has departed and the sun that warmed and brightened our lives has set and we shiver in the cold and dark. . . . He was perhaps the greatest symbol of the India of the past, and, may I say, of the India of the future. . . . We stand on this perilous edge of the present between that past and the future to be, and we face all manner of perils and the greatest peril is sometimes the lack of faith which comes to us, the sense of frustration that comes to us, the sinking of the heart and of the spirit that comes to us when we see ideals go overboard, when we see the great things that we talked about somehow pass into empty words."[92]

As Nehru reflected on Bapu's murder with a faltering heart and a weary spirit, he recognized that so many of the great things he had always talked about were mere empty words. But Nehru resolved at least to try now to turn Gandhi's death into a rallying cry for secular union, helping his government root out the dark forces of Hindu obscurantism that had almost buried him as well. "We have to consider and decide what attitude we should adopt towards communal organisations," Nehru badgered a crowd of mourners that night. "Recently a cry for Hindu Rashtra [Hindu nation] was raised by some organisations [R.S.S. and Hindu Mahasabha]. It was one of the votaries of this demand . . . who killed the greatest living Hindu. Was Hindu dharma [religion] protected by this foul deed?" These communal organizations were swiftly banned, and many of their leaders were arrested. Then a number of innocent Brahmans were killed in riots around India, especially in Pune and other Maharashtrian strongholds of the Hindu Right. Nehru admonished such zealots of secularism not to forget "Gandhiji's teachings."[93]

Mahatma Gandhi's ashes were divided into seven parts, so that all waters of India might share in their powerful blessings upon their immersion. In Bombay a beautiful white ship, flags covering its superstructure fore and aft, bore the urn with one portion of the sacred ashes to Back Bay, with hundreds among the half-million mourners who had gathered on Bombay's sands that afternoon in early February, wading in after the vessel, trying to swim close enough to the ship's fantail to touch a speck of the Mahatma's ashen remains as they fluttered down in softest of farewell flurries.

On February 12, 1948, Bapu's ashes were immersed in Allahabad's Triveni (three rivers). Again Nehru spoke: "The last journey has ended. The final pilgrimage has been made. . . . Some grieve over his passing. . . . But why should we grieve? Do we grieve for him or . . . for ourselves, for our own weaknesses, for the ill-will in our hearts, for our conflicts with others? We have to remember that it was to remove all these that Mahatma Gandhi sacrificed his life. . . . Democracy demands discipline, tolerance and mutual regard. Freedom demands respect for the freedom of others. This great tragedy has happened because many persons, including some in high places, have poisoned the atmosphere of this country of ours. It is the duty of the Government as well as the people to root out this poison."[94]

29

Pandit Prime Minister

1948–1950

GANDHI'S DEATH reunited Nehru and Patel. Their reconciliation not only saved Congress and India's central government from collapse, but it kept Nehru in power. Without the Sardar's strength and support Nehru might have broken down or been forced out of high office. Vallabhbhai ran India's administration for the next two years while Nehru indulged mostly in foreign affairs and high Himalayan adventures. Nehru also inveighed brilliantly against "Brahmanism" and Hindu "casteism," insisting that India was a "secular democracy," but every Kashmiri pandit and most of his adoring followers believed that "Panditji," as Nehru was now commonly called, was more the Brahman maharaja than any of India's dethroned princes.

Nehru flew back to Kashmir in mid-February 1948. "Whatever we have done in Kashmir has been based on the principle of truth and honesty," he informed his audience in Jammu. "India has nothing to hide. It was our duty to help the people of the State when they asked for help. At every step we have taken so far we had consulted Gandhiji and secured the approval of the saint of truth and nonviolence."[1] He also told his troops that same day, "We have not come here to rule; we have been called in here by the people . . . to help them against the enemy. . . . You have raised our prestige and that of the Indian Army. The world's eyes are on us. . . . We shall be successful."[2]

Back in Delhi the next day, he wrote to his sister Nan that "I am oppressed by the multitude of our problems and overwhelmed by the faith and confidence of large numbers of people in my ability to handle these problems." He was angry at the United States and England for being so sympathetic to Pakistan's eloquent defense of itself in the Security Council debates on Kashmir, feeling that both powers had "played a dirty role" against India. He told Nan that he had strongly expressed his feelings to Attlee, that he considered England the chief actor behind the scenes in the United Nations' conspiracy against him in Kashmir. "The time for soft and meaningless talk has passed," Nehru informed his sister. "Our choice of Czechoslovakia for the Security Council's Commission for India must have given the Anglo-American group a bit of a shock." He was determined now to deal with the United Kingdom and the United States as he had with Motilal and then with Gandhi, giving all those elders who tried to patronize him "a bit of a shock."[3]

Nehru also wrote to unburden himself to Krishna Menon at this time of escalating conflicts and problems. His activist undercover agent, Aruna Asaf Ali, had just flown back from London to report on Krishna's own proliferating worries in trying to keep at bay the growing Anglo-Indian criticism about his inefficiency as well as his strange and notorious nocturnal ventures around London. "This has no significance," Nehru reassured Krishna Menon, but the Security Council's decision to expand its debate from India's aggression complaint, focusing only on the Kashmir question, to many broader Indo–Pakistani questions, including the murder of Muslims around Delhi, depressed and distressed him greatly. Nothing seemed to work as neatly as he had imagined it would, as he believed it should. First the army bogged down in Kashmir, now the debate had bogged down in New York. "I have felt very angry with the U.S.A. and the U.K. Most of the others do not count. . . . [Commonwealth Secretary of State P. J.] Noel-Baker has been the villain of the piece in spite of his pious professions. . . . I think it is about time that we made it perfectly plain to the U.K. Government . . . what we really think," he told Krishna. "I have heavy work ahead. . . . The U.S.A. have practically refused to sell arms to us. The petrol position is also critical."[4] Two days later he sent his army commander in chief, General Sir Roy Bucher, to London to purchase military equipment, especially arms and to consult Krishna Menon there to see how much India could increase its gasoline quota. That quota had been much reduced, for obvious political reasons, and unless it was increased substantially, "we shall be gravely embarrassed."[5]

The potential threat of losing Kashmir because of diminished oil imports made Nehru focus more intensely on developing as swiftly as possible India's independent military and industrial resources. Russia's rich oil fields were still beyond reach, Nehru felt, but he wanted Krishna Menon to explore ways of ensuring Indian "production of power alcohol and liquid fuel from coal." He also soon sent Dr. Homi Bhabha on a search for nuclear power. Economic planning and the development of enough iron and steel oil and electrical power potential to fuel any future military and civil requirements had long been high-priority concerns. Nehru wanted to be able to tell every Noel-Baker and (U.S. Secretary of State John Foster) Dulles of the Western world exactly what he thought of their "pious, poisonous" attitudes toward India. Nehru always sent his letters to Krishna Menon by private courier and told him now in confidence that "even Mountbatten has . . . hinted at [the] partition" of Kashmir, leaving Jammu for India and the rest, including the lovely vale and Srinagar, to Pakistan. "This is totally unacceptable to us," Nehru insisted, although "if the worst comes to the worst, I am prepared to accept Poonch and Gilgit area being partitioned off. . . . But we are not going to put all this forward in any way before the Security Council."[6] Thus the stonewalling continued and in Kashmir the fighting never stopped, though a cease-fire was agreed on and accepted by both sides, under U.N. auspices and with U.N. monitors, beginning on January 1, 1949.

Indira was living with Feroze and their boys in Lucknow at this time and kept feeling "unwell," Nehru told Krishna Menon.[7] She remaind quite sickly as long as she continued to reside with Feroze but recovered "remarkably" soon after she moved with both boys into Nehru's magnificent prime ministerial residence (built for British India's commander in chief) at Teen Murti in New Delhi. There she enjoyed the last decade and a half of her father's life at his beck and call, in constant training to succeed him. To one of his sick old expatriate comrades in Switzerland, "Nanu" (A. C. N.) Nambiar, who in a letter appealing for funds asked how he should address the prime minister he had known for more than a quarter century, Nehru replied: "You need not address me formally. The only change that has come in me, or the principal one, is that I am older and sadder."[8] His sadness in April 1948, was magnified by the imminence of Edwina's departure. Dickie was eager to return to the fleet, to win the title of first lord of the Admiralty from which his father had been unjustly scuttled during World War I because of the Mountbatten family's German roots. It took him another seven years to be promoted to that highest British nautical rank.

Jawahar tried to talk Edwina into staying on with him after Dickie flew home, for he knew by now that her heart belonged to him alone. Mountbatten, of course, also "knew that they were lovers," as did all of their close friends. Edwina's sister Mary hated Nehru for it, blaming him for having "hypnotized" her. Edwina, "had no will where he was concerned . . . like water."[9] But in her own way Edwina was as strong as Nehru. "This was the only promise we ever made," Edwina reminded him a fortnight before they parted, "that nothing we did or felt would ever be allowed to come between you and your work or me and mine—because that would spoil everything." Still he wanted her, needed her, pleaded with her. A week before her departure, "Nehru could not face the prospect. Would Edwina not remain to carry on her work with the refugees? She reminded him that he had told her to be practical. 'And so it will be and it has to be,' he said sadly. 'How wise and right you are, but wisdom brings little satisfaction. A feeling of acute malaise is creeping over me, and horror seizes me. . . . Dickie and you cannot bypass your fate, just as I cannot bypass mine.' "[10] But when the two parted, it was Edwina who "shed copious tears" as the notoriously cool Jawaharlal reported to dear old friend Sarojini Naidu, his new governor of Uttar Pradesh (formerly the United Provinces).[11]

At the farewell banquet for Dickie and Edwina on June 20, 1948, Nehru sat next to Edwina. "I have often wondered why the people of India put up with people like me who are connected with the governing of India after all that has happened during the last few months," he confessed, in what may have been the most honest part of his singularly sentimental speech that evening. "I am not quite sure that if I had not been in the Government, I would put up with my Government," he added. But then he turned to his adoring Edwina, whose eyes never left his still handsome, though weary, face. "To you, Madam . . . [t]he gods or some good fairy gave . . . beauty and high intelligence, and grace and charm and vitality, great gifts, and she who possesses them is a great lady wherever she goes. But unto those that have, even more shall be given, and they gave you something which was even rarer than those gifts, the human touch, the love of humanity, the urge to serve those who suffer and who are in distress. . . . Wherever you have gone, you have brought solace, you have brought hope and encouragement. Is it surprising, therefore, that the people of India should love you and look up to you as one of themselves and should grieve that you are going?"[12] It was as close as he ever came to a public confession of his love for Edwina. "Very sad and lost," she noted in her diary, heading home with her earl husband, their daughter Pamela, and their entourage of friends, helpers, hangers-on, servants, and strangers.[13]

Nehru recovered faster; he was never one to sulk over separations. A few days after Edwina left him, Clare Boothe Luce returned to fill his mind with happy thoughts of their brief but passionate interlude. Writing in response to her letter, which he received on the eve of Edwina's departure, Nehru began, "Need I say that I was happy to receive your letter, . . . I have often thought of you ever since we met six long years ago." Their affair in Delhi had been sweet but too short. "It is a little difficult to pick up old threads again," he cautioned her now, for as prime minister he was a bit busier than he had been as Congress's left-wing leader on the eve of his longest prison term. "You remember our last conversation. You were very tired then . . . late in the evening, and I was leaving early next morning. . . . I had stayed on an extra day in the hope of seeing you. That hope was fulfilled. But that meeting left me unsatisfied. That feeling had to remain, for I have not seen you again and I do not know when we might meet."[14]

She wrote about how she had survived the tragic death of her child through prayer and wondered whether he ever prayed. "Your great sorrow led you to a certain faith, and if you have found peace in it, it is well with you. I do not myself see any peaceful or safe anchorage for my mind anywhere. I have to wander through life . . . often doubting as to what I should do and what I should avoid. . . . A strange fate threw me into a kind of life for which I was not temperamentally made. In a sense I succeeded in that life and gained . . . the love and affection of large numbers of human beings. That very love and affection became a burden to me. . . . I am sorry I am writing about myself when really I should like to know about you. You told me not to work too hard and not to sacrifice my health, vitality and energy. Very true. And yet the only satisfaction I have is in working. . . . We live for the high moments which seldom come and when they come they pass too soon."[15]

Nehru received about fifteen hundred letters a day, as he told J. R. D. Tata, whose letter he also took time to answer personally, though not as intimately and at half the length of his message to Clare Luce. "My dear Jeh," he wrote the head of India's premier industrial family. "Thank you for the lovely cherries. . . . Were they brought from Europe in one of your Constellations? As a matter of fact I have been having plenty of cherries this summer . . . from Kashmir. . . . In the future if you wish any of your letters specially to reach me, please mark it on the envelope 'FOR HIMSELF.' . . . [A]bout the running of your air service to Europe. All accounts agree in speaking well of it. . . . Congratulations. Perhaps I might use it some time. . . . There is a chance of my going to Europe . . . in October . . . only a chance."[16] He wrote Lady Colleen Nye [Madras Governor Sir Archibald Nye's wife] that same July 4, 1948, to confirm that he, Indira, and Mac would be flying down for a brief overnight visit to Madras before month's end. "I should love to attend any party that you may give," he assured the lovely Lady Colleen. "Still I hope I shall have some little leisure, and I would like to visit the woods about which you spoke to me."[17]

The Labour Party's Professor Reginald Sorensen, still active in London's India League, was most concerned when he learned that hundreds of Indian communists had been kept in prison under preventive detention arrest for many months by "Comrade" Nehru's own government. So Sorensen wrote his old friend Nehru to express his shock and grave concern. "I do not complain at all that you have drawn my attention to events in

India which have caused you grave concern," Nehru responded that July. Few of the many things that had happened in India over the past year, Nehru insisted, "have distressed me so much as the action that Government has had to take in regard to individual and civil liberty." But the preservation of peace and law and order were largely provincial functions, he explained to Sorensen, who obviously did not appreciate how India's federal government was run. "It is difficult for the Central Government to . . . interfere with individual cases when the provincial government is in a position to judge the situation far better." It was a letter worthy of a British viceroy, like many he had crumpled in rage during the past few decades. "In this kind of situation normal conceptions of civil liberty have little place. . . . [S]pecial rules have to be framed . . . ordinary labour disputes might also become communal rather suddenly. . . . [B]ehind the industrial disputes lay a definite policy of sabotage . . . a deliberate attempt to produce chaos in the country . . . facing a most difficult communal situation . . . exceedingly dangerous." In case he had not made his position perfectly clear, however, Nehru added that "I have personally no ill-feeling towards Communism and Communists . . . and I count some of them as my friends." He was, however, rather shocked to learn that communists in India seemed to be involved in such things as "political revolt, sabotage and individual killing."[18]

Nehru kept trying to justify himself to old comrades, to win back the support of ardent socialists like Achyut Patwardhan, to whom he offered many ambassadorships, but in vain, for Achyut, like J.P., had lost faith in him. There was, Nehru insisted, never time to sit down quietly with disillusioned former followers like Achyut, to whom he frankly confessed that having reached the top and won the crown he had always coveted, "many things" continued to "oppress the mind," and "I find myself in a harder prison than I have ever been."[19]

Only Krishna Menon remained perfectly loyal, able to keep pace with all of Nehru's twists and shifts. "The Russian attitude towards India has become progressively one of condemning and running down the Government of India," Nehru complained to Krishna in confidence. He reported that Nan had written repeatedly that she could do nothing very useful in Moscow. Stalin refused to see her; none of his confidants spoke to her; and she was never permitted to visit any Asian part of the Soviet Union. Nan wrote to say she felt "a moral defeat" in Moscow, asking her brother to bring her home or to send her somewhere else. So that September Nehru sent her to the United Nations. Less than a year later, in early 1949, Madame Pandit became India's ambassador in Washington. Nehru also confessed to Krishna that "the U.S.A. Government is not too pleased with us for various reasons," but Nehru blamed India's own press for that breakdown in Indian–American relations, rather than recognizing that it was primarily his own fault for so stubbornly refusing to agree to hold a plebiscite in Kashmir. "Our newspapers are . . . drunk with the new freedom and seldom understand the intricacies of foreign affairs and policy."[20]

Nehru and his entourage flew from Bombay on October 5, 1948, stopping briefly in Cairo and reaching London late in the evening of October 6. Krishna was waiting at the airport and drove Nehru directly to Dickie and Edwina's house "just round the corner" from Claridge's, where Krishna Menon had reserved him a suite. Dickie discreetly left the reunited lovers alone for their first midnight rendezvous in months. "Too lovely," Edwina noted in her diary.[21]

Nehru had lunch the next day with Attlee at 10 Downing Street, and, tightly smiling, Mountbatten joined them to show the public that nothing had changed. "The family 'line' has always been that they were just good *friends,*" as Edwina's grandson Lord Romsey recently put it.[22] Edwina insisted on driving him alone that evening to Broadlands, her splendid mid-Georgian Palladian mansion near Romsey that her father, Wilfred Ashley, had left her when he died in 1939. Dickie spent his nights in London while she and Jawaha remained blissfully alone for four more nights at Broadlands. "A heavenly weekend," Edwina called it, driving him back up to London on October 11, taking him to see *Medea* that evening, perhaps to prove to London gossips that they had some interest in culture as well as in being alone together.[23]

Back in London, Nehru had many other pressing engagements. The rationale for this long journey was the meeting of Commonwealth prime ministers, which he and Liaquat Ali Khan both attended for the first time. "Nothing important was done today," Nehru wrote Deputy Prime Minister Vallabhbhai, after that first meeting ended. "Liaquat Ali came to me in a very friendly and effusive way."[24] The next day, Nehru spoke at a crowded reception in Kingsway Hall, organized by Krishna Menon's India League, to welcome him to London. Old Pethick-Lawrence, who chaired the meeting, spoke of Nehru as one of the world's "great minds." Harold Laski introduced him as a fellow socialist, the "champion" of India's "poor and downtrodden." Nehru then spoke at great length, finding it "rather embarrassing" to be called so many great and selfless things. "I am made much of because of certain circumstances and because, as I often tell my friends, I have the capacity for showmanship," he confessed. "I do not think I could have survived but for the fact that Gandhi was there. Nevertheless, we survived, not only as individuals but as a nation; survived not only that horror . . . but, much more so, that mental pain and anguish."[25]

Nehru now agreed to keep India in the British Commonwealth, although most members of his cabinet had expected him to notify Attlee that India must leave, for Purna Swaraj, after all, required the severance of all dominion ties. How could an independent India swear allegiance to the Crown of Britain? How could Nehru go back on the sacred pledge he had made a thousand times since that bitter cold morning in January 1930 in Lahore? But Nehru suddenly realized that for India, membership in the Commonwealth meant "independence plus, not independence minus."[26] How simple for him to appreciate that in the aftermath of his weekend at Broadlands. Cripps had come down for a day and they all sat around the fire, chatting about economic development and prospects for world peace and trade. Edwina saw to his every desire and need. Even Dickie drove down from London one day. It was all so obvious now, everything that Motilal and Gandhi and Edward Thompson and C.R. all had tried, time and time again to tell him.

Nehru informed his fellow Commonwealth prime ministers that India's greatest need was for large supplies of capital goods to increase production.[27] Inflation was rampant, the railroad infrastructure badly worn and in many places broken down by the overuse of equipment. Large-scale economic support for India would enhance "the common good of all the peoples of the world," Nehru argued, Western industrial givers as well as the needy new nations of Asia. Communism "flourished only where the economic standards of the people were indefensibly low," he added. Thus, it was especially in the best interests of the United States to aid and support India's development. "On the other hand, although many Indians might envy the material prosperity of the United States, few would welcome the consequences of applying to India the principles of American capitalism." He favored Great Britain's approach to its economic problems, he told them, as he had told Cripps and Mountbatten earlier, assuring India of the industrial and military support required to carry on for the next year without having to turn to Washington.

Two days later he flew to Paris, where the U.N. General Assembly was meeting, Nan leading the Indian delegation there. He met with U.S. Secretary of State George Marshall and with Soviet Foreign Minister Andrei Vyshinsky, determined to try carving out a separate "Third World" path between both superpowers as the cold war heated up. Nehru's commitment to the Commonwealth—betrayal though it seemed to most of his former colleagues in and out of Congress—had never closed his mind to the prospects and possibilities of leading a "nonaligned" bloc of Asian and African nations, thus allowing India to receive support for its much needed development and many five-year plans from East, West, and Commonwealth alike. He also thought of himself as the world's foremost leader committed to peace. In Paris he addressed the General Assembly:

"I am not afraid of the bigness of the great powers, and their armies, their fleets and their atom bombs. That is the lesson which my Master taught me. We stood as an unarmed people against a great country and a powerful empire. . . . [W]e decided not to submit to evil, and I think that is the lesson which I have before me. . . . I think if we banish . . . fear, if we have confidence, even though we may take risks of trust rather than to risk violent language . . . those risks are worth taking. . . . [T]he Indian people happen to be three hundred and thirty millions in number; it is well to remember that. We have had a year of freedom and a year of difficulty. . . . We propose to build and construct and be a power for peace and for the good of the world." [28]

Just as peace became his international mantra, planning for economic development became Nehru's highest internal priority. "The first problem was to get political freedom" Nehru informed his country in Delhi on January 26, 1949, whose annual celebration he had inaugurated nineteen years earlier with the slogan "Purna Swaraj." "But we must remember . . . that we are at a very crucial situation economically. We have to control the situation very quickly. . . . Urgent steps are required. . . . A government can function only so long as its mind is alert. . . . We are trying to lay the foundations of India's future." [29]

"I am worried often enough with a sense of my own inadequacy," he confessed two days later at the convocation in Lucknow University, where he was given an honorary doctorate. "I carry on, though sometimes there is a certain weariness of spirit, and sometimes a little sorrow, that the wide dreams that many of us had somehow do not take shape as we wanted. . . . Somehow when work is to be done, solid work, . . . our attention is distracted by petty conflicts, all kinds of wrong things . . . it amazes me. Here is a time when work is required, labour is required . . . what are we doing? . . . running down and finding fault, petty factions and the like. I see them in all grades, above and below. . . . It is the job and the work that counts, not the thinking and dreaming and shouting of our people. . . . And when I see our young men going along different lines, when I see young men and hysterical young women misbehave, I am angry. . . . Is this your idea of liberty and democracy and freedom? What is happening here? . . . Have you any idea of the burdens of those who have been put in the seats of authority?" Only the day before a "number of refugees" had tried to "stop my car," Nehru reported, shocked at that sort of thing. He was also surprised at how weak the provincial government of Uttar Pradesh had been in dealing with "the situation here," as he put it, for the teachers were on strike, and one "girl" had actually "slapped a policeman." [30] That sort of behavior now seemed uncivilized to Pandit Nehru, but it made Governor Sarojini Naidu, sitting on the dais behind him, smile as she recalled how quickly he used to jump from every Congress platform to "pummel" anyone who disagreed with him. "I want you to understand that there are some people whose main occupation is to instigate people to mischief," Nehru solemnly warned the people of Lucknow.[31]

Nehru's old comrades kept pressing him to nationalize all industry and transform India into a socialist economy, but he was shrewd enough to see that such radical action would hardly help or solve any of India's mounting real problems, faced as it was with a lack of capital for industrial development, growing food shortages, labor unrest, and fear among most investors from Bombay to Calcutta that India was losing its balance, drowning in a sea of red tape and incompetence. "Socialism in itself is a good thing," Nehru told Ahmedabad's Textile Labour Association in February 1949. "But socialism is not a consummation. . . . The necessary conditions precedent for socialism must be created. Otherwise it will mean starvation and misery for another twenty years or more."[32]

Nehru turned instead to Washington and the International Bank for Reconstruction and Development for capital and food grains enough to carry India over its harshest famine conditions throughout 1949 and 1950. "Our food deficit this year, because of famines . . . is about 9 per cent," Nehru told Rajaji [C.R.], India's new governor-General. "The Grow More Food campaign has not been a success. . . . In view of the loan . . . from the International Bank, it is particularly important that food production should grow."[33] He soon turned to the Ford Foundation as well for assistance in rural and community development and relied heavily on help from Douglas Ensminger, who headed that multifaceted Ford Foundation enterprise throughout India from his headquarters in New Delhi. Thanks to Ford's commitment to India's development, thousands of American agronomists and other scientific professionals and many young volunteers came to India during the next decade to help in the continuing battle against famine and rural endemic diseases and all the problems compounded by poverty, illiteracy, caste, class, communal, and gender discrimination, none of which disappeared with the advent of Purna Swaraj.

Nehru agreed to a U.N.-monitored cease-fire in Kashmir at the start of 1949, which Pakistan accepted as the precursor of a plebiscite to be held throughout the state of Jammu and Kashmir, under U.N. auspices. Sheikh Abdullah feared that any impartial plebiscite would mean an end to his government in Srinagar, but Nehru was quick to reassure him: "You know well that this business of Plebiscite is still far away and there is a possibility of the plebiscite not taking place at all. (I would suggest however that this should not be said in public, as our *bona fides* will then be challenged.) It is not a question of hiding anything from the public, but . . . at this early stage there might be a complete lack of agreement between Pakistan and us on several issues. . . . [L]et us think of the future calmly.

. . . You know that we are at least as intensely interested in the future of Kashmir as you are."[34]

Publicly, Nehru pretended for several more years to favor a U.N.-monitored plebiscite. The cease-fire gave India time to develop industrially and to strengthen its civil and military economic arsenals. It was also important to allow distinguished international visitors to fly into and out of Srinagar without fear of being shot down. "Lady Mountbatten is due to arrive here on the 15th February, Tuesday, and will stay on for some time," Nehru informed Sheikh Abdullah in early February. "I am sure she will be happy to meet with you." Edwina was accompanied by her daughter Pamela, but Dickie stayed with his fleet.[35]

Edwina stayed in India for five weeks, sharing Nehru's wing of Teen Murti while in Delhi, visiting Anand Bhawan with him in Allahabad and joining him and Indira and Betty for a holiday on Puri's glorious beach in Orissa. "Jawaha too wonderful," she wrote before flying out of Delhi on March 21, 1949. And when she was gone, he lamented of the "darkness" that had descended on his house again, his "life stealing away and leaving an emptiness behind." But this time it would only be one month until they were reunited.[36]

Nehru flew off again to London on April 19 to attend the Commonwealth Prime Ministers' Conference. Krishna Menon was waiting with the Rolls, as usual, at London's airport and drove him back to Edwina shortly before midnight.[37] Nehru had lunch with King George the next day and met with Attlee and Cripps later in the afternoon. Then Edwina drove him down again to Broadlands for a lovely wet weekend.

Indira was upset by her father's unrelenting obsession with "that Mountbatten Lady!" "I hope the weekend at Broadlands was refreshing and that you will return looking less tired than when you left," she wrote anxiously from Bombay.[38] Even Mathai started worrying about Nehru when he "accidently" read one of Lady Mountbatten's many love letters to her "Darling Jawaha." "I cannot understand how a woman of Lady Mountbatten's age could write such adolescent stuff." Padmaja Naidu felt so enraged and jealous when Nehru brought Edwina to Lucknow to visit her mother that she "locked herself up in her room and went into a tantrum," refusing absolutely to meet Lady Mountbatten.[39] Despite her tantrums, Nehru later appointed Padmaja the governor of West Bengal, and his family physician, Dr. B. C. Roy, served there as chief minister. Roy died before Nehru, but Padmaja outlived him, remaining in charge of Calcutta's splendid old governor's house for more than a decade.

"Winston [Churchill] said to me that he felt as if a friend whom he had given up for dead had suddenly come back to life," Nehru reported to Indira on his flight from London to Zurich in early May. "If Winston feels that way you can imagine how others feel. . . . I am sure that we have done right for India and the world. We have not given up an iota of our freedom, of anything that matters, and we have gained much," he insisted, rationalizing how wise and good it was for India to remain in the Commonwealth. "I return a little more light-hearted than I went, a little surer of myself, a little fresher in mind and body. . . . [M]y two visits to Broadlands were very delightful. It is a lovely place which fits in with my mood and I wish you could also visit it."[40]

Being with Edwina at Broadlands was always exhilarating to Nehru. In some ways it seemed more remarkable than being prime minister, for that had long been his destiny. He was not quite sixty, still fit enough physically to enjoy every moment of it, all the passionate pleasures he had never known with Kamala, all the talk and laughter he had lavished on Nan, all the wisdom often wasted on Indira. But although he loved it, he tired much more quickly than Edwina did, feeling distracted and bored, after a day or two. He was always too polite to show it and luckily, there were always more urgent matters that demanded his immediate attention.

Nehru flew off to Switzerland where he met with many bankers as well as President Ernst Nobs of the Swiss Confederation, with whom he ratified an Indo–Swiss treaty of friendship. After touring Switzerland, he went back to Geneva to meet with King Leopold of Belgium. At this time he contracted for Swiss and Belgian high-tech centrifuges and the other advanced machinery required for nuclear processing, to liberate plutonium from Indian mountains. Nehru personally retained control over India's top secret nuclear energy program, which was housed in Delhi's Department of Scientific Research. He recruited Dr. Homi C. Bhabha, a brilliant nuclear physicist whose early death in a plane crash in Switzerland was a serious setback for India's program. Drs. S. S. Bhatnagar and K. S. Krishnan carried on the work, however, but the first plutonium bomb was not tested until a full decade after Nehru's death, when Indira was prime minister. She, too, kept tight control over the development of India's nuclear weapons.

Nehru flew home to Bombay on the night of May 6, 1949. Indira was waiting for him with both boys, as was Betty, who joined him and Indira on his next trip abroad that October, this time to the United States, as well as to London and Broadlands. First, however, there were urgent matters to deal with back in Delhi and in Kashmir. Nehru had hoped to fly to Srinagar on May 20, but was reminded when he got to his office that the A.I.C.C. was meeting at Dehra Dun that day and the next, "and I must obviously attend it," he explained to Sheikh Abdullah, postponing his flight to Kashmir until the end of the following week.[41] He was furious with Abdullah for having told a British journalist that "accession to either side cannot bring peace," adding wisely that "an independent Kashmir must be guaranteed not only by India and Pakistan but also by Britain, the United States and other members of the United Nations."[42]

Sheikh Abdullah had lived through the last year of war in Srinagar and knew just how few of his friends—to say nothing of virtually all other Muslim Kashmiris—ever wanted to belong to India. Nehru resolved when he read that report to get rid of Abdullah, whose utility to India had now all but ended and would soon have to be replaced by his much tougher deputy, Bakshi Ghulam Mohammad. In fact, Nehru decided not to wait until month's end, ordering Abdullah to fly to Delhi on May 14. In his letter to Krishna Menon that day, Nehru noted that Abdullah "lacks political foresight and has a knack of saying the wrong thing. He is influenced greatly by odd groups."[43] Nehru and Sardar Patel told Sheikh Abdullah never again to give such an "outrageous" interview to any journalist, and after two days on the carpet in Delhi he promised to behave himself. Nehru decided to give him another chance—or more rope, for he was anxious to visit Srinagar for the last weekend in May and had promised Colleen Nye, his other lovely British lady, to show her the stunning vale with its flowers in bloom.[44]

"No plebiscite is possible in Kashmir before the refugees are rehabilitated," Nehru told his Congress followers in Dehra Dun that May, preparing the ground for retracting his promise, now that the U.N. commissioners were starting to arrive with detailed plebiscite proposals. "So long as the raiders, who have driven out Kashmiris from their hearths and homes, remain in parts of Kashmir, the return of refugees to those areas cannot be expected. Till they return and settle down in their homes, how can you expect them to express their free will? . . . The price of freedom and progress is blood, sweat and tears. We have paid the price of our freedom in blood. For our future progress we have to pay that price in tears and sweat. Our generation is condemned to hard labour . . . to make India great. . . . I cannot forget my jail life in Dehra Dun . . . I had more peace and quiet as a prisoner than I have now as the Prime Minister of India."[45] Now that he and Churchill were good friends, Nehru began to speak a bit more like him, addressing his troops in Srinagar: "The history of this epic struggle has been written by the people of Kashmir in their own blood. . . . The Kashmir operation is a fight for the freedom of India and Kashmir. The attack on Kashmir is against all laws of international relations. There are laws even of war. An attack by dacoits and thieves cannot be a war. . . . Our participation in the Kashmir struggle is a great and noble task."[46]

Nehru addressed a large crowd the next day in Srinagar, hailing "Sher-e-Kashmir" (Lion of Kashmir) Sheikh Abdullah and promising never to let down him or Kashmir. He denounced the Pakistan government for not "keeping their promises. On the contrary, they continue to be aggressive and at the slightest opportunity try to grab further." India's plan for the future of Kashmir, Nehru promised its people on that lovely May day in 1949, was "to solve the problems of Kashmir and remove its poverty . . . so that this State might become rich and every one of its citizens may walk with his head held high."[47] That noble promise remained an ever more distant dream. Nehru promised as often to remove poverty from Kashmir as he did to permit the holding of an impartial plebiscite throughout that poor and ever more war-torn state. "India is the only stable and progressive nation in the whole of Asia," Nehru told his army officers in Kashmir, "and as such is the natural leader of Asian countries. The potential power of India is well realised by the world."[48]

President Harry Truman invited Nehru to visit the United States in October 1949, and Nehru brought along Indira and Betty. Nan had already settled down in Washington as India's ambassador and helped plan his three-week visit. "The real reason for my going via London is to take advantage of Air India," he told Nan the day after returning to Delhi from Srinagar. "I see no reason to pay money to an American airline if I can help it. Dollars are a very precious commodity. . . . We had a hint a little while ago that our Embassy in Washington had not kept in intimate touch with the State Department in recent months. . . . The recent visit of our military mission to the U.S. has not proved successful, chiefly because of the State Department's attitude. . . . I think it is necessary that the State Department should be cultivated a little more and in informal ways."[49]

This was one reason that he had sent the charming Nan to Washington. He wrote her in early June to say how interesting he found the "inherent conflict between England and the U.S.A." in courting India. "If we deal with the U.S.A. in regard to the sale of certain atomic energy material, they frankly tell us that they do not want us to sell them to the U.K., although the U.K. happens to be their close friend and ally. In England of course there is not too much friendship in evidence for the U.S.A., partly because they feel themselves dependent on America and do not like it."[50] Nehru astutely managed to take full advantage of these differences, thereby enriching India's economy as well as its military and nuclear development from both sides of the mutually suspicious Atlantic alliance. He also insisted on saving enough time during his visit to the United States to spend at least three full days in Canada, where he concluded an agreement to ship heavy water to India's nuclear energy production plant on the island of Tarapura off Bombay.

Although Nehru was thus eager to get military and nuclear energy support and "financial help from America . . . on easy terms," he remained suspicious of private American investment in India's economy, putting on indefinite hold most expressions of interest, like one he sent to Nan from the president of Westinghouse. He distrusted American big business, viewing its good intentions as primarily selfish interest in private profit. "The suggestion that we should invite a large crowd of prominent Americans to India does not appear to me to be feasible or desirable," he confessed to his ambassador sister. "Psychologically speaking, it will have a bad effect on our people and they will think that we are selling India to the U.S.A."[51] Nehru's inability to rise above his deep-rooted Marxist equation of Western capitalism with imperialism, and his almost paranoid, partly aristocratic, distrust of free enterprise in its most successful forms as "vulgar," cost India dearly in retarding its overall development for the remaining years of his rule, as well as for the even longer reign of his more narrowly doctrinaire daughter.

"Two years have gone by in our lives and two years have been added to the thousands of years of India's history," Nehru told his nation from Delhi's Red Fort on August 15, 1949. "We are birds of passage. . . . But the work that we do will be enduring. . . . India too will go on while people come and go. There are grave problems before us and we are imprisoned and crushed by them. But we must face them and try to overcome them. . . . We must forget our personal problems. . . . Mahatmaji's voice used to reverberate in our ears and hearts, and millions of us in India left our homes and petty preoccupations, even forgetting our families . . . and jumped into the fray. . . . We dreamt a great dream, and very often there was a kind of madness, an obsession in all of us. . . . Well, we got political freedom for India but another great problem remained unresolved . . . to ensure that all the people enjoy the benefits of that freedom. In the meantime, another great disaster descended upon us. . . . We made mistakes." He sounded almost as though he were awakening from a long interlude of amnesia, shocked to see how much had changed, more depressed perhaps to find how much more remained the same. "Finally, we reached our goal. . . . Why then should we be afraid of anything today or give in to panic and anxiety? . . . The only drawback is that our minds are wandering a little and we get bogged down in petty issues. . . . [W]e have to be large-hearted and broad-minded. Small men cannot do big things. They cannot solve problems by shouting slogans or making a noise or complaining and abusing one another. . . . Everyone must do his duty."[52] Now he sounded almost like Motilal or Gandhi.

"You must bear in mind two or three important things," Prime Minister Nehru told his people. "One, no matter what our policies are, we cannot do anything unless there is peace in the country and an opportunity to work. I am saying this because many of our misguided youth indulge in hooliganism, rioting, killing and throwing bombs. I am amazed that any sensible thinking individual can do such unpatriotic things. . . . You have the right to change the laws, even the government. But it has to be done by peaceful, lawful methods. . . . I am amazed that some of our youth think that they can serve the country by hooliganism and rioting. What amazes me even more is that even people who say that they are opposed to lawlessness actually side with these elements in the political field. They do so for some small benefit in elections."[53]

"We want friendship with the whole world," Nehru explained, trying to clarify his foreign as well as his domestic policy. "We also want . . . amity among the millions of people in the country . . . there should be love and cooperation among the people. We will accept any help that comes from outside. But . . . [w]e cannot tolerate interference in our internal affairs or in our freedom. . . . We have friendly relations with both the blocs and yet maintain our freedom. . . . [T]his is the policy best suited for our country and also . . . the only one by which we can serve the cause of world peace."[54] Nehru long considered himself Asia's foremost champion of world peace, and one of the most bitter disappointments of his life was that he did not win the Nobel Peace Prize, although he often was nominated for it. Even Indira later hoped for that highest honor, but she too was considered less than worthy, especially after the atrocities committed by her 1975 "emergency" Raj.

"Although Nehru greeted me affably his whole attitude changed when he began discussing Kashmir," U.S. ambassador to India Loy W. Henderson wired the U.S. secretary of state, Dean Acheson, that same August. "He said he was tired of receiving moralistic advice from U.S. India did not need advice. . . . His own record and that of Indian foreign relations was one of integrity and honesty, which did not warrant admonitions. He did not care to receive lectures. . . . So far as Kashmir was concerned he would not give an inch. He would hold his ground even if Kashmir, India and the whole world would go to pieces. . . . He would not be swayed by talks or persuasion. He was under too deep obligations to Kashmir. He would give State up, only in case Kashmir people should freely express their desire not to remain a part of India. The Kashmir issue affected underlying philosophy of India which was that of a secular progressive State. . . . This philosophy was opposed to that of Pakistan which was a theocratic State. . . . In latter part of his talk, he used conversational tone of voice and turned on his well known charm. . . . He said he . . . would not like anything which he had said to injure our relations."[55]

Both Truman and Attlee urged Nehru to accept the U.N. commission's recommended arbitration of the deadlocked Kashmir dispute. Pakistan accepted that sensible solution unconditionally on September 6, 1949, but Nehru vehemently rejected it. Sir Archibald Nye, now the British high commissioner in Delhi, went to speak with Nehru about Kashmir on September 9, and reported having heard a long account from him on the entire history and background of the dispute, as he viewed it, including "no arguments with which we were not already familiar but he stated them cogently . . . and with obvious sincerity. . . . There is nothing machiavellian about this man," Nye concluded.[56]

Two days later Nehru wired Krishna Menon, soon to become his point man on Kashmir at the Security Council, always keeping him abreast of his latest thinking on the dispute that he was firmly resolved never to settle at any risk of losing Kashmir's vale. "Friends of India in England are always welcome but in this matter they should be told clearly that in our opinion they are completely in the wrong."[57] Nehru knew, of course, that he could always rely on Krishna Menon to support him completely, even against his oldest, once closest, radical English friends, including Agatha Harrison and Harold Laski, on this issue. Likewise, Krishna Menon knew he had the total support of his prime minister against every other officer in India's Foreign Service, many of whom reported on how "sick" or "incompetent" he was.

By June 1951 Nehru wrote to advise Krishna to "take three months' leave,"[58] knowing how very ill he had become. Krishna refused to step aside, however, so Nehru sent Mac to London to see if he could talk Krishna into accepting Nehru's sage advice. But the disgrace of Krishna Menon's "inner degradation and disintegration" had gone too far, "all the greater when it comes to a man of Krishna's brilliance,"[59] Nehru replied to Mac's negative assessment of their mutual friend. So in October 1951 Nehru ordered Krishna Menon, who only barely managed to "keep going by continually taking powerful tonics [drugs!],"[60] to take immediate "leave . . . and go off to Switzerland. . . . Do not trouble yourself about expense. . . . That will be our responsibility. . . . I have not said anything about this matter to anyone in the Ministry here. No one has had the slightest hint. Mathai has typed this letter. My love to you."[61] Six years later Nehru brought Krishna into his Cabinet as Minister of Defense. For Krishna Menon remained the only man whose mind was perfectly attuned to Nehru's own complex, often contradictory mental wavelengths.

Nehru also tried to keep Mountbatten with him on Kashmir, although he knew how negatively Dickie felt about much that he had done there and the costly intransigence of his position concerning Srinagar. "The Kashmir problem remains as insoluble as ever and perhaps there is more tension now than ever before," Nehru wrote Mountbatten two weeks before flying out of Bombay to London. "Meanwhile the threats and vulgarity of Pakistan continue on an ever-ascending scale. I am afraid Pakistan is a pathological case and it is becoming more and more difficult to discuss anything with them reasonably."[62]

Two days later he flew back to Srinagar to address Sheikh Abdullah's

National Conference, which was meeting there for the first time since the war had started. "Sher-e-Kashmir, Comrades and Compatriots," Nehru addressed them on September 24, 1949. "When I think about the events of the last two years . . . I cannot understand how, when and on what matters of policy the Government of India had erred. . . . There is a passion, a fire, in my heart. But that fire can be governed only by a calm mind. If the fire reaches the brain it reduces the capacity of the brain to think. So there should be a fire in the heart but the head should remain cool. . . .When the news reached us in Delhi that we were under attack . . . The heart pulled us in one direction but the mind cautioned that there is a danger in it and that no country can afford to get carried away. . . . I did not have any idea then that the matter would become so long-drawn or that we would need to send troops in such large numbers."[63] Forty-five years later, long after Nehru and Indira and Rajiv all were dead, India had to station more than half a million of its troops in Kashmir, vainly trying to pacify its troubled vale. When Nehru spoke in 1949 there were still far fewer than 100,000 Indian troops in all of Jammu and Kashmir.

"We sent in our troops not to attack or to occupy the state but for the protection of the people, and at their invitation," Nehru told his audience of mostly skeptical, war-weary Muslim members of the National Conference, tired of seeing so many Indian soldiers, tanks, and armored trucks on the streets of Srinagar. "We also promised that the troops would remain as long as Kashmir needed protection. We will fulfill our promise. When the Commission said in their resolution that we should withdraw our troops from here, we replied . . . that we would first assess if Kashmir was likely to face any danger . . . of another ruinous attack and destruction from inside or outside. . . . We said we do not wish to keep a single soldier here if Kashmir is safe. . . . But we still need to keep our forces here to ensure that Kashmir is not attacked again because the fact is that it is very easy for Pakistan . . . to attack Kashmir. They are sitting right at the doorstep and could be there in a matter of a few hours or a day or two. Therefore, unless we are fully convinced . . . [that] Kashmir is completely safe . . . we cannot withdraw our troops nor can we let the people of Kashmir face danger. . . . I am just trying to show where our duty lies. . . . So this is the situation at present," Nehru assured them, "that whatever is done in Kashmir will be done ultimately only by the consent of the people of Kashmir. . . . [R]emember our policy that we have no desire to rule by the sword or to pressurize anyone. We want to maintain freedom in all corners of India."[64]

Nehru flew with Indira and Mac from Bombay on October 7, and they picked up Betty and her boys in Cairo next day, for the second lap to London, where Krishna Menon was waiting. President Truman sent his own plane to London to take Nehru and his party to the United States. Krishna Menon, more anti-American than Nehru, urged him to turn down that generous offer, but Nehru was delighted to fly in such comfort and security, without spending any precious dollars, just when the Indian rupee had been devalued and hard currency remained scarce. He met with Attlee and Cripps in London, stayed overnight at Broadlands, flew to Washington on October 10, and was welcomed by President Truman, his cabinet, and Nan there on October 11, 1949, for his first visit to the United Sates. "I come to you, Mr. President, bringing to you and to this great Republic the cordial greetings of my Government and my people of the new Republic to be of India."[65]

The Constituent Assembly had not quite finished drafting the world's longest constitution, under which the Republic of India would be born on January 26, 1950. India's eclectic constitution borrowed from many models, primarily Britain's parliamentary one, but instead of a crown there was to be a president, as in the United States, except that in times of "national emergency" the usually nominal head of state could suspend the constitution, as Indira ordered her president to do in June 1975, after Uttar Pradesh's high court found her guilty of electoral malpractice.

Nehru met with Dean Acheson on October 12 and told him of his desire to stockpile one million tons of wheat, in order to reduce the price of wheat in India by about 10 percent. He was eager to procure that wheat from the United States, and Acheson explained that it would require congressional approval, which he would try to expedite and also propose legislation to permit the U.S. government to "store wheat in India and sell it as withdrawals were necessary." Secretary Acheson asked Nehru to talk "fully and frankly" about Kashmir, which Nehru did at length. "He spoke bitterly of Pakistan deception and intrigue," Acheson noted. "I said that I understood what he had told me but I was not clear as to his idea for getting forward with a peaceful solution of the matter."[66] Nehru talked on but never clarified his "peaceful" solution.

The next day Nehru met with Truman in the White House. "The President referred to the Kashmir situation as one of the problems which he hoped fell within the Prime Minister's determination to solve without conflict. The Prime Minister assured the President that this was so."[67] Acheson hosted the dinner that evening, and Nehru said he felt "rather overwhelmed by the kindness and friendship" with which he had been welcomed to Washington. "I have come here not with a blind eye, but with a friendly eye and a friendly outlook," he assured his debonair host.[68] But Nehru's intransigence on Kashmir diminished the effectiveness of everything else he said, at least in Washington. He tried to convince Acheson of the wisdom of recognizing China, which India was now ready to do, and against supporting Bao Dai in Vietnam, although "he was convinced that Ho Chi Minh was a Communist," understanding, nonetheless, that his was a true "nationalist movement."[69] Had Nehru's brilliant mind been open and reasonable on Kashmir, his words of wisdom about other parts of Asia might have carried more weight in the State Department and Oval Office. He addressed the U.S. Congress as well, rightly noting that "what the world today lacks most is, perhaps, understanding and appreciation of one another among nations and people. I have come here, therefore, on a voyage of discovery of the mind and heart of America."[70]

From Washington Nehru flew to New York on October 15 and enjoyed that evening alone with Clare Boothe Luce in her New York apartment. "I had a telegram from Clare Luce saying that she had given up attending public functions, but she hoped to have a chance of seeing me," he wrote Nan before leaving for this most busy trip. "I should like to meet her somewhere."[71] He had turned down many New York invitations but relished that brief interlude of private reunion with Clare Luce. Nehru received a hero's welcome from New York City the next day and thanked Mayor Bill O'Dwyer for the big parade, insisting he was just "a simple man, from a country of simple ways . . . rather dazzled by the wealth and prosperity of this country . . . but I have been attracted even more by that background of freedom which this country had and by its struggle for freedom, and its championship of the cause of freedom in other countries."[72] Dwight D. Eisenhower, president of Columbia University, conferred on Nehru an honorary doctor of laws degree that afternoon. "I have come to you not so much in my capacity as Prime Minister of a great country or as a politician but rather as an humble seeker after truth and as one who has continuously struggled to find the way, not always with success, to fit action to the objectives and ideals that he has held. The process is always difficult but it becomes increasingly so in this world of conflict and passion."[73]

Nehru addressed the U.N. General Assembly the next day at Lake Success, New York, calling the United Nations "the only real hope for the future."[74] He spoke after dinner the next night to four major foreign policy organizations. "We have no atom bomb at our disposal. We have no great forces at our command, military or other. Economically we are weak. . . . But if I may say so, we rejoice in not having the atom bomb. . . . We have lost all fear of external aggression. . . . What has pained us is our own inner weakness, because that has sometimes made us doubt ourselves. . . . Some of you may be acquainted with wild animals. I have had some little acquaintance . . . and am convinced that no animal attacks man, except very rarely, unless the animal is afraid. . . . Man becomes afraid of the animal and then the animal becomes afraid of him and, between them, they make a mess of it." His animal lesson might have served as a parable for Indo–Pakistani relations, although he probably meant it more for U.S.–Soviet problems. "We have no desire for leadership anywhere," Nehru insisted. "Our greatest anxiety and yearning today is to build up India and to solve somehow the problems that face us . . . and to cooperate with other countries in the United Nations and elsewhere."[75]

He flew to Boston on October 21, and President Margaret Clapp of Wellesley College was at the airport to greet him and bring him to her campus, where Nan's daughters had studied. The next morning he left for Niagara Falls and Canada. Prime Minister Louis St. Laurent greeted Nehru at Ottawa's airport. Indira and Nan went with him to Canada, as did his ubiquitous private secretary-valet, Mac. Nehru's old Harrow friend Harold Alexander, then viscount (later earl) of Tunis was the governor-general of Canada at this time, and Nehru stayed with him in Ottawa. Thanks to Alexander, Nehru managed to expedite agreements to ship Canadian heavy water to India, as well as more urgently needed Canadian wheat. He addressed both houses of Canada's parliament on October 24 and flew to Chicago the next evening.

"I am being worked pretty hard here and by no stretch of imagination can this be called a holiday," Nehru wrote C.R. from Chicago. I am beginning to feel more and more stale. . . . I ramble on and on which is unfortunate. . . . I am continually being surprised at the type of popular welcome I am getting . . . another twelve days. . . . I am rather looking forward to the end of this tour. On my return to India I shall immediately be immersed in arrears of work and other problems. In between, I hope to get two days' rest at Broadlands."[76] He brought sugar from the United States for Edwina, since it was still rationed in Britain. "I think I am not interested in sex as sex," Edwina wrote him. "There must be so much more to it, beauty of spirit and form and in its conception. But I think you and I are in the minority! Yet another treasured bond." She felt "you understand as no one else has ever done."[77]

He flew home to Bombay on his sixtieth birthday in mid-November. "Brothers and Sisters," Nehru told the large crowd assembled to welcome him, "as you know, generally people, at least in government service, retire at the age of sixty. . . . So in such circumstances to remind a man that he is sixty years old is not being very kind to him. . . . Well, this is not the first time when you have celebrated my birthday with great love and affection. . . . You have always forgiven me for my mistakes. . . . [T]he one great task before us is the economic upliftment of the people because freedom is not complete till this task is done. . . . Secondly we are, as you know, caught up in a strange world of tensions, quarrels and of tilting imbalances. We do not wish to get involved in other people's quarrels. We want to be left alone to do our work. . . . But whether we like it or not, we cannot isolate ourselves from the world. . . . [T]he moment freedom comes, it brings responsibilities." Nehru profusely thanked the crowd that cheered him. "I wonder what I can do in return for the enormous love that I have received from the people of India, for there can be nothing more valuable than love. . . . Well, anyhow, whatever remains of my life will be spent in such tasks as I feel will do good to the country. . . . [I]t becomes a kind of a disease almost[,] . . . very difficult to get away from it. But why call it a disease? . . . [H]ow does a man achieve real happiness in life? Mere worldly pleasures can only bring superficial happiness. Real happiness can come only by doing big tasks. I have become addicted to it. . . . [S]o long as I have strength, I will continue to do my work."[78]

"I did not go there [to America] for deals and bargains or for intrigues," Nehru told his Constituent Assembly soon after he reached Delhi. "I did go there to create a friendly impression, if I may say so, a friendly interest in our problems, and . . . I believe we succeeded in doing that, and I believe that the responsible people in the United States thoroughly appreciated, if I may say so, the frank way in which I explained our position in world affairs. . . . That is a position in which India pursues her foreign policy without any commitments to other countries, without any binding ties, but in friendly cooperation with all the countries that cooperate with her. Indeed, it was largely in consonance, to give a historical parallel, with the policy which the great founder of their own nation had pursued in the early days of the American Republic. . . . I am not conceited enough to imagine that we can control the fortunes of the world or prevent something happening that otherwise could happen. But . . . we can occasionally at least make a difference. Well, I hope that this country will make that difference . . . in favour of peace."[79]

Not too much "pacifism," however, as Nehru explained to the first World Pacifist Meeting held that December in Mahatma Gandhi's last ashram in Sevagram. "Horace Alexander [chairman of the organization] said . . . I was not a pacifist and he was right in saying so. . . . I have to compromise all the time. . . . [T]he question is what do you compromise with? Are you compromising with evil? It is bad. Well, you can compromise holding on to the truth by not going too far . . . perhaps because you cannot carry others far. . . . I am just putting to you some of the difficulties . . . one has to face. . . . In a country like India, which is poverty-stricken and industrially backward, people are more interested in the primary needs of life than in theories. There may be democracy or capitalism or communism or socialism, but they want their primary needs of life satisfied. . . . It is just fantastic nonsense for people to talk of India being the leader of Asia. It irritates me. . . . It only flatters the vanity of the Indian people. . . . If you think that by sitting in Sevagram you have seen India, you are wrong. . . . Mahatma Gandhi is of course almost worshipped all over India . . . and may be some of the persons who worked in the closest companionship with him have got the quaintest notions of what he stood for. Quaintest, I say in the sense of very narrow notions. . . . What exactly is violence? It is not an easy question for me to answer. . . . I refer to the individual violence . . . behaviour in a most uncontrolled way. . . . It is not what a man utters that matters. Slogans do not make a man. What the man is counts for something."[80]

Nehru flew to Colombo on January 8, 1950, to attend the Commonwealth Foreign Ministers Conference held there, for he remained his own foreign minister. After what had been a busy month in Delhi, he enjoyed that week's respite. On January 12 he received another doctor of laws degree from the British vice-chancellor of the University of Ceylon. Indira and Krishna Menon joined him for that week in Sri Lanka. "My politics has always lain with the masses . . . and I always thought of freedom as something which will relieve the masses of their poverty," Nehru told his friends in Colombo on the eve of his departure. "I emphasise that Government policies can never have anything of trickery or jugglery about them. A wrong act by an individual or Government leads to wrong results . . . our great leader—father of our nation—Mahatma Gandhi taught us. . . . There was magic in his message. . . . I therefore urge that the Gandhian way should be applied to practical politics of the world today." That Sunday evening he returned to Delhi, and among the select group waiting to welcome him home were Sir Archibald and Lady Colleen Nye.[81]

Nehru signed the constitution of the Republic of India on January 24, 1950, and two days later Purna Swaraj Day was renamed Republic Day and has been celebrated as such ever since in New Delhi and throughout the Indian republic. "Twenty years ago we took the first pledge of independence. . . . The man who led us through apparent failure to achievement is no more with us, but the fruit of his labours is ours. What we do with this fruit depends upon many factors. . . . We are fortunate to witness the emergence of the Republic of India and our successors may well envy us this day; but fortune is a hostage which has to be zealously guarded by our own good work and which has a tendency to slip away if we slacken . . . or if we look in wrong directions."[82]

Four days later he addressed another large public meeting in Delhi on the second anniversary of Gandhi's assassination. "What is the true meaning of independence? . . . Just four days ago, we celebrated . . . it with some pomp . . . because when something big happens in the country, it should be marked by grandeur, dignity, and discipline. . . . It is equally right and proper that . . . we should observe today . . . in memory of Mahatma Gandhi. . . . It is not enough to take his name repeatedly or to praise him. . . . It would be very easy for you or me to criticize or accuse one another, but it would be totally futile to do so."[83]

Indira was always doing that, every time she wrote or spoke to him: "The politics of this province stink so much it is becoming difficult to ignore. . . . I cannot live in such an atmosphere. . . . You tolerate a lot of things. . . . It makes my heart bleed to hear everyone say that it is no use bringing anything to your notice since you don't do anything about righting things. . . . [I]t is very ugly. . . . I don't know what you can do but I do know that if you don't call Pantji to order . . . then please stop in future talking about democracy & the freedom of the press in India. With all our other ills let us not also have hypocrisy." She had always been high spirited and emotional.[84]

"Yes, the country will no doubt go along on its own slow momentum," Nehru told that huge crowd of mourning Gandhians. "Some work does get done and people continue to function in their own ways. But if you and I want to move fast . . . then it becomes necessary for all of us to create a new energy. . . . I am responsible anyhow—I cannot escape it. . . . Very few of you who are present here today may be able to remember the events of the last twenty or thirty years. . . . Our intellectual and the educated citizens would shake their heads and argue endlessly. . . . They could not understand what use it was to spin a *charkha* or sweep and do this and that. Their concept of politics was to pass long resolutions in large gatherings and assemblies. . . . It was very easy to mock at the idea of achieving swaraj by doing menial tasks. . . . Those of us who were young in those years can still remember the fresh awakening of minds and hearts and the spontaneous burst of enthusiasm. We forgot everything else that we are doing. . . . A great man had come among us and reminded us of great principles. . . . Had we learnt those lessons better and been more influenced by them perhaps we would have gone even further." A poignantly painful thought, but he brushed it swiftly aside, for Nehru was never one to linger on what might have been. He shed no tears at any funeral, brusquely adding, "Anyhow, we achieved a great deal. . . . Jai Hind."[85] So Nehru ended his speech memorializing the Mahatma with the "Victory to India" battle cry of Netaji Subhas Bose. His political genius was to continue to be able to attract somehow the support of such diametrically different groups of Indians—Gandhi's pacifist disciples and Bose's military devotees—while retaining the love and devotion of millions of ardent and passionate young idealists.

30

Himself at the Top

1950–1956

THROUGHOUT THE 1950s Nehru enjoyed unlimited, indeed, virtually unchallenged power over the Indian republic. He was the darling of India's people, the hero of his party, the unrivaled leader of his government. Deputy Prime Minister Sardar Patel had stomach cancer and weakened rapidly during his last pain-filled year of life, dying less than two months after his seventy-sixth birthday, in mid-December 1950 in Bombay. Soon after the Republic of India was born, Governor-General Rajagopalachari (C.R.), the only other leader strong enough to challenge Nehru, was obliged to step down from the high office he held only two years, with Dr. Rajendra Prasad emerging as its first chosen president. Perhaps Nehru's severest Indian critic in these years of his greatest triumph was hapless Feroze, the nation's son-in-law, doomed to an early death soon after losing his Indira to her father.

But in April 1950, Nehru took along both Feroze and Indira on his first visit to Pakistan, leaving for Karachi on the morning of April 26. Prime Minister Liaquat Ali Khan and his entire cabinet were waiting when Nehru landed. Hundreds of thousands of Pakistani schoolchildren lined the road to Karachi's Government House, where Jinnah had breathed his last in early September 1948. Cries of "Pandit Nehru Zindabad" (Victory to Pandit Nehru) reverberated as he was driven swiftly from the airport. He rode with Liaquat Ali, whose own fate was sealed now that he had welcomed Pakistan's enemy so warmly, gunned down eighteen months later by a hired assassin in Rawalpindi's army cantonment. For Nehru and

Liaquat Ali had reached an agreement of sorts, one that almost seemed destined to bring an era of peace and tranquillity to Indo–Pakistani relations.

Nehru had written Liaquat Ali in January to propose a "no-war" declaration, stating that "in no event are we going to war for a settlement of disputes."[1] Liaquat replied that Pakistan preferred both nations to agree to arbitration to settle each dispute—primarily Kashmir—rather than agreeing to a blanket declaration. An influx of Hindu refugees from East Pakistan to Calcutta in early 1950 led to deadly communal riots in retaliation against Muslims in Calcutta and Howrah, which in turn stimulated harsher and bloodier attacks against Hindus in Pakistan's East Bengal. "I am terribly sorry for what took place in Calcutta," Nehru wrote Liaquat. "But all the information at my disposal indicates that there is no comparison between [the] Calcutta happenings and [the] East Bengal happenings. It is little comfort, however, for either of us to measure and balance evil. . . . It is patent that we cannot wait and watch supinely for tragedy to descend upon us. . . . Both of us . . . have to bear a terrible responsibility."[2] Liaquat agreed to joint investigations of the atrocities to help "restore confidence in minority communities on either side."[3] Although fifteen years Nehru's junior, Nawabzada Liaquat Ali was as sensitive to human suffering as Nehru was.

Nehru wanted Liaquat to go with him to the troubled districts of both Bengals. He wrote, in fact, to Vallabhbhai Patel at this time, saying that in order best to serve Bengal, "I should get out of office."[4] Vallabhbhai calmed him down, urging him to do nothing in haste, and so Nehru agreed, "I shall not rush into any action." He felt frustrated, nonetheless: "I have largely exhausted my utility in New Delhi." He felt tired and stale. He longed for action: "First things must come first in a crisis." Time "we all shook ourselves up," he told the dying Vallabhbhai. "We grow too complacent and smug."[5] Was it frail Vallabhbhai he had in mind then or himself? At the end of February, Edwina flew in to soothe his nerves and troubled spirit for five days and nights. "It was delightful having her here," Jawaharlal wrote Dickie on March 5, when she left for Bombay and he prepared to fly to Calcutta the next morning. "Somehow we cannot put an end to the terrible consequences of partition. . . . Practically every Hindu in East Bengal is clamouring for evacuation. . . . There are 12 million."[6]

Four days in Calcutta left Nehru feeling exhausted, not from any physical strain, but rather from "mental and emotional exhaustion, having had to deal . . . with hysterical people in a highly supercharged atmosphere." Edwina was on her way to Colombo but planned to return to Delhi to spend another week with him.[7] Army intelligence informed Nehru that on just two days in March, 4,700 Hindus had fled from East Pakistan to

Calcutta, and 8,860 Muslims had fled the other way. "Whether we like it or not, I think it is time we dealt with this matter effectively," Nehru wrote Liaquat Ali the next day. "I am prepared to face the truth, whatever it is, and take the consequences. . . . Much evil has happened in East and West Bengal and in Assam."[8] Nehru returned to Calcutta the next day and learned from an old Muslim friend from Allahabad there that during the previous five weeks between 100,000 and 150,000 terrified Muslims had been forced out of Calcutta. "I was astonished at this number and did not believe it, but later official figures . . . corroborated it."[9]

Now Nehru confessed to Krishna Menon that he feared "a concerted effort, backed by strong forces, to drive us into war with Pakistan."[10] He told C.R. that same day, "I am pressed all round for what is called 'action' . . . a euphemism for war."[11] Edwina decided to come back on March 22 and planned to stay with him until April 1. He was sorely tempted at this time to give up his job and fly off with her, back to Broadlands, where he could live out whatever remained of his life in its beauty and splendor. The duke of Windsor had given up more for much less, after all, given all that Edwina offered compared with what Wallace Simpson possessed. He confessed to Nan on the eve of Edwina's return that he was seriously thinking "of resigning from the prime ministership." He blamed his discontent on Bengal and his critics in parliament, some of whom were rather noisy, but it was more the daily routine of public life he found so onerous, vulgar, and boring, for as he told Nan, "My whole nature rebels against sitting here in an office, when I should be up and doing. . . . I do not quite know what I shall do. But negatively, I feel sure that my mere going away will do good . . . I want to tell you not to worry at all. Also that you must stick to your post. . . . It would be improper and undesirable for you to take any action, simply because I have faded out of the Governmental picture."[12] First Punjab and Kashmir, now Bengal, all the horrors Bapu alone had been wise enough to anticipate as the aftermath of partition heaped on his weary head. Why not fly away with her? What happier escape from perdition to paradise? He felt sorely tempted.

Instead of flying away, however, Nehru waved Edwina off alone on April 2, and shortly after she left, he welcomed Liaquat Ali Khan and his advisers into Delhi seeking to calm the Bengali wave of communal violence with diplomatic words. Liaquat was eager to reach an agreement with Nehru and assured him that Pakistan "was an ordinary democratic State like England" and that the term "Islamic state" applied only to "personal law."[13] Nehru was happy to hear this, but when Pakistan's most militant maulanas learned of their prime minister's "betrayal" of his Muslim faith in Delhi, his tragic fate in Pindi the following year was sealed. On April 8, 1950, Nehru and Liaquat Ali Khan signed their agreement, affirming the determination of their governments to restore to minorities in India and

Pakistan some "sense of security" and to establish a "real equality of citizenship."[14] Many minor details were also agreed on. Committees of inquiry were to be appointed in all affected areas to resolve conflicts that special courts would adjudicate. Nehru promised to visit Pakistan before month's end, and the tide of war receded. "We have stopped ourselves at that edge of a precipice and turned our backs to it," Nehru told his Parliament a few days later. "It is now up to us . . . to face all our problems with sanity and goodwill."[15]

He and Indira spent two days in Karachi during the last week of April, visiting Jinnah's tomb, calling on his lonely sister Fatima Jinnah in Clifton, and dining with Liaquat and his wife, with whom Jinnah and Fati had so often played bridge. He spoke at length with Liaquat about many complaints and problems, from trade, water rights, and issues of evacuee property to the tone of each nation's often alarming press. They never shouted, rarely raising their cultivated Oxbridge voices, but said almost nothing about Kashmir, which Liaquat tried to raise but Nehru refused to discuss, insisting that since it was now on the agenda of the Security Council, he preferred to leave it there. Nor would he go to New York again to engage in that discussion, for nothing about Kashmir had changed in Nehru's mind. Acheson, Truman and Attlee all had tried again—in vain—to move him on the issue of arbitration.

"I must say that I just cannot understand the attitude of U.K. and U.S.A. about Kashmir," Nehru wrote his U.N. ambassador, Sir Benegal Rau. "As you say, there appears to be an almost invincible prejudice against India. We have tried to explain our position in the greatest detail. . . . All these people seem to . . . have lost their sense of balance and perspective."[16] After Nehru refused to talk to the U.N.'s first choice as arbitrator, Admiral Chester W. Nimitz, the Security Council suggested Australia's high court justice Sir Owen Dixon. "No other choice likely to be better," Nehru wired Abdullah. "We think we should accept him. Hope you agree."[17] Despite his impartiality and brilliance as a jurist, mediator, and friend of South Asia, Sir Owen failed to budge either Nehru or Abdullah from their hard-line positions on Kashmir. Dixon arrived in Delhi on May 27, 1950. Nehru met with him and found him "a patient listener."[18] Then Sir Owen flew from Delhi to Karachi and thence to Kashmir.

Nehru sailed on June 2, aboard India's cruiser I.N.S. *Delhi,* from Cochin to Indonesia. He took along Indira and his grandsons Rajiv and Sanjay but left Feroze home for this voyage to Southeast Asia as the guest of President Sukarno, his charismatic Indonesian counterpart. Nehru had strongly supported Indonesia's struggle against Dutch imperial rule and thought of Sukarno as one of his Asian "disciples," but Sukarno had ambitions for Third World leadership almost as grandiose as his own. "We must combine the spirit of the ancient civilisation as represented by the great continent of Asia with the mighty achievements of the present day which the countries of Europe and America represent. Actions divorced from reality are not likely to lead to good results."[19] But as the most populous Muslim nation in the world, Indonesia became one of Pakistan's major Asian allies, second only to China. Five years later, when Nehru returned to Bandung to attend the Afro-Asian Conference hosted by Sukarno, he found to his chagrin that China's popular Chou En-lai posed a formidable challenge to his ambitions of Third World leadership and his hope of Sino-Indian friendship, based on Sukarno's "five principles" of "peaceful coexistence."

Nehru and his family visited Singapore, Malaya, and Burma on their way home, after their ten-day holiday tour of Indonesia's islands. They flew into Calcutta from Rangoon, because the Hughli River tides were too treacherous for I.N.S. *Delhi* to navigate. Chief Minister Dr. B. C. Roy was waiting to greet India's First Family at Dum Dum airport. On June 26 they returned to Delhi, one day after North Korea's invasion of the South, and before the U.N. Security Council had resolved to support the United States' resolution to help South Korea repel that unprovoked invasion, to restore peace and security. India's ambassador, B. N. Rau, voted with the world majority against the North Korean aggression, alarming Krishna Menon enough to wire Nehru, urging no further support for what he considered a Western imperialist conspiracy against North Korea and China. "We are informing our representative at Lake Success not to commit ourselves further in any way," Nehru immediately wired back.[20] The next day Nehru called in U.S. Ambassador Loy Henderson to tell him that since India recognized China and was following a policy of nonalignment, it would hardly be proper to commit any Indian troops or ships to U.S. General Douglas MacArthur's United Nations command. He added that he was being called a "tool" of "Anglo-American imperialists" by various elements but did not mention Krishna Menon.[21]

Sheikh Abdullah was suddenly also proving much more difficult than Nehru had anticipated. He had ignored messages from Delhi's Foreign Ministry while Nehru was on his cruise, and now that Nehru was home and cabled him, he delayed responding. When he finally did write, Abdullah informed Nehru that he could not come to Delhi "because of Ramzan," and Nehru was also told that "Bakshi was ill."[22] Nehru understood too well how easy it was for Kashmiris to become invisible because of "illness."

In fact, Sheikh Abdullah had introduced a radical land reform act, expropriating all large estates (more than twenty-three acres) in Kashmir for conversion to cooperatives to be run by the state with landless laborers. The act was modeled on those introduced in China. "I knew of course that it was your programme to put an end to the *Jagirdari* and *Zamindari* systems and I entirely agreed with you," Nehru wrote his disciple in July, "but I did not know that you were tackling them immediately. . . . I was surprised to learn that you had practically committed yourself without any reference to us. . . . [Y]ou take objection to the Government of India interfering in this matter. . . . But I was not aware that mutual discussions on any subject were ruled out. . . . My own relations with you . . . have not been confined to the formal relations of Ministers of the Government of India. We were under the impression that . . . we had common objectives and there was mutual respect and friendship. . . . I greatly regret that you should have taken up a position which indicates that you do not attach any value to any friendly advice that we might give and indeed, consider it as improper interference," Nehru concluded his letter of July 4, 1950. "I have not thought of Kashmir or of you in that way and so I am rather at a loss how to act when the very foundation of my thought and action has been shaken up."[23] The die was cast, but Nehru would take three more years to decide just when his pet lion must be removed from the palace he had so generously given him in Srinagar, to be locked away for most of the rest of his life. Abdullah's tougher but more obedient deputy, Bakshi Ghulam Mohammad, was promoted to the thankless job of prime minister of Kashmir on August 9, 1953.

Nehru planned it all carefully, ever conscious of the political utility of adhering to the finer points of constitutional law in such matters. He had trained a young "tiger" to dispose of his no longer useful lion, Karan Singh, old Maharaja Hari Singh's son. A bright and pliable young man, Karan Singh was Nehru's choice to take over from his father. He had read books by Bertrand Russell and Aldous Huxley and agreed with everything Nehru told him, especially about how best to deal with Sheikh Abdullah, who lectured him for hours on end and treated him like a child rather than the future head of Kashmir state. For until the night he was deposed, Abdullah continued to think himself the most important person in Srinagar. He was so arrogant that he called Kashmir virtually "free and autonomous" of India, simply because the instrument of accession signed by the old maharaja had said that. A few months after "Tiger" (Karan Singh's nickname) turned twenty-one, Nehru invited him to his house in Delhi to explain "some of the perplexity that he was beginning to feel in dealing with Sheikh Abdullah." Karan Singh agreed after this long and frank "explanation" to accept, at Nehru's urging, his Constituent Assembly's invitation to become the head of state (Sadar-i-Riyasat) of Kashmir.[24]

"A new chapter opens now in the Jammu and Kashmir State," Nehru wrote Karen Singh after his election. "And yet, although it is new, it is a continuation of the old." Kashmir's monarchy was abolished, but the new head of state was the last maharaja of its dynasty. "My dear Tiger," Nehru appended as a personal note to his more formal letter, "You know that I shall often think of you and that you can always rely on such help and guidance as I can give."[25]

The following August 8, Tiger signed the order to "dismiss" Sheikh Abdullah. "Our gamble was a risky one," Karan Singh recalled, "for if the Sheikh got even an inkling of what was happening he would react ferociously, and our own lives may well have been in danger." But Nehru had trained him carefully. The order was served by police, who surrounded his house in the dead of night when Sheikh Abdullah and his wife were asleep. Abdullah angrily roared, "*I* made that chit of a boy Sadar-i-Riyasat."[26] He was wrong, of course, forgetting that the same person who had made him the prime minister also made the young Tiger, who trapped him, the head of state, and could remove them both whenever he wished.

Abdullah was jailed that night without being charged and was kept behind bars until January 8, 1958, when Nehru believed he had learned his lesson and would behave. But he still talked about self-determination for the people of Kashmir, who wanted nothing more than freedom from Indian occupation, and he said it not only in Srinagar but also in London. So he was rearrested on April 29, 1958, charged with treason, and thrown back into prison for another five years. Still Nehru continued to call Sheikh Abdullah his "friend," pretending to play no part in his unconscionable detention for almost eleven years.

Nehru kept up his busy travel schedule, flying abroad at least once a year and spending the first weekend of every trip to England at Broadlands. In January 1951, Nan accompanied him there, but Indira remained cool and aloof, preferring "*not* to go to Broadlands," as she told her father in May 1953, when he invited her to join them for the weekend.[27] Padmaja was then living with him in Edwina's bedroom at Teen Murti, "and will stay on, I suppose, as long as I am here," Nehru wrote his daughter, explaining how busy he had been since returning from a tour of Maharashtra. "Among other things, our honourable two Houses of Parliament nearly came to blows . . . over a relatively trivial matter. However, I have succeeded in pouring oil over these troubled waters!"[28] So his daily life went on, ever changing, yet always "a continuation of the old," much like Kashmir, much like all of India, whose philosopher king he had become.

"I have a feeling that another stage in our journey has been reached and a duty done—well done, if I may say so," he told Parliament, extolling the economic-planning process he had recently launched. "At the same time . . . a still more difficult duty is ahead of us. Another journey has immediately to be undertaken in which there are no resting places. . . ." No rest for him now that he longer went to jail. "The Planning Commission has worked very hard, very conscientiously . . . with a true crusading spirit in preparing this Plan. . . . They had to deal with a very difficult problem . . . with a complicated federal structure and with an economy which in many ways is very backward. . . . They had to keep in view the great ambition to progress rapidly, which we all share and, at the same time, to work with very limited resources . . . a stormy period of history, a period of trial and crisis and change. . . . We are dealing with India and not any other country. . . . The ultimate aim is economic democracy . . . to put an end to the differences between the rich and the poor, between the people who have opportunities and those who have very few or none. . . . Naturally, the Plan is not perfect."[29] Nehru never ceased to resort to such socialist and radical rhetoric, knowing how popular it was, especially on the eve of elections.

But wherever he looked, there were more problems. "Freedom is always accompanied by responsibility," he tried to explain to them, scolding a national conference of newspaper editors. "I want you to recognize and put before the public our great achievements in India. . . . Our fight will be relatively easy if our newspapers throw themselves into the fray on the right side."[30] He always felt rather cranky in India, except when Edwina came. She returned his visits every year on her Asian inspection tours of the St. John Ambulance Service. "Those were the only times Bhai recovered his youthful gaiety," his sister Betty recalled. "Nothing Edwina did was wrong . . . she was such a wonderful companion for my brother . . . always talked as a friend would, laughing and putting everyone at ease. . . . Every so often she would tell Bhai he didn't know how to cut his soft-boiled egg. That was a standing joke. He would say, 'Now don't tell me, Edwina. I know I do it wrong.' "[31]

Feroze, whose presence always reminded Nehru of his two greatest frustrations and failures, became more troublesome with time, more of a nuisance and misfit in his father-in-law's house. For a while he had managed the Nehru family newspaper, the *National Herald,* and was appointed the managing director of Associated Journals Limited, its parent company. Then he was elected to Lok Sabha (house of the people), India's House of Commons, in the first general elections of 1951–1952, from Rae Bareilly, one of the Nehru family's pocket-borough districts in Uttar Pradesh. He lived in a small government bungalow reserved for members of Parliament rather than suffer the humiliation of staying at Teen Murti, where his father-in-law treated him coldly and his wife, the nation's official "hostess," with contempt or disdain. Feroze sought female consolation elsewhere, in much-discussed affairs, one with a movie star in Lucknow, but Indira was never jealous. Rather, she seemed relieved to be rid of his importunate demands. Shortly before his death at the age of forty-eight, Feroze made a name for himself by rising in Parliament to expose embezzlement in an insurance business scandal that triggered the resignation of Nehru's trusted finance minister, T. T. Krishnamachari. A year later Feroze suffered the first of three heart attacks that finally killed him in September 1960.[32]

Indira now flew everywhere with her father. In 1953 they attended the coronation of Queen Elizabeth in London, driving from Buckingham Palace to Westminster Abbey in a royal gold carriage, joining the queen's procession inside with the other Commonwealth prime ministers. As the prime minister's consort, Indira participated in all public functions, meeting every world leader her father met, and she was privy to all his relations, public, private, and top secret. From London they flew to Bern, Switzerland, and then to Rome and Cairo en route home to Bombay. The next year they flew together to Burma and China, stopping briefly in Vietnam, Laos, and Cambodia. They were greeted warmly by Chou En-lai and met with Mao Tse-tung for more than an hour. "Mao didn't come to India," Indira recalled. "I was told later that he was waiting for an invitation. If only we had called him, many things would have been different . . . I liked Chou En-lai."[33] Nehru also liked him, spending three hours on his last night in China talking alone with Chou, mistakenly believing then—indeed, almost until the end of his own life—that "Hindi-Chini Bhai-Bhai!" (Indians and Chinese are brothers!), one of his sadly shattered fantasies.

Nehru's youthful memory of Moscow and the Soviet system also remained—if not untarnished in his mind, then something of a role model for India's economic development. He flew from Bombay to Prague with Indira, who inherited his love for many things Soviet, and to Moscow in June 1955. All of Moscow's leaders turned out to greet him, including Molotov.

Many Russian school girls waited to cheer him and Indu at the airport with bouquets of flowers as well as kisses. Marshal Bulganin's guard of honor snapped to attention for his review, and the Soviet Army band played India's national anthem. Crowds of workers and children cheered him wherever he went in the Soviet Union, from Leningrad to Tashkent and Yalta over the next busy fortnight, which included stellar evenings at the Bolshoi and Leningrad ballet. Nehru said, before boarding the plane to Poland, "I am leaving a part of my heart" in the Soviet Union. Five months later Nehru warmly toasted Khrushchev and Bulganin at a state banquet in India's Rashtrapati Bhavan, expressing "my deep gratitude to your Excellencies and to the people of the Soviet Union for their affection. . . . I saw in the Soviet Union mighty tasks undertaken and many accomplished for the well-being of the people. I saw, above all, the urgent and widespread desire for peace."[34] Indian–Soviet friendship and economic cooperation remained a major theme of Nehru's foreign policy and became an even more important pillar of Indira's policy. Her twenty-year Treaty of Indo-Soviet Friendship signed in 1971 effectively replaced Nehru's policy of nonalignment with an Indo-Soviet military pact.

Nehru's intransigence regarding Kashmir, his courtship of China, and his revival of Indian–Soviet friendship during this peak period of his power chilled Indian–U.S. relations to cold war lows. With John Foster Dulles as the U.S. secretary of state from 1953 to 1959, the United States' global policy focused on containing Soviet expansion and bolstering NATO. Pakistan was eager to join every alliance at Washington's behest and started to receive U.S. military aid in 1953 under the Mutual Security Act, soon to be linked to CENTO (Central Treaty Organization) and SEATO (Southeast Asia Treaty Organization), becoming the United States' "most allied Ally" in South Asia. Nehru, however, refused all such Washington overtures to join up and viewed every shipment of American planes, tanks, and guns to Pakistan as at least potentially aimed at India's "head" in Kashmir, if not at its heart in New Delhi. A year after his death, Nehru's worst fears came true when Pakistan launched a Patton-tank attack against Indian outposts in the Rann of Kutch and a few months later rolled into Kashmir.[35]

Nehru focused his foreign policy on trying to rally a Third World alternative to superpower confrontations. His articulate advocacy of peaceful, rather than nuclear, solutions to world conflicts in Korea and Vietnam enhanced his global reputation just at the time that his sister Nan, whom he had earlier appointed to lead India's delegation in New York, was elected the first woman to preside over the United Nations General Assembly. The Nehru family thus was identified in the minds of millions of people in the United States, the United Kingdom, the Soviet Union, China, all of Asia, and most of Africa with peace, freedom, and world progress. In honoring Nan with a gavel on U.N. Day that October, Eleanor Roosevelt stated that she had "lighted a lamp of understanding and goodwill in our hearts," and most of those who heard this tribute felt much the same way about her older brother.[36] Unfortunately, Nan's elevation left Krishna Menon to head the Indian delegation in New York, defending Kashmir against every rational attempt at peaceful solution. President Eisenhower addressed the General Assembly during Madame Pandit's presidency and in 1956 invited Nehru to Washington to help him avert global nuclear war in the aftermath of the Soviet invasion of Hungary and the Israeli–British–French attacks on Egypt in the wake of Nasser's nationalization of the Suez Canal.

Nehru was loudly criticized by many Indians in Lok Sabha for saying nothing immediately after the Soviet's crushing of Hungary's nationalist uprising, though he wasted no time in denouncing British–French support of Israel's intervention in Egypt. "In regard to Hungary, the difficulty was that the broad facts were not clear to us," Nehru replied from the floor of

Parliament, whereas in Egypt "we were in very intimate touch . . . [but] in regard to Hungary or Egypt or anywhere else, any kind of suppression by violent elements of the freedom of the people was an outrage on liberty."[37] A month later he and Indira flew to Washington and Vice President Richard Nixon escorted them to the White House, where President and Mrs. Eisenhower waited with a family lunch. Eisenhower drove Nehru down to his Gettysburg farm for two days of top secret talks that helped defuse the world conflagration and turned the tide of Indian–U.S. relations. Nehru was never again called "immoral" by Dulles for preaching "neutralism." His close association with Colonel Nasser of Egypt and Marshal Tito of Yugoslavia, moreover, gave him easy access to both Eastern European and Islamic revolutionary leaders in North Africa and Indonesia. Indeed, he attracted more than six hundred world correspondents when he spoke to the National Press Club in December 1956, and he was introduced as "one of the most important" men in the world.

At the age of sixty-seven then, after ten years in power, Nehru might well have chosen to retire gracefully, perhaps advising his Congress successors from the wings or simply withdrawing to the calm and quiet of Anand Bhawan or Broadlands, to write his memoirs of the years since he had revised his autobiography. Edwina's doctor had warned her, after she had suffered "mild angina" that year that "if she did not slow down she would be dead in three years."[38] But for those who ride the tiger, dismounting is never easy; and neither of them ever found a way. She lived only one more year than her doctor had predicted. Nehru's constitution proved itself four years stronger than that. But holding on until the bitter end hardly enhanced Nehru's reputation, nor did it really help India overcome the worst of its economic misery or emerge from the glacial crevass of Kashmir. The major beneficiaries of his tenacity were darling Indu, dear Krishna Menon, and grandson Rajiv. But only for a while.

31

Decline and Fall

1957–1962

NEHRU REMAINED Congress's premier political campaigner throughout 1957. Indira was catapulted that year to Congress's Central Election Committee, selecting candidates for Lok Sabha and campaigning for those most loyal to her father. But many of Nehru's old friends, to his left and right alike, now lost faith in him. J.P. no longer called him Bhai, and few of Nehru's former socialist comrades considered him much to the left of Winston Churchill. At the other end of the spectrum, C.R. viewed Nehru's stubborn refusal to listen to South Indian warnings about how dangerous it would be to replace English with Hindi-language examinations for central services in Tamil Nadu (land of the Tamils), as proof of his authoritarianism. Nehru's only truly loyal old friend was Krishna Menon, who collapsed on the floor of the Security Council after talking nonstop for nine hours in defense of India's position on Kashmir.[1]

"There are some people in this country and some people in other countries whose job in life appears to be to try to run down Krishna Menon," Nehru told his compatriots a few days after Krishna Menon collapsed, "because he is far cleverer than they are, because his record of service for Indian freedom is far longer than theirs, and because he has worn himself out in the service of India." Nehru found a Lok Sabha constituency for Krishna Menon in Bombay and campaigned hard for him, ensuring his victory, then made him India's minister of defense.[2]

In regard to Kashmir, Nehru continued to insist, as Krishna Menon had so exhaustively argued, that "the major consideration for me and for my Government has been the good of the people of Kashmir. Some people . . . say: 'Mr. Nehru who pretends to be a high moral figure, doling out moral advice to everybody in the world, forgets his own morality when he deals with Kashmir; he has a double standard.' Well, it is very difficult for me to know my own failings, but . . . I think that on moral issues India stands rather well over the Kashmir matter. . . . Fortunately we had Gandhiji with us at that time. I am not using Gandhiji's name to entangle him in this matter. I do so merely to tell you that as usual I ran to him for some advice and light. . . . He told us it was our duty to go to the help of the people of Kashmir. He, a man of peace, told us so. We went to Kashmir and we found that it was not an attack by mere raiders only, but that the Pakistan Army had entered Kashmir. . . . They had come in and had committed aggression."[3]

Nehru flew to Scandinavia that summer and in Stockholm explained India's foreign policy: "I want neither Western nor Eastern domination. If anybody tries to interfere with us we may be terribly weak, but nobody is going to succeed against us in spite of atom bombs. . . . If we have no weapons we will fight with sticks, but will not submit to being bullied by anybody."[4] A month later he responded to questions in Parliament about his Atomic Energy Department's budget demands. "The subject is naturally one which excites the imagination of everyone, and there is a feeling that . . . we should not lag behind . . . we have no intention of lagging. . . . [T]he fact remains that the development of atomic energy work in India has been remarkably rapid and good. . . . I may inform the House that nobody in the Government of India, anxious as we are to economize and save money, has ever refused any urgent demand of the Department or come in the way of its development. . . . The Prime Minister's being in charge merely shows how much importance has been given to work on atomic energy."[5] Three years earlier the Atomic Energy Commission had been upgraded to a department.

During his June month on tour in Scandinavia, Indira reported to him that "the familiar green-eyed monster jealousy" had begun to rear its head in opposition to his personal expenditures and powers of appointment.[6] Nehru had started to build his own secluded prime minister's retreat in Simla, and as Indira reported, "it seems now that our motives for building the small house are in doubt."[7] The estimates had grown much higher than anticipated, but Nehru wrote Indira from Oslo, before he had received her cautionary warnings, "In the new house that is being built for us, I think the study room should have built-in bookshelves, as many as possible."[8] Edwina's sitting room at Broadlands had many such built-in shelves. No matter how busy he was, Nehru's mind could always focus on fine points of detail where his pleasures were concerned. To some of his closest Congress colleagues, however, that predisposition seemed wasteful of their prime minister's time or was even excessive personal interest and vanity. Others said he was simply tired of the job or bored, which he usually was.

Nehru flew from Stockholm to London for the Commonwealth Prime Ministers' Conference in the last week of June. On his first night in London he met Queen Elizabeth and Prince Philip and Dickie and Edwina. The next day he called on Prime Minister Harold Macmillan at 10 Downing Street and returned to Buckingham Palace for another evening with the royal family, with whom he now felt quite at home. Nan was now his high commissioner in London, and Defense Minister Krishna Menon had flown over to brief him on their military hardware needs to keep Kashmir's people happy. But Krishna Menon was never happy with Nan, who complained to her brother about him, but as Nehru wrote to her: "If I speak to him, he [Krishna] has an emotional breakdown. He is always on the verge of some such nervous collapse." [9]

"Our Embassy residence in Kensington Palace Gardens . . . was a beautiful house," Nan recalled. "The main salon was paneled with boiserie from one of the famous French chateaux on the Loire. It was completely French. . . . There was an Italian fireplace in the dining room and two authentic Adam fireplaces in the bedrooms. Krishna had not wanted to live in this house and had had a bathroom added to a small room behind his office at the Chancery. This gave the image of austerity he wished to create about himself." [10] Nan loved the residence at 9 Kensington Palace Gardens, however, as did Nehru, who stayed there during his London visits and enjoyed pointing out to Edwina its Loire architectural charms and Adam fireplaces. "I have now been here for two days," he wrote Indira from the residence that June. "Tomorrow evening I go to Chequers for the night. The next day I shall proceed to Broadlands for another day and night. . . . It is very warm here, almost like early October weather in Delhi. Last night I had to go to Windsor for a banquet there. . . . I must say I was impressed by the collection of treasure there. The rooms are very gorgeous indeed." [11]

Nehru could not bring himself to return directly to Delhi and so stopped for three days in the Netherlands on his way back in July. He said nothing negative about Dutch imperialism in Indonesia, instead seeking Dutch assistance in a number of economic development fields. He then spent several days with President Nasser and flew to the Sudan from Cairo, where he received a "hero's welcome" from thousands of chanting tribals in Khartoum. Nehru flew home in mid-July. Indira had planned to be there to greet him, but remained on holiday in Kashmir, where she had been enjoying a few weeks of sunshine with her boys, after a month of heavy political touring.

Soon after returning to Delhi, Nehru addressed his upper house of Parliament on Kashmir. "Hon. Members will remember that when this matter came up before the Security Council last year, we had to face a very considerable opposition," he said. "It was an astonishing opposition. . . . I have seldom come across anything so astounding as the attitude of some . . . great powers. . . . [T]hey passed a resolution about the accession of Kashmir not taking place and nothing being done with regard to it on the 2th January, 1957. They were told repeatedly that the accession of the State . . . had taken place in October 1947. . . . They talk about a plebiscite. Again and again we have pointed out that . . . the first thing to be done was for Pakistan to get out . . . and until Pakistan goes out nothing else is going to be done. Instead of going out, Pakistan has entrenched itself. . . . Pakistan has got enormous aid from the United States. . . . I should like our friends concerned to realize how by some of their policies of military alliances and military aid, they have added to the burdens of India a feeling of insecurity and thereby come in the way of our working out our Five Year Plan." [12]

The second five-year plan had run out of funding and had to be abandoned in midcourse, which Nehru now blamed on U.S. aid to Pakistan. He also spoke in Lok Sabha on Kashmir, in April 1958, trying to explain why he had totally rejected the efforts of Dr. Frank Graham, the Security Council's third mediator on Kashmir, who had labored as valiantly but as futilely as did Owen Dixon from 1951 to 1953, on a plan for the demilitarization of Kashmir that would have allowed an impartial plebiscite to be held there. Pakistan was receptive, but India refused to consider the idea. Now Graham published his report, which Nehru found unpleasant reading. Graham suggested that both prime ministers meet with him to resolve their differences over Kashmir. "It places us in a position of, let us say, equality in this matter with Pakistan," Nehru protested. "We have always challenged that position. Pakistan is an aggressor country. . . . Secondly, for the two Prime Ministers who meet, it would almost appear as if they have to plead with Dr. Graham . . . as advocates. . . . This kind of thing does not lead to proper consideration of problems." [13] Nehru never liked pleading, nor did he recognize the standing of any U.N. mediator. His mind had long since closed down on Kashmir, and he no longer bothered so much as to pay lip service to his ancient promise to agree to holding a plebiscite there.

China's rapacity in Tibet forced the Dalai Lama and many of his followers to flee from Lhasa to India's Mussoorie on the last day of March 1959. The following month Nehru reported to Lok Sabha on his visit and long talk with that most revered Buddhist priest, to whom he gave Indian asylum. China never forgave Nehru for sheltering the twenty-four-year-old Dalai Lama, contributing to China's invasion of India three years later.

Mao and Chou and their Tibetan puppet, the Panchen Lama, charged India with kidnapping the Dalai Lama and keeping him prisoner in Mussoorie. They also hinted that the Dalai Lama was lured away by C.I.A. agents, who had infiltrated his order. "In these days of the cold war, there has been a tendency to use unrestrained language and often to make wild charges without any justification," Nehru wisely told his Lok Sabha the following month. "The matter is too serious to be dealt with in a trivial or excited way. . . . As the Panchen Lama had made himself . . . some strange statements, I have stated that we would welcome him to come to India and meet the Dalai Lama. . . . I have further said that the Chinese Ambassador or any other emissary of the Chinese Government can come to India for this purpose." They never accepted his offer, of course. The Chinese charged that India's northern border town of Kalimpong, near Darjeeling, was a center of Tibetan rebellion, a charge Nehru rejected as wholly unjustified.[14]

In his last frank and full talk with Chou, Nehru explained, the Chinese premier had spoken of Tibet as an autonomous region and said it was absurd to "imagine that China was going to force communism on Tibet."[15] Nehru had believed Chou to be sincere and called him "Brother," embracing him warmly when he had come to Delhi. Three months later a unit of Indian border police were captured by Chinese troops in Ladakh, "inside our territory," as Nehru told Parliament. The Chinese had established a camp at Spanggur, and when India "lodged a protest with the Chinese Government . . . the Chinese claimed that that part of the territory was theirs."[16] Nor had the Chinese ever recognized the McMahon line, which since 1914 had demarcated the international border farther east between British India and China, but no signposts or fences had ever been erected in the snow and ice of Ladakh. The Indian maps, inherited from the British cartographers who made them, showed Ladakh well within India's northern border, as part of Kashmir. But the Chinese believed their own maps. At this time the Chinese released their Indian prisoners unharmed, but now Nehru noted that they had also built a road across northern Ladakh, linking Tibet to China's Sinkiang Province.

"Does it mean that in parts of our country which are inaccessible, any nation can come and build roads and camps?" one shocked member of Lok Sabha asked his prime minister. "We just send out parties, they apprehend the parties and . . . [t]he road remains there, the occupation remains there and we do not do anything about it."[17] "I do not know," Nehru replied softly, equivocally, clearly growing more uncomfortable as the debate continued. "In regard to some parts of the border . . . it is rather difficult to say where the immediate border is. . . . [E]ven in a map . . . a big line is drawn, that line itself covers three or four miles. . . . Then there are parts still where there has been no demarcation. . . . Nobody was in that area." Many members were on their feet now, wanting to know what positive steps India was taking to secure its northern borders. Nehru's weak response was that "the border is 2,500 miles long."[18] Much of it remained unprotected, unguarded, unexplored, and uninhabited.

So began India's "cartographic war" with China. Neither Nehru nor his minister of defense, Krishna Menon, had any fears of Chinese aggression, for China was not Muslim Pakistan, after all. China was Communist, Tibet Buddhist, and to Krishna Menon and Nehru, Communists and Buddhists were India's brothers. But the Chinese kept sending more armed guards into the area, and members of both houses of India's Parliament became more and more alarmed, their questions to the prime minister–foreign minister growing more urgent and angry. "We have to face here a particular situation," Nehru responded vaguely. "There is no alternative for us but to defend our country's borders and integrity. Having said that, . . . we must not, as often happens in such cases, become alarmist and panicky."[19] Two weeks later Chinese border troops opened fire on Indians at Longju, and Chou En-lai charged India with aggression in crossing over China's border there. "I would venture to say that there appears to me to be a lack of understanding or recognition in China of the revolution in India," a weary and wary Nehru told his upper house. "But nothing can be more amazing folly than for two great countries like India and China to get into a major conflict and war for the possession of a few mountain peaks."[20]

Indira was elected Congress president in February 1959. She kept busy trying to prevent party loyalists from losing faith in her Papu, but the pace was exhausting, and the frustrations and compromises frazzled her nerves to the point that when she was urged to continue in that post for a second term, she politely but firmly refused. "Many reasons, all equally valid and cogent, can be given for my wishing to discontinue," Indira wrote him that October. "Some people are attaching importance to my presiding over a Congress session. . . . To me it has no special meaning or attraction. . . . I felt a burden on me. . . . I have worked harder and longer as the years went by, always feeling that I could never do enough. . . . By the time I became President, I just was not in the mood for this sort of work and I have felt like a bird in a very small cage, my wings hitting against the bars whichever way I move. . . . For the moment, I just want to be free as a piece of flotsam waiting for the waves to wash me up on some shore."[21] She enjoyed more real power and leisure, Indira well knew, as his hostess than as Congress president and, like her father in his earlier days, found the Congress crown to be one more of thorns than roses.

In 1958, Edwina had a growth in her parotid gland surgically removed, causing her face to sag "at one side, as if she had had a stroke . . . people stared at her lopsided cheek and drooping eye." Adding illness to that surgery, she caught chickenpox from a grandchild in January 1959, but flew to Delhi as soon as she recovered, seeking comfort from Jawaha.[22] He had little solace to offer her in his seventieth year, with so many angry representatives of his people snapping at him in Parliament and China moving down from the north with such glacial intransigence. Although she had flown to Delhi for a complete rest, Edwina returned home in late March feeling tired.[23] And by the end of June she wrote to her beloved sadly to report that she was still "not so good."[24]

To compound their frustrations, Dickie and Edwina's daughter Pamela was to be engaged at Broadlands that mid-November, so Edwina could not possibly fly back to Nehru for his birthday but delivered a charming tribute to him at a royal dinner for the young couple in London on November 16. He had, of course, been invited, but the frustrations of their last few trysts and the toll that age had taken on them both persuaded him that his many public commitments in India must take priority over any private urgency of flying to London. They both knew it was too late for either of them to escape their societal and political destinies. Not even Nehru, vivid though his romantic imagination remained, could any longer indulge in such fantasy. Still, Edwina, desperate, tried one last time. She flew to India in January 1960, her first free week since the previous March, and reached Delhi just when Soviet President Voroshilov landed to begin his first official state visit to India. Nehru was there with flowers for both of them. His old friend, the new mayor of Delhi, Aruna Asaf Ali, came with him to greet the Soviet president.

On February 5, 1960, Edwina left India for the last time. Jawahar took her to the airport, of course. She was headed east, the last lap of her mission, first to Malaya and then to north Borneo. She flew into Jesselton on the eighteenth, mortally ill, burning up with high fever and a splitting headache, and heartsore at the realization that she would never see Nehru again. She excused herself early on Saturday evening, February 20, and alone in that remote British Borneo residence bedroom, Edwina Lady Mountbatten "died in her sleep," rereading Nehru's love letters, which lay scattered about on the floor at her bedside where they had fallen from her fingers.[25] Her body was flown to Singapore the next day and then on to London, reaching Heathrow Airport on February 24, 1960, taken directly to Broadlands, and kept for half a day at Romsey Abbey, where the family and friends could pay respects. Dickie had hoped she would agree to be buried in the family crypt in Romsey, but Edwina's will requested that her remains be set free in the open sea, so her coffin was taken to Portsmouth on February 25, the tiny body released from the British navy's *Wakeful* into the wind-whipped waters onto which the Indian frigate *Trishul* dropped its prime minister's farewell garland of marigolds, sent to adorn her body, as every good Hindu garlands his goddess.

But Nehru shed no tears and showed no emotion. His face was "expressionless and self-contained," Marie Seton recalled, watching him enter Delhi's Sapru House to introduce Arnold Toynbee, who lectured there that Sunday evening, when a mournful Indian capital learned of Edwina's death. Four days later, Kingsley Martin's wife, Dorothy Woodman, dined with Nehru alone and told Marie how "surprised" she was to note with what "detachment" Nehru spoke to "Lord Mountbatten on the phone to England" that evening, the first time they had talked since Edwina's death. He had risen to his feet in silence at Sapru House with all the rest of them, as he did next day in Lok Sabha, but he canceled none of his appointments, having known in his heart and head long ago that she was lost to him, their lovely affair no more than a fragrant memory.[26]

That April Nehru flew off again to London for the Commonwealth Prime Ministers Conference. A colder prospect awaited him there, without her, and this was his first trip abroad in many years without Mac as well. Almost like Edwina, Mac Mathai had grown to be so intimate a part of his life that countless others—who wished they could take his place as Nehru's private secretary—spread wild rumors about him, from charges of petty theft to spying for the C.I.A. Nehru had long dismissed such talk as the product of "small-minded men," as he called those rumormongers, but the chorus grew louder and the documents sent up to his government's cabinet secretary grew weightier and the "top secret" file became thicker. Finally, after charges of corruption were made against him in Parliament, Mac offered to resign in early 1959. The then cabinet secretary, V. Sahay, quietly informed Nehru that "Mathai could not account for his great wealth and without doubt had received money from the C.I.A. as well as from businessmen in India."[27] Mac denied any receipt of improper or excessive funds, blaming "some Communists" for the "virulent attack on me."[28] He sent Nehru a written account of his personal assets, reminding him that he had worked without salary for some time until "you asked me to work with you in Government. . . . [B]eing a bachelor, I had enough to live on and I was not in need of paid employment. Since you thought that my joining Government would facilitate your work, I agreed to do so without payment. But you did not, as a matter of principle, approve of my not taking a salary." Poor Mac had become the lightning rod for several radical newspapers and journals, which found it much easier to attack him than the prime minister. So Mac resigned, closing his last letter to Nehru with "my love to you as always." Nehru silently accepted both.[29]

"I arrived here . . . a couple of hours ago," Nehru wrote Indira from 9 Kensington Palace Gardens on May Day, 1960. Both Nan and Betty were there with him, helping make that ambassadorial French chateau with its Adam fireplaces in the heart of London seem more like Anand Bhawan in the old days when the ghost haunted the gardens. He called on

Dickie next day in another house haunted by a ghost so recently buried at sea. Then he went to Harrow on May 3 for a special concert of the old school songs, sung in his honor by the entire school body, assembled in the grand Speech Room where his full-length painting still hangs, facing Churchill's, and he had tea with the headmaster, R. L. James, in his study in the very house in which he had first felt so chilly and alone. Finally, Nehru was driven from Harrow to Windsor Castle, where he dined that evening with Queen Elizabeth and all the other prime ministers of the Commonwealth, just as Motilal had once dreamed and hoped he might, many long years ago. The next day he spent at 10 Downing Street and in Buckingham Palace, for the queen's party and dance on the eve of Princess Margaret's wedding.

"I was glad to read of your visit to Harrow," Indira wrote him from Delhi, reporting that "the state of the Congress & Parliament is deteriorating. Firm guidance and attention to detail is imperative but you are not feeling fresh & alert enough to give it & your programme is so crowded with appointments & tours—good in themselves but secondary in importance to the other things. So many good ideas and programmes are being channelled in the wrong direction, and there is so much drift in our political life, only because no one has the time or inclination to keep an eye on them."[30] She was right but was about to fly off to London herself. In June she went to Srinagar with the boys, and he planned to join them there in July. "The real object is to go to Leh in Ladakh," he explained. "I would then go, let us say, for two days to Srinagar and one or two days to Ladakh. On my return journey, I should like to pay a visit to Kishtwar for a day or so. That would probably mean coming back by car up to Jammu."[31] Always restless. Never at ease with himself any longer, always planning to break away.

Delhi's political temperature was also rising that blistering June of 1960, with black-turbaned and -bearded Sikh Akali (Immortal) Party cadres marching, shouting, and demanding a Punjabi-language state for Punjab, which devout Sikhs alone would run. Nehru adamantly opposed the Sikhs' demand. Although he had just approved redrawing India's provincial map elsewhere along linguistic lines, he remained hostile to any partitions based on religion, viewing the popular Akali Sikh demand for Punjab a painful echo of Pakistan. "The Akali demonstrations . . . were a big show yesterday," he wrote Indira on June 13. "Our police behaved rather well and with great restraint. The Akalis were very violent . . . stone-throwing from houses, and especially from the Sisganj Gurudwara [largest Sikh temple in Old Delhi]. . . . [T]here was no firing and not much in the way of lathi charges. Tear gas was used. . . . In spite of the violence and aggressiveness of the Akali crowd, the police held them. . . . I hope you are having a good time."[32]

Many militant Sikh leaders now shouted that to win Sikh support against Pakistan before British India's partition Nehru had falsely promised the Sikhs their own Punjabi state. But once he secured power, he betrayed those promises, jailing Master Tara Singh and putting his own bright but faithless Sikh chief minister, Pratap Singh Kairon, in charge of Punjab's swift green-revolution provincial development. Since 1857, Sikh "lions" *(singhs)* had been the "sword arm" of British India and remained the bravest of India's soldiers, guarding its Punjab border with Pakistan. Indeed, the First Sikh Battalion had saved Kashmir for Pandit Nehru. But by 1960 more Sikhs recalled the question asked after partition by the disillusioned Tara Singh, and to some, the demand for a Punjabi state was nothing less than a national battle cry for a separate Sikhistan or Khalistan (land of the pure).

On the eastern extreme of north India, in Assam, the Nagas also demanded a state of their own, and many Muslim peasant émigrés from East Pakistan who had settled in that lush and sparsely populated region were violently expelled by angry Assamese tribals, a struggle destined to continue for at least thirty-five more years. At first, Nehru resisted the Nagas' demands but finally agreed to appoint a committee to look into the matter. Three years later a separate Nagaland emerged as India's sixteenth state. "It is amazing how all higher considerations are swept away when communal passions are roused," Nehru declaimed in Lok Sabha in early September 1960. "It is not only the Assamese or the Bengalis who are guilty in this regard; each one of us is a guilty party . . . each person's idea of nationalism is his own brand. . . . When two brands of nationalism come into conflict, there is trouble. We live in a closed society—not one closed society, but numerous closed societies . . . a Bengali closed society, a Marathi closed society. . . . When we bring in democracy and open the door of opportunity to everyone this narrow outlook brings about group conflict. . . . What is communalism itself? You may well have described Hindu communalism as Hindu nationalism and Muslim communalism as Muslim nationalism, and you would have been correct."[33]

More than a decade in power had taught him the accuracy of many of Jinnah's arguments. Nehru no longer blamed all of India's communal conflicts simply on British policies of divide and rule or Marxist economic forces. "In Assam . . . evil men flourish on occasions like this because they are in tune with the mind of the multitude. . . . The language question to the Assamese is only a symbol of this mind. . . . It goes above reason, and becomes an article of faith. When this happens, it is relatively easy for it to be exploited for wrong ends. . . . It is a very grave tragedy for people in one State to be driven out either by force or through sheer panic. Panic is so infectious that it is difficult to deal with. . . . It is a symbol of our weakness, of our failings . . . of our incapacity to function together and of a tendency to go to pieces. . . . In this matter everyone of us has to blame himself, and I include myself."[34] The horrible aftermath of partition refused to die, another ghost that continued to haunt his waking hours.

Nehru flew off again on September 19, to Karachi, where he signed a comprehensive Indus Valley canal waters agreement with Pakistan's military president, Ayub Khan, negotiated through the International Bank for Reconstruction and Development, which funded the construction of many high dams and power plants on both sides of the Punjab's international border. That week in Pakistan Nehru met for many hours with Ayub, whose coup two years earlier had swiftly ended Pakistan's political debates over what sort of constitution it wanted. Nehru also met with Ayub's fast-rising youngest minister and soon-to-be foreign minister, Zulfikar Ali Bhutto. Nehru had met "Zulfi" years before in Bombay, and they would meet again in New York during Nehru's last visit to the U.N. late in 1962.[35] Nehru flew back to Delhi from Lahore on September 23 and left the next night for New York, where he met again with Eisenhower and Khrushchev, as well as Tito, Ghana's President Nkrumah, Cuba's Castro, Nasser, and Sukarno. His policy of nonalignment gave him easy access to all heads of state, and many of the new African, revolutionary Latin American, and Asian leaders looked up to him as their mentor in diplomacy.

"Without peace all our dreams vanish and are reduced to ashes," Nehru told the General Assembly that October. "The main purpose of the United Nations is to build up a world without war, a world based on the co-operation of nations and peoples. . . . [U]ltimately it is necessary to bring about the change in our minds and to remove fears and apprehensions, hatreds and suspicions. Disarmament is a part of this process. . . . But it is only a step towards our objective. . . . But the sands of time run out. . . . We live in an age of great revolutionary changes brought about by the advance of science and technology. Therein lies the hope for the world and also the danger of sudden death. . . . [I]f we aim at right ends, right means must be employed. Good will not emerge out of evil methods. That was the lesson which our great leader Gandhi taught us, and though we in India have failed in many ways in following his advice, something of his message still clings to our minds and hearts."[36]

Nehru flew home via London and Bonn, meeting with Chancellor Konrad Adenauer, and on to Cairo, where he also touched down before returning to Bombay. None of India's communal, regional, economic, or social problems were solved during his long trips abroad, nor did any of the internal Congress Party disputes and grand or petty matters of corruption that continued to plague India's public life disappear. Indira had ventured off on her own to Mexico, returning to New York in early November, just in time to wait up to watch John Kennedy defeat Richard Nixon in the U.S. presidential election. Nehru went to Raipur for the A.I.C.C. meeting there at the time, reporting to Indira on Congress's interminable "squabbles."[37] Parliament was scheduled to launch its winter session on his birthday, and Padmaja would be coming for that seventy-first celebration, bringing as she always did, his largest cake. Burma's Prime Minister U Nu and Madame Nu also came for his party.[38]

President Kennedy sent one of Harvard's most distinguished economists, Professor John Kenneth Galbraith, to serve him as America's most scintillating ambassador to India. Galbraith and Nehru had Cambridge in common and quickly became friends despite many diplomatic differences, from Kashmir and Goa to Vietnam and China. They both were much too intelligent and sophisticated to let diplomatic differences undermine their mutual personal admiration, and Galbraith helped save Nehru and India from China's massive invasion little more than a year after he settled down in Delhi. Galbraith's lovely young wife, Catherine, also charmed Nehru from their first formal introduction. "There were about a dozen guests," Kitty recalled of that luncheon at the prime minister's house, "nearly all members of the Nehru family, to welcome us to our post. When the Prime Minister came in, he walked over to me and said, 'I have heard you are shy.' He then set out to put me at ease, even to cutting my mango at dessert. . . . After that we saw him often."[39]

Nehru returned to London for the Commonwealth conference in March and also attended a "nonaligned" conference in Tito's capital of Belgrade, to which he brought Krishna Menon, and his cousin R. K. Nehru, the secretary-general of his Ministry of External Affairs. "We call ourselves nonaligned countries," Nehru told that Belgrade meeting in the midst of the latest cold war crises in Berlin and Laos on September 2, 1961. "But if we give it a positive connotation it means nations which object to lining up for war purposes, to military blocs. . . . [W]e want to throw our weight in favour of peace. . . . I know that the key to the situation . . . lies essentially in the hands of the two great powers. . . . The power of nations assembled here is not military power or economic power; nevertheless it is power. Call it moral force. It does make a difference. . . . I am amazed that rigid and proud attitudes are taken up by the great countries as being too high and mighty to negotiate for peace. I submit that it is not their prestige which is involved . . . but the future of the human race. It is our duty and function to say that they must negotiate. . . . [T]he only possible way to solve many of these problems ultimately is complete disarmament."[40] Nehru's moral prestige had never been higher, a voice for the fears and conscience of humanity. From Belgrade Nehru flew to Moscow, where he met with Khrushchev. His mission for the nonaligned nations was to try to bring Khrushchev and Kennedy together to discuss their differences, seeking to lower the level of escalating confrontation around the world.

Ambassador Galbraith arranged for Nehru a November meeting with Kennedy. Nehru reached New York on the sixth, accompanied by Indira and his second cousin, B. K. Nehru, India's new ambassador to Washington. President and Mrs. Kennedy and Galbraith picked up Nehru and his party at Newport, taking them on the Kennedys' yacht *Honey Fitz* to their house at Newport. The long flight had wearied Nehru, but he perked up as soon as he saw Jackie and was most excited by the prospect of her imminent visit to India with her lovely sister, both of whom he invited to stay in his house, in the suite that Edwina had always occupied. But Kennedy found Nehru so unresponsive in their talks—which for the most part turned out to be Kennedy monologues—that he later rated his summit with Nehru as "the worst State visit" he had ever experienced.[41] Nehru's age and reluctance to "open up" in Washington proved most frustrating to his young host, who also found infuriating Nehru's focus on his wife and his inability to keep his hands from touching her.[42] Galbraith wrote of Kennedy's longest meeting with Nehru: "Nehru simply did not respond. Question after question he answered with monosyllables or a sentence or two. . . . [T]he President found it very discouraging."[43] Nehru's wisest advice to Kennedy was ignored, however, for he tried to make it clear that the United States should not send soldiers to Vietnam.

Immediately after Nehru's disappointing visit to Washington, he flew to New York, where he met with Governor Nelson Rockefeller. "I asked Nehru later how they got along," Galbraith recalled, to which Nehru responded: "A most extraordinary man. He talked to me about nothing but bomb shelters. Why does he think I am interested in bomb shelters?"[44] Indian–U.S. relations continued to fall even faster that December, shortly after Nehru flew home, when India's army closed in on Goa, taking that oldest Western enclave, which Portugal stubbornly considered one of its own provinces, after a single day of almost bloodless combat. It was not the seventeen Portuguese citizens killed that elicted so shocked and angry a response from the United States' U.N. ambassador, Adlai Stevenson, but India's resort to violence after so much high moralizing against it. Nehru's Minister of Defense Krishna Menon ordered the Indian Army to take the sort of action he had so righteously condemned on the part of Pakistan, insisting "action" had to be taken to prevent much more "dangerous violence" from impatient Goan Nationalists in India. Opposition members of Parliament charged that Aruna's jingoistic radical journal *Link* was paid for by Krishna Menon, who ordered thousands of copies for Western India's army bases surrounding Goa.

Galbraith had tried his best to derail the Indian attack. His long and eloquent pleading with Nehru may have postponed the army's move for a few days, but Nehru's mind was made up on Goa at least several months before his army marched in in late December. "Because Portugal has become a part of military alliances like the NATO," Nehru told an Indian Council Seminar on Portuguese Colonies in New Delhi that October, "Portugal gets special consideration sometimes from the big countries. . . . Perhaps very recently it is dawning on some of the big powers that they might be backing the wrong horse. They ought to have known that long ago. . . . Historically speaking, colonialism is a past phenomenon. . . . [T]he force which builds up the colonial domains is over. The odd thing is that today Portugal which is . . . a backward country socially, economically, politically . . . represents the biggest colonial empire. And it continues to try to hold on to this empire in spite of its amazing record of backwardness. . . . As you know, we have a little bit of the Portuguese colony in India. It is almost a dot in size. Yet it has created a strong feeling in India . . . it has occupied our mind a good deal. . . . All these years we were thinking not merely of solving a problem in the immediate present but solving it for good. . . . At no time did we in our minds . . . renounce or give up the possibility of military action. But we did not want to resort to it. . . . We were prepared to wait, as we have waited very long. . . . If we have to take some other action, we shall take it."[45]

Major General B. M. (Bijji) Kaul, another of Nehru's Kashmiri relatives, had been promoted and put in command of the Goan operation by Krishna Menon, over the strongest professional objections of the chief of army staff, General K. S. Thimayya. General Thimayya, like several other first rate Indian generals, had little more than contempt for Krishna Menon and tried his best to warn Nehru before it was too late of how incompetent a minister of defense he had appointed. But Nehru listened to no such criticism of Krishna Menon, or at least never acted on it, until after the disastrous Chinese invasion a year later. By then it was too late. Galbraith saw the prime minister on the eve of India's attack, asking Nehru to announce the suspension of any Indian action and promising that Washington was ready to "do something to bring the Portuguese around. . . . But, in the course of the discussion, it became plain to me that the zero hour had passed. [Nehru] cited more newsprint atrocities by the Portuguese that morning. . . . He said disorder was imminent and thousands of Indian volunteers were waiting to march on Goa." Saddened by that dark glimpse into the future, Galbraith concluded, "It is remarkable how little people want to be saved . . . the Indians have badly tarnished their reputation."[46] President Kennedy, who expected less of Nehru, was less disappointed by his sudden leap from the high ground of moral principle and inspiring advocacy of peace into the muddy pit of battle.

"On New Year's Day, 1962, I went to see Nehru," Marie Seton recalled. "What struck me the moment I saw him, was his unhappy and glum expression. A delegation of Goans had come to thank him. Yet there was no joy in him, only a struggle not to display too great a feeling of depression. . . . 'People have become more brutal in thought, speech and action,' " Nehru told his old friend sadly. " 'All the graciousness and gentleness of life seem to have ebbed away. . . . I am not saying that everything worthwhile is completely destroyed, but I do say that the process of coarsening is going on apace all over the world. We are being coarsened and vulgarized because of many things, but chiefly because of violence and the succession of wars.' "[47]

That March, in Lok Sabha, Nehru introduced the Twelfth Constitutional Amendment, fully integrating Goa and two other tiny former Portuguese enclaves into India. "When long ago we started our movement for independence . . . [w]e took it for granted that when British rule ceased in India, the other enclaves would also be freed. We never thought that there would be any difficulty about them. . . . [D]iscussions with the French . . . took a few years to settle. . . . We tried to do the same thing with the Portuguese . . . but they refused. . . . Portugal . . . maintained that Goa was Portugal, in fact. . . . Certainly we could not accept that position . . . I gave them notice. I gave them and other countries notice . . . rather suddenly our hands were forced. . . . [T]here was some firing on Indian shipping carrying on in the normal way and there were some actual incursions from the Goan territory into India proper. . . . We thereafter took steps and sent some military forces there. They hardly functioned in a military manner, and within 24 hours or 36 hours the whole action was over. . . . [T]he people of Goa welcomed the Indian forces." So ended the last phase of Western rule in India, which had also been its vanguard four and a half centuries earlier, and Nehru proudly informed Parliament that "the independence of India has become complete."[48]

Mrs. Kennedy and her sister Lee Radziwill flew into Delhi on March 13, 1962. Nehru was waiting with flowers at the foot of the Air India ramp as Jackie, "looking a million dollars in a suit of radioactive pink," Galbraith noted, descended for two weeks of "great fun" in India.[49] Nehru planned to send two small tigers—his "favorite breed of pet," Kitty recalled—to the White House for John and Caroline Kennedy, but soon after Jackie flew away his tiger pets died.[50] Was it an omen of their master's fate? Unlike his mother, of course, who had read so much into the death of her king cobra, Nehru was never really a superstitious person.

Yet just a few days after the beautiful Jackie Kennedy flew out of his life, headed for Pakistan, he almost collapsed on the floor of Lok Sabha, running a high fever. He had a kidney infection that kept him confined to his bed from March 30 through April 7, 1962. It was the first of a series of illnesses that plagued him for the last two years of his life with periods of pain, on the eve of India's most devastating foreign attack from the north.[51]

32

Good Night, Sweet Prince

1962–1964

MANY CIVIL AND military leaders in Delhi, except for Krishna Menon, were anxious about the increasingly powerful Chinese presence on India's northern border. But the massive medication prescribed for Nehru's painful kidney disease had distorted his face almost beyond recognition and left him so lethargic in step and speech that he seemed unaware of most of the problems growing around him in that summer of 1962. He rallied enough to fly off to London with Indira, however, on September 7 for the Commonwealth Prime Ministers Conference, which remained the high point of his diplomatic year. It was in London a week later that he was obliged to confirm news reports of a Chinese "incursion" into India's North East Frontier Agency (NEFA). Nehru now invited Krishna Menon to join him in London and then sent him on to New York to serve as his voice in the U.N. during the Cuban missile crisis. "So if the Chinese decide to take Delhi," ever-trenchant Galbraith noted, "they will find few important captives."[1]

Shortly before leaving for London, Nehru assured his upper house of Parliament that "the one border which we protected more or less adequately was the NEFA border."[2] A month later that optimistic assessment was shattered by Chinese fire. Before flying out of India, Krishna Menon ordered his new army chief of staff, General P. N. Thapar, to "hold firm" but not to "fire except in defense." Nehru was examined by the queen's personal physician in London and then flew to Paris on September 20 with Indira to help support him and his cousin R. K. Nehru to handle the diplomatic talks with France's foreign minister, Couve de Murville. President Georges Pompidou welcomed Nehru, calling him "a great statesman." But Mao now claimed more of NEFA for China, and Chinese soldiers kept moving in on all the Indian outposts along the entire northern tier. Nehru was in Paris when Indian patrols fired back at Chinese invaders for the first time. "We cannot negotiate as long as daily acts of aggression are committed," he told the press, still trying to play down the grave dangers ahead, calling what happened only "petty conflicts between patrols."[3] Krishna Menon, of course, echoed his words in New York.

From Paris Nehru flew to Nigeria and confidently informed the press that if the Chinese pushed into India, they would be ousted by force. Leaving Lagos he flew to Rome, where he met with President Antonio Segni, and went on to Cairo the next day, where he talked with Nasser about nonalignment. President Radhakrishnan and Vice President Zakir Husain—destined to be India's first Muslim president—were at Palam Airport to welcome Nehru home. Krishna Menon also was there, looking almost as worried as he felt, as was tiny Lal Bahadur Shastri, Nehru's right hand in Uttar Pradesh in getting out the vote. He was now home minister, soon to be India's next prime minister, but only very briefly, for though his spirit was mighty, Shastri's heart was very weak.

The dire reports Nehru received from Krishna Menon convinced him that the Chinese were no longer India's brothers but a new imperialist menace that only looked Asian and dishonestly claimed to be communist. Nehru thus ordered two divisions of troops sent to NEFA, under the command of no less trusted a general than his cousin B. M. Kaul, promoted to commander of the Eastern Corps. Nehru gave Kaul his marching orders in early October and a few days later flew with Indira south to Ceylon, informing the assembled press that he had ordered the army to push the Chinese out. "The sorrow clouding Nehru's face as he came on to the plane was the saddest thing I ever saw," Marie Seton recalled.[4]

None of the doomed Indian troops sent so swiftly and perilously to altitudes three miles high—without adequate physical training or proper gear—were able to stop high altitude–hardened Chinese divisions toughened by years of training in subduing Tibet. Pulmonary edema choked and immobilized most of the Indian troops thrown into Chinese harm's way without time for their bodies to adapt to such high altitudes.[5] General Kaul also fell victim to that breathless, paralyzing illness soon after he was flown up to examine the front line that Nehru expected him to hold. Kaul was immediately flown back to Delhi and hospitalized, allowing him to recover his health, though not his ruined reputation.[6] He did not fly back to his high command until it was too late. Most of his corps soon collapsed in the face of wave after wave of Chinese attackers. The battle-hardened and brilliant Sikh general Harbaksh Singh, briefly put in command of the NEFA front, might have saved thousands of soldiers' lives had he been permitted to fall back to strategically defensible lower ground, but Nehru judged that too humiliating, listening instead to his Kashmiri cousin Kaul, returning him to a front that his troops were not equipped to defend.[7]

On October 20, some twenty thousand Chinese troops poured over the NEFA border, attacking all along a twelve-mile front with heavy mortars and machine guns. More than four thousand Indian troops were overwhelmed and could not be accounted for. Fewer than four hundred of those Indian soldiers subsequently emerged from behind Chinese lines. The Chinese government had stated from Beijing that morning that "Indian forces had attacked them in large numbers," Nehru explained. "It has become a habit for the Chinese to blame others for what they propose to do."[8] Every other member of Nehru's cabinet and most members of Parliament now demanded Krishna Menon's resignation. But Nehru continued to shield him.

"Comrades, Friends and Fellow-Countrymen," Prime Minister Nehru broadcast to his nation on that bleak October 22, 1962. "We are men and women of peace in this country . . . unused to the necessities of war. . . . But all our efforts have been in vain in so far as our own frontier is concerned. . . . The time has, therefore, come for us to realize fully this menace, which threatens the freedom of our people. . . . [T]o conserve that freedom and integrity of our territory we must gird up our loins and face this greatest menace. . . . The price of freedom will have to be paid in full measure, and no price is too great. . . . I earnestly trust and I believe that all parties and groups in the country will unite in this great enterprise and put aside their controversies and arguments . . . and present a solid united front."[9]

Ambassador Galbraith had just reached London, en route to New York and Washington, and was fast asleep when he was called by the embassy and ordered by the president to fly back to Delhi immediately. The next day he met with Nehru, assuring him of American sympathy and military support whenever India requested it.[10] But Krishna Menon was still unable to bring himself to ask for U.S. military aid, despite how desperately India's troops needed help. And Nehru was as yet unable to fire him. But on October 31, he finally did so when demands within his own party became universal, as was the clamor against Krishna Menon outside. After his dismissal, the United States flew military equipment and high-altitude gear of every variety into Calcutta from November 3 on, around the clock, until enough U.S. supplies to arm and clothe no less than ten Indian mountain divisions had reached India.

"China, which claimed and still claims to be anti-imperialist, is pursuing a course today for which comparisons can only be sought in the eighteenth and nineteenth centuries," Nehru told his lower house of Parliament on November 8. "It is sad to think that we in India, who have pleaded for peace all over the world, sought the friendship of China, and . . . pleaded their cause in the councils of the world, should now ourselves be victims of a new imperialism and expansionism. . . . History has taken a new turn in Asia and perhaps the world, and we have to fight with all our might this menace to our freedom and integrity." [11] As American military aid moved up the line, the Chinese realized that unless they withdrew before winter closed every northern pass, their victorious army would soon be doomed to death or capture. Beijing wisely opted to withdraw, therefore, shortly after Galbraith called for aircraft carriers of the U.S. Seventh Fleet to steam into the Bay of Bengal to provide air cover and support for Kaul's shattered NEFA Corps. On November 19, 1962, the Chinese decided to advance no farther toward the plains and capital of Assam, which lay within reach but would have been a trap for their entire force. A unilateral Chinese cease-fire was announced from Beijing on November 21, and by month's end all Chinese troops had pulled back to the McMahon line they had so long claimed not to recognize.

Krishna Menon left the cabinet in early November yet was the only nonfamily member to breakfast with Nehru on his seventy-third birthday that November 14. "This is of course not a parting," Nehru had written his one remaining close friend upon accepting his formal resignation, "as both you and I are dedicated to serve our country in whatever position either of us may be placed." [12] But Krishna knew he had become a heavy rock round the neck of his elder brother, dragging him down in the eyes of most of the world, as well as most members of his own party and Parliament. Whatever else might be said of Krishna Menon, he was not stupid. So he carried his tea pot and toaster off to a house near Nehru's and spent his days at the Supreme Court, where he tried, unsuccessfully, to busy himself pleading the law. No one in the government but Nehru and Indira missed him much, and although he outlived Nehru by a decade, he never served again in any office of high responsibility. "Radhakrishnan always thought that Krishna Menon's was a diseased mind," Mac wrote of him, also obviously agreeing with the many negative Western press assessments of "His Grey Eminence; India's Rasputin; the venomous cobra; Hindu Vishinsky; tea-fed tiger." [13]

Nehru's own reputation was badly scarred by the ease with which Chinese forces had invaded and virtually destroyed India's NEFA army. Those who had long considered him infallible—or if not always perfect, at least as close to perfection as a human leader could be—never quite viewed him in that way after October–November 1962. Deep shadows had fallen on

Nehru, etching rings under his deeper set, sadder eyes, and the puffiness of his now flaccid cheeks and neck added the ever present look of mortality to the once handsome face that had seemed to defy age and death itself.

By 1963 everyone knew that India's king was dying, yet few could agree on the successor to his crown and throne. Indira, after all, was too young—was she not? Most of the older men in the cabinet believed so at any rate, and some of Nehru's former socialist supporters, like Ram Manohar Lohia, considered her no more than a "dumb doll." Indira, much like Krishna Menon, was many things, but never dumb. Yet at forty-five she had never been in the cabinet and had served as Congress president for only a year. The family knew, of course, how close she was to her father, and many officials knew how important she was to all daily decision making now, from such minor things as which petitioners Nehru would see and which meetings he would address, to major socioeconomic and political issues. If only he could last five more years! That would give her ample time to grow into the succession naturally, without raising too many "green-eyed monsters" among those ambitious elders in the cabinet and Congress who had waited so patiently all their mature lives for Nehru to step down or die. "In another ten years or so you will probably be thinking of retiring," Dickie had written him in 1952. "I know that you and I have always seen eye-to-eye about this, and as long ago as 1948 you promised me to find someone who you would train up. The fact remains that four years have passed since then, and I have yet to hear that you have anybody in sight!"[14] What even Dickie never realized, of course, was that Nehru had always had his very own Darling Indu in sight.

Sixty-six-year-old Morarji Desai, Nehru's finance minister at this time, was considered by many of his colleagues and most of his business supporters in Gujarat and Bombay to be the man best qualified to succeed Nehru as prime minister. Young Morarji had been Sardar Patel's disciple, as well as Gandhi's. He was as solidly rooted in India's conservative Hindu traditions and basically cautious landed-mercantile nature as Nehru was mercurial and predisposed to flights of foreign and radical fancy. Morarji knew most of Nehru's weaknesses as well as Vallabhbhai had, but Morarji was a patient person, confident that his turn would come. But even Morarji never imagined that he would wait another fourteen years rather than the fourteen more months left for Nehru's tenure in power. Nehru, of course, recognized Morarji's ambitions and appreciated his willpower and administrative strengths for he was cut from the same solid Gujarati mold that had formed Bapu and the Sardar. But Nehru had never liked Morarji, and Indira feared and hated him, as did Krishna Menon, always whispering about how "right-wing" he was and how much "the tool of Washington." So Nehru had to devise some scheme to get rid of him, to remove him from the vortex of power, from the press and public eye alike, before it was too late and Indira was swept aside by the rush of Congress leaders racing to hail Morarji Desai as the premier.

Nehru never liked associating himself directly with any unscrupulous act, anything as immoral as throwing his "friend" Sheikh Abdullah behind bars or forcing his most likely and best-qualified successor Morarji out of his cabinet into the political wilderness virtually on the eve of his demise. That would have looked awful and might have jeopardized Indira's chances of stepping into his shoes, for he never admitted wanting her to take power, always insisting, in all humility, that "in a Democracy the People will decide" such things. So just as he had found his Tiger to take care of Lion Abdullah, he now discovered a South Indian "king of love," a *kama-raj,* to do the tricky job of toppling Morarji Desai, shaming him into stepping down, resigning from his high office in the best Hindu tradition of withdrawal and retreat, in much the way that Gandhi himself had done so often.

The sixty-year-old Tamil leader Kumaraswami Kamaraj Nadar was the chief minister of Madras until August 1963, when Nehru talked him into resigning that post, by promising him the presidency of Congress as part of his reward for devoting himself full time to building up the party, which had sustained three serious bielection defeats in the wake of the Chinese debacle. Kamaraj swiftly emerged from the obscurity of his south Indian provincial base to the center stage of India's national politics. He spoke little English and hardly any Hindi, but he managed not only the political eclipse of Morarji Desai but also the brief elevation to power of Lal Bahadur Shastri and, less than two years later, the much more important anointment of Indira Gandhi as prime minister, thereby finishing the job he was hired to do from the start. For Nehru always planned ahead. And like his earlier five-year plans, the clever trick he devised in mid-1963 was called the Kamaraj plan. Nehru modestly called it "an idea taking hold of the mind and growing by itself."[15] A few months later, however, when nominating Kamaraj for president of Congress, Nehru assured most of his puzzled colleagues that "his name has got engraved in the history of India."[16]

The original plan was for Kamaraj and a few other chief ministers, primarily Biju Patnaik of Orissa, to announce their resignations to return to their grass roots to revitalize enthusiasm and support for Congress among the common people and help lift the sagging morale of the nation. But if that would work in the provinces, why not try the same thing at the center? So Nehru called his senior cabinet ministers to meet with Kamaraj and announced that he considered this so good an idea that he was inspired by Kamaraj's example to submit his own resignation, in order to go back to private work at his Allahabad roots. "When he said this," Morarji recalled, "Kamaraj and two or three other colleagues said that that was an impossible proposition. . . . Jawaharlalji did not need much persuasion to give up his suggestion."[17] Home Minister Lal Bahadur Shastri, who was famed for his loyalty to Nehru and may well have been his "decoy elephant," then offered his resignation, and immediately told Morarji about it.[18] "I suspected that some game was being played behind the scenes," Morarji reflected, but by then he had been a minister for two decades and readily submitted his letter of resignation, which Nehru quickly accepted. Morarji later charged that there had been an understanding between Nehru and Shastri to bring Lal Bahadur back into the Cabinet after a few months. Shastri was, indeed, appointed minister without portfolio early in 1964, shortly before Nehru died.[19]

In the wake of the Chinese debacle, most of those opposed to Congress in Lok Sabha tabled a no-confidence motion in Nehru's government, and forty members of Parliament criticized Nehru's foreign policy and demanded his resignation in late August 1963. "Personally, I have welcomed this motion and this debate," Nehru responded. "What has brought together these various Members in such a curious array? It is obvious that what has brought them together is . . . not only a dislike of our Government, but—I am sorry to say so—perhaps a personal attitude against me. . . . I must confess, and I say so with all respect, that the Members, the leaders of the Opposition . . . have not done justice to this motion, nor to themselves."[20] His rhetoric reverted to its clever Cambridge Union style, although the forty members of Parliament who spoke had raised many serious charges, including those of corruption and incompetence.

"I do not mean to say that all the charges they made had no substance. . . . But, generally, the debate has proceeded on rather personal grounds, personal likes and dislikes. . . . I do not wish to argue about myself; it is unbecoming of me to do so, and it would be wrong. . . . We were dealing with the future of India, not of Jawaharlal Nehru. . . . In [the] course of time, we shall go and others will take our place. . . . I felt that at a moment like this, to talk in this petty and small-minded way was not becoming . . . it showed an absence of a larger vision. . . . When many years ago most of us here . . . were participating in the struggle for freedom under the leadership of Gandhiji, we had that larger vision . . . a vision of the future which we were going to build, and that gave us a certain vitality, a certain measure of a crusading spirit. Now, most of us are perhaps lost, or are tied up in humdrum politics and petty matters."[21] Yes, indeed, nodded many of those who had tabled this motion, thumping their seats, muttering "Quit, Nehru, quit!" It was the chant of some ten thousand opposition party supporters outside, who marched around the circular stone mausoleum of Parliament on that hot and wet day.

"I do not think India is going to perish," Nehru droned on. "It has not perished for five thousand years or more. . . . But, I do not want India merely to exist; I want it to live a full life. . . . We want real freedom . . . not merely political freedom; it is economic freedom in two senses. One is that you do not have to rely on other countries. You are friends with them . . . you take their help, but you are not dependent on them to carry on. . . . And the second . . . is economic freedom for the vast masses of our country . . . putting an end, in stages, if you like, to gross differences in wealth and opportunities."[22] But many opposition members, to the left as well as the right of Congress, no longer believed Nehru's rhetoric and promises.

Still he persisted, arguing that "the whole philosophy of Gandhiji, although he did not talk perhaps in the modern language, was not only one of social justice, but of social reform and land reform. . . . It was inevitable that Congress should begin to think that way. . . . This Constitution talks of social justice. . . . Later, this Parliament definitely adopted the ideal of socialism, and so did the Planning Commission. . . . We have been slow for a variety of reasons. . . . But I am convinced that there is no choice for India. No party, whatever it may feel, can stop this country's march to socialism . . . to achieve this ideal of social democracy through planning."[23] Nehru spoke slowly, slurring some of his words, "an old man, looking frail and fatigued," as H. V. Kamath noted with surprise whenever he saw the prime minister enter Parliament nowadays, "with a marked stoop in his gait, coming down the gangway . . . with slow, faltering steps, and clutching the backrests of benches for support as he descended."[24]

Hundreds of guests and neatly dressed schoolchildren assembled on "Chacha" (Uncle) Nehru's lush lawn on the morning of his seventy-fourth birthday, for which Governor Padmaja Naidu of West Bengal had sent a thirty-pound cake, topped with a huge candle. Sister Betty spent more time with him now, but Nan was in Bombay, for her brother had appointed her governor of Maharashtra, her last post of power. Indira, of course, presided over all his affairs at Teen Murti. She was his sole heir and would soon inherit the Nehru Memorial Trust Fund as well as all his personal possessions. He had written out his "last will and testament" a decade earlier, on June 21, 1954: "I have received so much love and affection from the Indian people that nothing that I can do can repay even a small fraction of it. . . . I can only express the hope that in the remaining years I may live, I shall not be unworthy of my people. . . . I do not want any religious ceremonies performed for me after my death. I do not believe in any such ceremonies and to submit to them, even as a matter of form, would be hypocrisy."[25] But his cremation and funeral would be that of a traditional Hindu Brahman, just as his marriage had been.

"When I die, I should like my body to be cremated," he had written in his will, wanting "my ashes sent to Allahabad. A small handful of these ashes should be thrown into the Ganga . . . no religious significance, so far as I am concerned. . . . I have been attached to the Ganga and the Jumna rivers in Allahabad ever since my childhood and, as I have grown older, this attachment has also grown. I have watched their varying moods as the seasons changed, and have often thought of the history and myth and tradition and song and story that have become attached to them through the long ages and become part of their flowing waters. The Ganga, especially, is the river of India, beloved of her people, round which are intertwined her racial memories, her hopes and fears, her songs of triumph, her victories and her defeats. She has been a symbol of India's age-long culture and civilization, ever-changing, ever-flowing, and yet ever the same Ganga. She reminds me of the snow-covered peaks and deep valleys of the Himalayas, which I have loved so much. . . . Smiling and dancing in the morning sunlight, and dark and gloomy and full of mystery as the evening shadows fall. . . . And though I have discarded much of past tradition and custom, and am anxious that India should rid herself of all shackles that bind and constrain her and divide her people, and suppress vast numbers of them, and prevent the free development of the body and the spirit . . . yet I do not wish to cut myself off from that past completely. I am proud of that great inheritance that has been, and is, ours, and I am conscious that I too, like all of us, am a link in that unbroken chain which goes back to the dawn of history in the immemorial past of India."[26]

Nehru addressed the International Congress of Orientalists, which met in Delhi early in January 1964. "I must confess to you that I do not claim to be a scholar or historian," he told that gathering of scholars. "What I am, it is difficult for me to say—a dabbler in many things—but I certainly feel a certain feeling of embarrassment standing before this distinguished audience . . . because, apart from dabbling in many things, I have not studied carefully the work of Orientalists. I have always thought their work important . . . I feel grateful to the many eminent scholars in Europe who have studied these subjects and shed a great deal of light on them. . . . It is extraordinary how in . . . India, these old ideas and thoughts have clung to the people through the ups and downs of history . . . various customs and attitudes which we find a little difficult to discard. . . . We cannot entirely discard the old and uproot ourselves," he reaffirmed more emphatically, not wanting to "become rootless. . . . Many of you, ladies and gentlemen, are interested in finding out facts about the old. . . . The chief concern that fills my mind is how to find a synthesis between the old and the new. . . . The two have to be brought together."[27]

Even though his government had easily managed to survive that vote of no confidence, Nehru knew that many of the brightest young members of Lok Sabha opposed him, and he could hear those angry shouts of "Quit, Nehru!" They reminded him of Gandhi's "Quit India," which he too had shouted once, and he remembered how often Gandhi had tried to restrain him, to teach him greater moderation and more patience, trying to convince him of the value of coordinating their steps, the old and the new.

"There was a certain depth in the traditional way of living," he told the Oriental scholars, who nodded appreciatively as he added, "With life today, with its rush and hurry and technical developments which are, of course, very important . . . we are apt to lose something. . . . And that is why I have tried to think of how the two can be harmonized. . . . But all the progress which we make is essentially in our knowledge of the external world . . . not very much . . . in the knowledge of ourselves. We go back to our ancient saying, whether Greek or Indian . . . which always laid stress on a person knowing himself. . . . These two . . . the external approach and the internal approach, have to be, I suppose, combined in order to make us realize what we are and how we are to face our problems. . . . Perhaps . . . it would be helpful if we thought quietly about ourselves. . . . and not merely in terms of the atom bomb and how to escape from it."[28]

The next day Nehru flew with Indira to the capital of Orissa, Bhubaneshwar, to attend the Congress's annual session held there. Kamaraj was nervous about his new position, and needed reassurance from Nehru, as well as three interpreters to help him understand what was being said. Many others were anxious to speak to Nehru as well, pressing close with their questions and problems.

In addition to the weariness of the flight, therefore, Nehru found no rest on arrival and little sleep, for his nights were always plagued with those dreams of terror that left him more exhausted upon waking than he had been when he closed his eyes. So on January 6, 1964, as he tried to stand up on that crowded platform of seated Congress pundits, the left side of his body went numb, a cerebral stroke flooding the right side of his brain with blood. Indira jumped up, catching hold of him, and, with the aid of a security guard, managed to lead him off stage with surprisingly little fuss or attention from the crowd that continued earnestly to debate the major socialist resolution of this last session of Congress that Nehru was ever to attend. They spirited him away and secured him in the State House, where no one but Dr. B. C. Roy and his most trusted medical assistants were permitted to examine him as he lay dying. Too soon! The Kamaraj plan had only just been activated, yet now Lal Bahadur Shastri had to be called back into harness. Kamaraj pleaded with Indira to join the cabinet as well, for how else could he function?

But Indira wisely waited. She knew that her most important job was to stay at her father's side as long as he remained in bed and then to help him stand up and walk once he was on his feet again. Luckily, as it soon became clear, this was not as massive a stroke as initially feared, and Nehru clung to life with that same tigerlike tenacity with which he had held and guarded his throne. He held on, as once he had to a slender rope that kept him from sinking into early oblivion, as he had held in those vermin-infested cells of solitude where British magistrates had locked him away for a decade, thinking he was easily broken. Although now half of him felt nothing at all and he could not speak, Dr. Roy's glucose feeding and morphine drip helped revive him enough to fly him back to Delhi by week's end.

The press was handled well, of course, merely reporting that the prime minister was slightly ill, but was making a remarkably swift recovery, and would soon be able to take up his normal schedule. For the interim, Lal Bahadurji would take some of the burden off his shoulders. Nehru's chair remained vacant for the remaining sessions of that Bhubaneshwar Congress, and the crowds who had come hoping to see Panditji never glimpsed him again. Only Kamaraj and Indira saw him, reporting that he was getting better and better each day, as indeed he was. For the stroke did not kill him, although he remained mortally wounded.

By April 3, 1964, Nehru was strong enough to attend the budget session debate in Lok Sabha but sat slumped over, mostly silent, and when he tried to rise in response to a remark made by opposition member H. V. Kamath, he fell back, "obviously angry, rattled . . . he was frothing at the corner of his mouth . . . fagged out."[29] Indu took him up to Dehra Dun, where he could see the mountain peaks he loved. "The major portion of my ashes," he had written in his will, after noting his desire to have a handful thrown into Mother Ganga, "should . . . be carried high up into the air in an aeroplane and scattered from that height over the fields where the peasants of India toil, so that they might mingle with the dust and soil of India and become an indistinguishable part of India."[30] A few weeks in the cooler mountain air helped revive him.

But the calls and wires kept coming in from Delhi, asking Indira what to tell the press and how to keep hiding the news from the entire world. So many urgent problems had been left unresolved, and Kashmir was heating up again. Now Bakshi was out, for even he could not endure his job any longer and had to be replaced with some one less independence minded. Why not let poor Sheikh Abdullah out of prison? Maybe he could help make Ayub and Zulfi feel less angry over Kashmir, before everything blew up. Pakistani U.S. fighters kept flying sorties over Jammu airspace, and there were no Indian planes to stop them or shoot them down, for now the Pakistanis had new high-speed American jets. Indira finally agreed to fly back with her father to the hellish heat of Delhi in mid-April and to the firing line of India's political battlefront, where Congress itself was in daily danger of sinking under the rising tides of opposition, both internal and external. The once pressing question "After Nehru, who?" had become "After Nehru, what?"

Indira decided to fly to Washington on April 15, to deliver a letter from her father to President Lyndon B. Johnson, who also had just recently taken up the burdens of a fallen leader. Nan and Padmaja moved temporarily into Teen Murti to take over Indira's nursing and guarding duties, keeping Nehru happy and the press at bay. Although Indira had taken no cabinet post, this was her first actual mission as foreign minister, viewed by many observers in Delhi as the trial balloon of India's next prime minister. Indira's mission was to reassure President Johnson of her father's good health and his most sincere desire to reach a peaceful agreement with Pakistan on a permanent solution to the Kashmir problem, by launching new three-way talks with the just released Sheikh Abdullah and President Ayub, a summit in Delhi to which Nehru would invite them both as soon as possible. She had heard how susceptible Lyndon Johnson was to lovely ladies and so decided to end her three years of mourning for Feroze a bit early, dressing elegantly for her Washington minisummit. She hoped to be able to charm Johnson enough to win his promise of supersonic U.S. jets to ward off further Pakistani incursions over the Jammu–Kashmir border. But that required a special trip to Washington a month later by Defense Minister Y. B. Chavan, whose mission was cut short by an urgent call to return to Delhi for a state funeral.

Indira met with President Johnson on April 27, and flew home herself directly from Washington, for nothing in Delhi had changed for the better, and Shastri, whom she never trusted, was using her time abroad, she had learned, to consolidate his own power base. "The desire to be out of India and the malice, jealousies and envy, with which one is surrounded, are now overwhelming," Indira wrote to her closest American friend, Dorothy Norman, a week after flying home—"also the fact that there isn't one single person to whom one can talk or ask advice."[31] Like her father, Indira never felt fully at ease in India, never really trusted anyone.

Indira arrived home early on April 29, and later that same day returned to the airport to meet Sheikh Abdullah. He had flown down from Srinagar for his summit with Nehru at Teen Murti, where he stayed. The Lion of Kashmir smiled for photographers and embraced the frail prime minister, who took him to his lounge for a private cup of tea and talk, which led nowhere and changed nothing. Abdullah enjoyed his vacation, however, free to go wherever he wished, to see whomever he wanted, including his friend J.P. and old C.R. (Rajaji) in Madras, who had started an opposition Swatantra (Freedom) Party of his own, rallying some of the brightest young members of Lok Sabha to stand up and speak out against Congress corruption, calling for a more rational policy in Kashmir. But

Nehru ignored Rajaji's sound advice, much as he always had, for in some respects Rajaji reminded him of Gandhi, at times of Motilal.

Two weeks before Nehru died, old Syed Mahmud, his closest Cambridge friend in India, came in from Patna to say good-bye. "I loved him," Mahmud confessed tearfully, unashamedly.[32] Nehru knew it, of course, and felt much the same way, though at times he had yelled harshly at his lethargic, almost blind old friend, to whom he had read aloud so many newspaper reports of Churchill's speeches in Parliament during the years they spent together in Ahmadnagar Fort, sharing a prison cell. Nehru had loved Mahmud, not as he had loved Edwina or Padmaja, but the Old School way. Loving brothers they had always remained.

In May, Nehru forced himself to fly once again to Bombay, where the A.I.C.C. was meeting. He took the "Chair," a soft chair on the platform—he could no longer manage the lotus position of a Congress-yogi on the floor mat, as the rest of them did, as Bapu had done till the day he was murdered. He sat through it all, silent, glum, and barely awake, that final painful party meeting at which President Kamaraj, faced with so many strange languages, found it almost as hard as Nehru did to follow what was happening. Another mass exodus of Hindus from East Pakistan was the focus of that meeting. Bitter communal reprisals against Muslims in India ensued. Nehru's summit with the long-dead Liaquat had solved nothing, yet he reaffirmed his faith in secularism and refused to abandon the hope of trying to come to at least some understanding with Pakistan. The only alternative, after all, was war, a legacy left for Indira to fight and win, with Soviet support, in 1971. They flew back to Delhi on May 17, where Nehru collapsed from another stroke while walking in his garden.

On May 22, 1964, he had to face the press, for everyone kept clamoring for a chance to ask him questions. "In regard to Kashmir," Nehru told them, speaking slowly, his voice thick, slurred, hesitant, "Sheikh Abdullah is going to Pakistan the day after tomorrow, I think . . . best that these talks take place without any inhibition." How did he propose to promote greater Indo–Pak amity? "I did not propose anything," Nehru replied. "It is difficult to say what the . . . approach might be." What type of relationship did he envisage for both countries? "I do not have anything very definite in my mind. It depends upon . . . Pakistan. . . . There was no particular purpose." He would now be willing to accept an agreement that left Pakistan "holding on to that part of Kashmir which they have occupied." But Sheikh Abdullah had "stated strongly that he does not want a division of Kashmir." Nehru was asked about the fourth plan and rising prices. How did he plan to tackle that question? "I am afraid I cannot deal with the price question in detail here. It is a difficult . . . question." More difficult questions were asked, none of which he could answer with

NOTES

ABBREVIATIONS OF WORKS FREQUENTLY CITED

Auto	Jawaharlal Nehru, *An Autobiography* (London: John Lane, 1936).
Campbell-Johnson's *Mission*	Alan Campbell-Johnson, *Mission with Mountbatten* (New York: Dutton, 1953).
CWMG	M. K. Gandhi, *Collected Works of Mahatma Gandhi*, vols. 23–90 (Ahmedabad: Navajivan Trust, 1967–1984).
Freedom's Daughter	Sonia Gandhi, *Freedom's Daughter: Letters Between Indira Gandhi and Jawaharlal Nehru, 1922–39* (London: Hodder & Stoughton, 1989).
Galbraith's *Journal*	John Kenneth Galbraith, *Ambassador's Journal: A Personal Account of Kennedy Years* (Boston: Houghton Mifflin, 1969).
George's *Krishna Menon*	T. J. S. George, *Krishna Menon: A Biography* (London: Jonathan Cape, 1964).
Hough's *Edwina*	Richard Hough, *Edwina: Countess Mountbatten of Burma* (London: Weidenfeld and Nicolson, 1983).
Hough's *Mountbatten*	Richard Hough, *Mountbatten: A Biography* (New York: Random House, 1981).
INC	*Indian National Congress,* 2nd ed. (Mandras: G. A. Nateson, 1917).
Indira's *MT*	Indira Gandhi, *My Truth,* presented by Emmanuel Pouchpadass (New Delhi: Vision Books, 1981).
Jinnah of Pakistan	Stanley Wolpert, *Jinnah of Pakistan* (New York: Oxford University Press, 1984).
JN's Speeches	*Jawaharlal Nehru's Speeches,* 1919–1953, 2nd impression (Delhi: Publications Division, 1954).
Mathai's *Days*	M. O. Mathai, *My Days with Nehru* (New Delhi: Vikas, 1979).
Mathai's *Reminiscences*	M. O. Mathai, *Reminiscences of the Nehru Age* (New Delhi: Vikas, 1978).
Morgan's *EM*	Janet Morgan, *Edwina Mountbatten: A Life of Her Own* (New York: Scribner, 1991).
Nehru's *Discovery*	Jawaharlal Nehru, *The Discovery of India* (New York: John Day, 1946).
Norman's *Nehru*	Dorothy Norman, *Nehru: The First Sixty Years* (London: The Bodley Head, 1965), vol. 2.
Old Letters	Jawaharlal Nehru, *A Bunch of Old Letters: Written mostly to Jawaharlal Nehru and some written by him* (New York: Asia Publishing House, 1960).
PD	Nehru's prison diary
Scope of Happiness	Vijaya Lakshmi Pandit, *The Scope of Happiness: a Personal Memoir* (New York: Crown Books, 1979).
Seton's *Panditji*	Marie Seton, *Panditji: A Portrait of Jawaharlal Nehru* (London: Dennis Dobson, 1967).
SWJN	S. Gopal, ed., *Selected Works of Jawaharlal Nehru,* vols. 1–15 (New Delhi: Orient Longman, 1972–1982).

SWJN (2)	S. Gopal, ed., *Selected Works of Jawaharlal Nehru,* 2nd series, vols. 1–16 (New Delhi: Orient Longman, 1984–1994).
Tendulkar, *Mahatma*	D. G. Tendulkar, *Mahatma: Life of Mohandas Karamchand Gandhi,* vols. 2–7 (Bombay: V. K. Jhaveri & D. G. Tendulkar, 1951–1954).
TF	Jawaharlal Nehru, *Toward Freedom* (New York: John Day, 1941).
TP	Nicholas Mansergh and E. W. R. Lumby, eds., *The Transfer of Power, 1942–1947,* vols. 1–12 (London: HMSO, 19 – 1983).
Two Alone	Sonia Gandhi, ed., *Two Alone, Two Together: Letters Between Indira Gandhi and Jawaharlal Nehru, 1940–1964* (London: Hodder & Stoughton, 1992).
Voice of Freedom	K. M. Panikkar and A. Pershad, eds., *The Voice of Freedom: The Speeches of Pandit Motilal Nehru* (London: Asia Publishing House, 1961).
We Nehrus	Krishna Nehru Hutheesing, *We Nehrus* (New York: Holt, Rinehart and Winston, 1967).
With No Regrets	Krishna Nehru, *With No Regrets: An Autobiography* (New York: John Day, 1945).
YI	M. K. Gandhi, *Young India, 1919–1922* (New York: B. W. Huebsch, 1924).

CHAPTER 1

1. Quotations from Nehru's "Tryst with Destiny" speech to the Constituent Assembly in New Delhi on August 14, 1947, are from Norman's *Nehru,* 2:336–40. *Tryst* was "Mac" (M. O.) Mathai's word. See Mathai, *Reminiscences,* p. 11.
2. I am indebted to Inder K. Gujral for this quotation from Nehru, who said it of Indira when many friends insisted that artist Satish Gujral's portrait of her made her look "too old." Nehru, who liked the painting that Inder's brother had made of his daughter, silenced its critics saying, "They don't know her. Hers is an old soul." Mr. Gujral told me this in Los Angeles in 1993.
3. Norman, *Nehru,* p. 336.
4. Ibid.
5. Lord Mountbatten told me this during a morning interview I had with him in London in the summer of 1978.
6. *We Nehrus,* p. 20.
7. Nehru dedicated his autobiography "To KAMALA, Who is No More."
8. *Auto,* p. 7; *TF,* p. 22.
9. Ibid.
10. Ibid.
11. Ibid.; *Auto,* pp. 14–16; *TF,* pp. 27–29.
12. *Auto,* pp. 14–15. *TF,* 28–29.
13. Gregory Tillett, *The Elder Brother: A Biography of Charles Webster Leadbeater* (London: Routledge & Kegan Paul, 1982). Also see Annie Besant and Charles Webster Leadbeater, *Occult Chemistry: Clairvoyant Observations on the Chemical Elements,* 2nd. ed. (Adyar: Theosophical Publishing House, 1919).
14. Annie Besant; *An Autobiography* (London: T. Fisher Unwin, 1893); and Anne Taylor, *Annie Besant: A Biography* (Oxford: Oxford University Press, 1992).
15. George Bernard Shaw's judgment of the passionately eloquent Annie.
16. Tillett, *The Elder Brother,* pp. 78–86.

17. Ibid., pp. 103–13.
18. Walter Crocker, *Nehru: A Contemporary's Estimate* (London: Allen & Unwin, 1966), p. 60 and n. 1. Frank Moraes, *Jawaharlal Nehru: A Biography* (New York: Macmillan, 1956), pp. 20–21.
19. *Auto,* p. 15; *TF,* p. 29.
20. *Auto,* p. 16; *TF,* p. 29.
21. George's *Krishna Menon.*
22. Norman's, *Nehru,* p. 337.
23. *Auto,* p. 597.

CHAPTER 2

1. Harrow School was founded by John Lyon on Harrow Hill in 1572. See E. D. Laborde, *Harrow School: Yesterday and To-Day* (London: Winchester, 1948).
2. *TF,* p. 30.
3. I am indebted to Harrow's archivist A. D. K. Hawkyard for this information. When I visited Harrow School in 1992, Mr. Hawkyard kindly showed me the old register in which Nehru's "last school" was inscribed in ink as "J. Atkins, Esq., 154 Dalston Lane." Nehru also mentions the Tanners in a letter to Motilal Nehru, written from London on March 28, 1907, *SWJN,* vol. 1 (1972), p. 22.
4. Tikka Paramjit Singh, heir to Punjab's maharaja of Kapurthala and the son of the Maharashtrian gaekwar of Baroda, Maharaj Kumar Jaisinghrao, are listed in the Harrow School Registrar, 1800–1911, pp. 230 and 272.
5. *TF,* p. 32.
6. *SWJN,* 1: 3–29.
7. B. R. Nanda, *The Nehrus: Motilal and Jawaharlal* (New York: John Day, 1963), p. 70.
8. Jawahar to father, Highgate, January 3, 1906, *SWJN,* 1: 10.
9. Nanda, *The Nehrus,* pp. 76–77.
10. Jawahar to father, Harrow, January 30, 1907, *SWJN,* 1: 18.
11. Ibid., p. 19.
12. Jawahar to father, Harrow, February 8, 1907, ibid., p. 20.
13. Stanley Wolpert, *Morley & India, 1906–1910* (Berkeley and Los Angeles: University of California Press, 1967).
14. Jawahar to father, February 8, 1907, *SWJN,* 1: 19 and n. 1.
15. Jawahar to father, May 17, 1907, ibid., p. 24.
16. Jawahar to father, Highgate, August 1, 1907, ibid., p. 30; Also see his last letter from Harrow, July 25, 1907, ibid., p. 29. The quotation is from *Auto,* p. 19.
17. Written in 1872 by Edward Bowen and set to music by John Farmer. The musical score and words are in John Leaf's *Harrow School* (Eastleigh, Hants: Pitkin Pictorials, 1990), p. 9.
18. Jawahar to father, Dublin, September 12, 1907, *SWJN,* 1: 32.
19. Jawahar to father, Trinity College, Cambridge, October 24, 1907, ibid., p. 35.
20. Mathai, *Reminiscences,* pp. 58–61.
21. Jawahar to father, October 31, 1907, *SWJN,* 1: 36.
22. Jawahar to father, November 7, 1907, ibid., p. 37.
23. Winston Churchill also wrote for that most conservative British paper in Allahabad.
24. Jawahar to father, London, December 20, 1907, SWJN, 1: 39.
25. Jawahar to father, Cairn Hydro, Harrogate, January 2, 1908, ibid., pp. 40–41.
26. *Auto,* pp. 19–20.
27. Jawahar explains his monetary problems to his father in several letters written from late January through February 1908, *SWJN,* 1: 43–46.
28. Jawahar to father, London, April 3, 1908, ibid., p. 48.

29. Jawahar to mother, London, April 5, 1908, ibid., pp. 48–49.
30. Ibid., p. 49.
31. Jawahar to father, London, April 10, 1908, ibid., p. 50.
32. On January 10, 1908, Motilal wrote him, "You know me and my views well enough to understand that I do not approve of the opinions expressed by you." On January 30, Jawahar replied, "I am sorry you don't approve of my opinions but really I can hardly help holding them. . . . The government must be feeling very pleased with you." Ibid., p. 44 and n. 1; p. 50.
33. Jawahar to father, Cambridge, May 7, 1908, ibid., p. 51.
34. Nehru's own reflection on himself in *Auto,* p. 26.
35. Jawaharlal Nehru's application for admission on January 11, 1909, to the Inner Temple in the Hilary term of that year is preserved in the record book in the basement archives of the Inner Temple. Fletcher's letters of introduction to Barristers Bond and Ball, dated December 5, 1908, are also there, as are the certificates by both barristers of "five years' standing," repeating his statement in support of young Nehru's good character.
36. J. H. Baker, *The Inner Temple: A Brief Historical Description* (London: Honourable Society of the Inner Temple, 1991).
37. Jawahar to mother, Cambridge, May 7, 1909, *SWJN,* 1: 67.
38. Jawahar to father, Cambridge, May 14, 1909, ibid., p. 68.
39. Ibid.
40. Syed Mahmud's vivid recollection is in Rafiq Zakaria, ed., *A Study of Nehru* (Calcutta: Rupa, 1989), p. 173.
41. Fenner Brockway's record of that first encounter with Nehru in 1911 is preserved in London's National Sound Archive in South Kensington on the Series 3 record, "Personality and Power."
42. Lord Wyatt's statement to me at his home in London in 1993.
43. Jawahar to father, Cambridge, May 12, 1910, *SWJN,* 1: 71.
44. Jawahar to father, 60, Elgin Crescent, Holland Park, June 17, 1910, ibid., p. 74.
45. Jawahar to father, London, June 24, 1910, ibid., p. 75.
46. Jawahar to mother, London, June 24, 1910, ibid.
47. Jawahar to father, London, July 15, 1910, ibid., pp. 76–77.
48. I am indebted to Professor John Kenneth Galbraith for this recollection, reported to me at his home in Cambridge, Massachusetts, in 1993.
49. Jawahar to father, Maidenhead, August 5, 1910, *SWJN,* 1: 77–78.
50. *Auto,* p. 26.
51. Jawahar to father, London, August 19, 1910, *SWJN,* 1: 78–79.
52. Jawahar to father, Casino de Dieppe, September 22, 1910, ibid., p. 79.
53. Jawahar to father, London, November 11, 1910, ibid., p. 82.
54. *Auto,* pp. 20, 26.
55. Jawahar to father, London, December 22, 1911, *SWJN,* 1: 92–93.
56. Promilla Kalhan, *Kamala Nehru: An Intimate Biography* (Delhi: Vikas, 1973).
57. Jawahar to mother, Beacon Hotel, Crowborough, Sussex, March 14, 1912, *SWJN,* 1: 97.
58. Jawahar to father, London, June 21, 1912, ibid., p. 98.

CHAPTER 3

1. *We Nehrus,* pp. 2, 3, 4.
2. *Scope of Happiness,* p. 55.
3. *Auto,* p. 28. The paragraph about the races was deleted from the American edition. See *TF,* p. 40.
4. *Auto,* p. 27.

5. Stanley A. Wolpert, *Tilak and Gokhale: Revolution and Reform in the Making of Modern India* (Berkeley and Los Angeles: University of California Press, 1962).
6. Mohandas K. Gandhi wrote a memoir of Gokhale entitled *Gokhale: My Political Guru* (Ahmedabad: Navajivan Trust, 1955), in which he acknowledges his deep and abiding debt to "Mahatma" Gokhale.
7. *Auto,* p. 27.
8. R. N. Mudholkar's presidential address at Bankipore is reprinted in *INC,* p. 1065.
9. *Auto,* p. 30.
10. *We Nehrus,* p. 5.
11. DeWitt C. Ellinwood and S. D. Pradhan, eds., *India and World War I* (New Delhi: Manohar, 1978).
12. *Auto,* p. 31.
13. *With No Regrets,* p. 27.
14. *Auto,* p. 34.
15. *Scope of Happiness,* pp. 55–56.
16. *We Nehrus,* p. 12.
17. *Auto,* p. 37.
18. Ibid., pp. 37–38.
19. Kalhan, *Kamala Nehru,* pp. 12, 13.
20. *Jinnah of Pakistan,* p. 45.
21. A. C. Mazumdar's presidential address to Congress at Lucknow in 1916, *INC,* p. 1274.
22. *Auto,* p. 35.
23. Arthur H. Nethercot, *The First Five Lives of Annie Besant* (London: Rupert Hart-Davis, 1961); Arthur H. Nethercot, *The Last Four Lives of Annie Besant* (Chicago: University of Chicago Press, 1963).
24. Nehru's letter to the editor, June 21, 1917, is reprinted in *SWJN,* 1: 106.
25. Ibid., p. 107.
26. *Auto,* p. 34.

CHAPTER 4

1. Erik H. Erikson, *Gandhi's Truth: On the Origins of Militant Nonviolence* (New York: Norton, 1969), p. 283.
2. M. K. Gandhi, *The Story of My Experiments with Truth* [Gandhi's' Autobiography], trans. Mahadev Desai (Washington, D.C.: Public Affairs Press, 1948), pp. 545, 551–554.
3. *Jinnah of Pakistan,* p. 61.
4. *Auto,* p. 41.
5. Ibid., pp. 41–42.
6. *We Nehrus,* pp. 40–41.
7. *Auto,* p. 42.
8. *We Nehrus,* p. 32.
9. *Auto,* p. 42.
10. *We Nehrus,* p. 34.
11. Stanley Wolpert, *Massacre at Jallianwala Bagh* (New Delhi: Penguin Books, 1988).
12. Motilal Nehru's presidential address to Congress, Amritsar, December 27, 1919, *Voice of Freedom,* pp. 6–8.
13. *Jinnah of Pakistan,* p. 67.
14. Ibid., p. 69.
15. Ibid., p. 70.
16. Shan Muhammad, *Unpublished Letters of the Ali Brothers* (Delhi: Idarah-i Adabiyat, 1979); and Gail Minault, *The Khilafat Movement: Religious Symbolism and Political Mobilization in India* (New York: Columbia University Press, 1982).

17. *Auto,* pp. 46–47.
18. Ibid., p. 47.
19. Jawaharlal Nehru to the editor, *The Leader,* September 4, 1920, *SWJN,* 1: 167.
20. *We Nehrus,* p. 34.
21. *Auto,* p. 52.

CHAPTER 5

1. *Auto,* p. 57.
2. *We Nehrus,* p. 45.
3. *Scope of Happiness,* p. 65.
4. Ibid., pp. 72–73.
5. *Auto,* p. 69.
6. Ibid., p. 77.
7. Ibid., p. 78.
8. Ibid.
9. Hough's *Edwina:* p. 68. Hough also wrote *Mountbatten: A Biography* (New York: Random House, 1981). See, too, Morgan's *EM* and Hough's *Mountbatten.*
10. This quotation is from the first of many prison diaries that Nehru wrote during his years of incarceration, dated December 6, 1921, *SWJN,* 1: 232.
11. *We Nehrus,* p. 35.
12. *No Regrets,* p. 36.
13. *SWJN,* 1: 235.
14. *We Nehrus,* p. 53.
15. *SWJN,* 1: 235.
16. *We Nehrus,* p. 56.
17. Ibid., p. 54.
18. M. K. Gandhi, "The Crime of Chauri Chaura," February 16, 1922, *YI,* pp. 993–94.
19. *Auto,* p. 81.
20. *YI,* pp. 995–96.
21. *Auto,* p. 82.
22. *SWJN,* 1: 239, 242.
23. Nehru's "Statement at Trial," from *The Leader,* May 19, 1922, *SWJN,* 1: 252–57.

CHAPTER 6

1. *Auto,* p. 98.
2. Ibid., pp. 108, 109.
3. The Maharaja of Nabha lived in New Delhi with his maharani till his death in 1995. He served in World War II with great distinction as a major general.
4. *Auto,* p. 110. Also see Nehru's "Draft statement to be read in Court at Nabha," *SWJN,* 1: 369.
5. *Auto,* pp. 110–11.
6. M. N. to J., September 28, 1923, in *Old Letters,* pp. 28–29.
7. Swaraj Party Manifesto, issued from Allahabad, October 14, 1923, *Voice of Freedom,* p. 506.
8. Jawaharlal Nehru, "On a Reception to Lord Reading," October 25, 1923, *SWJN,* vol. 2 (1972), p. 33.
9. *Auto,* p. 101.
10. *SWJN,* 2: 33.
11. *Auto,* pp. 101–2.

12. Nehru's presidential address at Kakinada, December 25, 1923, *SWJN,* 2: 83–86.
13. Ibid., pp. 87–89.
14. *Auto,* p. 124.
15. M. K. Gandhi, "Thoughts on Council Entry," *CWMG,* vol. 23 (1967), pp. 413–14.
16. In *Navajivan,* April 13, 1924, *CWMG,* pp. 429, 431.
17. April 18, 1924, *Voice of Freedom,* pp. 514–15.

CHAPTER 7

1. Gandhi to Motilalji, Bombay, September 2, 1924, *CWMG,* vol. 25 (1967), p. 65.
2. Gandhi to Jawaharlal, September 6, 1924, ibid., p. 98.
3. *Auto,* p. 130.
4. B. N. Pandey, *Nehru* (London: Macmillan, 1976), p. 111.
5. *Scope of Happiness,* pp. 89–90.
6. *We Nehrus,* p. 55.
7. Indira's *MT,* p. 17.
8. *No Regrets,* p. 43.
9. Indira's *MT,* p. 18.
10. Jawahar to Mahmud, Geneva, May 24, 1926, *SWJN,* 2: 233.
11. Jawahar to Mahmud, Geneva, July 15, 1926, ibid., p. 235.
12. Jawahar to Mahmud, Geneva, August 11, 1926, ibid., p. 241.
13. Jawahar to My dear Mahmud, September 12, 1926, ibid., p. 243.
14. *Auto,* p. 156.
15. Isaac Kramnick and Barry Sheerman, *Harold Laski: A Life of the Left* (Harmondsworth: Allen Lane, Penguin Books, 1993). See also George's *Krishna Menon.*
16. Nehru's speech at the Brussels Congress, February 10, 1927, *SWJN,* 2: 272–76.
17. Nehru's touching "message" when he learned of Toller's death is in ibid., p. 279, n. 9.
18. Father to Jawahar, Anand Bhawan, December 2, 1926, *Old Letters,* p. 51.
19. Ibid., pp. 51–52.
20. Ibid., p. 52.
21. Gandhi to Jawaharlal, Mysore, March (not May) 25, 1927, *Old Letters,* p. 57.
22. Jawaharlal to Gandhi, London, April 22, 1927, *SWJN,* 2: 325–26.
23. Jawaharlal to Mahmud, Montreux, July 14, 1927, ibid., pp. 328–29.
24. *We Nehrus,* p. 72.
25. Jawahar to Nan, Moscow, November 10, 1927, *SWJN,* 2: 369–70.
26. Jawahar to Nan, Poland, November 12, 1927, ibid., pp. 371–72.
27. Ibid., p. 374.
28. *Auto,* p. 165.

CHAPTER 8

1. Subjects Committee of Madras Congress, December 25, 1927, *SWJN,* vol. 3 (1972), p. 3.
2. Ibid., p. 4.
3. Ibid.
4. Nehru's presidential address to the Republican Congress, which met in Madras on December 28, 1927, *SWJN,* 3: 7–8.
5. Gandhi to Jawaharlal, Sabarmati, January 4, 1928, *Old Letters,* p. 58.
6. *SWJN,* 3: 10, n. 2.
7. Jawaharlal to Gandhi, Allahabad, January 11, 1928, ibid., p. 11.
8. Ibid., pp. 12–15.

9. Gandhi to Jawaharlal, Sabarmati, January 17, 1928, *Old Letters,* pp. 59–60.
10. Nehru's letter to the editor, *The Leader,* January 25, 1928, *SWJN,* 3: 17.
11. Jawaharlal to Gandhi, Allahabad, January 23, 1928, ibid., pp. 18–19.
12. Jawaharlal to Gandhi, February 23, 1928, ibid., p. 36.
13. *Jinnah of Pakistan,* p. 95.
14. Nehru's statement at the All Parties Conference, Delhi, March 8, 1928, *SWJN,* 3: 37.
15. Jawaharlal to Mahmud, March 17, 1928, ibid.
16. Gandhi to Indira, *Freedom's Daughter,* p. 38.
17. Subhas to Motilal Nehru, Calcutta, July 18, 1928, *Old Letters,* p. 62.
18. Motilal Nehru Gandhi, July 11, 1928, ibid., pp. 60–61.
19. Motilal Nehru to Annie Besant, September 30, 1928, ibid., p. 67.
20. Motilal Nehru to Gandhi, July 11, 1928, ibid., p. 61.
21. Jawahar to Mahmud, June 30, 1928, *SWJN,* 3: 51.
22. Nehru's speech at the All Parties Conference, Lucknow, August 29, 1928, ibid., pp. 57–60.
23. *Jinnah of Pakistan,* p. 101.
24. Annie Besant's statement to the press, *SWJN,* 3: 76, n. 2.
25. Nehru's interview, November 16, 1928, ibid., p. 77.
26. *Auto,* p. 174.
27. Nehru's message to the students of Lucknow, ibid., pp. 104–5.
28. Ibid., pp. 177–78.
29. Ibid., p. 178.
30. Ibid., pp. 179–80.
31. Ibid., p. 181.
32. Nehru's statement to the press, Allahabad, December 1, 1928, ibid., pp. 108–15.
33. Nehru's presidential address to the Bombay Youth Conference, December 12, 1928, ibid., pp. 203–4.
34. Ibid., pp. 205–6.
35. Ibid., pp. 207–10.
36. Jawaharlal on "independence" in the Subjects Committee of the Calcutta Congress, December 27, 1928, ibid., pp. 270–74.
37. D. G. Tendulkar, *Mahatma,* vol. 2 (1951), p. 441.
38. Motilal Nehru's presidential address to the forty-third Congress in Calcutta, December 29, 1928, *Voice of Freedom,* pp. 54–69.
39. B. R. Nanda, *The Nehrus, Motilal and Jawaharlal* (New York: John Day, 1963) p. 337.
40. Jawaharlal's amendment to the resolution on independence, *SWJN,* 3: 270–73.
41. Motilal's letter to Gandhi on January 12, 1929 is quoted in Nanda, *The Nehrus,* p. 307.
42. Ibid.

CHAPTER 9

1. Nehru's speech in Delhi, February 5, 1929, *SWJN,* vol. 4 (1973), pp. 1–2.
2. B. R. Nanda, *The Nehrus: Motilal and Jawaharlal* (New York: John Day, 1963), p. 310.
3. Motilal speaking against the finance bill on March 19, 1929, *Voice of Freedom,* p. 461.
4. *Jinnah of Pakistan,* p. 105.
5. Nehru in Lahore, February 8, 1929, *SWJN,* 4: 2–3.
6. Deva to Nehru, Benares, February 9, 1929, *Old Letters,* pp. 72–73.
7. Jawaharlal to Gandhi, Allahabad, July 13, 1929, *SWJN,* 4: 156.
8. Promilla Kalhan, "A Husband's Testimony," in her *Kamala Nehru: An Intimate Biography* (Delhi: Vikas, 1973), p. 107.
9. Indira's *MT,* p. 19.
10. *Auto,* p. 189.
11. Nehru to Comrade D. B. Kulkarni, Allahabad, September 10, 1929, *SWJN,* 4: 38–40.

12. Nehru to press, June 2, 1929, ibid., p. 101.
13. Nehru on Britain's Labour government, Calcutta, June 7, 1929, ibid., p. 102.
14. Nehru to Fenner Brockway, August 1, 1929, ibid., p. 109.
15. Nanda, *The Nehrus,* pp. 312–13.
16. Jawaharlal to Gandhi, July 13, 1929, ibid., p. 156.
17. Ibid., p. 157.
18. Jawahar to father, Sabarmati, August 30, 1929, ibid., pp. 158–59.
19. *Auto,* pp. 194–95.
20. Sarojini Naidu to Jawahar, Lucknow, September 29, 1929, *Old Letters,* p. 75.
21. *Jinnah of Pakistan,* p. 107.
22. Ibid., p. 109.
23. Ibid., p. 110.
24. Delhi Manifesto, November 1, 1929, *SWJN,* 4: 165–66.
25. Jawaharlal to Gandhi, November 4, 1929, ibid., pp. 166–68.
26. Gandhi to Jawaharlal, Aligarh, November 4, 1929, *Old Letters,* p. 78.
27. Gandhi to Jawaharlal, November 8, 1929, ibid., p. 80.
28. V. Chattopadhyaya to Jawahar, Berlin, December 4, 1929, ibid., pp. 81–83.
29. Nehru's presidential address to the All-India Trade Union Congress, Nagpur, November 30, 1929, *SWJN,* 4: 49–51.
30. *Auto,* p. 197.
31. *Jinnah of Pakistan,* p. 111.
32. Nehru's speech in Lahore, December 29, 1929, *SWJN,* 4: 183.
33. Indira's *MT,* p. 22.
34. Nehru's presidential address to the Lahore Congress, December 29, 1929, *SWJN,* 4: 184–98.
35. Ibid.
36. "Independence Day Pledge," in H. Y. Sharada Prasad et al., eds., *Jawaharlal Nehru* (New Delhi: Jawaharlal Nehru Memorial Fund, 1983), p. 92.
37. *Auto,* p. 202.
38. "Independence Day Pledge," p. 92.
39. *Auto,* p. 203.
40. Ibid., pp. 204–6.
41. *We Nehrus,* p. 83.
42. M. K. Gandhi to Viceroy Irwin, Sabarmati, March 2, 1930, in P. Sitaramayya, *The History of the Indian National Congress,* vol. 1, 1885–1935 (Bombay: Padma, 1946), pp. 372–76.
43. Gandhi to Jawaharlal, March 11, 1930, *Old Letters,* p. 86.
44. Nehru's statement of March 14, 1930, *SWJN,* 4: 259.
45. *Auto,* p. 212.
46. Sitaramayya, *History,* p. 388.
47. Nehru, leaving the courtroom, April 1930, *SWJN,* 4: 314.
48. Sitaramayya, *History,* p. 395.
49. PD, May 5, 1930, *SWJN,* 4: 341.

CHAPTER 10

1. Jawahar to father, Naini Prison, May 14, 1930, *SWJN,* 4: 345.
2. Ibid., p. 347.
3. Jawahar to Nan, June 25, 1930, ibid., pp. 361–63.
4. *Auto,* pp. 207–8.
5. B. R. Nanda, *The Nehrus: Motilal and Jawaharal* (New York: John Day, 1963), p. 333.
6. P. Sitaramayya, *The History of the Indian National Congress,* vol. 1, *1885–1935* (Bombay: Padma, 1946), p. 401.

7. *PD,* June 28, 1930, *SWJN,* 4: 367.
8. Ibid., June 7, 1930, pp. 367–68.
9. Motilal to Krishna (Betty), July 30, 1930, *Old Letters,* p. 92.
10. Jawaharlal to Gandhi, July 28, 1930, *SWJN,* 4: 369–71.
11. PD, August 1, 1930, ibid., p. 373.
12. PD, August 9, 1930, ibid., p. 374.
13. Both Nehrus and Gandhi as well as other Congress leaders signed this letter to Sapru and Jayakar, addressed "Dear Friends." The letter was drafted by Jawaharlal, August 15, 1930, ibid., pp. 375, 376.
14. Jawahar to father, September 10, 1930, ibid., p. 383.
15. October 11, 1930, ibid., p. 391.
16. Nehru's speech at Allahabad, October, 12, 1930, ibid., pp. 392–95.
17. "Message to Comrades," October 24, 1930, ibid., pp. 413–14.
18. Nehru's statement at his trial, October 24, 1930, ibid., pp. 414–16.
19. PD, December 7, 1930, ibid., pp. 425–26.
20. PD, New Year's Day, 1931, ibid., p. 451.
21. PD, January 12, 1931, ibid., p. 452.
22. Jawahar to father, January 12, 1931, ibid., pp. 452–53.
23. *Jinnah of Pakistan,* p. 121.
24. Ibid., p. 123.
25. The prime minister's declaration at the R.T.C., January 19, 1931, is Appendix I, *CWMG,* vol. 45 (December 1930–April 1931) (1971), pp. 424, 426.
26. Nehru's note in response to the declaration, *SWJN,* 4: 453–54.
27. Jawahar to father, January 22, 1931, ibid., pp. 454–56, 457.
28. Viceroy's statement, January 26, 1931, Appendix II, *CWMG,* 45: 427.
29. Gandhi's interview to the Associated Press of India, January 26, 1931, ibid., p. 125.
30. Gandhi's Bombay interview, January 27, 1931, ibid., p. 128.
31. Tendulkar, *Mahatma,* vol. 3 (1952), p. 66.
32. *Scope of Happiness,* p. 99.
33. Tendulkar, *Mahatma,* 3: 66.
34. *Auto,* p. 247.

CHAPTER 11

1. *Auto,* pp. 249–50.
2. Ibid., p. 251.
3. Gandhi's report of his first interview with the viceroy, February 17, 1931, *CWMG,* 45: 188–91.
4. Gandhi's report of his second interview, February 18, 1931, ibid., p. 197.
5. Ibid., p. 201.
6. Irwin's cable to Secretary of State Wedgwood Benn, Appendix IV, ibid., p. 429.
7. February 19, 1931, ibid., p. 206.
8. Tendulkar, *Mahatma,* 3: 69.
9. February 27, 1931, *CWMG,* 45: 235.
10. March 2, 1931, ibid., p. 242.
11. Nehru's note on the provisional settlement, Appendix V, ibid., p. 431.
12. Gandhi's statement to the press, March 5, 1931, ibid., p. 255.
13. *Auto,* pp. 257–59.
14. Gandhi's statement to the press, March 5, 1931, ibid., p. 255.
15. Nehru's speech to the Students Convention, Karachi, March 27, 1931, *SWJN,* 4: 503.
16. Tendulkar, *Mahatma,* 3: 92–93.
17. Nehru in Karachi Congress, March 29, 1931, *SWJN,* 4: 505.

18. *Auto,* p. 266.
19. Nehru, Karachi Congress, March 31, 1931, *SWJN,* 4: 507.
20. Nehru on "fundamental rights" in Karachi, ibid., pp. 511–12.
21. *We Nehrus,* p. 100.
22. *Auto,* p. 265.
23. Jawahar to Nan, April 22, 1931, *SWJN,*4: 520.
24. Jawahar to Betty, May 22, 1931, ibid., pp. 532–33.
25. Nehru to Fenner Brockway, August 20, 1931, in *SWJN,* vol. 5 (1973), p. 17.
26. Tendulkar, *Mahatma,* 3: 139.
27. Jawaharlal to Gandhi, September 1, 1931, *SWJN,* 5: 29.
28. Jawahar to Gandhi, September 27, 1931, ibid., pp. 46–47.
29. *Jinnah of Pakistan,* p. 125.
30. Jawaharlal to Ansari, October 4, 1931, *SWJN,* 5: 48.
31. Nehru's speech, Delhi, September 24, 1931, ibid., p. 128.
32. Jawaharlal to Gandhi, October 1, 1931, ibid., p. 135.
33. Nehru to Chief Secretary Prasad, October 15, 1931, ibid., p. 151.
34. Jawaharlal to Gandhi, October 16, 1931, ibid., pp. 156–57.
35. Tendulkar, *Mahatma,* 3: 160.
36. Jawaharlal to Vallabhbhai, November 2, 1931, *SWJN,* 5: 169.
37. Nehru's interview to the press, Bombay, November 7, 1931, ibid., p. 175.
38. Nehru to Kisans Conference, Allahabad, November 25, 1931, ibid., p. 180.
39. Nehru to Mieville, November 28, 1931, ibid., p. 182.
40. *Auto,* pp. 319–20.

CHAPTER 12

1. Nehru's statement in District Magistrate's Court, Allahabad, January 4, 1932, *SWJN,* 5: 348.
2. PD, January 2, 1932, ibid., p. 349.
3. *Scope of Happiness,* pp. 102–3.
4. PD, January 27, 1932, *SWJN,* 5: 351.
5. *We Nehrus,* pp. 102–5.
6. PD, February 7, 1932, *SWJN,* 5: 354–55.
7. PD, February 8, 1932, ibid., pp. 355–56.
8. PD, March 15, 1932, ibid., p. 369.
9. PD, March 20, 1932, ibid., p. 370.
10. PD, March 29, 1932, ibid., pp. 371, 372.
11. Nehru to Bidhan (Roy), April 6, 1932, ibid., pp. 376–77.
12. PD, April 23, 1932, ibid., p. 381.
13. *Auto,* pp. 334–35.
14. *We Nehrus,* p. 106.
15. PD, May 10, 1932, *SWJN,* 5: 382.
16. Jawahar to Nan and Betty, May 19, 1932, ibid., pp. 383–84.
17. PD, June 6, 1932, ibid., p. 387.
18. PD, June 8, 1932, ibid., pp. 388–89.
19. PD, June 13, 1932, ibid., pp. 393–94.
20. Promilla Kalhan, *Kamala Nehru: An Intimate Biography* (Delhi: Vikas, 1973), p. 85.
21. Kamala Nehru to Bhai Abhayandaji, November 26, 1933, ibid., p. 86.
22. Gandhi to Indira, June 15, 1932, *SWJN,* 5: 394–95.
23. PD, June 23, 1932, ibid., p. 396.
24. PD, July 17, 1932, ibid., p. 399.
25. PD, July 29, 1932, ibid., p. 400.

26. PD, August 13, 1932, ibid., p. 401.
27. PD, August 16, 1932, ibid., p. 402.
28. PD, August 18, 1932, ibid., pp. 402, 403.
29. Gandhi to MacDonald, August 18, 1932, *CWMG,* vol. 50 (June–August 1932) (1972), pp. 383–84.
30. Gandhi to MacDonald, September 9, 1932, *CWMG,* vol. 51 (September 1–November 15, 1932) (1972), p. 31.
31. PD, September 22, 1932, *SWJN,* 5: 407–8.
32. Gandhi's statement to the press, September 16, 1932, *CWMG,* 51: 62–64.
33. PD, September 22, 1932, *SWJN,* 5: 408.
34. Tagore to Gandhi, September 19, 1932, *CWMG,* 51: 109, n. 2.
35. Gandhi's statement on September 20, 1932, ibid., pp. 116, 118–19.
36. Gandhi's discussion with Ambedkar, September 22, 1932, ibid., pp. 458–60.
37. Gandhi's message to Great Britain, September 25, 1932, ibid., pp. 140–41.
38. Gandhi's statement to the press, September 26, 1932, ibid., pp. 143–44.
39. PD, September 27, 1932, *SWJN,* 5: 410.
40. Indira's *MT,* p. 24.
41. Nehru to Indira, October 3, 1932, *SWJN,* 5: 410.
42. PD, October 16, 1932, ibid., p. 417.
43. PD, October 25, 1932, pp. 419–20.
44. PD, October 26, 1932, ibid., pp. 420–21.
45. PD, October 29, 1932, ibid., p. 422.
46. Nehru to Indira, November 1, 1932, ibid., pp. 422–23.
47. PD, November 13, 1932, ibid., pp. 425, 426.
48. PD, November 14, 1932, ibid., p. 426.
49. PD, November 17, 1932, ibid., p. 427.
50. PD, November 21, 1932, ibid., p. 433.
51. Kalhan, *Kamala Nehru,* pp. 88–89.
52. Jawaharlal to mother, December 14, 1932, *SWJN,* 5: 439–41.
53. PD, December 20, 1932, ibid., pp. 442–43.
54. Quoted in *Times of India,* November 10, 1932, *CWMG,* 51: 467–68.
55. PD, December 20, 1932, *SWJN,* 5: 443.
56. Jawaharlal to Gandhi, January 5, 1933, ibid., pp. 450–51.
57. PD, January 8, 1933, ibid., p. 452.
58. PD, January 14, 1933, ibid., p. 454.
59. Nehru to Wood, January 23, 1933, ibid., p. 456.
60. PD, February 5, 1933, ibid., p. 457.
61. Gandhi to Jawaharlal, Poona, February 5, 1933, *Old Letters,* p. 111.
62. Jawaharlal to Gandhi, March 7, 1933, *SWJN,* 5: 459–62.
63. PD, March 12, 1933, ibid., pp. 463–64.
64. Nehru to Girdhari Lal, March 20, 1933, ibid., pp. 465–66.
65. Nehru to Indira, February 21, 1933, *Freedom's Daughter,* p. 81.
66. Nehru to Indira, March 20, 1933, ibid., p. 84.
67. PD, March 30, 1933, *SWJN,* 5: 470.
68. Indira to Nehru, March 17, 1933, *Freedom's Daughter,* pp. 82–83.
69. Nehru to Indira, April 3, 1933, ibid., pp. 84–85.
70. PD, April 1, 1933, *SWJN,* 5: 470–71.
71. Jawaharlal to mother, April 4, 1933, ibid., p. 471.
72. Tendulkar, *Mahatma,* 3: 243.
73. Jawaharlal to Gandhi, May 5, 1933, *SWJN,* 5: 473–74.
74. Nehru to Desai, May 23, 1933, ibid., pp. 475–76, 477.
75. PD, June 4, 1933, ibid., pp. 478–79.
76. Jawahar to Betty, June 13, 1933, ibid., pp. 481–83.

77. PD, June 18, 1933, ibid., pp. 483–84.
78. PD, June 19, 1933, ibid., p. 484.
79. PD, July 18, 1933, ibid., pp. 489, 499.
80. Nehru's interview in Allahabad, August 30, 1933, ibid., pp. 505–6.

CHAPTER 13

1. Nehru's interview in *The Pioneer,* Lucknow, August 31, 1933, *SWJN,* 5: 506–7.
2. Ibid., p. 507.
3. Nehru to Desmond Young, Lucknow, September 1, 1933, ibid., p. 515.
4. Nehru on Congress, Allahabad, September 4, 1933, ibid., pp. 517–19.
5. Nehru on the world situation, ibid., p. 519.
6. *We Nehrus,* p. 116.
7. Ibid., pp. 113, 114, 115, 116.
8. Jawaharlal to Gandhi, September 13, 1933, *SWJN,* 5: 526.
9. Ibid., pp. 527–28.
10. Gandhi to Jawaharlal, Poona, September 14, 1933, *CWMG,* vol. 55 (April–September 1933) (1973), pp. 427–30.
11. *Auto,* p. 403.
12. Jawaharlal to Gandhi, *SWJN,* 5: 529–30.
13. Interview at Kanpur, September 19, 1933, ibid., p. 538.
14. "Whither India?" *SWJN,* vol. 6 (1974), pp. 1–16, 2–3.
15. Ibid., pp. 4–5.
16. Ibid., p. 12.
17. Ibid., p. 16.
18. Nehru on "some criticisms," ibid., pp. 18–19.
19. Indira to Nehru, November 11, 1933, *Freedom's Daughter,* p. 102.
20. Nehru to Indira, November 14, 1933, ibid., p. 103.
21. Gandhi to Jawaharlal, October 9, 1933, *CWMG,* vol. 56 (1933–1934) (1973), p. 79.
22. Nehru to Juddelpore Municipality, December 5, 1933, ibid., p. 80.
23. Nehru's speech in Delhi, December 12, 1933, *SWJN,* 6: 129–31.
24. Nehru to Indira, Santiniketan, January 19, 1934, *Freedom's Daughter,* pp. 108–9.
25. Tendulkar, *Mahatma,* 3: 303–4.
26. Nehru, "Earthquakes—Natural and Political," January 31, 1934, *SWJN,* 6: 151–53, 154.
27. Nehru, Allahabad, January 29, 1934, ibid., p. 191.
28. Nehru to Patna press, February 10, 1934, ibid., pp. 195–96.
29. Nehru, February 12, 1934, ibid., pp. 196–98.
30. Ibid., p. 227.
31. Allahabad, February 12, 1934, *Freedom's Daughter,* p. 111.
32. PD, *SWJN,* 6: 228.

CHAPTER 14

1. Nehru's court statement, February 15, 1934, *SWJN,* 6: 233–35.
2. Nehru to Indira, March 30, 1934, ibid., p. 246.
3. Tendulkar, *Mahatma,* 3: 306, 307–8.
4. Ibid., p. 318.
5. PD, April 13, 1934, *SWJN,* 6: 247–48.
6. *Auto,* pp. 504, 505.
7. Ibid., pp. 505–6.

8. Ibid., p. 506.
9. Ibid., p. 507.
10. Ibid., pp. 509–12.
11. PD, May 9, 1934, *SWJN,* 6: 251.
12. PD, May 25, 1934, ibid., p. 254.
13. PD, June 3, 1934, ibid., pp. 255–56.
14. Indira to Nehru, Srinagar, May 28, 1934, *Freedom's Daughter,* pp. 115–16.
15. Nehru to Indira, June 15, 1934, *SWJN,* 6: 256–58.
16. PD, June 20, 1934, ibid., p. 259.
17. PD, June 25, 1934, ibid., p. 260.
18. PD, June 26, 1934, ibid.
19. PD, June 30, 1934, ibid., p. 262.
20. Indira to Nehru, July 7, 1934, *Freedom's Daughter,* p. 122.
21. Nehru to Indira, July 12, 1934, *SWJN,* 6: 263.
22. Jawaharlal to Gandhi, Anand Bhawan, August 13, 1934, Tendulkar, *Mahatma,* 3: 379–80.
23. Ibid., pp. 380–81.
24. Ibid., p. 382.
25. Ibid.
26. Gandhi to Jawaharlal, Wardha, August 17, 1934, ibid., pp. 384–85.
27. Gandhi to Vallabhbhai, September 1934, ibid., p. 386.
28. Ibid., pp. 387–88.
29. PD, November 14, 1934, *SWJN,* 6: 302–3.
30. Nehru to Indira, November 26, 1934, ibid., p. 305.
31. PD, December 6, 1934, ibid., p. 306.
32. PD, December 21, 1934, ibid., p. 307.
33. PD, January 1, 1935, ibid., pp. 307, 308.
34. PD, January 12, 1935, ibid., p. 309.
35. PD, January 16, 1935, ibid., pp. 310–11.
36. PD, January 22, 1935, ibid.
37. PD, February 1, 1935, ibid., pp. 312, 313.
38. Nehru to Indira, February 4, 1935, ibid., pp. 314–15.
39. Indira to Nehru, January 12, 1935, *Freedom's Daughter,* p. 136.
40. Nehru to Indira, February 4, 1935, *SWJN,* 6: 316–17.
41. Ibid., p. 318.
42. PD, February 6, 1935, ibid., p. 319.
43. *Jinnah of Pakistan,* pp. 137–38.
44. Nehru to Indira, February 22, 1935, *SWJN,* 6: 323–24.
45. PD, March 3, 1935, ibid., pp. 326–27.
46. PD, March 5, 1935, ibid., pp. 327–28.
47. PD, March 10, 1935, ibid., pp. 329–30.
48. PD, March 12, 1935, ibid., p. 330.
49. PD, March 19, 1935, ibid., pp. 331–32.
50. Indira to Nehru, March 27, 1935, *Freedom's Daughter,* p. 151.
51. Nehru to Indira, April 4, 1935, *SWJN,* 6: 342.
52. Jawaharlal to Gurudev (Tagore), April 16, 1935, ibid., pp. 346–47.
53. PD, April 17, 1935, ibid., p. 347.
54. PD, April 30, 1935, ibid., p. 349.
55. Ibid., pp. 348–51.
56. Nehru to Indira, May 6, 1935, ibid., pp. 352–55.
57. Ibid., pp. 356–57.
58. PD, May 16, 1935, ibid., pp. 364–65.
59. Nehru to Indira, June 7, 1935, ibid., pp. 368–69.

60. Ibid., p. 370.
61. PD, June 9, 1935, ibid., pp. 371–72.
62. PD, June 13, 1935, ibid., pp. 372–73.
63. Nehru to Indira, July 5, 1935, *Freedom's Daughter*, pp. 182–84.
64. Jawahar to My dear Madan (Atal), July 5, 1935, *SWJN*, 6: 384.
65. Nehru to Indira, July 5, 1935, ibid., pp. 385–86.
66. Ibid., pp. 391–92.
67. PD, September 2, 1935, ibid., p. 416.

CHAPTER 15

1. PD, September 2, 1935, *SWJN*, 6: 417–18.
2. Nehru's statement to the press, September 3, 1935, ibid., p. 418.
3. Nehru's diary, September 5, 1935, *SWJN*, vol. 7 (1975), p. 1.
4. Jawaharlal to My dear Gurudev (Tagore), September 19, 1935, ibid., p. 3.
5. Kamala Nehru to Bhai Abhayanandaji, Bhowali, March 28, 1935, in Promilla Kalhan, *Kamala Nehru: An Intimate Biography*(Delhi: Vikas, 1973) pp. 93–94.
6. Kamala to Abhayananda, Berlin, June 13, 1935, ibid., pp. 96–97.
7. Agatha Harrison to Mahatmaji, September 13, 1935, Agatha Harrison Papers, Friends House, London, TEMP MSS 46/12.
8. Nehru to Alexander, September 17, 1935, *SWJN*, 7: 2.
9. Nehru to Bharati, October 19, 1935, ibid., p. 5.
10. Jawaharlal Nehru to Bharati, October 24, 1935, ibid., p. 5.
11. Jawahar to Bharati, November 21, 1935, ibid., pp. 6–8.
12. Ibid., p. 8.
13. Jawahar to Bharati, November 25, 1935, ibid., pp. 9–10.
14. Seton's, *Panditji*, p. 88.
15. Gandhi to Jawaharlal, October 3, 1935, *Old Letters*, p. 123.
16. Nehru to Agatha, September 25, 1935, *SWJN*, 7: 25–29.
17. Nehru to Amiya Chakravarty, November 29, 1935, ibid., p. 13.
18. Nehru to Rajendra Prasad, November 20, 1935, ibid., pp. 38–40.
19. Nehru to Krishna Menon, December 9, 1935, ibid., pp. 15–16.
20. Nehru to Lord Lothian, December 9, 1935, Agatha Harrison Papers.
21. Agatha's confidential memo, December 12, 1935, ibid.
22. Nehru to Lothian, January 17, 1936, *SWJN*, 7: 62–64.
23. Ibid., pp. 65–67.
24. Nehru's interview on his return to London, January 26, 1936, ibid., pp. 78–79.
25. Nehru's speech in Caxton Hall, London, February 3, 1936, ibid., pp. 88–90.
26. Nehru to the India Conciliation Group, February 4, 1936, ibid., pp. 90–97.
27. Nehru to the press, London, February 6, 1936, ibid., p. 113.
28. Nehru to Swami Abhayanandaji, Lausanne, February 26, 1936, in Kalhan, *Kamala Nehru*, p. 122.
29. Seton's *Panditji*, p. 86.
30. Nehru to Indira, Cairo, March 8, 1936, *Freedom's Daughter*, pp. 241–42, 240.
31. Nehru to Indira, Baghdad, March 9, 1936, ibid., pp. 243–45.
32. Gandhi to Jawaharlal, Delhi, March 9, 1936, *Old Letters*, p. 174.

CHAPTER 16

1. Nehru to Indira, Allahabad, March 12, 1936, *Freedom's Daughter*, p. 245.
2. Rajendra Prasad to Jawaharlal, Wardha, December 19, 1935, *Old Letters*, pp. 159–60.

3. Ibid., p. 160.
4. Subhas Bose to My dear Jawahar, March 4, 1936, ibid., pp. 172–73.
5. Tagore's tribute to Kamala, March 8, 1936, ibid., pp. 179–80.
6. Nehru to youth, March 15, 1936, SWJN, 7: 163.
7. Nehru to Indira, Delhi, March 22, 1936, *Freedom's Daughter,* p. 247.
8. Nehru to Indira, March 26, 1936, ibid., p. 248.
9. Nehru to Indira, March 30, 1936, ibid., p. 249.
10. Indira to Nehru, Bex, April 1, 1936, ibid., pp. 250–251.
11. Nehru to Krishna Menon, September 28, 1936, *SWJN,* 7: 471.
12. Nehru to Subjects Committee, Lucknow, April 11, 1936, ibid., pp. 167–69.
13. Nehru's presidential address, Lucknow, April 12, 1936, ibid., pp. 170, 171, 172–73.
14. Ibid., p. 173.
15. Gandhi to Agatha, Wardha, April 30, 1936, *Old Letters,* p. 182.
16. Nehru's address, Lucknow, *SWJN,* 7: 179–80.
17. Ibid., pp. 182–85.
18. Nehru in Bombay, May 15, 1936, ibid., pp. 219–22.
19. Nehru, Bombay, May 18, 1936, ibid., p. 235.
20. Gandhi to Jawaharlal, May 21, 1936, *Old Letters,* p. 183.
21. Gandhi to Jawaharlal, May 29, 1936, ibid., p. 184.
22. Rajendra Prasad et al. to Jawaharlalji, Wardha, June 29, 1936, ibid., p. 191.
23. Nehru to Indira, June 29, 1936, *Freedom's Daughter,* p. 261.
24. Rajendra Prasad to Jawaharlalji, July 1, 1936, *Old Letters,* pp. 192–94.
25. Nehru to Indira, Wardha, July 30, 1936, *Freedom's Daughter,* pp. 261–62.
26. Jawaharlal to Gandhi, Allahabad, July 5, 1936, *Old Letters,* pp. 194–98.
27. Gandhi to Jawaharlal, July 8, 1936, ibid., p. 198.
28. Nehru to Indira, July 28, 1936, *Freedom's Daughter,* p. 268.
29. Ibid., pp. 268–69.
30. Nehru's press conference on Punjab, August 3, 1936, *SWJN,* 7: 330–31.
31. Nehru to Agatha, August 6, 1936, ibid., p. 332.
32. Nehru's statement in Allahabad, August 9, 1936, ibid., pp. 333–34.
33. Nehru to Students Conference, August 12, 1936, ibid., p. 335.
34. Ibid., pp. 336–37.
35. Ibid., p. 338.
36. Ibid., p. 339.
37. Ibid, pp. 339, 342–43.
38. Nehru to Agatha, September 3, 1936, ibid., pp. 346–47.
39. August 8, 1936, *CWMG,* vol. 63 (1976), pp. 206–7.
40. Gandhi to Jawaharlal, August 28, 1936, ibid., p. 249.
41. Gandhi to Indira, Allahabad, September 3, 1936, *Freedom's Daughter,* p. 278.
42. Nehru to the press, Allahabad, September 16, 1936, *SWJN,* 7: 350.
43. Nehru's account of his tour, September 16, 1936, ibid., pp. 348–49.
44. Nehru to the editor, *The Leader,* September 20, 1936, ibid., p. 352.
45. *CWMG,* 63: 347.
46. Nehru to Prema Kantak, September 25, 1936, *SWJN,* 7: 353.
47. Nehru to Chimanlal Shah, September 25, 1936, ibid., p. 354.
48. Nehru to Krishna Menon, November 19, 1936, ibid., p. 356.
49. Nehru to Krishna Menon, October 29, 1936, ibid., pp. 438–39.
50. Nehru to Indira, October 15, 1936, *Freedom's Daughter,* pp. 286–87.
51. Nehru to Indira, October 29, 1936, ibid., p. 293.
52. Indira to Nehru, November 7, 1936, ibid., pp. 293–94.
53. Nehru to Indira, November 5, 1936, ibid., p. 293.
54. Nehru's statement to the press, Bareilly, November 20, 1936, *SWJN,* 7: 591; 594, n. 4.
55. Nehru to Indira, January 4, 1937, *Freedom's Daughter,* p. 307.

56. Nehru to A.I.C.C, Faizpur, December 25, 1936, *SWJN,* 7: 595–96.
57. Nehru's presidential address, Faizpur, December 27, 1936, ibid., pp. 598–99.
58. Indira to Nehru, December 6, 1936, *Freedom's Daughter,* pp. 300–301.
59. Nehru to Indira, December 4, 1936, ibid., pp. 302–3.
60. Presidential address, Faizpur, *SWJN,* 7: 599–600, 602–3.
61. Ibid., pp. 604–5, 608.
62. Ibid., pp. 611–12.
63. Nehru's concluding address, Faizpur, December 28, 1936, ibid., pp. 614–15.
64. Ibid., p. 615.
65. Faizpur, December 29, 1936, ibid., p. 618.

CHAPTER 17

1. Nehru to Indira, January 14, 1937, *Freedom's Daughter,* pp. 309–10.
2. Jawaharlal to Krishna Menon, January 4, 1937, *SWJN,* vol 8 (1976), p. 3.
3. Nehru at Ghazipur, January 13, 1937, ibid., pp. 4–5.
4. Nehru to the press, Fyzabad, January 15, 1937, ibid., p. 6.
5. Nehru in Bombay, February 10, 1937, ibid., p. 25.
6. *Scope of Happiness,* p. 129.
7. Nehru's message, February 16, 1937, *SWJN,* 8: 28.
8. Jawaharlal to Krishna Menon, February 22, 1937, ibid., pp. 29–30.
9. Nehru to Cripps, February 22, 1937, ibid., p. 31.
10. Ibid., pp. 32–34.
11. Nehru to the press, Purnea, January 10, 1937, ibid., pp. 119–21.
12. Nehru to "Dear Comrade," March 31, 1937, ibid., p. 123.
13. Jawaharlal to Krishna Menon, March 28, 1937, ibid., p. 76.
14. Gandhi to Jawaharlal, April 5, 1937, *CWMG,* vol. 65 (1976), p. 55.
15. Gandhi to Agatha Harrison, April 5, 1937, ibid.
16. Nehru to Subhas Bose, April 16, 1937, *SWJN,* 8: 408.
17. Indira's *MT,* p. 35.
18. Nehru in Rangoon, May 7, 1937, *SWJN,* 8: 651–52.
19. Nehru to the press, Rangoon, May 20, 1937, ibid., pp. 657–58.
20. Jawaharlal to Krishna Menon, May 22, 1937, ibid., p. 659.
21. Ibid., p. 660.
22. Nehru to Indian Ladies Union, May 27, 1937, ibid., pp. 676–77.
23. Nehru to Rajendra Prasad, May 29, 1937, ibid., p. 678.
24. Nehru in Allahabad, June 24, 1937, ibid., p. 250.
25. Working Committee resolution, Wardha, July 7, 1937, *CWMG,* 65: 373–74.
26. *Scope of Happiness,* pp. 133, 134, 137–78.
27. Nehru to Indira, September 23, 1937, *Freedom's Daughter,* pp. 332, 333.
28. Jawahar to Bebee, Allahabad, October 5, 1937, *SWJN,* 8: 519.
29. "The Rashtrapati," ibid., pp. 520–21.
30. Ibid., p. 523.
31. Ibid.
32. Ibid., p. 523.
33. Ibid.
34. Jawahar to Padmaja, October 20, 1937, ibid., pp. 523–24.
35. Ibid., pp. 524–25.
36. Gandhi to Jawaharlal, Wardha, October 12, 1937, *Old Letters,* p. 254.
37. Sarojini Naidu to Jawahar, Calcutta, November 13, 1937, ibid., p. 255.
38. Indira to Nehru, Oxford, November 8, 1937, *Freedom's Daughter,* p. 342.
39. Nehru to Indira, November 10, 1937, ibid., p. 346.

40. Nehru to Indira, January 30, 1938, ibid., p. 368.
41. Indira to Nehru, Oxford, February 3, 1938, ibid., pp. 373–74.
42. Indira to Nehru, November 21, 1937, ibid., p. 346.
43. Nehru to Govind Pant, November 25, 1937, *Old Letters,* pp. 263, 264.
44. Nehru to Indira, January 14, 1938, *Freedom's Daughter,* p. 357.
45. Nehru's report to A.I.C.C., Haripura, February 16, 1938, *SWJN,* 8: 751–54.
46. President Bose in Haripura, Tendulkar, *Mahatma,* vol. 4 (1952), pp. 266–67.
47. *Jinnah of Pakistan,* p. 153.
48. Ibid., pp. 153–54.
49. Ibid., p. 157.
50. Nehru to Jinnah, February 4, 1938, *SWJN,* 8: 214–16.
51. Nehru, April 7, 1938, ibid., pp. 873–74.
52. Nehru to Indira, March 15, 1938, *Freedom's Daughter,* pp. 385–87.
53. *SWJN,* 8: 874.
54. Indira to Nehru, Oxford, April 23, 1938, *Freedom's Daughter,* pp. 389–90.
55. Nehru to Indira, Allahabad, April 30, 1938, ibid., pp. 391–92.

CHAPTER 18

1. Aboard the SS *Biancamano,* June 11–13, 1938, *SWJN,* vol. 9 (1976), p. 8.
2. Nehru to Indira, Lucknow, April 24, 1938, *Freedom's Daughter,* p. 391.
3. Aboard the SS *Biancamano,* June 11–13, 1938, *SWJN,* 9: 17.
4. Nehru to Indira, Marseilles, June 14, 1938, *Freedom's Daughter,* p. 398.
5. London, June 19, 1938, *Manchester Guardian* June 24, 1938, *SWJN,* 9: 17–18.
6. Nehru over Paris Radio, June 20, 1938, ibid., p. 19.
7. Nehru at Hall of Nations, June 21, 1938, ibid., pp. 21–22.
8. Jawahar to Bharati, Houlgate, August 3, 1938, ibid., pp. 105–6.
9. Nehru to the press, London, June 23, 1938, ibid., p. 25.
10. Nehru to Commons reception, June 23, 1938, ibid., pp. 24–25.
11. Lothian to Nehru, Aylsham, June 24, 1938, *Old Letters,* pp. 287–88.
12. Nehru's note to Working Committee, August 1, 1938, *SWJN,* 9: 94–95.
13. Ibid., pp. 96–98.
14. Ibid., pp. 100–101.
15. Ibid., p. 101.
16. Ibid., pp. 102–3.
17. Ibid., pp. 103; 102, n. 11.
18. Nehru in Caxton Hall, London, June 27, 1938, ibid., pp. 27–29.
19. Nehru to Jivat, Prague, August 16, 1938, ibid., pp. 108–9.
20. Nehru to *National Herald,* Budapest, August 21, 1938, ibid., pp. 111–12.
21. *Scope of Happiness,* p. 146.
22. Indira's *MT,* p. 36.
23. Nehru to Kripalani, August 30, 1938, *SWJN,* 9: 117–18.
24. Kripalani to Nehru, September 9, 1938, *Old Letters,* p. 297.
25. Nehru's note to Working Committee, London, September 6, 1938, *SWJN,* 9: 132–34.
26. Nehru to Indira, London, September 15, 1938, *Freedom's Daughter,* pp. 398–99.
27. Indira to Nehru, Brentford, September 15, 1938, ibid., p. 399.
28. Nehru's "Munich Crisis," Geneva, September 20, 1938, *SWJN,* 9: 150–52.
29. Nehru's "Great Betrayal," Paris, September 22, 1938, ibid., pp. 153, 154.
30. Nehru to the press, London, September 26, 1938, ibid., p. 160.
31. Nehru's report from London, September 28, 1938, ibid., p. 163, n. 3.
32. Ibid., p. 164.
33. Ibid., pp. 164–65.

34. Nehru to Pantji, London, October 1, 1938, ibid., pp. 172–73.
35. Nehru to Nahas Pasha, London, October 1, 1938, ibid., pp. 175–78.
36. Nehru to Friends House, London, October 4, 1938, ibid., pp. 187–88.
37. Nehru to India League, November 10, 1938, ibid., p. 200.
38. Nehru's report, Bombay, November 17, 1938, ibid., pp. 202–3.
39. Jawahar to Krishna Menon, Wardha, November 21, 1938, ibid., p. 211.
40. Jaya Prakash Narayan to Nehru, Calicut, November 23, 1938, *Old Letters,* pp. 304–7.
41. Haig to Linlithgow, "secret," November 23, 1938, United Provinces governor's reports to the viceroy, L/PJ/5/266 (London: India Office Library).
42. Haig to Linlithgow, "secret," December 6, 1938, ibid.
43. Tagore to Jawaharlal, November 28, 1938, *Old Letters,* pp. 307–8.
44. Jawaharlal Nehru, *The Unity of India: Collected Writings, 1937–1940* (New York: John Day, 1942), pp. 126–27.
45. Nehru to Indira, Wardha, December 10, 1938, *Freedom's Daughter,* pp. 401–2.
46. Nehru to Indira, December 22, 1938, ibid., p. 402.
47. Ibid., p. 403.
48. Haig to Linlithgow, "secret," January 10, 1939, "secret," L/PJ/5/266 (London: India Office Library).
49. Jinnah, January 1, 1939, quoted in *SWJN,* 9: 354, n. 2.
50. Nehru to the press, Calcutta, January 3, 1939, ibid.
51. Gandhi in *Harijan,* December 17, 1938, *CWMG,* vol. 68 (1977), p. 208.
52. Nehru to Subhas, February 4, 1939, *Old Letters,* pp. 317–21.
53. Bose to Nehru, March 28, 1939, ibid., pp. 329–49.
54. Nehru's "Where Are We?" *SWJN,* 9: 488–89.
55. Nehru to Krishna Menon, March 16, 1939, ibid., p. 524.
56. Indira to Nehru, April 18, 1939, *Freedom's Daughter,* p. 408.
57. Nehru to Indira, April 26, 1939, ibid., p. 409.

CHAPTER 19

1. Nehru to Indira, September 30, 1939, *Freedom's Daughter,* p. 433.
2. Nehru to the press, Rangoon, September 8, 1939, *SWJN,* vol. 10 (1977) p. 119.
3. Tendulkar, *Mahatma,* vol. 5 (1952), pp. 196–97.
4. *Jinnah of Pakistan,* p. 171.
5. Working Committee resolution on the war, September 14, 1939, *SWJN,* 10: 122–35.
6. Ibid., p. 130, n. 8.
7. Nehru to *National Herald,* September 20, 1939, ibid., pp. 144–45.
8. Nehru's speech in Lucknow, September 25, 1939, ibid., p. 163.
9. Jawaharlal to Krishna Menon, September 26, 1939, ibid., p. 164.
10. Nehru's speech in Allahabad, September 28, 1939, ibid., pp. 164–65.
11. Nehru's cable to London, October 5, 1939, ibid., p. 169.
12. Nehru to Indira, Wardha, October 6, 1939, *Freedom's Daughter,* p. 436.
13. Nehru to Linlithgow, Wardha, October 6, 1939, *SWJN,* 10: 173.
14. Jawaharlal to Krishna Menon, October 6, 1939, ibid., pp. 174–76.
15. Ibid., p. 177.
16. Nehru's statement to A.I.C.C., Wardha, October 10, 1939, ibid., pp. 178–81.
17. Cripps to Nehru, October 11, 1939, *Old Letters,* pp. 395–96.
18. Raghu Nandan Saran to My dear Panditji, Delhi, October 14, 1939, ibid., pp. 398–99.
19. Nandan (Saran) to Panditji, October 17, 1939, ibid., pp. 400–401.
20. Nehru's "answer," Lucknow, October 17, 1939, *SWJN,* 10: 190–93.
21. Cable to *News Chronicle,* October 18, 1939, ibid., p. 194.
22. Lucknow, October 19, 1939, ibid., p. 201.

23. Nehru's editorial, Lucknow, October 20, 1939, ibid., pp. 203–4, 205–6.
24. Jawaharlal to Krishna Menon, October 25, 1939, ibid., p. 207.
25. Nehru's speech in Bombay, October 26, 1939, ibid., pp. 208–9.
26. Gandhi to Jawaharlal, Segaon, October 26, 1939, *Old Letters,* pp. 403–4.
27. Gandhi to Jawaharlal, November 4, 1939, ibid., p. 404.
28. Nehru to Indira, November 6, 1939, *Freedom's Daughter,* p. 443.
29. Jawaharlal to Krishna Menon, November 8, 1939, *SWJN,* 10: 230–31.
30. *Jinnah of Pakistan,* p. 176.
31. Sarojini Naidu to Jawahar, Diwali, November 1939, *Old Letters,* p. 407.
32. Edward Thompson to Jawaharlal, Aylesbury, December 3, 1939, ibid., pp. 409–12.
33. Jawaharlal to Edward Thompson, January 5, 1940, *SWJN,* 10: 602–3.

CHAPTER 20

1. Nehru in Bombay, December 14, 1939, *SWJN,* 10: 401–3.
2. Jawaharlal to Krishna Menon, Wardha, December 21, 1939, ibid., pp. 412, 413.
3. Ibid., pp. 413–14.
4. Gandhi in Segaon Ashram, January 5, 1940, *CWMG,* vol. 71 (1978), pp. 109–10.
5. Nehru to Indira, January 16, 1940, *Two Alone,* pp. 6–9.
6. Nehru to Indira, January 23, 1940, ibid., pp. 9–10.
7. Jawaharlal to Gandhi, January 24, 1940, *Old Letters,* p. 424.
8. Jawaharlal to Gandhi, February 4, 1940, ibid., pp. 425–26.
9. Ibid., pp. 425–26, 427.
10. Nehru to Indira, February 4, 1940, *Two Alone,* pp. 11–12.
11. Nehru to Indira, February 27, 1940, ibid., pp. 25–26.
12. Jawaharlal to Bharati, February 22, 1940, *SWJN,* 10: 614.
13. Maulana Abul Kalam Azad, *India Wins Freedom: The Complete Version* (Hyderabad: Orient Longman, 1988), p. 29.
14. Ibid., p. 30.
15. Appendix IV, Government's Communique, February 5, 1940, *CWMG,* 71: 438–39.
16. Gandhi's "Task Before Us," *Harijan,* February 10, 1940, written February 6, ibid., pp. 109–11.
17. Nehru to Indira, March 11, 1940, *Two Alone,* pp. 35–36.
18. Nehru's speech at Ramgarh, March 20, 1940, *SWJN,* vol. 11 (1978), pp. 11, 12.
19. Jinnah's presidential address to the Muslim League, Lahore, *Jinnah of Pakistan,* p. 182.
20. Nehru on Pakistan, Allahabad, April 13, 1940, *SWJN,* 11: 17.
21. Nehru to Indira, Bombay, April 20, 1940, *Two Alone,* pp. 53–54.
22. Nehru to Indira, May 10, 1940, ibid., p. 60.
23. Nehru to Indira, Kashmir, May 29, 1940, ibid., pp. 63–64.
24. Nehru to Indira, June 16, 1940, ibid., p. 65.
25. Working Committee resolution, passed June 21, 1940, *SWJN,* 11: 55–57.
26. Nehru to the press, Bombay, June 23, 1940, ibid., p. 65.
27. Tendulkar, *Mahatma,* 5: 352, 254–55.
28. Azad, *India Wins Freedom,* p. 35.
29. Tendulkar, *Mahatma,* 5: 365.
30. Nehru to the press, Delhi, July 8, 1940, *SWJN,* 11: 68.
31. Nehru, *National Herald,* July 16, 1940, ibid., p. 71.
32. Nehru to A.I.C.C, Poona, July 28, 1940, ibid., pp. 92, 93.
33. Ibid., p. 94.
34. Azad, *India Wins Freedom,* p. 34.
35. Jawaharlal to Krishna Menon, July 29, 1940, *SWJN,* 11: 96.
36. Nehru's *National Herald* editorial, August 9, 1940, ibid., pp. 99–101.
37. Nehru to Indira, Cawnpore, August 11, 1940, *Two Alone,* pp. 69–70.

38. Nehru to students, August 12, 1940, ibid., *SWJN,* 11: 119–20.
39. Nehru, August 16, 1940, ibid., pp. 120–21.
40. Nehru to the press, August 21, 1940, ibid., pp. 122–23.
41. Nehru to Indira, Wardha, August 20, 1940, *Two Alone,* p. 71.
42. Tendulkar, *Mahatma,* 5: 395.
43. Nehru to the press, Bombay, August 27, 1940, *SWJN,* 11: 127–28.
44. Nehru's address in Bombay, September 1, 1940, ibid., pp. 130–31.
45. Ibid., p. 133.
46. Nehru's speech at Gorakhpur, October 6, 1940, ibid., pp. 150–55.
47. Hallim Tennyson, *India's Walking Saint: The Story of Vinoba Bhave* (Garden City, N.Y.: Doubleday, 1995).
48. Gandhi's introduction of Vinoba Bhave, Tendulkar, *Mahatma,* 5: 427–29.
49. Nehru's report on his tour of Lucknow, October 22, 1940, *SWJN,* 11: 184–85.
50. Gandhi to Nehru, ibid., p. 186, n. 3.
51. Jawaharlal to Gandhi, October 24, 1940, ibid.
52. Nehru to Indira, October 25, 1940, *Two Alone,* pp. 74–75.
53. Nehru to Indira, District Jail, Gorakhpur, November 5, 1940, ibid., pp. 76–77.
54. Ibid.

CHAPTER 21

1. PD, November 14, 1940, *SWJN,* 11: 493.
2. Tendulkar, *Mahatma,* vol. 6 (1953), p. 4.
3. PD, January 26, 1941, *SWJN,* 11: 535.
4. *Scope of Happiness,* pp. 150–51.
5. Nehru to Indira, January 31, 1941, *SWJN,* 11: 539.
6. PD, April, 7, 1941, ibid., p. 571.
7. PD, April 8, 1941, ibid., p. 574.
8. Jawahar to Bebee (Padmaja), April 30, 1941, ibid., pp. 582–84.
9. PD, April 28, 1941, ibid., p. 586.
10. PD, May 2, 1941, ibid., p. 586.
11. PD, April 19, 1941, ibid., p. 575.
12. Nehru to Indira, May 3, 1941, ibid., pp. 589–90, 587.
13. Nehru to Indira, May 4, 1941, ibid., pp. 589–90.
14. Nehru to Indira, May 15, 1941, ibid., pp. 591–92, 593–95.
15. Ibid., pp. 595–96.
16. Indira to Nehru, Mussoorie, May 22, 1941, *Two Alone,* pp. 107–8.
17. Nehru to Indira, May 16, 1941, *SWJN,* 11: 597.
18. PD, May 29, 1941, ibid., pp. 604–5.
19. Indira to Nehru, Mussoorie, June 2, 1941, *Two Alone,* pp. 112–13.
20. *We Nehrus,* p. 143.
21. PD, June 23, 1941, *SWJN,* 11: 637.
22. Nehru to Indira, June 29, 1941, ibid., pp. 638–39.
23. Nehru to Indira, July 9, 1941, ibid., pp. 643–45.
24. Ibid., pp. 646, 647.
25. *We Nehrus,* p. 143.
26. Nehru to Indira, July 28, 1941, *Two Alone,* pp. 123–25.
27. PD, August 7, 1941, *SWJN,* 11: 670–71.
28. PD, August 14, 1941, ibid., pp. 674–76.
29. Ibid., pp. 676–77.
30. PD, September 4, 1941, ibid., p. 692.
31. PD, September 19, 1941, ibid., pp. 696–97.
32. PD, October 3, 1941, ibid., p. 708.

33. PD, October 4, 1941, ibid., p. 709.
34. Nehru to Indira, October 10, 1941, ibid., p. 710.
35. Nehru to Indira, October 26, 1941, ibid., p. 715.
36. Jawahar to Betty, October 28, 1941, ibid., p. 716.
37. Ibid., p. 717.
38. Ibid., p. 718.
39. Nehru to Rajan Nehru, November 16, 1941, ibid., p. 737.
40. PD, November 18, 1941, ibid., pp. 738–39.
41. Ibid., p. 744, n. 117.
42. PD, November 22, 1941, ibid., pp. 744, 745.
43. PD, December 2, 1941, ibid., p. 750.
44. Nehru to the press, December 5, 1941, *SWJN,* vol. 12 (1979), p. 1.
45. Jawahar to Bebee, December 16, 1941, ibid., p. 622.
46. Nehru to the press, Allahabad, February 26, 1941, ibid., p. 623.
47. Gandhi in Sevagaon, March 2, 1942, in *CWMG,* vol. 75 (1979), pp. 375–76.
48. Indira's *MT,* p. 50.
49. Nehru to Lachhmi Dharji, March 16, 1942, *SWJN,* 12: 626–27.
50. *Scope of Happiness,* p. 155.
51. *We Nehrus,* p. 145.
52. Indira's *MT,* p. 50.

CHAPTER 22

1. Nehru to the A.I.C.C, Wardha, January 15, 1942, *SWJN,* 12:75–76.
2. Nehru to C.R., January 26, 1942, ibid., pp. 91–92.
3. Nehru's speech, Gorakhpur, January 31, 1942, ibid., pp. 102–3, 105.
4. Nehru's article in *Fortune,* March 1942, ibid., pp. 169–70.
5. Linlithgow to Amery, New Delhi, February 16, 1942, *TP,* vol. 1 (1970), p. 185.
6. *Jinnah of Pakistan,* p. 198.
7. Colin Cooke, *The Life of Richard Stafford Cripps* (London: Hodder & Stoughton, 1957), p. 285. Also see Cripps to the press, Delhi, March 23, 1942, *TP,* 1: 462.
8. Cripps's note, March 24, 1942, ibid., p. 464–65.
9. Cripps's interview with Azad, March 25, 1942, ibid., p. 479.
10. Draft declaration of Cripps's offer, ibid., pp. 292–93.
11. Cripps's note on interview with Jinnah, March 25, 1942, ibid., pp. 480–81.
12. Cripps's interview with Gandhi, March 27, 1942, ibid., pp. 498–500.
13. Ibid., p. 500.
14. Cripps's interview with C.R., March 28, 1942, ibid., pp. 511–12.
15. Cripps's note on Congress members, March 29, 1942, ibid., pp. 527–28.
16. Cripps's second interview with Azad, March 28, 1942, ibid., p. 514.
17. Ibid., March 29, 1942, p. 528.
18. Cripps's interview with Nehru and Azad, ibid., pp. 530–31.
19. Cripps's interview with Nehru, March 30, 1942, ibid., pp. 557–58.
20. Cripps's broadcast, March 30, 1942, ibid., pp. 566–71.
21. Resolution by Congress's Working Committee, April 2, 1942, *SWJN,* 12: 188–89.
22. Sir G. S. Bajpai to Lord Linlithgow, "most secret," Washington, D.C., April 2, 1942, *TP,* 1: 619.
23. Amery to Linlithgow, April 3, 1942, ibid., pp. 632–33.
24. Caroe's "most secret" memo, April 6, 1942, ibid., p. 665–66.
25. Nehru to Colonel Johnson, Delhi, April 6, 1942, *SWJN,* 12: 197–99.
26. Note by Linlithgow on conversation with Cripps and Johnson, April 8, 1942, *TP,* 1: 694–95.

27. Linlithgow to Amery, April 9, 1942, ibid., pp. 699–700.
28. Churchill to Cripps, 10 Downing Street, April 10, 1942, ibid., p. 721.
29. Cripps to war cabinet, April 10, 1942, ibid., p. 717.
30. Azad to Cripps, April 10, 1942, ibid., pp. 726–30.
31. Cripps to Churchill, April 10, 1942, ibid., pp. 726–30.
32. Cripps to Azad, April 10, 1942, ibid., pp. 732–33.
33. Nehru to President Roosevelt, April 12, 1942, *SWJN,* 12: 212–13.
34. Cripps's broadcast, Delhi, April 11, 1942, *TP,* 1: 755.
35. Roosevelt to Harry Hopkins, April 12, 1942, ibid., p. 759.
36. Churchill to Roosevelt, Chequers, April 12, 1942, ibid., p. 764.
37. Nehru to the press, April 15, 1942, *SWJN,* 12: 234.
38. Nehru's speech, Gauhati, April 24, 1942, ibid., pp. 237–38.
39. Nehru to Evelyn Wood, June 5, 1942, ibid., pp. 241–42.
40. A.I.C.C. final resolution, drafted by Nehru and passed May 1, 1942, ibid., pp. 277–85.
41. Gandhi in *Harijan,* May 3, 1942, Tendulkar, *Mahatma,* 6: 80.
42. Nehru's speech in Lahore, May 21, 1942, *SWJN,* 12: 321–22.
43. Glancy to Linlithgow, Lahore, May 1, 1942, *TP,* vol. 2 (1971), pp. 7–8.
44. Nehru's speech at UP PCC United Provinces (Provincial Congress Committee), Lucknow, May 31, 1942, *SWJN,* 12: 332.
45. Nehru's public speech, Lucknow, May 31, 1942, ibid., pp. 333–34.
46. Gandhi in *Harijan,* May 31, 1942, Tendulkar, *Mahatma,* 6: 90.
47. In *Harijan,* May 24, 1942, quoted in *TP,* 2: 219.
48. Sir H. Twynan to Linlithgow, May 25, 1942, ibid., p. 117.
49. Amery to Linlithgow, May 27, 1942, ibid., p. 141.
50. Nehru to Yunus, June 22, 1942, *SWJN,* 12: 373.
51. Gandhi to Roosevelt, Sevagram, July 1, 1942, ibid., pp. 378–79.
52. Maulana Abul Azad, *India Wins Freedom: The Complete Version,* (Hyderabad; Orient Longman, 1988), pp. 74–75.
53. Nehru on Working Committee resolution, July 15, 1942, *SWJN,* 12: 398–99.
54. Nehru's speech in Meerut, July 19, 1942, ibid., p. 413.
55. Nehru to Krishna Menon, July 23, 1942, ibid., p. 415.
56. Nehru's speech in Parel, Bombay, August 5, 1942, ibid., pp. 428–31.
57. Nehru on "Quit India" resolution, Bombay, August 7, 1942, ibid., p. 457.
58. Gandhi at A.I.C.C., Bombay, August 8, 1942, Tendulkar, *Mahatma,* 6: 157–59.
59. *Jinnah of Pakistan,* p. 208.
60. Madeleine Slade, *The Spirit's Pilgrimage* (New York: Coward McCann, 1960), pp. 241, 245.
61. PD, September 10, 1942, *SWJN,* vol. 13 (1980), pp. 4, 5, 6.
62. Ibid., pp. 1, 8–9.
63. PD, September 12, 1942, ibid., p. 10.

CHAPTER 23

1. Nehru to Indira, somewhere in India, September 18, 1942, *SWJN,* 13: 13–16.
2. Nehru to Indira, October 15, 1942, ibid., pp. 20–21.
3. PD, November 13, 1942, ibid., p. 28.
4. PD, November 16, 1942, ibid., p. 29.
5. PD, January 26, 1943, ibid., pp. 47–48.
6. Mohan Lal Saksena to My dear Lord Linlithgow, September 10, 1942, TP, 2: 938–42.
7. Governor R. Lumley to Linlithgow, Bombay, December 30, 1942, ibid., pp. 436–37.
8. Linlithgow to Amery, January 13, 1943, *TP,* vol. 3 (1971), p. 500.
9. PD, January 26, 1943, *SWJN,* 13: 49–50.

10. Gandhi to Dear Lord Linlithgow, New Year's Eve 1942, *TP,* 3: 439–40.
11. Linlithgow to Amery, January 6, 1943, ibid., pp. 462–63.
12. Gandhi to Linlithgow, January 19, 1943, ibid., pp. 517–19.
13. Linlithgow to Gandhi, January 25, 1943, ibid., pp. 517–19.
14. Churchill to Linlithgow, "most secret," February 8, 1943, ibid., p. 619.
15. Amery to Linlithgow, February 8, 1943, ibid., pp. 631–32.
16. Azad to Linlithgow, February 13, 1943, TP, 3: 660–61.
17. PD, January 26, 1943, *SWJN,* 13: 49.
18. PD, February 15, 1943, ibid., pp. 62–63.
19. PD, February 17, 1943, ibid., pp. 66, 68.
20. Linlithgow to Amery, February 19, 1943, *TP,* 3: 687–88.
21. Linlithgow to Amery, ibid., p. 691.
22. Amery to Linlithgow, February 19, 1943, ibid., p. 695.
23. Amery to Linlithgow, "private and secret," February 19, 1943, ibid., pp. 698–99.
24. PD, February 25, 1943, *SWJN,* 13: 69.
25. Sir T. B. Sapru to Sir G. Laithwaite, February 20, 1943, *TP,* 3: 705–6.
26. Sapru to Churchill, February 21, 1943, ibid., pp. 711–12.
27. Churchill to Linlithgow, "important, most secret," February 25, 1943, ibid., p. 730.
28. Churchill to Smuts, February 26, 1943, ibid., p. 738.
29. Quoted in Tendulkar, *Mahatma,* 4: 201.
30. PD, March 16, 1943, *SWJN,* 13: 81.
31. PD, April 2, 1943, ibid., pp. 92–93.
32. PD, April 17, 1943, ibid., pp. 114–15, 115–16.
33. PD, April 28, 1943, ibid., p. 122.
34. Nehru to Indira, May 7–8, 1943, ibid., pp. 126–27.
35. PD, May 9, 1943, ibid., pp. 130–31.
36. Halifax to Amery, Washington, May 5, 1943, *TP,* 3: 945–46.
37. Amery to Churchill, April 16, 1943, ibid., pp. 895–96.
38. Amery to Linlithgow, May 6, 1943, ibid., p. 950.
39. Amery to Linlithgow, May 7, 1943, ibid., p. 950.
40. Jinnah's invitation in his speech of April 24, 1943, Appendix II, ibid., p. 982.
41. *Jinnah of Pakistan,* p. 221.
42. Churchill to Attlee and Amery, Washington, May 14, 1943, *TP,* 3: 978.
43. PD, May 15, 1943, *SWJN,* 13: 142–43.
44. PD, May 22, 1943, ibid., p. 145.
45. PD, May 28, 1943, ibid., p. 154, n. 198.
46. PD, May 31, 1943, ibid., pp. 154–55.
47. PD, July 2, 1943, ibid., p. 174.
48. PD, July 14, 1943, ibid., pp. 185–86.
49. Note by Military Intelligence, "most secret," July 14, 1943, TP, vol. 4 (1973), pp. 74–75.
50. Linlithgow to Amery, July 15, 1943, ibid., p. 78.
51. Nehru to Indira, August 14, 1943, *SWJN,* 13: 221.
52. PD, August 30, 1943, ibid., p. 228.
53. Nehru to Indira, September 4, 1943, ibid., pp. 231–32.
54. Hough's, *Edwina,* p. 178. Also see Morgan's *EM.*
55. PD, October 8, 1943, *SWJN,* 13: 252 and n. 351.
56. PD, October 17, 1943, ibid., p. 261.
57. PD, November 8, 1943, ibid., p. 274.
58. Madame Pandit to me in an interview at the India International Centre, New Delhi, 1985.
59. PD, November 8, 1943, *SWJM,* 13: 275–76.
60. PD, November 10, 1943, ibid., pp. 278–79.

61. PD, November 11, 1943, ibid., p. 284.
62. PD, November 20, 1943, ibid., pp. 291, 292.
63. PD, December 17, 1943, ibid. pp. 311–12.
64. PD, December 28, 1943, ibid., pp. 322–23, 324.
65. PD, January 14, 1944, ibid., p. 332.
66. PD, February 18, 1944, ibid., p. 352.
67. Gandhi to Wavell, March 9, 1944, *TP,* 4: 789–93.
68. PD, February 27, 1944, *SWJN,* 13: 361.
69. Nehru to Indira, February 29, 1944, ibid., pp. 361–64.
70. PD, April 1, 1944, ibid., p. 383.
71. PD, April 3, 1944, ibid., p. 388.
72. Wavell to Amery, April 18, 1944, *TP,* 4: 895.
73. Indira to Nehru, April 18, 1944, *Two Alone,* p. 363.
74. Nehru to Indira, April 29, 1944, *SWJN,* 13: 403.
75. Wavell to Amery, April 18, 1944, *TP,* 4: 895.
76. Amery to Wavell, May 11, 1944, ibid., pp. 965–66.
77. Nehru to Indira, May 13, 1944, *SWJN,* 13: 410.
78. Indira to Nehru, May 8, 1944, *Two Alone,* p. 368.
79. Nehru to Indira, May 27, 1944, *SWJN,* 13: 417.
80. PD, June 17, 1944, ibid., pp. 429–31.
81. Nehru to Indira, June 17, 1944, ibid., p. 432.
82. *Jinnah of Pakistan,* pp. 230–31.
83. PD, August 5, 1944, *SWJN,* 13: 456.
84. Nehru to Indira, August 5, 1944, ibid., pp. 458–59.
85. Nehru to Indira, August 21, 1944, ibid., p. 466.
86. Indira to Nehru, September 3, 1944, *Two Alone,* p. 414.
87. Ibid., pp. 414–15.
88. Indira to Nehru, September 15, 1944, p. 421.
89. Nehru's *Discovery,* p. 25.
90. PD, September 20, 1944, ibid., p. 478.
91. Wavell to Amery, September 27, 1944, *TP,* vol. 5 (1974), p. 47.
92. PD, October 5, 1944, *SWJN,* 13: 486.
93. PD, December 5, 1944, ibid., p. 523.
94. Jawahar to Nan, October 23, 1944, ibid., p. 501.
95. *Scope of Happiness,* pp. 189–90.
96. PD, December 31, 1944, *SWJN,* 13: 532.
97. Nehru to Indira, January 5, 1945, ibid., p. 534.
98. Nehru to Indira, February 17, 1945, ibid., p. 562.
99. Nehru to Indira, February 24, 1945, ibid., p. 563.
100. PD, March 9, 1945, ibid., pp. 571–72.
101. PD, March 16, 1945, ibid., p. 580.
102. PD, March 19, 1945, ibid., p. 585.
103. Ibid., pp. 586–87.
104. Nehru to Indira, March 24, 1945, ibid., p. 590.
105. Ibid., p. 591.
106. Nehru to Indira, April 3, 1945, ibid., p. 600.
107. PD, April 30, 1945, ibid., p. 615.
108. Nehru to Indira, May 11, 1945, ibid., p. 620.
109. *Scope of Happiness,* p. 196.
110. Nehru to Indira, May 25, 1945, *SWJN,* 13: 625–26.
111. Nehru to Indira, June 12, 1945, ibid., pp. 635–36.
112. PD, June 14, 1945, ibid., pp. 638–39.

CHAPTER 24

1. Indira's *MT,* p. 55.
2. Nehru's speech at Almora, June 15, 1945, *SWJN,* vol. 14 (1981), pp. 1–2.
3. Wavell's broadcast, June 14, 1945, *TP,* 5: 1122.
4. Nehru's speech at Ranikhet, June 17, 1945, *SWJN,* 14: 4–5.
5. George's *Krishna Menon,* p. 137.
6. Wavell to Amery, "secret," Simla, June 25, 1945, *TP,* 5: 1151–52.
7. Maulana Abul Azad, *India Wins Freedom: The Complete Version* (Hyderabad: Orient Longman, 1988), p. 112.
8. Wavell to Amery, June 25, 1945, *TP,* 5: 1153–54.
9. Wavell to Amery, "secret," June 26, 1945, ibid., pp. 1162–63.
10. Wavell's note on Nehru, July 2, 1945, ibid., pp. 1192–93.
11. Wavell to Amery, "secret," July 9, 1945, ibid., p. 1210.
12. Wavell to earl of Scarbrough, July 9, 1945, ibid., p. 1218.
13. Amery to Wavell, "secret," July 12, 1945, ibid., p. 1237.
14. Minutes of final meeting, Simla Conference, July 14, 1945, ibid., pp. 1243–45.
15. Wavell's note on Nehru, July 14, 1945, ibid., p. 1249.
16. Wavell to Amery, "secret," July 15, 1945, ibid., pp. 1262–63.
17. Ibid., p. 1263.
18. Nehru to Indira, Lahore, August 25–26, 1945, *Two Alone,* p. 517.
19. Ibid., p. 518.
20. Wavell to Pethick-Lawrence, August 12, 1945, *TP,* vol. 6 (1976), p. 60.
21. Poona, September 14, 1945, Enclosure 1 in Wavell to Pethick-Lawrence, September 18, 1945, ibid., pp. 274–75.
22. Ibid.
23. Wavell to cabinet, August 29, 1945, ibid., p. 174.
24. Cabinet, August 31, 1945, ibid., p. 190.
25. *Sunday Statesman,* New Delhi, September 16, 1945, Enclosure 4, ibid., p. 279.
26. Wavell's broadcast, New Delhi, September 19, 1945, ibid., pp. 282, 283.
27. Nehru to Indira, November 17, 1945, *Two Alone,* pp. 522–23.
28. Wavell to Pethick-Lawrence, October 9, 1945, *TP,* 6: 319–20.
29. Pethick-Lawrence to Wavell, October 19, 1945, ibid., pp. 362–63.
30. Pethick-Lawrence to Wavell, October 26, 1945, ibid., p. 413.
31. Wavell to Pethick-Lawrence, November 4, 1945, ibid., p. 439.
32. Wavell's note on Nehru, Enclosure 188, ibid., p. 440.
33. Nehru's speech in Lahore, November 18, 1945, *SWJN,* 14: 175.
34. "Top secret" Enclosure 194 with Wavell to Pethick-Lawrence, November 6, 1945, *TP,* 6: 451–53.
35. Pethick-Lawrence to cabinet, November 14, 1945, ibid., p. 482.
36. Nehru's speech in Annexure III to ibid., p. 484.
37. "Secret" Pethick-Lawrence to Wavell, "secret," November 16, 1945, ibid., p. 490.
38. Cabinet meeting of India and Burma Committee, November 19, 1945, ibid., pp. 501–3.
39. Wavell to Pethick-Lawrence, November 27, 1945, ibid., pp. 552, 554.
40. Auchinleck to Wavell, November 26, 1945, ibid., pp. 544–45.
41. Casey to Wavell, December 8, 1945, ibid., p. 623.
42. *National Herald,* December 4, 1965, Annex 276, Wavell to Pethick-Lawrence, December 9, 1945, ibid., pp. 628–29.
43. Hope to Wavell, December 10, 1945, ibid., p. 631.
44. Wavell to Pethick-Lawrence, Calcutta, December 11, 1945, ibid., p. 633.
45. Minute 2 in India and Burma Committee meeting, December 19, 1945, ibid., p. 663.
46. Woodrow Wyatt, *Confessions of an Optimist* (London: Collins, 1985), pp. 122, 128–29.
47. *Jinnah of Pakistan,* p. 254.
48. "Top secret" draft telegram, Pethick-Lawrence to Viceroy, *TP,* 6: 810.

49. Jawaharlal to Stafford, January 27, 1946, ibid., pp. 856–57.
50. Seton's *Panditji,* pp. 117–18.
51. Hough's *Edwina,* p. 179.
52. Colville to Wavell, Bombay, February 27, 1946, *TP,* 6: 1079–81.
53. *We Nehrus,* pp. 179–80.
54. Colville to Wavell, Bombay, February 27, 1946, *TP,* 6: 1084.
55. *We Nehrus,* p. 180.
56. Governor Sir F. Mudie (Sind) to Wavell, February 27, 1946, *TP,* 6: 1073.
57. Jawaharlal to Cripps, Anand Bhawan, March 5, 1946, ibid., pp. 1107–8.
58. Wavell to Pethick-Lawrence, March 5, 1946, ibid., pp. 1110–11.
59. *The Statesman,* New Delhi, March 4, 1946, Enclosure 2, No. 499, ibid., p. 1117.
60. Nehru to Indira, March 10, 1946, *Two Alone,* pp. 524–25.
61. Nehru to Indira, March 12, 1946, ibid., p. 526.
62. Morgan's *EM;* Hough's *Edwina;* and Hough's *Mountbatten.*
63. Hough's *Edwina,* p. 180.
64. Mountbatten to earl of Listowel, Broadlands, October 3, 1978, MSS. EUR. C. 357, IOL, London.
65. Mathai's *Reminiscences,* pp. 1, 248–49.
66. Pethick-Lawrence's statement, March 23, 1946, *TP,* vol. 7 (1977), p. 1.
67. Cripp's note, March 30, 1946, ibid., p. 59.
68. Cabinet delegation's meeting with Azad, April 3, 1946, ibid., p. 115.
69. Note of the delegation's meeting with Gandhi, April 3, 1946, ibid., pp. 116, 117–18.
70. Delegation's interview with Jinnah, April 4, 1946, ibid., pp. 119–20, 123, 124.
71. Azad, *India Wins Freedom,* pp. 137–38.
72. Ibid., p. 162.
73. Wavell's note on Nehru, May 2, 1946, *TP,* 7: 395.
74. Nehru to Luce, May 4, 1946, *SWJN,* vol. 15 (1982), pp. 597–99.
75. "Top secret" record of meeting, May 16, 1946, ibid., pp. 571–72.
76. Pethick-Lawrence's broadcast, May 16, 1946, ibid., pp. 592, 594.
77. Gandhi in *Harijan,* May 17, 1946, ibid., p. 615.
78. Note to Abell on message from Jinnah, May 18, 1946, ibid., p. 619.
79. Wavell to delegation, May 22, 1946, ibid., pp. 656–57.
80. Nehru to Wavell, May 25, 1946, ibid., pp. 692–95.
81. Nehru to Indira, June 16, 1946, *Two Alone,* p. 531.
82. Nehru to Indira, Kashmir, June 19, 1946, ibid., p. 532.
83. Nehru to Indira, Kashmir, June 21, 1946, ibid., pp. 533–35.
84. Nehru to Begum Sahiba, June 4, 1947, *SWJN* (2), vol. 3 (1985), p. 197.

CHAPTER 25

1. Wavell to King George, July 6, 1946, *TP,* 7: 1091–93.
2. Tendulkar, *Mahatma,* 7: 148.
3. Nehru's press conference, July 10, 1946, *TP,* vol. 8 (1979), p. 25.
4. *Jinnah of Pakistan,* p. 280.
5. Ibid., pp. 282–83.
6. Nehru's editorial in *National Herald, SWJN,* 15: 255–56.
7. Nehru's speech in Delhi, July 20, 1946, ibid., p. 264.
8. Wavell's interview with Nehru, July 22, 1946, ibid., pp. 265–66.
9. Nehru's note on his visit to Kashmir, ibid., pp. 413, 415.
10. Nehru to Rani of Jhansi Brigade, August 4, 1946, ibid., p. 477.
11. Indira to Nehru, August 13, 1946, *Two Alone,* p. 537.
12. *Jinnah of Pakistan,* p. 285.
13. Wavell's note on Nehru, September 1, 1946, *TP,* 8: 382–83, 384.

14. Gandhi on the interim government, September 2, 1946, ibid., p. 386.
15. Wavell to Pethick-Lawrence, September 3, 1946, ibid., pp. 398, 399.
16. Nehru to Pethick-Lawrence, September 27, 1946, *SWJN* (2), vol. 1 (1984), p. 164.
17. Jawaharlal to Krishna Menon, October 6, 1946, ibid., pp. 168–72.
18. Wavell to Pethick-Lawrence, October 1, 1946, *TP*, 8: 636.
19. Wavell's Note on Nehru, October 14, 1946, ibid., p. 720.
20. Nehru to Wavell, October 15, 1946, ibid., pp. 732–33.
21. Caroe to Wavell, October 23, 1946, ibid., pp. 786–88, 792.
22. Nehru to Wavell, October 30, 1946, ibid., pp. 835–37.
23. Nehru to Indira, Patna, November 3, 1946, *Two Alone*, p. 540.
24. Pethick-Lawrence, to Wavell, November 8, 1946, *TP*, vol. 9 (1980), p. 34.
25. Reported to me by a Cabinet Minister of India in my home in Los Angeles, 1993.
26. Pethick-Lawrence, to Wavell, November 8, 1946, *TP*, 9: 36.
27. Wavell to Pethick-Lawrence, November 11, 1946, ibid., pp. 41–42.
28. Abell to Wavell, November 18, 1946, ibid., p. 96.
29. Wavell's note on Nehru, November 19, 1946, ibid., pp. 111–12.
30. *Hindustan Times*, November 22, 1946, ibid., pp. 131–32.
31. Major Short to Cripps, November 28, 1946, ibid., pp. 208–9.
32. Ian Scott to Wavell, November 29, 1946, ibid., p. 220.
33. Morgan's *EM*, p. 373.
34. Pethick-Lawrence's note on Nehru, December 3, 1946, *TP*, 9: 249.
35. Cabinet meeting at 10 Downing Street, December 4, 1946, ibid., p. 253.
36. Cabinet meeting, December 10, 1946, ibid., p. 319.
37. Acting Viceroy Colville to Pethick-Lawrence, December 10, 1946, ibid., p. 321.
38. Nehru's speech to the Constituent Assembly, December 13, 1946, *SWJN* (2) 1: 240–44.
39. Prof. John Kenneth Galbraith to me in his home in Cambridge, Massachusetts, 1993.
40. Hough's *Mountbatten*, p. 215.
41. Dickie to Bertie, January 4, 1947, *TP*, 9: 453.
42. Campbell-Johnson's, *Mission*.
43. Mountbatten to the king, January 4, 1947, *TP*, 9: 453.
44. Mountbatten to prime minister, January 3, 1947, ibid., p. 452.
45. See R. J. Moore, *Escape from Empire: The Attlee Government and the Indian Problem* (Oxford: Clarendon Press, 1983).
46. King George to Mountbatten, January 5, 1947, *TP*, 9: 454, n. 6.
47. Woodrow Wyatt, *Confessions of an Optimist* (London: Collins, 1985), pp. 162–63.
48. Nehru to Wavell, January 23, 1947, *TP*, 9: 541.
49. Nehru to Constituent Assembly, January 22, 1947, *SWJN* (2) 1: 256.
50. Jawaharlal to Krishna Menon, January 13, 1947, ibid., pp. 562–63.
51. Indira's *MT*, p. 55.
52. *We Nehrus*, pp. 193–94.
53. Wavell's note on Nehru, March 10, 1947, *TP*, 9: 907.
54. Governor Jenkins (Punjab) to Wavell, March 10, 1947, ibid., pp. 912–13.
55. Nehru to Wavell, March 13, 1947, ibid., p. 929.
56. Krishna Menon's plan to Mountbatten, March 13, 1947, ibid., pp. 946–51.
57. Jenkins' note on Nehru, March 14, 1947, ibid., pp. 952, 953, and 953, n. 2.
58. Campbell-Johnson's *Mission*, p. 35.
59. Ibid., p. 41.
60. Ibid., p. 45.

CHAPTER 26

1. Mountbatten's interview with Azad, March 27, 1947, *TP*, vol. 10 (1981), p. 34.
2. Mounbatten's "top secret" interview with Gandhi, March 31, 1947, ibid., p. 55.

3. Mountbatten and Gandhi, April 1, 1947, ibid., p. 69.
4. Mountbatten's interview with Nehru, April 1, 1947, ibid., pp. 70–71.
5. Mountbatten's interview with Gandhi, April 2, 1947, ibid., pp. 83–84.
6. Mountbatten's interview with Azad, April 2, 1947, ibid., p. 86.
7. Viceroy's personal report no. 1, April 2, 1947, ibid., pp. 90–91.
8. Mountbatten's interview with Gandhi, April 3, 1947, ibid., p. 103.
9. Mountbatten's interview with Krishna Menon, April 5, 1947, ibid., p. 133.
10. Mountbatten's meeting with Jinnah, April 5–6, 1947, ibid., pp. 137–38.
11. Uncirculated record of staff discussion, "top secret," April 5, 1947, p. 126.
12. Gandhi to Mountbatten, April 7, 1947, ibid., p. 146.
13. Mountbatten's meeting with Jinnah, April 9, 1947, ibid., pp. 163–64.
14. "Top secret," April 11, 1947, ibid., p. 190.
15. April 15, 1947, ibid., p. 262.
16. Jenkins's note, April 16, 1947, ibid., p. 282.
17. Sir T. Shone's note, April 17, 1947, ibid., pp. 293, 294.
18. Nehru to Lord Mountbatten, April 17, 1947, ibid., pp. 305–6.
19. Nehru to the States People's Conference, Gwalior, *SWJN* (2), vol. 2 (1984), pp. 268–71.
20. Mountbatten's meeting with Krishna, April 17, 1947, "top secret," "specially restricted circulation," *TP,* 10: 310–12.
21. Lord Mountbatten told me this when I interviewed him at his flat in London in 1978.
22. "Top secret" interview with Krishna, April 17, 1947, *TP,* 10: 312–13.
23. Ibid., p. 313.
24. Mountbatten's interview with Krishna, April 22, 1947, ibid., pp. 371–74.
25. Jawaharlal to Gandhi, April 23, 1947, *SWJN* (2) 3: 498.
26. Nehru's remarks during his meeting with Mountbatten, April 22, 1947, *SWJN* (2) 2: 101, 102.
27. Nehru's Working Committee letter to Mountbatten, May 1, 1947, ibid., pp. 106–7.
28. Nehru to Huxley, May 14, 1947, *SWJN,* (2) 3: 499.
29. Mathai's *Reminiscences,* p. 4.
30. Mountbatten's meeting with Krishna, April 22, 1947, *TP,* 10: 673–75.
31. Morgan's *EM,* p. 393.
32. Minutes of Simla meeting, May 8, 1947, "Nehru's Plan," *TP,* 10: 673–75.
33. Minutes of viceroy's meeting in Simla, May 10, 1947, ibid., p. 734.
34. Ibid., p. 735.
35. Minutes of viceroy's meeting in Simla, May 11, 1947, ibid., p. 759.
36. Item 3, ibid., p. 761.
37. "Top secret" note by Nehru for Mountbatten, May 11, 1947, ibid., pp. 766–67.
38. Burrows to Mountbatten, May 11, 1947, ibid., pp. 772, 773.
39. Mountbatten to Ismay, May 11, 1947, "immediate top secret," ibid., pp. 774–75.
40. Mieville to Ismay, May 12, 1947, ibid., p. 780.
41. Mieville to Mountbatten, May 20, 1947, ibid., p. 916.
42. Cabinet meeting, 10 Downing Street, May 20, 1947, ibid., p. 922.
43. Nehru to Gandhi, May 16, 1947, *SWJN* (2) 2: 151.
44. Tendulkar, *Mahatma,* vol. 7 (1962), pp. 397, 401.
45. Mountbatten's record of his interview with Churchill, May 22, 1947, *TP,* 10: 95–96.
46. Krishna Menon to Lord Mountbatten, May 21, 1947, ibid., p. 940.

CHAPTER 27

1. Nehru's broadcast, June 3, 1947, *TP,* vol. 11 (1982), pp. 94–95, 96–97.
2. Krishna to Mountbatten, June 4, 1947, ibid., p. 109.
3. Mountbatten's press conference, New Delhi, June 4, 1947, ibid., p. 121.
4. Campbell-Johnson to Ronnie Brockman, June 4, 1947, ibid., p. 128.

5. Mathai's *Days,* pp. 54, 55.
6. *Scope of Happiness,* pp. 227–28, 235.
7. Cabinet meeting, June 19, 1947, 10 Downing Street, *TP,* 11: 515–16.
8. Nehru to Mountbatten, June 22, 1947, ibid., pp. 561–62.
9. Quoted by Mountbatten to Jenkins, June 24, 1947, ibid., p. 594.
10. Jenkins to Mountbatten, June 24, 1947, ibid., pp. 605–6.
11. Mountbatten's record or interview with Nehru, June 24, 1947, ibid., pp. 591–92, 593.
12. Jenkins to Mountbatten, Lahore, June 25, 1947, ibid., pp. 623–24.
13. Abell to Abbott, June 25, 1947, ibid., pp. 633–34.
14. Mountbatten to Attlee, June 30, 1947, ibid., p. 804.
15. Attlee to Mountbatten, July 1, 1947, ibid., p. 807.
16. Churchill to Attlee, July 1, 1947, ibid., pp. 812–13.
17. Mountbatten's report of that incident to me in his London flat in 1978.
18. Attlee to Churchill, July 1, 1947, *TP,* 11: 812.
19. Mountbatten to Attlee, July 2, 1947, ibid., p. 826.
20. Reuters report, July 8, 1947, *TP,* vol. 12 (1983), pp. 17–18.
21. Mountbatten's meeting with Gandhi, July 9, 1947, ibid., pp. 50–51.
22. Jenkins to Mountbatten, July 10, 1947, ibid., p. 71.
23. Krishna Menon to Mountbatten, July 11, 1947, ibid., p. 59.
24. Mountbatten to Krishna Menon, July 10, 1947, ibid., p. 70.
25. Nehru to Attlee, July 11, 1947, ibid., pp. 110–11.
26. Attlee to Mountbatten, July 17, 1947, ibid., p. 215.
27. Mountbatten to Gandhi, July 17, 1947, ibid., p. 211.
28. Mountbatten's meeting with Gandhi, Nehru, and Patel, July 29, 1947, ibid., p. 398.
29. Ibid., pp. 398–99.
30. Mountbatten's "top secret" report, August 1, 1947, ibid., pp. 443–52.
31. Mountbatten to Listowel, "private and top secret," August 8, 1947, ibid., p. 590.
32. Viceroy's personal report no. 16, August 8, 1947, ibid., p. 594.
33. Ibid., p. 595.
34. "Secret" minutes of Mountbatten's staff meeting, *TP,* 12: 611.
35. Campbell-Johnson's *Mission,* p. 152.
36. Quoted in *TP,* 12: 611, n. 3.
37. Nehru to Mountbatten, August 13, 1947, ibid., pp. 695–96.
38. Jenkins to Mountbatten, August 13, 1947, ibid., pp. 702–4.
39. "A Tryst with Destiny," August 14–15, 1947, *SWJN* (2) 3: 135.
40. Campbell-Johnson's *Mission,* pp. 156–57.
41. Appendix I to no. 489, *TP,* 12: 776–80.
42. Viceroy's personal report no. 17, August 16, 1947, ibid., pp. 772, 773.
43. Nehru's broadcast to the nation, August 15, 1947, *SWJN* (2) 3: 137–38.

CHAPTER 28

1. Nehru's speech at the Red Fort, August 16, 1947, *SWJN* (2), vol. 4 (1986), p. 2.
2. Bapu to Amrit Kaur, August 16, 1947, *CWMG,* vol. 89 (1983), p. 50.
3. Nehru's statement in Amritsar, August 18, 1947, *SWJN* (2) 4: 5–6.
4. Nehru's broadcast to the nation, August 19, 1947, ibid., pp. 7–8.
5. Jawaharlal to Gandhi, August 22, 1947, ibid., pp. 14–15.
6. Indira to Nehru, August 21, 1947, *Two Alone,* p. 547.
7. Ibid.
8. Indira to Nehru, Lucknow, December 5, 1947, ibid., pp. 547–49.
9. Nehru to Mountbatten, August 27, 1947, *SWJN* (2) 4: 25–26.
10. Jawaharlal to Vallabhbhai Patel, August 31, 1947, ibid., p. 42.

11. Nehru to Mountbatten, August 31, 1947, ibid., pp. 44–45.
12. Nehru to Gandhi, August 31, 1947, ibid., pp. 45–46.
13. Gandhi to Vallabhbhai, September 1, 1947, *CWMG,* 89: 126–27.
14. Gandhi's statement to the press, September 1, 1947, ibid., p. 132.
15. Gandhi's speech to Calcutta, September 4, 1947, ibid., p. 154.
16. Gandhi to the press, Delhi, September 8, 1947, ibid., p. 166.
17. Gandhi at his prayer meeting, September 10, 1947, ibid., pp. 167–68.
18. Ibid., pp. 168–70.
19. Nehru's broadcast to the nation, September 9, 1947, *SWJN* (2) 4: 56.
20. To Cabinet Emergency Committee, New Delhi, September 15, 1947, ibid., p. 75.
21. Nehru to Rajendra Prasad, September 17, 1947, ibid., pp. 83–84.
22. Gandhi's speech at prayers, September 14, 1947, *CWMG,* 89: 183–86.
23. Note 2, September 17, 1947, prayer meeting, ibid., p. 195.
24. Nehru to mill workers in Delhi, September 30, 1947, *SWJN* (2) 4: 107–9.
25. Nehru to Patel, September 30, 1947, ibid., pp. 113, 114.
26. Nehru to Nan, September 28, 1947, ibid., pp. 235–36, and n. 2.
27. Nehru to Patel, September 27, 1947, ibid., pp. 263, 265.
28. Nehru to My dear Sheikh Saheb, September 30, 1947, ibid., pp. 265–66.
29. Nehru to Sheikh Abdullah, October 10, 1947, ibid., pp. 268–69.
30. Ibid., pp. 270–71.
31. Nehru to Mahajan, October 21, 1947, ibid., pp. 272–73.
32. Nehru to Attlee, October 25, 1947, ibid., pp. 274–75 and 286, n. 2.
33. *Jinnah of Pakistan,* pp. 348, 349.
34. Nehru to M. C. Mahajan, October 26, 1947, *SWJN* (2) 4: 276.
35. *Jinnah of Pakistan,* p. 349.
36. Nehru to Mahajan, October 27, 1947, *SWJN* (2) 4: 277.
37. Nehru to Maharaja Saheb, October 27, 1947, ibid., pp. 278–79.
38. Nehru to Abdullah, October 27, 1947, ibid., pp. 279–82.
39. Campbell-Johnson's *Mission,* p. 225.
40. Nehru to Atal, October 27, 1947, *SWJN* (2) 4: 283–84.
41. Gandhi at prayer meeting, October 26, 1947, *CWMG,* 89: 413–14.
42. Gandhi's prayer meeting speech, October 29, 1947, ibid., pp. 432–33.
43. Nehru to Atal, October 27, 1947, *SWJN* (2) 4: 285–86.
44. Nehru's cable to Attlee, October 28, 1947, ibid., pp. 286–88a.
45. Nehru to Liaquat, October 28, 1947, ibid., p. 289.
46. *Jinnah of Pakistan,* pp. 350–51.
47. Campbell-Johnson's *Mission,* p. 226.
48. *Jinnah of Pakistan,* p. 351.
49. Ibid., p. 352.
50. Nehru to Patel, October 30, 1947, *SWJN* (2) 4: 290–91.
51. Nehru to Mahajan, October 30, 1947, ibid., pp. 292–93.
52. Gandhi at prayer meeting, November 2, 1947, *CWMG,* 89: 460–62.
53. Gandhi at prayer meeting, November 5, 1947, ibid., pp. 480–81.
54. Nehru to Abdullah, November 4, 1947, *SWJN* (2) 4: 318–19.
55. Nehru to Abdullah, November 1, 1947, ibid., p. 300.
56. Nehru to My dear Dwarkanath (Kachru), November 7, 1947, ibid., pp. 320d–e.
57. Nehru's speech in Srinagar, November 11, 1947, ibid., pp. 321–22.
58. Nehru to maharaja of Kashmir, November 13, 1947, ibid., pp. 324–25.
59. Nehru to Kachru, November 21, 1947, ibid., p. 329.
60. Nehru to Dalip Singh, November 21, 1947, pp. 330–31.
61. Nehru to Abdullah, November 21, 1947, ibid., pp. 336–37.
62. Nehru to Attlee, November 23, 1947, ibid., pp. 338–39.
63. Nehru to Constituent Assembly, November 25, 1947, ibid., p. 345.

64. Nehru to Sri Prakasa, November 25, 1947, ibid., pp. 346, 347.
65. Ibid., November 26, 1947, p. 348, n. 2.
66. Nehru to Abdullah, December 3, 1947, ibid., p. 355.
67. Nehru to Indira, *Two Alone,* pp. 549–50.
68. Campbell-Johnson's *Mission,* December 11, 1947, p. 250.
69. Record of meeting in Lahore, December 8, 1947, *SWJN* (2) 4: 361, 362.
70. Ibid., pp. 366–67.
71. Nehru to Abdullah, December 12, 1947, ibid., pp. 368–69.
72. Gandhi to Constructive Workers, New Delhi, December 11–12, 1947, *CWMG,* vol. 90 (1984), pp. 215–17, 223–24.
73. Nehru to My dear Maharaja Saheb, December 12, 1947, *SWJN* (2) 4: 371–73.
74. Nehru to Attlee, December 12, 1947, ibid., p. 374.
75. Nehru's note on Kashmir, December 19, 1947, ibid., pp. 375–78.
76. Mountbatten's talk with Nehru, December 21, 1947, ibid., pp. 381, 382.
77. Nehru to maharaja of Kashmir, December 21, 1947, ibid., pp. 387–88.
78. Nehru's December 28, 1947, cable of his U.N. appeal to Attlee, ibid., p. 407.
79. Mountbatten's record of his meeting with Nehru, December 31, 1947, ibid., p. 421.
80. Gandhi's prayer meeting speech, January 12, 1948, *CWMG,* 90: 408–9, 409–10.
81. Nehru's wire to Ayyangar, January 15, 1948, *SWJN* (2) 5: 7.
82. *CWMG,* 90: 444, n. 2.
83. Gandhi in Birla House, January 18, 1948, ibid., pp. 445–46.
84. Gandhi to prayer meeting, January 20, 1948, ibid., p. 465 and n. 1.
85. Gandhi's speech to prayer meeting, January 21, 1948, ibid., p. 472.
86. Gandhi to prayer meeting, January 29, 1948, ibid., pp. 524, 525.
87. Mathai's *Reminiscences,* pp. 36–37.
88. Campbell-Johnson's *Mission,* p. 275.
89. Mathai's *Reminiscences,* p. 37.
90. Nehru's broadcast to the nation, January 30, 1948, Norman's *Nehru,* 2: 364–65.
91. Nehru's resolution on Gandhi's death, February 2, 1948, *SWJN* (2), vol. 5 (1987), pp. 37–38.
92. Nehru to Constituent Assembly, February 2, 1948, ibid., pp. 40–41.
93. Nehru in New Delhi, February 2, 1948, ibid., p. 44.
94. Nehru's speech at Triveni, February 12, 1948, ibid., pp. 54–57.

CHAPTER 29

1. Nehru's speech in Jammu, February 15, 1948, *SWJN,* (2) 5: 216.
2. Nehru to Indian troops, February 15, 1948, ibid., p. 217.
3. Jawahar to Nan, February 16, 1948, ibid., p. 218.
4. Nehru to Krishna Menon, February 16, 1948, ibid., pp. 219, 220.
5. Nehru to Krishna Menon, February 20, 1948, ibid., p. 221.
6. Ibid., pp. 222–23.
7. Nehru to Krishna Menon, April 6, 1948, ibid., p. 579.
8. Nehru to Nan, April 6, 1948, ibid., p. 581.
9. Hough's *Edwina,* p. 182.
10. Morgan's *EM,* pp. 423–24.
11. Nehru to Sarojini, July 6, 1948, *SWJN* (2), vol. 7 (1988), p. 706.
12. Nehru's speech on June 20, 1948, Norman's *Nehru,* 2: 412.
13. Morgan's *EM,* p. 425.
14. Nehru to Clare Boothe Luce, July 1, 1948, *SWJN* (2) 7: 703.
15. Ibid., p. 704.
16. Jawaharlal to J. R. D. Tata, July 4, 1948, ibid., pp. 704–5.

17. Nehru to Lady Nye, July 4, 1948, ibid., p. 705.
18. Nehru to Sorensen, July 10, 1948, ibid., pp. 587, 588.
19. Nehru to Achyut, June 6, 1948, *SWJN* (2), vo. 6 (1987), p. 462.
20. Jawaharlal to Krishna Menon, June 26, 1948, ibid., pp. 463–64, 465.
21. Morgan's *EM,* p. 430.
22. Lord Romsey told me this during high tea with him at Broadlands in 1993.
23. Morgan's *EM,* p. 431.
24. Nehru to Patel, London, October 11, 1948, *SWJN* (2), vol. 8 (1989), p. 359.
25. Nehru in Kingsway Hall, October 12, 1948, ibid., pp. 360–61.
26. Norman's *Nehru,* 2: 414.
27. Nehru to Commonwealth Prime Minister's Conference, October 13, 1948, *SWJN* (2) 8: 279.
28. Nehru to U.N. General Assembly, Paris, November 3, 1948, ibid., pp. 290–94.
29. Nehru's speech in Delhi, January 26, 1949, *SWJN* (2), vol. 9 (1990), pp. 6 and 17.
30. Nehru's speech in Lucknow, January 28, 1949, ibid., pp. 20–22, 23, 28–32.
31. Nehru's address at a public meeting, Lucknow, January 28, 1949, ibid., p. 38.
32. Nehru to textile workers, Ahmedabad, February 12, 1949, ibid., p. 63.
33. Nehru to Rajaji (C.R.), February 11, 1949, ibid., pp. 71–72.
34. Nehru to Abdullah, January 12, 1949, ibid., pp. 198–99.
35. Nehru to Abdullah, February 10, 1949, ibid., p. 200.
36. Morgan's *EM,* p. 434.
37. O. P. Ralhan, *Jawaharlal Nehru Abroad: A Chronological Study* (Delhi: S.S. Publishers, 1983), p. 50.
38. Indira to Nehru, April 28, 1949, *Two Alone,* p. 563.
39. Mathai's *Reminiscences,* pp. 209, 204.
40. Nehru to Indira, May 3, 1949, *Two Alone,* p. 564.
41. Nehru to Abdullah, May 10, 1949, *SWJN* (2), vol. 11 (1991), p. 115.
42. Abdullah's interview with Michael Davidson in the *Scotsman,* April 14, 1949, ibid., p. 115, n. 3.
43. Nehru to Krishna Menon, May 14, 1949, ibid., p. 117.
44. Nehru to Abdullah, May 18, 1949, ibid., p. 122.
45. Nehru's speech in Dehra Dun, May 22, 1949, ibid., p. 126.
46. Nehru to the Indian army in Srinagar, May 28, 1949, ibid., p. 130.
47. Nehru at public meeting, Srinagar, May 29, 1949, ibid., pp. 138, 140.
48. Nehru to Army officers in Srinagar, May 29, 1949, ibid., p. 141.
49. Jawahar to Nan, May 31, 1949, ibid., pp. 354–55.
50. Jawahar to Nan, June 8, 1949, ibid., p. 356.
51. Nehru to Nan, June 9, 1949, ibid., p. 360.
52. Nehru to the nation, August 15, 1949, *SWJN* (2), vol. 12 (1991), pp. 34–35.
53. Ibid., pp. 36–37.
54. Ibid., p. 37.
55. Henderson to Acheson, August 15, 1949, ibid., pp. 342–43.
56. Nye's minute on his meeting with Nehru, September 9, 1949, ibid., pp. 223–24.
57. Nehru to Krishna Menon, September 11, 1949, ibid., p. 226.
58. Jawaharlal to Krishna, June 5, 1951, *SWJN* (2), vol. 16, part 1 (1994), pp. 636–37.
59. Jawaharlal to Mac (M. O. Mathai), September 29, 1951, *SWJN* (2), vol. 16, part 2 (1994), pp. 751–52.
60. Jawaharlal to Krishna, October 14, 1951, ibid., pp. 752–53.
61. Ibid., pp. 754–55.
62. Nehru to Mountbatten, September 22, 1949, ibid., p. 230.
63. Nehru to Kashmir's National Conference, Srinagar, September 24, 1949, ibid., pp. 231–32.
64. Ibid., pp. 235–38, 239.

65. Nehru on his arrival in Washington, October 11, 1949, ibid., p. 295.
66. Minutes of Acheson's meeting with Nehru, October 12, 1949, ibid., pp. 295–96, 298.
67. Minutes of Truman's meeting with Nehru, October 13, 1949, ibid., p. 299.
68. Nehru's speech in Washington, October 13, 1949, ibid., p. 300.
69. Nehru's meeting with Acheson at the State Department, October 12, 1949, ibid., p. 296.
70. Nehru's address to the U.S. Congress, October 13, 1949, ibid., p. 301.
71. Jawahar to Nan, August 24, 1949, ibid., p. 287, n. 2.
72. Nehru's speech at New York City reception, October 17, 1949, ibid., p. 313.
73. Nehru at Columbia University, October 17, 1949, ibid., pp. 314–15.
74. Nehru to U.N. General Assembly, October 19, 1949, ibid., p. 328.
75. Nehru to Foreign Policy Associations, October 19, 1949, pp. 332–35, 338.
76. Nehru to Rajaji (C.R.), October 26, 1949, pp. 345–46.
77. Morgan's *EM,* pp. 440–41.
78. Nehru's speech in Bombay, November 14, 1949, ibid., pp. 460–61, 462.
79. Nehru to Constituent Assembly, November 28, 1949, *SWJN* (2), vol. 14, pt. 1 (1992), p. 234, 235.
80. Nehru to World Pacifist meeting, Wardha, December 31, 1949, ibid., pp. 467–73.
81. Nehru in Colombo, January 15, 1950, ibid., pp. 561–62.
82. Norman's *Nehru,* 2: 537.
83. Nehru on death anniversary of Gandhi, January 30, 1950, *SWJN* (2) 14, pt. 1: 261–64.
84. Indira to Nehru, December 16, 1949, *Two Alone,* pp. 566–68.
85. Nehru at public meeting, January 30, 1950, *SWJN* (2) 14, pt. 1: 264–70.

CHAPTER 30

1. Nehru to Liaquat, January 18, 1950, *SWJN* (2) 14, pt. 1: 33.
2. Nehru to Liaquat Ali Khan, February 17, 1950, ibid., pp. 40–41.
3. Liaquat to Nehru, February 18, 1950, ibid., p. 45, n. 2.
4. Nehru to Patel, February 20, 1950, ibid., p. 48.
5. Nehru to Vallabhbhai, February 21, 1950, ibid., pp. 50, 51.
6. Jawaharlal to Dickie, March 5, 1950, ibid., p. 91.
7. Nehru to Rajaji, March 10, 1950, ibid., pp. 97, 98.
8. Nehru to Liquat, March 13, 1950, ibid., pp. 111–13.
9. Nehru's Note on Calcutta, March 15, 1950, ibid., p. 120.
10. Nehru to Krishna Menon, March 16, 1950, ibid., pp. 125–26.
11. Nehru to C.R., March 16, 1950, ibid., p. 126.
12. Jawahar to Nan, March 20, 1950, ibid., pp. 132–33.
13. Nehru to B. C. Roy, April 4, 1950, ibid., p. 171.
14. Nehru to Krishna Menon, April 8, 1950, ibid., p. 178.
15. Nehru to Parliament, April 10, 1950, *SWJN* (2), vol. 14, pt. 2 (1993), p. 9.
16. Nehru to Rau, February 14, 1950, *SWJN* (2) 14, pt. 1: 201–2.
17. Nehru to Abdullah, April 6, 1950, ibid., p. 205.
18. Nehru to Ghulam Mohammad, May 30, 1950, *SWJN* (2) 14, pt. 2: 153.
19. Nehru to Indonesia's Parliament, June 7, 1950, ibid., p. 387.
20. Nehru to Krishna Menon, June 27, 1950, ibid., p. 307.
21. Loy Henderson's record of his talk with Nehru, June 28, 1950, ibid., pp. 307–8.
22. Nehru to Abdullah, July 4, 1950, ibid., pp. 154–55.
23. Ibid., pp. 156, 157.
24. Karan Singh, *Autobiography (1931–1967)* (Delhi: Oxford University Press, 1989), p. 145.
25. Ibid., pp. 145, 146.
26. Ibid., pp. 163, 164.

27. Indira to Nehru, May 11, 1953, *Two Alone,* p. 587.
28. Nehru to Indira, May 8, 1953, ibid., pp. 585–86.
29. Nehru to Parliament, December 15, 1952, *JN's Speeches,* vol. 1: *1919–1953* (1954), pp. 89, 91–92, 95–96.
30. Nehru to newspaper editors, August 13, 1954, *JN's Speeches,* vol. 3: March 1953–August 1957 (1958), pp. 454–55.
31. *We Nehrus,* pp. 247–48.
32. Mathai's *Reminiscences,* pp. 93–98; and Shashi Bhushan, *Feroze Gandhi: A Political Biography* (Delhi: Progressive People's Sector Publications, 1977), pp. 129–73.
33. Indira's *MT,* p. 75.
34. Nehru in Rashtrapati Bhavan, November 20, 1955, *JN's Speeches,* 3: 311.
35. Stanley Wolpert, *Zulfi Bhutto of Pakistan* (New York: Oxford University Press, 1993), pp. 84–99.
36. *Scope of Happiness,* p. 277.
37. Nehru in Lok Sabha, November 19, 1956, *JN's Speeches,* 3: 321–22.
38. Morgan's *EM,* p. 469.

CHAPTER 31

1. Sarvepalli Gopal, *Jawaharlal Nehru: A Biography,* vol. 3: *1956–1964* (London: Jonathan Cape, 1984), p. 25.
2. Nehru's speech in Madras, January 431, 1957, *JN's Speeches,* 3: 228.
3. Ibid., pp. 229–30.
4. Nehru in Sweden, June 24, 1957, in O. P. Ralhan, *Jawaharlal Nehru Abroad: A Chronological Study* (Delhi: S.S. Publishers, 1983), pp. 231–32.
5. Nehru's reply to the debate on atomic energy in Lok Sabha, July 24, 1957, *JN's Speeches,* 3: 513–15.
6. Indira to Nehru, June 14, 1957, *Two Alone,* p. 619.
7. Indira to Nehru, Simla, June 18, 1957, ibid., pp. 619–20.
8. Nehru to Indira, Oslo, June 21, 1957, ibid., p. 621.
9. *Scope of Happiness,* p. 287.
10. Ibid., p. 288.
11. Nehru to Indira, London, June 27, 1957, *Two Alone,* p. 622.
12. Nehru to Rajya Sabha, September 9, 1957, *JN's Speeches,* vol. 4: *1957–1963* (1964), pp. 275–77.
13. Nehru to Lok Sabha, April 9, 1958, ibid., pp. 278–80.
14. Nehru to Lok Sabha, April 27, 1959, ibid., pp. 185–89.
15. Ibid., p. 190.
16. Nehru to Lok Sabha, August 28, 1959, ibid., p. 193.
17. N. G. Goray to Nehru in Lok Sabha, ibid., p. 195.
18. Nehru in Lok Sabha, ibid., pp. 195–96.
19. Nehru to Lok Sabha, August 28, 1959, ibid., p. 204.
20. Nehru to Rajya Sabha, September 10, 1959, ibid., pp. 205–7.
21. Indira to Nehru, 1959, *Two Alone,* pp. 627–28.
22. Morgan's *EM,* pp. 469–70.
23. Hough's *Edwina,* p. 2112.
24. Morgan's *EM,* p. 470.
25. Ibid., p. 471.
26. Seton's *Panditji,* pp. 284–85.
27. Gopal, *Jawaharlal Nehru,* p. 122, n. 78.
28. Mathai's *Reminiscences,* p. 16.
29. Mathai to Panditji, January 12, 1959, Appendix III, ibid., pp. 278–79.

30. Indira to Nehru, May 4, 1960, *Two Alone,* pp. 629–30.
31. Nehru to Indira, June 11, 1960, ibid., p. 631.
32. Nehru to Indira, 13, 1960, ibid., pp. 632–33.
33. Nehru to Lok Sabha, September 3, 1960, *JN's Speeches,* 4: 7–9.
34. Ibid., p. 9.
35. Stanley Wolpert, *Zulfi Bhutto of Pakistan* (New York: Oxford University Press, 1993), pp. 21, 72.
36. Nehru to U.N. General Assembly, October 3, 1960, *JN's Speeches,* 4: 314–24.
37. Nehru to Indira, November 4, 1960, *Two Alone,* pp. 645–46.
38. Nehru to Indira, November 14, 1960, ibid., p. 649.
39. Catherine A. Galbraith, "Nehru: A View from the Embassy," *Harper's Magazine,* July 1965, pp. 76–80.
40. Nehru to Non-aligned Conference, Belgrade, September 2, 1961, *JN's Speeches,* 4: 361–62.
41. Galbraith's *Journal,* p. 248.
42. Janki Ganju reported this to me at his home in the suburbs of Washington, D.C., in 1993.
43. Galbraith's *Journal,* pp. 248, 246.
44. Ibid., p. 252.
45. Nehru's speech on the Portuguese colonies, October 20, 1961, *JN's Speeches,* 4: 365–68.
46. Galbraith's *Journal,* pp. 284–85.
47. Seton's *Panditji,* p. 317.
48. Nehru's speech in Lok Sabha, March 14, 1962, *JN's Speeches,* 4: 43–45, 46.
49. Galbraith's *Journal.* "Great Fun" is the title of his Chapter 16 on the visit, and the quotation is from p. 317.
50. Catherine A. Galbraith, "Nehru: A View from the Embassy."
51. Seton's *Panditji,* pp. 320–21.

CHAPTER 32

1. Galbraith's *Journal,* September 18, 1962, p. 405.
2. Nehru to Rajya Sabha, August 22, 1962, *JN's Speeches,* 4: 225.
3. Seton's *Panditji,* p. 329.
4. Ibid., p. 333.
5. I first learned of this from my colleague in cardiology, UCLA's former dean of medicine, Dr. Kenneth Shine, who had flown to India to examine soldiers suffering from this dreadfully debilitating illness.
6. B. M. Kaul, *The Untold Story* (Bombay: Allied Publishers, 1967) is his attempt to explain that debacle.
7. Patwant Singh, *Of Dreams and Demons: An Indian Memoir* (London: Duckworth, 1994), p. 73.
8. Nehru to chief minister, October 21, 1962, in G. Parthasarathi, ed., *Jawaharlal Nehru's Letters to Chief Ministers, 1947–1964,* vol. 5: *1958–1964* (New Delhi: Teen Murti House, 1989), p. 535.
9. Nehru's broadcast, October 22, 1962, *JN's Speeches,* 4: 226–29.
10. Galbraith's *Journal,* October 23, 1962, pp. 429–30.
11. Nehru in Lok Sabha, November 8, 1962, *JN's Speeches,* 4: 230–31.
12. George's *Krishna Menon,* p. 256.
13. Mathai's *Reminiscences,* p. 184.
14. Dickie to Jawaharlal, "*Personal,*" February 18, 1952, S/P (India Files), MB 1/928/F1, Mountbatten Papers, University of Southampton Library.
15. Sarvepalli Gopal, *Jawaharlal Nehru: A Biography,* vol. 3: *1956–1964* (London: Jonathan Cape, 1984) p. 245.

16. Nehru's speech to A.I.C.C., November 3, 1963, ibid.
17. Morarji Desai, *The Story of My Life* (Madras: Macmillan of India, 1974), vol. 2, p. 199.
18. Gopal, *Jawaharlal Nehru,* p. 245.
19. Desai, *The Story of My Life,* p. 200, 204.
20. Nehru's speech on no-confidence motion, August 22, 1963, *JN's Speeches,* vol. 5: *1963–1964* (1968), pp. 75–76.
21. Ibid., pp. 76–78.
22. Ibid., pp. 79–80.
23. Ibid., pp. 80–81.
24. H. V. Kamath, *Last Days of Jawaharlal Nehru* (Calcutta: Jayasree Prakashan, 1977), p. 1.
25. Nehru's "last will and testament," Norman's *Nehru,* 2: pp. 573–75.
26. Ibid., pp. 574–75.
27. Nehru to Orientalists, January 4, 1964, *JN's Speeches,* 5: pp. 152–53.
28. Ibid., pp. 154–55.
29. Kamath, *Last Days,* p. 24.
30. Norman's *Nehru,* 2: 574–75.
31. Indira to Dorothy, May 8, 1964, in Dorothy Norman, *Indira Gandhi: Letters to an American Friend, 1950–1984* (San Diego: Harcourt Brace Jovanovich, 1985), p. 103.
32. Seton's *Panditji,* p. 442.
33. Nehru's last press conference, May 22, 1964, *JN's Speeches,* 5: 220–21, 224, 228.
34. *Scope of Happiness,* p. 315.
35. "Hamlet," in *Shakespeare: Ten Great Plays* (New York: Golden Press, 1962), p. 348.
36. N. K. Seshan, *With Three Prime Ministers, Nehru, Indira and Rajiv* (New Delhi: Wiley Eastern, 1993), p. 87.

BIBLIOGRAPHY

The most important sources of information about Jawaharlal Nehru's life are his own books, letters, and prison diaries. Nehru wrote in so engaging a style and with such seeming frankness, however, that his writings hide as much as they reveal of his brilliant mind and troubled heart. Nehru was not without powers of self-criticism and analysis, but he was clever enough to cloak whatever he did—no matter how foolish or selfish his motives—in high-minded, politically correct rationalizations, most of which he probably believed. His *Autobiography,* first published in 1936 by John Lane, London, is the most revealing of his works. An altered and updated version was published in New York by John Day in 1941, entitled *Toward Freedom.* Nehru wrote all three of his major books in prison. *Glimpses of World History,* published in London in 1934, is a series of letters first written to his daughter Indira, begun behind bars in 1930, Nehru's initial global briefing to his heir. Nehru's most ambitious book, *The Discovery of India,* was written in jail between 1942 and 1945 and published by John Day in New York in 1946. His literary agent was his close friend V. K. Krishna Menon, who arranged for all his major publications, including the much slimmer volume of Nehru's collected writing from 1937 to 1940, entitled *The Unity of India* (New York: John Day, 1942).

Nehru's letters and prison diaries are even more interesting and revealing than his books. In 1958 Nehru released *A Bunch of Old Letters: Written mostly to Jawaharlal Nehru and some written by him* for publication by the Asia Publishing House in Bombay and two years later by the same publisher in New York. These letters span three decades, from 1917 through 1948. Other important letters written by and to Nehru were published by both his sisters. Krishna Nehru published the first version of her autobiography, entitled *With No Regrets* (Bombay: Padma Publications, 1944). She (Krishna Hutheesing) later published *Nehru's Letters to His Sister* (London: Faber, 1963), and her expanded autobiography, Krishna Nehru Hutheesing, *We Nehrus* (New York: Holt, Rinehart and Winston, 1967). Nehru's sister Vijaya Lakshmi Pandit's autobiography is entitled *The Scope of Happiness: A Personal Memoir* (New York: Crown Books, 1979). Two volumes of Nehru's letters to and from his daughter Indira were collected by her daughter-in-law Sonia Gandhi, ed., *Freedom's Daughter: Letters Between Indira Gandhi and Jawaharlal Nehru, 1922–39* (London: Hodder & Stoughton, 1989), and *Two Alone, Two Together: Letters Between Indira Gandhi and Jawaharlal Nehru, 1940–1964* (London: Hodder & Stoughton, 1992).

Nehru's official biographer, S. Gopal, edited over thirty volumes of Nehru's letters, editorial writings, and prison diaries, published in two series of fifteen and sixteen volumes each, entitled *Selected Works of Jawaharlal Nehru.* The first series was published in New Delhi by Orient Longman between 1972, 1982, and the second series by the same publisher between 1984 and 1994. Gopal, the son of India's second president, Dr. S. Radhakrishnan, is the only scholar to date who has enjoyed full access to the Nehru papers locked away in the Nehru Museum and Library at Teen Murti in New Delhi. He used the same papers for his three-volume official *Jawaharlal Nehru,* published in London by Jonathan Cape between 1975 and 1983, and in Cambridge, Massachusetts, by Harvard University Press, between 1976 and 1984.

Many of Nehru's letters are among the last fifty or so volumes of *The Collected Works of Mahatma Gandhi,* more than one hundred volumes of which have now been published in Ahmedabad by the Publications Division of the Indian government, with permission of the Navajivan Trust (1958–1994). P. N. Chopra edited six volumes of *The Collected Works of Sardar Vallabhbhai Patel,* which also contain Nehru's letters and were published in Delhi by the Sardar Patel Society between 1990 and 1995.

I also used the twelve volumes edited by Nicholas Mansergh and E. W. R. Lumby, *Constitutional Relations Between Britain and India: The Transfer of Power* (London: Her Majesty's Stationery Office, 1970–83), many of which contain important statements by or about Nehru made during the Cripps and cabinet missions and in the period of Mountbatten's viceroyalty. Most of the Mountbatten Papers were transferred from Broadlands to the University of Southampton Library, where I used them, and I found many of Nehru's letters among the other materials so well preserved there. Many private collections of papers contain references to Nehru, and those that I found most useful are the Agatha Harrison Papers at the Friends House Library in London; the Lord Attlee Papers in the Bodleian Library, Oxford, and in Churchill College, Cambridge, as well as his Cabinet Office Papers in the Public Record Office at Kew; the Sir Richard Stafford Cripps Papers, Nuffield College Library, Oxford; the Lord Pethick-Lawrence Papers, Trinity College, Cambridge; and India Office Libray (IOL), London; the Lord (Richard Austen) Butler Papers, Trinity College, Cambridge; the Lionel George Curtis Papers, Bodleian Library, Oxford; the Geoffrey Dawson Papers, Bodleian Library, Oxford; the Ernest Bevin Papers, Churchill College, Cambridge; the Olaf K. Caroe papers, IOL, London; and the Archer Collection, IOL, London.

Hundreds of books have been written about Nehru, most of them uncritical if not hagiographic, for Nehru was, after all, the George Washington of independent India. In recent years, a number of more critical works have emerged, especially as it became clear that Nehru's foreign policy with respect to the Soviet Union as well as China was as flawed as his economic development plans proved to be. Few of the many works written about Nehru, however, probe beneath the veiled surface of Nehru's complex personality. The man who perhaps knew him best was his closest personal secretary, M. O. (Mac) Mathai, who published two thin but insightful volumes: *Reminiscences of the Nehru Age* (New Delhi: Vikas, 1978), and *My Days with Nehru* (New Delhi: Vikas, 1979). Another sensitively empathetic memoir of Nehru was written by a foreign diplomat, Walter Crocker, *Nehru: A Contemporary's Estimate* (London: Allen & Unwin, 1966). B. R. Nanda wrote a most helpful study of both Jawaharlal Nehru and his father, *The Nehrus: Motilal and Jawaharlal* (New York: John Day, 1963), and has now also published a volume of essays that he calls its sequel, *Jawaharlal Nehru: Rebel and Statesman* (Delhi: Oxford University Press, 1995).

Other biographies, monographs, and anthologies concerning Nehru that are worth reading include the following: Frank Moraes, *Jawaharlal Nehru: A Biography* (New York: Macmillan, 1956); Marie Seton, *Panditji: A Portrait of Jawaharlal Nehru* (London: Dobson, 1967); B. N. Pandey, *Nehru* (Briarcliff, N.Y.: Stein & Day, 1976); Michael Brecher, *Nehru: A Political Biography* (London: Oxford

University Press, 1959); D. R. Sar Desai and Anand Mohan, eds., *The Legacy of Nehru: A Centennial Assessment* (Springfield, MA.: Nataraj Books, 1992); Dorothy Norman, ed., *Nehru: The First Sixty Years,* 2 vols. (London: The Bodley Head, 1965); M. N. Das, *The Political Philosophy of Jawaharlal Nehru* (New York: John Day, 1961); H. V. Kamath, *Last Days of Jawaharlal Nehru* (Calcutta: Jayasree Prakashan, 1977); Vincent Sheean, *Nehru: The Years of Power* (London: Victor Gollancz, 1960); Norman Cousins, *Talks with Nehru* (London: Gollancz, 1951); Michael Edwardes, *Nehru: A Political Biography* (London: Allen Lane, Penguin Books, 1971); Aruna Asaf Ali, *Private Face of a Public Personality: A Study of Jawaharlal Nehru* (New Delhi: Radiant, 1989); John Grigg, ed., *Nehru Memorial Lectures, 1966–1991* (Delhi: Oxford University Press, 1992); Rafiq Zakaria, ed., *A Study of Nehru* (Calcutta: Rupa, 1989); Y. D. Gundevia, *Outside the Archives* (Hyderabad: Sangam, 1984); B. N. Mullik, *My Years with Nehru* (New Delhi: Allied, 1971); Subimal Dutt, *With Nehru in the Foreign Office* (Calcutta: Minerva, 1977); and H. Y. Sharada Prasad, ed., *Jawaharlal Nehru* (New Delhi: Jawaharlal Nehru Memorial Fund, 1983).

Many of the autobiographies, memoirs, and biographies of Nehru's contemporaries and comrades shed more light on his life and times than do some of the books that focus exclusively on him. The most valuable among these are Abul Kalam Azad, *India Wins Freedom: The Complete Version* (Hyderabad: Orient Longman, 1988); D. G. Tendulkar, *Mahatma: Life of Mohandas Karamchand Gandhi,* 8 vols. (Bombay: Jhaveri & Tendulkar, 1952–54); Pyarelal, *Mahatma Gandhi: The Last Phase,* 2 vols. (Ahmedabad: Navajivan Trust, 1958); T. J. S. George, *Krishna Menon: a Biography* (London: Jonathan Cape, 1964); Promilla Kalhan, *Kamala Nehru: An Intimate Biography* (Delhi: Vikas, 1973); K. M. Panikkar and A. Pershad, eds., *The Voice of Freedom: The Speeches of Pandit Motilal Nehru* (London: Asia Publishing House, 1961); P. N. Chopra, *The Sardar of India: Biography of Vallabhbhai Patel* (New Delhi: Allied Publishers Ltd., 1995); Richard Hough, *Mountbatten: A Biography* (New York: Random House, 1981); Janet Morgan, *Edwina Mountbatten: A Life of Her Own* (New York: Scribner, 1991); Isaac Kramnick and Barry Sheerman, *Harold Laski: A Life of the Left* (Harmondsworth: Allen Lane, Penguin Books, 1993); Indira Gandhi, *My Truth,* presented by Emmanual Pouchpadass (New Delhi: Vision books, 1981); Shashi Bhushan, *Feroze Gandhi: A Political Biography* (New Delhi: Progressive People's Section, 1977); Colin Cooke, *The Life of Richard Stafford Cripps* (London: Hodder & Stoughton, 1957); R. J. Moore, *Escape from Empire: The Attlee Government and the Indian Problem* (Oxford: Clarendon Press, 1983); Chester Bowles, *Ambassador's Report* (New York: Harper, 1954); John Kenneth Galbraith, *Ambassador's Journal: A Personal Account of the Kennedy Years* (Boston: Houghton Mifflin, 1969); Moraji Desai, *The Story of My Life,* 2 vols. (Madras: Macmillan of India, 1974); Karan Singh, *Autobiography (1931–1967)* (Delhi: Oxford University Press, 1989); Alan Campbell-Johnson, *Mission with Mountbatten* (New York: Dutton, 1953); Woodrow Wyatt, *Confessions of an Optimist* (London: Collins, 1985); Steven A. Hoffman, *India and the China Crisis* (Berkeley and Los Angeles: University of California Press, 1990); and Stanley Wolpert, *Jinnah of Pakistan* (New York: Oxford University Press, 1984).

INDEX

Abdullah, Sheikh Mohammad, 269, 366, 368, 372, 400, 421, 425ff., 449–50, 461, 463, 494–95, 496
Abell, George, 383, 407
Abhayananda, Swami, 144, 189
Abyssinia, 197, 219
Acheson, Dean, 448, 451, 460
Afghanistan, 313
Aga Khan's Palace, 316
Ahimsa, 147, 156
Ahmadnagar Fort, 316, 318
A.I.C.C. *See* Congress, Indian National
Ajanta, 218
Alexander, A. V., 357
Alexander, Horace, 190, 453
Alexander, Lord Harold, 453
Ali, Aruna Asaf, 322, 474
Ali, Asaf, 322
Ali, Maulana Mohammad, 66, 85
Ali, Maulana Shaukat, 44, 85, 250
Allahabad, 6, 40, 94, 120, 152, 163, 170, 181, 201, 211, 259, 299, 491
Almora, 174, 183, 187, 282, 343
Ambedkar, Dr. B. R., 141–42, 266, 348, 378
America. *See* United States of America
Amery, Leo S., 275, 277, 301, 307–8, 314, 320ff., 326, 335, 343
Amritsar, 42, 381, 419
Anand Bhawan, 6, 40, 45, 94, 112, 119, 152, 163, 170, 185, 188, 203, 227, 232–33, 237, 249, 294, 299, 331
Anderson, Sir John, 301
Andrews, C. F., 190
Aney, Dr. M. S., 323, 348
Ansari, Dr. M. A., 69, 80
Ashley, Edwina. *See* Mountbatten, Edwina Lady
Ashoka, 143
Ashram, 48, 171, 278
Assam, 314, 341
Astor, Lady, 238
Atal, Dr. Madan, 179, 180–81, 186–87, 190, 199, 284–85
Atal, Pandit Kishan Lall, 26
Atlantic Charter, 301
Atomic Energy Commission, 469
Attlee, Clement, 240, 301, 307, 343, 346, 356, 377, 378, 381, 401, 415, 448, 460
Auchinleck, Field Marshal Claude, 330, 355, 365, 390, 419
Austria, 235
Ayodhya, 249
Azad, Maulana Abul Kalam, 70, 82, 160, 226, 234, 246ff., 251, 256, 270–71, 273–74, 278, 281, 303, 314, 319–20, 325, 332, 334, 341, 346, 348, 355, 364, 384, 386

Badenweiler, 185, 189
Badshah Khan. *See* Khan, Abdul Ghaffar
Baghdad, 200
Bajpai, Sir G. S., 308
Baldwin, Stanley, 76, 217, 238
Ball, W. Rowse, 20
Baltistan, 272
Bande Mataram, 250
Bangladesh, 389
Barcelona, 237, 241, 280
Bardoli, 98, 249, 298
Batlivala, Bee, 237–38
Belgium, 272
Benaras (Varanasi), 276
Bengal, 248, 251ff., 274, 304; famine in, 320, 330, 395
Berlin, 185, 189, 342
Besant, Annie, 9, 36, 37–38, 76, 83
Betty. *See* Nehru, Krishna Hutheesing
Bevan, Aneurin, 240
Bex, 175, 199, 204
Bey, Fouad, 200
Bhabha, Dr. Homi, 434, 444
Bhargava, Dr., 151
Bhatnagar, Dr. S. S., 444
Bhave, Vinoba, 278
Bhubaneshwar, 493
Bhutto, Zulfikar Ali, 478, 494
Bihar, 161, 165, 221
Bikaner, 244
Birju Bhai. *See* Nehru, Brijlal

Index

Birkenhead, Secretary of State, 99
Birla, G. D., 72, 141, 151
Birla House, 306
Birmingham, 190
Blake, 179
Blavatsky, Madame, 36
Blicking, 196, 238
Bodley Head, 190
Bombay, 89, 168, 184, 207, 237, 246, 272, 298, 315, 319–20, 335, 346, 358
Bose, "*Netaji*" Subhas Chandra, 82, 84, 91, 109, 185, 190, 200, 202, 203, 219, 230, 233, 247ff., 251, 256; on Nehru, 252–53, 329, 350
Broadlands, 391, 439, 459, 470, 474
Brockway, Fenner, 21
Bruce, General, J. G., 399
Bucher, General Sir Roy, 434
Buck, Pearl, 339
Budapest, 242
Bulganin, Marshal, 465
Burma, 224–25, 255, 314, 358
Burrows, Governor Sir Frederick, 395
Butler, Sir Harcourt, 52

Cairo, 200, 246
Calcutta, 92, 161–62, 231, 255, 301, 319, 323, 355, 372, 381, 395, 458–59
Calicut, 145, 247
Cambridge, 16, 19, 20, 56, 149, 192, 212
Campbell-Johnson, Alan, 379, 382
Canton, 241
Caroe, Sir Olaf, 309
Casey, Governor Richard, 355
Cawnpore (Kanpur), 250, 275
Caxton Hall, 198, 241
Chamberlain, Neville, 235, 238–39, 242–43, 272
Chanakya, 227, 230
Chattopadhyaya, Chatto, 99, 103
Chaupati, 44, 122
Chavan, Y. B., 495
Chiang Kai-shek, Madame, 255, 282, 295–96, 302–3, 323, 325–26, 330
Chiang Kai-shek, Marshal, 255, 295, 302–3, 330
China, 255, 281, 295, 302, 313, 485
Chittagong, 235, 238, 381
Chou En-lai, 465, 473
Chungking, 255, 280
Churchill, Winston, 12, 15, 239–40, 245–46, 258, 272, 275, 301, 305, 307, 309, 311, 317–18, 321–22, 330, 333, 346, 358, 396, 400, 443, 468
Civil disobedience (C.D.) 262, 272, 278, 320
Civil Liberties Union, 216
Cliveden Set, 235, 238
Commonwealth, British, 303, 382, 390
Communal Award, 177, 202
Communal question, 213–14, 477
Communalism, 222, 389
Congress, Indian National, 30, 58, 77, 166, 204; A.I.C.C., 68, 168, 180, 210, 218, 222, 274, 299, 312, 315; Working Committee of, 169, 171, 180, 202, 204, 209–10, 223, 226, 238ff., 249ff., 256, 258, 272, 305, 308, 314, 346
Constituent Assembly, 3, 240, 267, 271, 273, 275, 369–70, 371, 376ff., 382
Coward, Noel, 382
C.R. (Chakravarti Rajagopalachari), 57, 181, 207, 226, 302, 305, 312, 324, 337, 440, 442, 453, 457, 468, 495
Cripps, Stafford, 199, 222, 239–40, 258, 266, 275, 301–2; Mission, 303ff., 324, 333–34, 354–55, 357, 379, 395
Crossman, R. H. S., 240
Czechoslovakia, 242–43, 245

Dalton, Hugh, 240
Das, C. R., 57ff., 61, 72, 82, 106, 202
Dehra Dun, 28, 168, 281, 283, 497
Delhi, 188, 203, 259, 319, 344
Desai, Bhulabhai, 165, 169, 292, 315
Desai, Mahadeva, 150, 294, 316
Desai, Morarji, 488–49
Deva, Narendra, 95, 172, 204, 252
Dickens, Charles, 178
Dickinson, G. Lowes, 17, 29
Discovery of India, 105, 293–94, 312, 338
Dixon, Sir Owen, 460
Dominion status, 61, 100, 391, 400
Dulles, John Foster, 435, 467
Dutt, R. Palme, 241, 346

Economic development and planning, 245
Eden, Anthony, 217, 235, 323, 326, 401
Edward, Prince of Wales, 169, 331
Egypt, 246

Index

Eisenhower, President Dwight D., 452, 466
Elizabeth II, Queen, 476
Ensminger, Douglas, 442

Fabian, 156
Faizpur, 218
Fascist, 228–29, 235, 246, 256
Finland, 268
Fischer, Louis, 314
Five-Year Plan, 256
Ford Foundation, 442
Forster, E. M., 29
Forward Bloc, 256
France, 242, 246
Franco, Francisco, 238
Friends House, 246
Fyzabad, 249

Galbraith, John Kenneth, 23, 379, 479ff., 486
Gandhi, Devadas, 141
Gandhi, Feroze, 119, 160, 175, 181, 195, 224, 226, 242, 284, 287–88, 291, 294, 298, 315–16, 329, 342, 345, 457
Gandhi, Indira. *See* Nehru, Indira
Gandhi, Kasturbai, 316, 334
Gandhi, *Mahatma* Mohandas Karamchand, 30, 32, 39ff., 41, 48, 54, 67, 71, 73, 77, 78ff., 98, 110ff., 140, 145, 149–50, 156, 162, 165, 170–71, 173, 204, 207–8, 214, 218, 223–24, 231, 241, 251–52, 256, 262, 273, 291, 294, 298, 313, 314ff., 320ff., 327, 335, 337, 362, 366, 384ff., 402, 412ff., 428–29, 431
Gandhi, Rajiv, 333, 337–38, 339, 342, 345, 349
Gauhati, 312
Geneva, 175, 243, 282
Genoa, 237
Germany, 31, 180, 187, 239, 242, 272, 289, 293
Ghadr Party, 60
Ghose, Aurobindo, 326
Gitanjali, 152, 324
Glancy, Governor Sir B., 303, 313
Glimpses of World History, 112, 152, 165, 183
Goa, 479, 481
Gokhale, Gopal Krishna, 30, 37
Gorakhpur, 277, 280, 302
Government of India Act (1935), 202, 204, 206, 220, 260
Graham, Dr. Frank, 471
Grigg, Sir James, 301
Gulmarg, 349

Haig, Sir Harry, 160, 226, 247
Halifax, Lord, 326
Hallett, Sir Maurice, 276, 355
Harijan, 147, 160–61, 165, 203, 313
Haripura, 227, 233, 235
Harrison, Agatha, 165, 187, 189, 193–94, 199, 206, 211, 214, 246, 320, 335
Harrow, 8, 11ff., 20, 25, 56, 143, 169, 184, 245, 264, 453, 476
Heath, Carl, 195, 199
Heidelberg, 185
Hemmerlin, Mlle., 183
Henderson, Loy W., 449
Hindu Mahasabha, 81, 177, 260, 410
Hindu-Muslim conflicts and riots, 180, 234, 249, 278, 332, 353, 410ff., 458–59, 477
Hindu-Muslim question, 36, 259
Hindu-Muslim unity, 65, 78, 95, 250, 316
Hindu Raj, 266
Hitler, Adolf, 235, 242–43, 251, 273, 329, 342
Hoare, Sir Samuel, 145, 163, 174, 194, 326
Hong Kong, 255
Hope, Governor Arthur, 355
Hopkins, Harry, 311
Hossain, Syed, 398
Hull, Cordell, 323
Human rights, 345, 392
Husain, Dr. Zakir, 409, 485
Hutheesing, Raja, 154, 248, 358
Huxley, Aldous, 181, 340, 462
Hyderabad, 241, 393

Iftikharuddin, Mian, 293
Imphal, 329
Inner Temple, 20
India Conciliation Group, 190, 199
India League, 10, 195, 353
India Office, 264, 301, 308, 310, 314
Indian National Army, 256, 329, 355
Indian National Congress. *See* Congress, Indian National
Indira Gate, 211
Iqbal, Dr. Muhammad, 120

Index

Iran, 313
Ireland, 240, 322
Ironside, General, 238
Irwin, Lord, 99, 100, 109, 114, 117
Ismay, Sir H. L. ("Pug"), 379, 392
Italy, 239, 272

James, R. L., 476
Jammu and Kashmir, 269. *See also* Kashmir
Japan, 256, 301, 311, 350
Jayakar, M. R., 15, 141, 324
Jenkins, Governor Sir Evan, 382, 389, 399
Jinnah, *Quaid-i-Azam* M. A., 6, 35, 42, 44, 64, 168–69, 176, 212–13, 222–23, 234, 239, 242–43, 246, 250, 256–57, 259, 261, 263, 266, 271, 274, 303–4, 316, 325, 333–34, 337, 347, 356, 363, 370, 377, 384–85, 386, 388–89, 395–96, 419–20, 477
Joan of Arc, 68, 178, 211
Jodhpur, 189, 200
Johnson, J. Wilson, 60
Johnson, Louis Arthur, 308, 311
Johnson, President Lyndon B., 495
Jones, Thomas, 238
J. P. *See* Narayan, Jaya Prakash
Junagadh, 393
Justice Party, 267

Kachru, Dwarkanath, 422–23
Kairon, Pratap Singh, 477
Kak, Pandit Ram Chandra, 366, 403
Kalidas, 177
Kamaraj, Kumaraswami, 489ff.
Kamath, H. V., 491, 494
Karachi, 200, 359, 457, 478
Kashmir, 34–35, 164–65, 168–69, 215, 241, 246, 269, 272, 300, 315, 340ff., 345, 349, 367, 371, 393, 413ff., 422ff., 433ff., 445ff., 449–50, 460, 462ff., 468, 496
Katju, Jivan Lal, 21
Kaul, General B. M., 481, 485
Kaur, Rajkumari Amrit, 12, 267, 348, 354
Kelappan, 145
Kennedy, Jacqueline, 480, 482
Kennedy, President John F., 479–80
Khaliquzzaman, 348
Khalistan, 313, 349, 389
Khalsa, 313
Khan, Abdul Ghaffar, 177, 218, 274, 314, 350, 375
Khan, Ayub, 478, 494–95
Khan, Liaquat Ali, 259, 348, 375, 408, 457
Khan Saheb, Dr., 177, 350, 375, 382
Khilafat movement, 44, 65, 85
Khrushchev, Premier, 465, 478
Kremlin, 166
Kripalani, Jivan B., 208, 241, 243, 253, 262, 320, 377, 380
Krishna. *See* Menon, V. K. Krishna
Krishnamachari, T. T., 464
Krishnan, K. S., 444
Kulu Valley, 299–300

Labour Party, British, 96, 195, 198, 240, 246, 260, 333
Ladakh, 272, 472, 476
Lahore, 193, 261, 271, 350, 354, 399
Lama, Dalai, 471–72
Lane, Allen, 190, 195, 198
Lanka, Sri, 215, 455
Laski, Harold, 70, 195, 198, 240–41, 275, 346, 439
Lausanne, 176, 193, 199
Lawrence, T. E., 229, 283
Leh, 476
Lenin, V. I., 74, 150, 157, 166–67, 233
Leysin, 263
Linlithgow, Lord, 196, 226, 236, 239, 247, 256–57, 259, 270, 273, 301, 308–9, 320–21, 323
Lisbon, 282
Listowel, Lord, 354, 395, 399, 404
Lohia, Ram Manohar, 204, 488
London, 145, 168, 192, 198, 239, 256, 275, 280, 315, 377
London School of Economics, 25, 190, 195
Lothian, Lord, 196, 235–36, 238
Luce, Clare Boothe, 364, 436, 452
Lucknow, 88, 153, 169, 184, 193, 204, 212, 226, 242, 292
Lumley, Governor, 320

MacDonald, Ramsay, 65, 96, 99, 120, 177
Madras, 227, 267, 312, 355
Madrid, 217
Madurai, 216
Mahajan, M. C., 414, 416
Mahmud, Syed, 69, 74, 114, 144, 174, 270, 294, 319, 337, 496
Maidenhead, 24
Malaviya, Pandit Madan Mohan, 72, 80

Index

Manipur, 329
Mao Tse-tung, 465, 485
Mare, Walter de la, 177, 192
Marseilles, 199
Marshall, General George, 440
Marx, Karl, 198
Marxists, 83, 156, 197, 213
Masaryk, Jan, 242
Mathai, M. O. ("Mac"), 16, 361, 392, 398, 429, 449, 475
McMahon line, 472
Mears, Sir Grimwood, 63
Meerut, 315
Mehta, Dr. Jivraj, 294
Menon, V. K. Krishna, 10, 70, 141, 165, 190, 195, 198, 200, 204, 216, 219, 222, 225, 232, 237–38, 240–41, 246, 253–54, 258, 261, 266, 275, 315, 331, 346, 353, 358, 374, 381, 387, 391, 395, 398, 439, 449, 468, 473, 480–81, 485, 487, 497
Menon, V. P., 393, 401, 416
Mieville, Sir Eric, 379, 395
Mission, British Cabinet, 360, 362ff.
Mody, Sir Homi, 323
Molotov, 343, 374
Mohammad, Bakshi Ghulam, 445
Morrison, Herbert, 240
Moscow, 74–75, 343, 374, 398
Mountbatten, Edwina Lady, 51, 331, 358, 360–61, 382–83, 435–36, 439, 443–44, 453, 459, 464, 467, 470, 473–74
Mountbatten, Louis Lord ("Dickie"), 50, 273, 302, 330, 358, 360–61, 377ff., 381–82, 384ff., 388, 390–91, 396, 401, 404, 410, 470
Mountbatten, Pamela Lady, 382, 407, 474
Mudie, Sir Francis, 359
Mukerji, Dr. Shyama Prasad, 348
Munich, 242, 245–46
Muslim League, 35–36, 42, 64, 75, 120, 169, 212, 222–23, 242–43, 250–51, 257, 261, 263, 271, 293, 303, 306, 312, 325, 347, 349, 356, 370
Mussolini, Benito, 185, 200, 202, 235, 251, 273, 329, 377
Muzaffarabad, 418

Nabha, 60, 366
Naidu, Padmaja ("Bebee"), 144, 149, 174, 185, 203, 208, 227, 230, 267, 283, 298, 331, 365, 443, 491
Naidu, Sarojini, 95, 101, 173, 181, 207, 216, 231, 263, 267, 315–16, 365, 440, 443
Nambiar, A. C. N. ("Nanu"), 435
Nan. *See* Nehru, Vijaya Lakshmi Pandit
Naples, 237
Napoleon, 215, 274
Narayan, Jaya Prakash (J.P.), 131, 160, 172, 204, 247, 252, 331, 365, 468, 495
Naval mutinies, 358–59
Nehru, Brijlal ("Birju"), 17, 18, 148, 151, 345
Nehru, B. K., 480
Nehru, Indira, 53–54, 68, 142–43, 146, 148, 152, 158, 161, 165, 169–70, 175, 180ff., 211, 217, 224–25, 232, 238, 242, 248, 263, 279, 282ff., 287–88, 291, 297–98, 299, 315–16, 318, 329, 331, 335–36, 337, 339, 342, 457, 473, 493
Nehru, *Panditji* Jawaharlal: ancestry, 13; and atomic energy, 444ff., 469; and Azad, 332, 364; and books, 177ff.; and Bose, 251ff.; in Cambridge, 16ff.; and Churchill, 333; and Congress, 30ff., 58ff., 65ff., 68, 71, 73, 77ff., 83ff., 89ff., 93ff., 121ff., 160ff.; as Congress President, 98ff., 104ff., 107ff., 210ff., 221ff.; and Cripps, 305ff., 333; and education, in Harrow, 11ff., 14ff.; at home, 6ff.; and human rights, 164; and Indira, 143ff., 148ff., 159, 279, 282ff., 287, 289ff.; at Inner Temple, 20ff.; and Jinnah, 223, 266, 325, 333, 356–57; on Lenin, 166–67; and Mahatma Gandhi, 36ff., 40ff., 44ff., 54ff., 67ff., 78ff., 90ff., 101ff., 114ff., 125ff., 130, 150ff., 166ff.; and peasants, 47ff., 278ff.; in prison, 51ff., 60ff., 110ff., 117ff., 133ff., 163ff.; as "Rashtrapati," 227ff.; and revolt, 262, 354; on science, 170; on Third World, 276; on Wavell, 347; on women, 225–26
Nehru, Kamala, 7, 26, 32ff., 41, 53, 96, 108, 118, 143–44, 149, 160–61, 165, 168, 170, 174–75, 179, 183, 189, 199–200
Nehru, Krishna Hutheesing ("Betty"), 33, 108, 146, 154, 161, 185, 233, 248, 272, 289, 291, 295, 299, 316, 339, 446, 475, 491
Nehru, Motilal, 6–7, 11ff., 13–14, 17, 27, 28ff., 31–32, 36, 41ff., 45, 48, 52, 57ff.,

Nehru, Motilal (*continued*) 60, 67ff., 73, 74, 84, 87–88, 97, 106, 112, 116; death of, 121ff.
Nehru, Rajan, 296
Nehru, Ratan (R.K.), 362, 374, 484–85
Nehru, Shridhar, 18, 19, 185
Nehru, Swarup Rani, 8, 11, 18, 27, 41, 45, 53, 144, 149, 161, 174–75, 233
Nehru, Vijaya Lakshmi Pandit ("Nan"), 29, 33, 41, 48, 75, 146, 169–70, 174, 185, 222, 226, 233, 235, 242, 247, 282–83, 289, 297, 299, 331, 339–40, 341, 343, 396, 398, 438, 446, 459, 466, 475, 497
Nehru Report, 93ff.
New York, 452, 478, 486
Nichol, Muriel, 356
Nietzsche, 17, 330
Noel-Baker, P. J., 434
Noon, Sir Feroze Khan, 244
North-West Frontier, 314, 350, 374–75
Norway, 24
Nye, Sir Archibald, 437, 448
Nye, Lady Colleen, 437
Nuclear power and plutonium, India's, 444, 446, 469

Orissa, 145
Oxford, 23, 25, 48, 190–91, 263–64

Pakistan, 234, 244, 259, 303, 312–13, 316, 333, 340, 349, 354, 356, 363, 365, 370, 457–58, 466, 478
Pandit, Ranjit, 48, 149, 168, 185, 235, 281, 297, 333
Pandit, Vijaya Lakshmi. *See* Nehru, "Nan"
Panditji. *See* Nehru, Jawaharlal
Pant, Govind Vallabh, 227, 233, 245, 306, 319–20
Paris, 25, 226, 238, 242, 272
Pasha, Nahas, 246
Patel, *Sardar* Vallabhbhai, 57, 82–83, 97, 98, 109, 172, 181, 201, 204, 207, 218, 252, 256, 281, 292, 316, 319, 320, 327, 332, 341, 348, 355, 358, 364, 377–78, 386, 396, 401, 426–27, 458, 488
Pathans, 314
Patil, S. K., 359
Patna, 161, 168
Patnaik, Biju, 489
Patwardhan, Achyut, 204, 438
Pavlova, Anna, 174
Pethick-Lawrence, Lord, 346, 351, 353, 362ff., 387, 439
Phillips, William, 323
Pirpur Report, 250
Plebiscite, 420, 445
Poona, 141, 165, 274, 316, 335
Portugal, 481
Prague, 242, 244
Prakasa, Sri, 409
Prasad, Dr. Rajendra ("Babu"), 57, 77, 161–62, 165, 176, 180, 195, 201, 204, 208, 218, 256, 263, 274, 348, 378, 406, 428
Princely States, 365–66
Punjab, 208, 210–11, 303, 350, 408ff.
Purna Swaraj, 106–7, 259, 306, 440, 455

Quakers, 165, 190, 195
Quetta, 184
Quit India. *See* Satyagraha

Radcliffe, Sir Cyril, 402, 404ff.
Radhakrishnan, President, 485
Rai, Lala Lajpat, 72, 80, 86
Rajagopalachari. *See* C.R.
Rama, King, 249
Ramkrishna Mission, 96, 143–44, 175
Rama Rajya, 249
Ramgarh, 270
Rangoon, 255
Rau, Sir Benegal N., 348, 460, 461
Rashtrapati Bhavan, 382, 465
Rawalpindi, 367
Reading, Lord, 51, 60, 62–63
Revolt, 275, 278, 302, 314, 319, 322, 354
Rhodes Trust, 264
Richards, Robert, 356
Robeson, Paul, 240
Rolland, Mlle., 183
Rolland, Romain, 69
Rollier, Dr., 263
Roosevelt, Eleanor, 323, 340, 466
Roosevelt, President F. D., 152, 244, 301, 308, 310, 314, 323, 326, 330, 333, 340
Round Table Conferences, London, 109, 120, 145, 155, 165, 202, 213
Roy, Dr. B. C., 165, 323, 335, 360, 443, 493
Russell, Bertrand, 70, 178, 462
Russia, 185, 214, 289

Sahay, V., 475
Saksena, Mohan Lal, 319–20
San Francisco, 60, 343
Santiniketan, 161, 169–70, 176, 180–81, 248, 251
Sappho, 192
Sapru, Sir Tej Bahadur, 115, 176, 181, 264, 324, 339
Sarabhai, Bharati, 190–91, 192, 235, 238, 242, 246, 269, 292
Saran, Nandan, 259
Sarasvati, 201
Sarkar, N. R., 323
Satyagraha, 40, 54, 110ff., 165, 202, 273, 277–78, 313ff., 320
Satyagraha Sabha, 40
Satya, 156
Self-determination, 246, 255, 269, 308, 313
Seshan, N. K., 497
Sevagram (Segaon), 171, 226, 253
Shakuntala, 177
Shankaracharya, 145
Shastri, Lachhmi Dhar, 299
Shastri, Lal Bahadur, 131, 485, 489–90, 493
Shaw, George Bernard, 16, 30, 36, 147, 158, 178, 324
Sikhs, 59, 65, 257, 303, 476
Simla, 234, 256, 346, 364, 393
Simon, Lord, 76, 301
Simon Commission, 76, 85–86, 153, 202
Sind, 145, 210–11
Singapore, 224, 225, 301, 358, 360
Singh, Baldev, 313, 377, 402, 416
Singh, Buta, 313
Singh, General Harbaksh, 485
Singh, Karan, 462
Singh, Maharaja Hari, 400, 403, 425–26, 462
Singh, Master Tara, 313, 348–49, 477
Sino-Indian "brotherhood", 303
Sitaramayya, Pattabhi, 249, 252
Sivananda, Swami, 96
Slade, Madeleine (Mirabehn), 316
Smuts, Marshal Jan, 324
Socialism, 171, 197, 206, 347
Socialist Book Club, 247
Socialist Party of India, 95, 172
Sorensen, Reginald, 356, 438
Southeast Asia, 223, 460–61
Soviet Union, 75–76, 248, 293, 315, 465
Spain, 237–38, 245
Spanish Civil War, 219, 246
Srinagar, 35, 342, 366, 415, 476
Srivastava, Dr. Ram Sarup, 284
States Peoples' Conference, 269, 367
Stephens, Ian, 21
Stilwell, General Joe, 330
Suhrawardy, H. S., 389
Sukarno, President, 460
St. John Ambulance Brigade, 360
Swadeshi, 204
Swaraj, 56, 61, 111, 150, 222, 252, 306
Swaraj Bhawan, 294
Swaraj Party, 58ff., 67, 106
Switzerland, 175, 180–81, 183, 263, 282

Tagore, *Gurudev* Rabindranath, 14, 140, 152, 161, 165, 170, 180–81, 189, 202, 216–17, 248, 264, 292
Tandon, Purushottamdas, 181
Thackersey, Lady, 335
Thapar, General P. N., 484
Thimayya, General K. S., 481
Thompson, Edward, 263, 440
Tibet, 472, 485
Tilak, *Lokamanya* B. G., 31, 37, 315
Tito, Marshal, 467, 478
Toller, Ernst, 71–72
Tolstoy, Leo, 177
Toward Freedom, 199
Trieste, 184
Trinity College, Cambridge, 16, 20
Tripuri, 252
Trotsky, Leon, 255
Trotskyists, 156, 247
Truman, President Harry S., 343, 446, 448, 450ff., 460

United Nations (UNO), 309, 343, 392, 420, 452
United Provinces (U.P.), 47, 222, 226, 248–49, 274, 276, 292, 302, 348
United States of America, 260, 295, 391–92, 446, 460

Vallabhbhai. *See* Patel, *Sardar* Vallabhbhai
Vedas, 156
Venice, 184
Victoria and Albert Museum, 198
Vienna, 185

Index

Vir, Dharma, 362
Vishnu, Lord, 249

Wafd Party, 246
Wardha, 171, 173, 193, 208–9, 226, 246, 248, 253, 258, 272, 298, 302, 313
Washington, D.C., 486, 495
Wavell, Field Marshal Archibald Percival, 303, 309, 318, 326, 330, 334–35, 350–51, 353, 358, 364, 377, 381
Wells, H. G., 178
Wilde, Oscar, 18
Wilkie, Wendell, 333
Wilkinson, Ellen, 198
Willingdon, Lord, 181
Windsor Castle, 476
Women's Conference, All-India, 267
Wood, Evelyn, 146
Wood, Joseph, 11
Woolf, Virginia, 330, 337
Working Committee. *See* Congress, Indian National
World Peace Congress, 219
World War I, 31ff., 35, 39, 71, 93
World War II, 255
Wyatt, Woodrow (Lord Wyatt of Weeford), 21, 356, 379

Yalta, 340
Yamuna, 123, 199, 201
Youth Conference, Bombay, 89
Yunus, Mohammad, 314, 349, 367

Zamorin, 145
Zetland, Lord, 223, 226, 239, 257–58, 260

DATE			